AF734661

The Laguna Madre of Texas and Tamaulipas

REVISED EDITION

TEXAS A&M UNIVERSITY CORPUS CHRISTI | HARTE RESEARCH INSTITUTE FOR GULF OF MEXICO STUDIES

Number Thirty-Six

GULF COAST BOOKS
Sponsored by Texas A&M University–Corpus Christi
Larry McKinney, General Editor
John W. Tunnell Jr., Founding Editor

The Laguna Madre of Texas and Tamaulipas

REVISED EDITION

EDITED BY
Kim Withers
Brian R. Chapman
John W. Tunnell Jr.
Frank W. Judd

Texas A&M University Press
College Station

Copyright © 2023 by Kim Withers
All rights reserved. First edition 2002.
Second revised edition 2023

∞ This paper meets the requirements of ANSI/NISO Z39.48–1992 (Permanence of Paper).
Binding materials have been chosen for durability.
Manufactured in China through Martin Book Management

Library of Congress Cataloging-in-Publication Data

Names: Withers, Kim, editor. | Chapman, Brian R., editor. | Tunnell, John W., Jr., editor. | Judd, Frank W., editor.
Title: The Laguna Madre of Texas and Tamaulipas / edited by Kim Withers, Brian R. Chapman, John W. Tunnell Jr. and Frank W. Judd.
Other titles: Gulf Coast books; no. 36.
Description: Second revised edition. | College Station: Texas A&M University Press, 2023. | Series: Gulf Coast books; number thirty-six | Includes bibliographical references and index.
Identifiers: LCCN 2022023714 (print) | LCCN 2022023715 (ebook) | ISBN 9781623499914 (hardcover) | ISBN 9781623499921 (ebook)
Subjects: LCSH: Natural history—Texas—Laguna Madre Region. | Nature conservation—Texas—Laguna Madre Region. | Lagoon ecology—Texas—Gulf Coast. | Lagoon ecology—Mexico—Tamaulipas (State) | BISAC: NATURE / Ecosystems & Habitats / Coastal Regions & Shorelines | NATURE / Environmental Conservation & Protection
Classification: LCC QH105.T4 L34 2023 (print) | LCC QH105.T4 (ebook) | DDC 577.7/80916364—dc23/eng/20220831
LC record available at https://lccn.loc.gov/2022023714
LC ebook record available at https://lccn.loc.gov/2022023715

Wes Tunnell

Frank Judd

Brian Chapman

This book is dedicated to my coeditors, Wes Tunnell, Frank Judd, and Brian R. Chapman. Unfortunately, none of them lived to see this book brought to fruition. Wes was the driving force behind both incarnations of *The Laguna Madre of Texas and Tamaulipas*—the original volume published in 2002—and this new updated and revised edition. Frank played a significant role in the development of the original book and this edition. Always students, Wes and Frank modeled the roles of scientist and scholar, and they meant a great deal to all of us who were privileged to know and work with them. Wes lost his life to cancer in July 2018 while we were putting this book together, and he left behind some really big shoes to fill. Luckily, Brian, who was a longtime colleague and dear friend of Wes's, Frank's, and mine, agreed to help us finish the book. Brian was able to reenergize the effort, and he, Frank, and I worked hard to produce a book that Wes would have been proud of. Shortly after the completed manuscript for the book was submitted to Texas A&M University Press, Frank Judd died suddenly and unexpectedly (December 2020). Then, in June 2021, Brian Chapman also died suddenly, devastating all who knew him. This book is dedicated to the memory of these three outstanding biologists whose teaching, research, and writings expanded the knowledge and interest in the unique area known as the "Mother Lagoon."

Kim Withers
November 2021

Contents

Foreword

The Laguna Madre is a place of superlatives, and it is a natural treasure on a global scale. Dick Bartlett, formerly a trustee with The Nature Conservancy (TNC) and now deceased, wrote an eloquent foreword to the first edition of this book (Tunnell and Judd 2002). Dick's passion for the TNC, as well as the science and conservation of the Laguna Madre, was evident in his words that are still relevant today. Before conserving an ecosystem, TNC always "does the science first"; "one has to know how all the pieces fit," he stated. Natural protection is provided to the east by long barrier islands and peninsulas. In Texas, Padre Island National Seashore and Laguna Atascosa National Wildlife Refuge protect most of these barrier island lands, and vast historical ranches, such as the King, Kenedy, and Yturria, protect the mainland shores. In Tamaulipas, we were concerned about a proposed intracoastal waterway to be dredged throughout the length of the Laguna Madre when Dick wrote the first foreword; today, the Mexican lagoon is federally protected in its entirety.

As series editor of this book in our Harte Research Institute for Gulf of Mexico Studies series, I have the great honor and privilege to briefly introduce you to this amazing ecosystem. It is the only coastal, hypersaline lagoon system on the North American continent, comprising two of only five such lagoons worldwide. Together they form the largest hypersaline system in the world. Extending along 455 km (277 mi) of shoreline in South Texas and northeastern Mexico, the Laguna Madre is separated into two bodies of water by 75 km (47 mi) of Rio Grande Delta. Each lagoon is about 185 km (115 mi) in length, and each is further divided into subunits; the Upper and Lower Laguna Madre in Texas are separated by the Land-Cut tidal flats, and the Northern and Southern Laguna Madre de Tamaulipas are separated by the El Carrizal Flats.

The superlatives that these two lagoons are best known for include their hypersaline condition, a large overwintering Redhead duck population, numerous protected species, vast seagrass meadows, and great fishery productivity. The commonly recorded extreme salinities of over 100 parts per thousand (ppt) in both systems have been moderated in recent decades due to changes in water circulation (Gulf Intracoastal Waterway [GIWW] and Mansfield Pass in Texas and five passes connecting the Gulf with the lagoon in Tamaulipas). The two lagoons together contain over 77% of the North American overwintering population of the Redhead and numerous protected species such as the Piping Plover, Snowy Plover, and Reddish Egret. Recently delisted species, such as the Brown Pelican and Peregrine Falcon, are found in both lagoons, and colonial waterbird rookeries abound. In Texas, almost 80% of all seagrass beds in the state are found in the Laguna Madre, and the historically highly productive commercial fisheries have now given way to some of the best recreational fishing in North America for Red Drum, Black Drum, and Spotted Seatrout.

In Tamaulipas, a "boom and bust" cycle of great fishery production alternated with briny, almost sterile waters before the 1970s. The cycle included a highly productive lagoon for several years after a wet hurricane, then dwindling fisheries and species diversity as the barrier island passes closed and the completely enclosed lagoon progressed to higher and higher salinities. During extended droughts or long periods between hurricanes, the lagoon began to dry up, with water level decreasing and salinities increasing due to evaporation. When salinities reached 150 ppt, only brine shrimp existed, and at the highest recorded salinity of 295 ppt, salt had precipitated along the shoreline and in the bottom of the lagoon. Then a hurricane would open Gulf passes, flush the system with floodwaters from rain, and the cycle would begin again. Major hurricanes of 1909, 1933, and 1967 all demonstrated this cycle in Tamaulipas, and it is likely that a similar cycle existed in Texas before channelization of that system (GIWW completed in 1949; Mansfield Pass completed in 1962).

In addition to these best-known features of the Laguna Madre, there are other unique lesser-known, but important characteristics, such as the most extensive wind-tidal flats and clay dunes in North America; the only strain of high-salinity-adapted oysters in North America; the only natural rocky shoreline in Texas; the only serpulid worm reefs in Texas; and the only locality of oolite (calcium carbonate) and gypsum crystal formation in Texas. Wind-tidal flats occupy ~631 km^2 (244 mi^2; estimated in early 2000s) of shoreline in the Texas Laguna Madre and 508 km^2 (196 mi^2) in Tamaulipas. Clay dunes form a mainland transition zone between the wind-tidal flats and the upland in many areas. A unique population of Eastern Oysters, adapted to the high-salinity conditions of the Laguna Madre, are found in South Bay, the southernmost portion of the Lower Laguna Madre in Texas. Oysters in the Southern Laguna Madre de Tamaulipas may be similarly adapted, but they have not been studied. Penascal Point between Baffin Bay and Upper Laguna Madre contains the only natural rocky shoreline (coquina, beach rock) in Texas, and a similar shoreline exists at Punta Piedras between Laguna el Catán and the Southern Laguna Madre de Tamaulipas. Relict serpulid worm reefs occupy 16 km^2 (6 mi^2) of Baffin Bay, providing otherwise absent, hard-substrate bottom habitat and an indication of a wetter climate in the past. Calcium carbonate is deposited as irregular sand grains in the quiet waters of the Upper Laguna Madre near Baffin Bay and within Baffin Bay, a process typically relegated to tropical regions of the world, and gypsum crystals and rosettes are formed below the surface of wind-tidal flats.

New or elevated conservation and management issues since the first edition of our book include Bahia Grande restoration, Texas Seagrass Conservation Plan, protection of South Padre Island land, dredging of Packery Channel, potential desalinization in uppermost and lowermost Texas lagoons, wind farms, water-quality issues, sea-level rise, cold stunning of sea turtles, potential causeways in uppermost and lowermost Texas lagoons, and continued barrier island development.

In summary, over 100 years of literature and more than 1,765 references were compiled and examined for the first edition and this second edition (32 references cited in the first edition were deleted because they were superseded by later or more comprehensive works). The unique biological resources and geological features within this world-renowned ecosystem are clearly worthy of conservation and protection. Monitoring its resources and research on understanding the physical and biological processes that govern its functioning will ensure that sound scientific information is provided for the long-term management and conservation of the Laguna Madre.

John W. Tunnell Jr.
June 2018

Not long after penning this "Foreword" and the following "Preface," Wes Tunnell lost his painful battle with cancer, leaving his family, friends, and professional colleagues devastated. Although he is still mourned, the authors and editors who contributed to this book picked up the pace, determined to complete the project to which Wes had dedicated himself. Kim Withers assumed the mantle of senior editor, Frank Judd continued his valuable contributions as coeditor, and Brian R. Chapman was asked to assist with editorial duties. Having been a fellow undergraduate student with Wes at Texas A&I University, a faculty member with Wes during the early years at Texas A&M University–Corpus Christi, and friend for more than 50 years, Brian was honored to participate. Although many worked to make this book possible, Wes was the source of inspiration to all participants. He is missed.

Kim Withers, Frank W. Judd,
and Brian R. Chapman
February 2020

Preface

The Laguna Madre of Texas and Tamaulipas (Tunnell and Judd 2002) was my first book with Texas A&M University Press, and it was the book that initiated the first Texas A&M University–Corpus Christi (TAMUCC) book series. The Gulf Coast Books (formerly Gulf Coast Studies) series now has over 30 titles on subjects as broad as archaeology, art, history, natural history, and science of the Gulf of Mexico. That series led to the development of the Harte Research Institute (HRI) for Gulf of Mexico Studies book series in 2007 on the science of major selected topics of the entire Gulf or selected regions of the Gulf.

The original book holds a special place in my heart, since it began our TAMUCC book series, which became an important part of my later career at HRI and TAMUCC. In addition, the subject is one of my favorite natural ecosystems, and I was able to write it with close friends and colleagues in the Center for Coastal Studies for my favorite nature organization, The Nature Conservancy (TNC). However, the first edition sold out after about 12 years, so it has been unavailable for the past five years or so. In addition, a number of new scientific and conservation issues have arisen that need to be addressed, most chapters need to be updated with new and current data, and a number of selected topics need to be added as new, relevant chapters.

As noted in the preface to the first edition, that book was a contract project to produce a report for TNC, who wanted to gather all scientific information and conservation issues into one document before starting a focused 20-year conservation effort on the Laguna Madre in the United States and Mexico. TNC trustee Dick Bartlett, who championed the project, remarked when he saw the report: "This is a book; come with me to TAMU Press, Wes, to meet Noel Parsons, the editor-in-chief about it." Noel agreed, and I began my first book project working with Shannon Davies, natural science editor at TAMU Press. As a delightful and knowledgeable person to work with, Shannon and I did many books together in our first book series, and then she encouraged the development and formation of our second book series just after HRI opened. Although Shannon (now retired) moved up to editor-in-chief and then became director of TAMU Press, we both acknowledge that it was the Laguna Madre book that started our long affiliation between TAMUCC and TAMU Press.

The preferred alternative, as noted in my preface to the first edition, would have been to get the experts on each topic in the book to write those chapters. However, the short nine-month timeline set by TNC would not allow that, so my colleagues in the Center for Coastal Studies, Liz Smith and Kim Withers, agreed to write the majority of all chapters with me. This time, there was no rush. David Newstead, one of our former master's degree students, now working on his PhD on the Red Knot, had first suggested that we update, or do a second edition, of the Laguna Madre book. However, that was around the time of my part-time retirement (2013), and I did not want to take on any new, big projects. Then, in 2016, Ed Proffitt, our new chair of the Department of Life Sciences, encouraged me to consider updating the book again, and he volunteered to do a chapter on the spread of Black Mangroves into and around the Laguna Madre. Because I have always enjoyed doing projects, and because I had a serious cancer diagnosis in 2015, I wanted something important to keep me occupied, so I decided to accept the challenge.

Since Frank Judd, my former coeditor, was retired but still active, I invited him to be coeditor with me again. I also invited Kim Withers, who is a great writer, copyeditor, and document organizer, to join us as a third coeditor. Kim had done this with my second TAMU Press book (Tunnell et al. 2007), and I knew she would be a great help with this project, especially since she knows the Laguna Madre and its biota/habitats so well. We then collectively asked the experts on various Laguna Madre topics to join us in either updating the first edition chapters or adding new chapters. We had a great response to our request, going from 8 to 26 authors and from 19 to 22 chapters. New chapters are included on the Coastal Sand Plain, adjacent and west of the Laguna Madre in Texas, and Black Mangroves. Because of travel restrictions to Tamaulipas, I asked Mexican colleagues to write a chapter providing detail on the Tamaulipas lagoon. These chapters broaden the scope of information about both lagoons, and the addition of experts to other chapters increases the depth of coverage on those previous topics.

The book begins with an overview of the physical environment of the Laguna Madre, the surrounding Tamaulipan Biotic Province, and important private ranches that protect the western shoreline in Texas and part of Tamaulipas. Then the book covers the natural resources of the area, including seagrass meadows, open bay, wind-tidal flats, barrier islands, various bird resources, and fisheries organisms. Finally, we cover special issues and concerns, such as Red Tide and Texas Brown Tide, recreational fisheries, Black Mangrove expansion northward, environmental stressors, a conservation framework for the Laguna, information gaps and needs, and conservation issues and recommendations. We also added an afterword written by longtime Laguna Madre cabin and houseboat owners.

As it was for the first edition, the purpose of this compendium is intended to be a synthetic overview of the Laguna Madre, and it is for a middle-of-the-road audience, both scientist and nonscientist. It is a book based in science, but one for use by an educated audience who want to learn about and take care of the Laguna Madre. The Nature Conservancy intended to use the first book as an information tool to spark interest and raise awareness for the long-term conservation and preservation of Laguna Madre natural resources. For an agency person or student, it should put everything known about the Laguna Madre in one place and serve as a resource information guide. To the scientist, the thorough literature review and listing should point out where the strengths and weaknesses are in our knowledge concerning the Laguna Madre. In a few areas, there is enough information to write a solid, scientific review or book, for example, on seagrass, colonial waterbirds, or fisheries. On the other hand, it points out great knowledge gaps in some areas where substantial research is still needed. This work, as with any of its size and scope, will undoubtedly contain some errors and unanswered questions. However, if we can prompt or challenge students and scientists to answer some of these questions and seek new knowledge, then we have achieved an important goal. Additionally, if we have introduced new habitats, biota, or issues within the region that need attention and that will further the cause of protection and conservation, we have achieved another goal.

Finally, the literature cited section contains 1,765 citations that are also listed in the comprehensive bibliography, a companion project of writing the book. The book synthesizes over 100 years of literature on the Laguna Madre and surrounding environments. The journals, proceedings, and other literature titles within the literature cited are spelled out to make acquisition in either English or Spanish easier. Due to its length, as well as the need to have it in a "searchable" format, the comprehensive bibliography is available separately in electronic format from the Harte Research Institute for Gulf of Mexico Studies at Texas A&M University–Corpus Christi.

John W. Tunnell Jr.
June 2018

Acknowledgments

Well, as they say, "it takes a village." If you read the foreword and the preface, you understand the impetus for the original book and the reasons for this update. In the original, a few of us at the Center for Coastal Studies at TAMUCC, where Wes Tunnell was the director, wrote the chapters for a report to The Nature Conservancy. When it was decided that the report should be turned into a full-fledged book, those chapters were reviewed by 35 experts, who were gratefully acknowledged in the first edition and who are listed below. In this update, the experts were asked to join us as coauthors on revisions of many of the chapters included in the first edition or as authors of new chapters covering topics that were either beyond the scope of the first edition or that were not even on the radar in the early 2000s. I am grateful that these scientists and experts were willing to spend their time contributing to this project that was so near and dear to Wes's heart. I also appreciate their patience, as this process played out over a much longer time than was expected.

Wes's untimely death caused a great deal of upheaval in the progress toward the completion of this edition. But a few people stepped bravely into the fray to do any number of tasks that had been left partially or completely undone. I gratefully acknowledge Mike Wetz (Harte Research Institute) for spearheading revision of chapter 22, "Conservation Issues and Recommendations," and Hannah Ehrmann (Harte Research Institute) for sifting through all the files that resided on several computers and file-sharing platforms and compiling them into an organized, useful archive. Shannon Davies at Texas A&M University Press, now retired, asked Brian Chapman to offer to join the editorial team, and this proved to be critical to the completion of the book. Brian had already helped update several chapters in the book, and his addition to the editorial team provided the rudder that was needed to get the book back on track and finished. Thanks to Shannon for seeing what was needed and then giving the push to make it happen.

One of the new chapters in this revised edition of the book is the afterword, "Cabins, Floaters, and Lagunatics," which provides a brief look at some of the "culture" of the Laguna Madre. The wonderful watercolors that enliven this chapter were painted by Dr. John Pettigrove, and I thank him for sharing them.

Funding for this edition of the book was provided by the Robert J. and Helen C. Kleberg Foundation, the Harte Research Institute for Gulf of Mexico Studies, and Texas A&M University–Corpus Christi. We thank Gail Sutton and Larry McKinney at the Harte Research Institute for the administrative and moral support that they provided.

Between this revised edition of the book and the first are myriad people who contributed time, knowledge, and expertise, and I am grateful to them all:

Reviewers: Russ Miget, Warren Pulich, Bill Longley, Junji Matsumoto, Jim Norwine, Gary Powell, Ruben Solis, Richard Watson, Jennifer Smith-Engle, Bob Lonard, Bruce Cheeseman, Keith Gutherie, Tom Krenek, Jerry Thompson, Mike Beck, Chris Onuf, Pete Sheridan, Ed Buskey, Tracy Villarreal, Hudson DeYoe, Lynn Drawe, Marc Woodin, Allan Chaney, Tim Brush, Mike Farmer, Curt Zonick, Maurice Renaud, Bob Edwards,

Larry McEachron, Kyle Spiller, Randy Blankenship, A. Raz-Guzman, Forrest S. Smith, Dan L. Duncan, Roberto Perez.

Assistants: Leticia Contreras, Sandra Alvarado, Gabriella de la Fuente, Mikaela Zieglar, Hannah Ehrmann, Kathryn (Harvey) Tunnell, Jace Tunnell, Suzanne (Dilworth) Bates, Susan Childs, Gloria Krause, Linda Simmons, Nancy Hilbun.

Technical contributions (data, logistics): Chris Onuf, Richard Watson, Tim Brush, Lee Elliott, Paul Montagna, Felipe San Martin, Ing. Marco Antonio Chapa Martinez, Lic. Gustavo Barroso Hernandez, Lic. Monica Castilo Sagastegul, Ing. Nora E. Macia Perez, Ernesto Enkerlin Hoeflich, Diego Fabian Lozano Garcia, Ing. Jorge Brenner Guillermo, Ing. Julie Noriega, Eduardo Carrera Gonzalez, Gabriela de la Fuente de Leon, Salvador Contreras Balderas, Jim Gibeaut, Brach Lupher, Charles Sassine.

Other supporters: James Bergan, Dick Bartlett, Noel Parsons.

Finally, although I have dedicated this book to Wes, Frank, and Brian, Wes dedicated the first book to Henry H. Hildebrand, a pioneer in the study of the Laguna Madre in both Texas and Tamaulipas. I quote Wes from the acknowledgments section of the first edition:

> We recognize and commend the untiring efforts and long-term dedication of Henry H. Hildebrand to the study of the Laguna Madre ecosystem, particularly in Tamaulipas. His early studies there, starting in 1951, form the basis of our scientific knowledge of that lagoon today. His many trips to the Tamaulipas system, usually self-financed, depict a naturalist dedicated to a unique North American natural resource.

Kim Withers
November 2021

The Laguna Madre of Texas and Tamaulipas

REVISED EDITION

Part 1 Overview

1 Introduction

JOHN W. TUNNELL JR.

The rich natural resources of the Laguna Madre of Texas and Tamaulipas are widely known (Tunnell and Judd 2002). Competition for these resources, or varied uses of these highly productive and unique ecosystems, continues to generate a greater need for resource management and conservation. Consequently, there are a number of compelling reasons for continuing to compile and synthesize all known and new information:

1. Although considerable scientific literature is available, it is important to continue to compile and synthesize information into a single source for ready access and use.
2. Because the two lagoons are politically separated in two different countries, the literature is in two languages.
3. Because the two systems are similar, studies completed in one may be applicable to the other.
4. Information gaps need to be identified.
5. Conservation and management issues need to be identified and recommendations presented.

Because effective management must be based on knowledge provided by sound science, the purposes of this compendium are to provide a literature synthesis of major natural resource and conservation issues facing the Laguna Madre of Texas and Tamaulipas as well as a comprehensive bibliography to stimulate further investigation (figs. 1.1 and 1.2).

Some historical concepts about the Laguna Madre of Texas have proven false yet continue to persist. For instance, popular lore accounts for the infilling of the Laguna Madre of Texas in the Land-Cut region by "a tropical storm and a tidal wave in 1836 which swept over Padre Island carrying much of the island into the Laguna" (Burr 1945a). This story was related by Captain Mifflin Kenedy, who heard it from a Mexican general. Marine scientists studying the Laguna Madre changed the story somewhat in time but perpetuated the concept by noting that the Laguna Madre was effectively divided in two by the hurricane of 1919 (Gunter 1945a; Hedgpeth 1947). However, more recent research revealed that this land-bridge area was filled by a much slower, long-term process (Miller 1975; Morton and Garner 1993).

Another hotly debated topic is the John F. Kennedy Causeway (Padre Island Causeway until 1968), an earthen or landfill causeway extending from Corpus Christi to Padre Island, and its impact on water circulation and the "health" of the Upper Laguna Madre. Even before its construction in 1950, marine scientists argued that it would likely do great harm or even cause death to the entire ecosystem. In regard to the

Figure 1.1. *Satellite image of South Texas and northeastern Mexico coastline from Copano-Aransas Bays to Northern Laguna Madre de Tamaulipas. Prepared by Brach Lupher (Satellite: Sentinel-2A; Product: 12A; Image Date: 1 May 2020; Image Type: Natural Color; Spatial Resolution: 10 m; Band Combination: Bands 4 3 2; Cloud Cover: <10%). (Copernicus Open Access Hub, European Space Agency [ESA], https://scihub.copernicus.eu)*

Figure 1.2. *Satellite image of South Texas and northeastern Mexico coastline showing Lower Laguna Madre of Texas, Rio Grande Delta, and entire Laguna Madre de Tamaulipas. Prepared by Brach Lupher (Satellite: Landsat 8; Product: L1TP; Image Date: 8 June 2020; Image Type: Pan-sharpened Natural-Color; Spatial Resolution: 30 m + 15 m [pan chromatic]; Band Combination: Bands 4 3 2 + Band 8; Cloud Cover: <20%). (US Geological Survey, https://landlook.usgs.gov/viewer.html)*

proposed causeway, Gunter (1946) stated that "this proposal is quite likely to turn the Laguna into lifeless brine to the lasting injury of the people of the State." Hedgpeth (1947) also noted that in terms of geological time the Laguna Madre "is already dying." Fifty years later, arguments still abound on the effects of the causeway on the overall ecosystem, yet the Laguna continues to be one of the most productive coastal embayments on the Texas coast.

Although the two Laguna Madre systems are separated by the Rio Grande Delta, interesting similarities exist between them:

1. Hypersalinity (40 to 80 ppt; more in the past)
2. Semiarid climate
3. Shallow, elongate coastal lagoons
4. Highly productive fisheries (cyclical highs and lows in the past)
5. Extensive wind-tidal flats
6. Clay dunes and/or *lomas* (hills) along mainland shores and in Rio Grande Delta area
7. Limited freshwater inflow
8. Divided into upper and lower sections by wind-tidal flats
9. Single embayment within each (Baffin Bay in Texas, Laguna el Catán in Tamaulipas)
10. Predominantly remote, publicly inaccessible shorelines
11. Extensive overwintering populations of Redheads and other waterfowl
12. One of most important shorebird overwintering areas in the New World
13. Essential habitat for many colonial waterbirds and shorebirds, including protected species, such as Piping Plover, Snowy Plover, and Reddish Egret

Historical studies on the Laguna Madre of Texas have focused on unique physical features (wind-tidal flats, tidal-flat geochemistry, clay dunes, hypersalinity) or biological resources (blue-green or cyanobacteria algal mats, aquatic birds, fisheries productivity, seagrass ecology). In recent times, attention has turned to anthropogenic impacts and concerns, such as dredging of the US Gulf Intracoastal Waterway (GIWW), dredged-material disposal, water circulation, new causeways, agricultural and other nonpoint source runoff, water quality, boat propeller scarring in seagrass beds, sea-level rise, continued barrier island development, and vehicular impacts on tidal flats. Natural perturbations such as fish kills from hypersalinity and severe cold spells are well documented, and Texas Brown Tide algal blooms have received much attention.

The Laguna Madre de Tamaulipas has received much less attention in the past, with early studies focusing on fisheries, hydrography, sediments, and micromolluscs. This lagoon system is the largest in Mexico and one of the most productive and economically important. Ducks Unlimited de México (DUMAC) ranks it first out of 28 key wetland areas in Mexico that harbor 84% of the wintering waterfowl (DUMAC 1990b). However, during the 1990s, considerable attention focused on an enormous dredging project that would potentially extend the GIWW from Texas southward through the coastal lagoons to Tampico (and perhaps beyond). Greater focus then turned to the extensive natural resources and environmental condition of the region, and DUMAC submitted a proposal to the federal government of Mexico to establish the entire Laguna Madre de Tamaulipas as a protected area (Carrera and de la Fuente 1996). On 14 April 2005, the National Commission of Natural Protected Areas (Comisión Nacional de Áreas Naturales Protegidas, CONANP) announced that a new flora and fauna protected area, the Laguna Madre y Delta del Río Bravo (Laguna Madre and Delta of the Río Bravo, i.e., the US Rio Grande) had been established and published in the *Diario Oficial de la Federación* (Official Federal Register). This new natural protected area encompasses 572,808 ha (1.4 million ac) of barrier island, lagoon, delta, and mainland habitats in the *municipios* (counties) of Matamoros, San Fernando, and Soto la Marina.

Among all the studies completed on both lagoonal systems, hypersalinity is a common theme. From the earliest writings of Burr (1930, 1945a, 1945b, 1947, 1950), to those of Gunter (1945a, 1945b), Hedgpeth (1947, 1967), Hildebrand (1958, 1969, 1980), Simmons (1957), and Breuer (1962), and most recently Quammen and Onuf (1993) and Onuf (1996a, 1999, 2007), the continuing scientific interest has been hypersalinity or changes related to salinity. Generally, one associates high salinity with low numbers of species and low productivity (Copeland and Jones 1965; Copeland 1967). Yet quite uniquely, and seemingly in contrast, many of the studies on the Laguna Madre emphasize the high productivity of seagrasses, fisheries, or waterfowl. This enigma of high productivity in a hypersaline environment appears to be governed by water circulation: navigation channels in Texas and hurricane-opened barrier island passes in Tamaulipas. Before the GIWW was dredged, fisheries productivity was high in the Texas Laguna Madre (20% of bay surface in Texas but over 50% of catch; Hedgpeth 1947). However, when salinity reached levels up to two or three times that of oceanic salinity, or when summer or winter temperatures became extreme, there was no escape and massive fish kills occurred.

Today, the GIWW and Mansfield Pass bisecting Padre Island allow for improved circulation and moderated salinities, so extremes have been eliminated or reduced (Quammen and Onuf 1993). Changes in hypersalinity and productivity have been affected by changes in water circulation. The high productivity also seems to be linked to the extensive seagrass meadows. This seagrass-based ecosystem provides much in the way of food, shelter, and nutrients for the semi-enclosed system; however, unlike a normal or positive estuary, which depends on freshwater inflow, salt marshes, and phytoplankton for nutrient input, the Laguna Madre appears to recycle its own.

In Mexico, however, Hildebrand (1980) demonstrated that "boom and bust" fishery cycles were linked to wet hurricanes driving the Laguna Madre de Tamaulipas system. Between hurricanes, passes connecting the Gulf of Mexico with the Laguna closed, and with little to no freshwater inflow, the Laguna became increasingly saline. Several highly productive years were followed by almost no fisheries production in a briny Laguna Madre. The next hurricane would open the Gulf passes, bring much-needed rain and flushing of the system, and subsequently stimulate another cycle of high fishery productivity. Hildebrand (pers. comm.) believed that hurricane rainfall floodwaters and El Niño wet cycles, coupled with long-term drought cycles, are more important regulators of change than channelized circulation in either system. He emphasized, however, that water-storage structures, such as Falcon Lake, greatly restrict freshwater inflow into either system.

Accessibility and remoteness have played major roles in the study and protection of these two lagoons. Although both systems are somewhat remote, close proximity and accessibility to marine biologists from the Texas Game, Fish, and Oyster Commission (now Texas Parks and Wildlife Department, TPWD), Marine Laboratory in Rockport, and marine scientists from the University of Texas Marine Science Institute in Port Aransas aided early studies in the 1940s and 1950s in the upper Laguna Madre. A small field station on the Lower Laguna Madre aided TPWD work there in those early years. In the 1970s, a small Coastal Studies Laboratory on South Padre Island associated with Pan American University in Edinburg (later University of Texas–Pan American and now University of Texas–Rio Grande Valley) increased studies in the Lower Laguna. Creation of the Center for Coastal Studies and Conrad Blucher Institute for Surveying and Science at Texas A&M University–Corpus Christi in the 1980s led to increased access and research in the Upper Laguna.

Accessibility continues to make research difficult in Mexico, and since the early 2000s increased drug trafficking and activity have made studies challenging and dangerous in Tamaulipas. The earliest studies in the Laguna Madre de Tamaulipas necessitated multiday field trips from the University of Texas Marine Science Institute (Hildebrand 1958; Copeland and Jones 1965; Copeland 1967) and the University of Corpus Christi (Hildebrand 1969) or longer expeditions from the Universidad Nacional Autónoma de México in Mexico City (García-Cubas 1968; Yañez and Schlaepfer 1968). Most recently, in the absence of a laboratory facility, marine scientists and biologists continue to make expeditions to the Laguna from Universidad Autónoma de Tamaulipas in Ciudad Victoria and Universidad Autónoma de Nuevo León, Instituto Tecnológico y de Estudios Superiores de Monterrey, Pronatura Noreste, and DUMAC, all in Monterrey. These institutions are all many hours' drive from the Laguna Madre de Tamaulipas, and when scientists arrive at the Laguna, accommodations are distant and few.

Remoteness of both Laguna Madre ecosystems has surely limited their scientific study. However, this same remoteness has contributed to the protection and conservation of some species and habitats. In Texas, large, private ranches along the mainland and two federally protected areas, Padre Island National Seashore and Laguna Atascosa National Wildlife Refuge, limit access and development. Large private ranches and few paved roads in Mexico limit access and settlement along the shores of the Laguna.

The points introduced in this chapter are developed more fully in the pages to follow. Most of the literature compiled for this compendium concerns the Texas Laguna Madre and its surrounding area. Unfortunately, much comprehensive information on the Laguna Madre de Tamaulipas, which was generated in regard to the potential GIWW during the 1990s, is owned by the Mexican federal government and is unavailable except by permission. However, the new Laguna Madre de Tamaulipas protected area (Laguna Madre y Delta del Río Bravo) is generating much new and past information about the lagoon. A separate comprehensive bibliography, available in PDF from the Harte Research Institute for Gulf of Mexico Studies at Texas A&M University–Corpus Christi, will lead interested persons to further information about these two distinctive and dynamic ecosystems.

2 Geography, Climate, and Hydrography

JOHN W. TUNNELL JR.

The Laguna Madre is a bar-built, coastal lagoon bordered by barrier islands and peninsulas to the east and the mainland to the west. Geomorphically, the Laguna Madre is divided into two separate lagunas by the Rio Grande Delta: the Laguna Madre of Texas to the north and the Laguna Madre de Tamaulipas, Mexico, to the south (fig. 2.1). These two coastal ecosystems form one of only five hypersaline ecosystems in the world (Javor 1989). The other four are (1) the Putrid Sea, or the Sivash, adjacent to the Sea of Azov on the Crimean Peninsula, Ukraine; (2) Laguna Ojo de Liebre, on the Pacific coast of Baja California, Mexico; (3) Spencer Gulf, South Australia; and (4) Shark Bay, Western Australia.

Geography

The Laguna Madre of Texas is the largest of seven estuarine systems along the Texas coast (Diener 1975; Montagna et al. 2013). The Laguna Madre de Tamaulipas is one of only three coastal lagoons in the State of Tamaulipas and the second largest of all coastal lagoons in Mexico. Together, the Laguna Madre of Texas and Laguna Madre de Tamaulipas form the largest hypersaline system in the world.

Extending along 445 km (277 mi) of coastline, Laguna Madre reaches from Corpus Christi Bay on the north (27°40.5′ N) to Río Soto la Marina on the south (23°47′N). Each of the Texas and Tamaulipas lagunas is approximately 185 km (115 mi) in length, but they vary in width, averaging 7 km (4.3 mi) and 12 km (7.5 mi), respectively (table 2.1). The lagunas are separated by 75 km (47 mi) of Rio Grande Delta—12 km (7.5 mi) in Texas and 63 km (39 mi) in Tamaulipas—which bulges the Texas-Tamaulipan shoreline over 35 km (28 mi) eastward into the Gulf of Mexico.

Each Laguna is further divided physiographically into subunits. The Upper Laguna Madre in Texas (fig. 2.2) is separated from the Lower Laguna Madre (fig. 2.3) by a land bridge extending from Padre Island to the mainland, variously known as the Salt Flats, Saltillo Flats, Kenedy Flats, Laguna Madre Flats, or most commonly, the Land-Cut. The latter name was derived when the Gulf Intracoastal Waterway (GIWW) was cut across the extensive tidal flats of the land bridge in the late 1940s. The upper and lower lagoons in Texas are fairly similar in size, 76 km (47 mi) long by 6 km (3.7 mi) average width and 91 km (57 mi) long by 8 km (5 mi) average width, respectively. In Tamaulipas, however, a shorter, wider northern basin (fig. 2.4) is divided from a longer, narrower southern basin (fig. 2.5) by the Río San Fernando (Río Conchos on some maps) Delta, a barren mudflat sometimes referred to as El Carrizal (Hildebrand 1969). The Northern Laguna in Tamaulipas is approximately 58 km (36 mi) long and averages 20 km (12 mi) in width, and the Southern Laguna is about 127 km (79 mi)

Figure 2.1. *Laguna Madre of Texas and Tamaulipas*

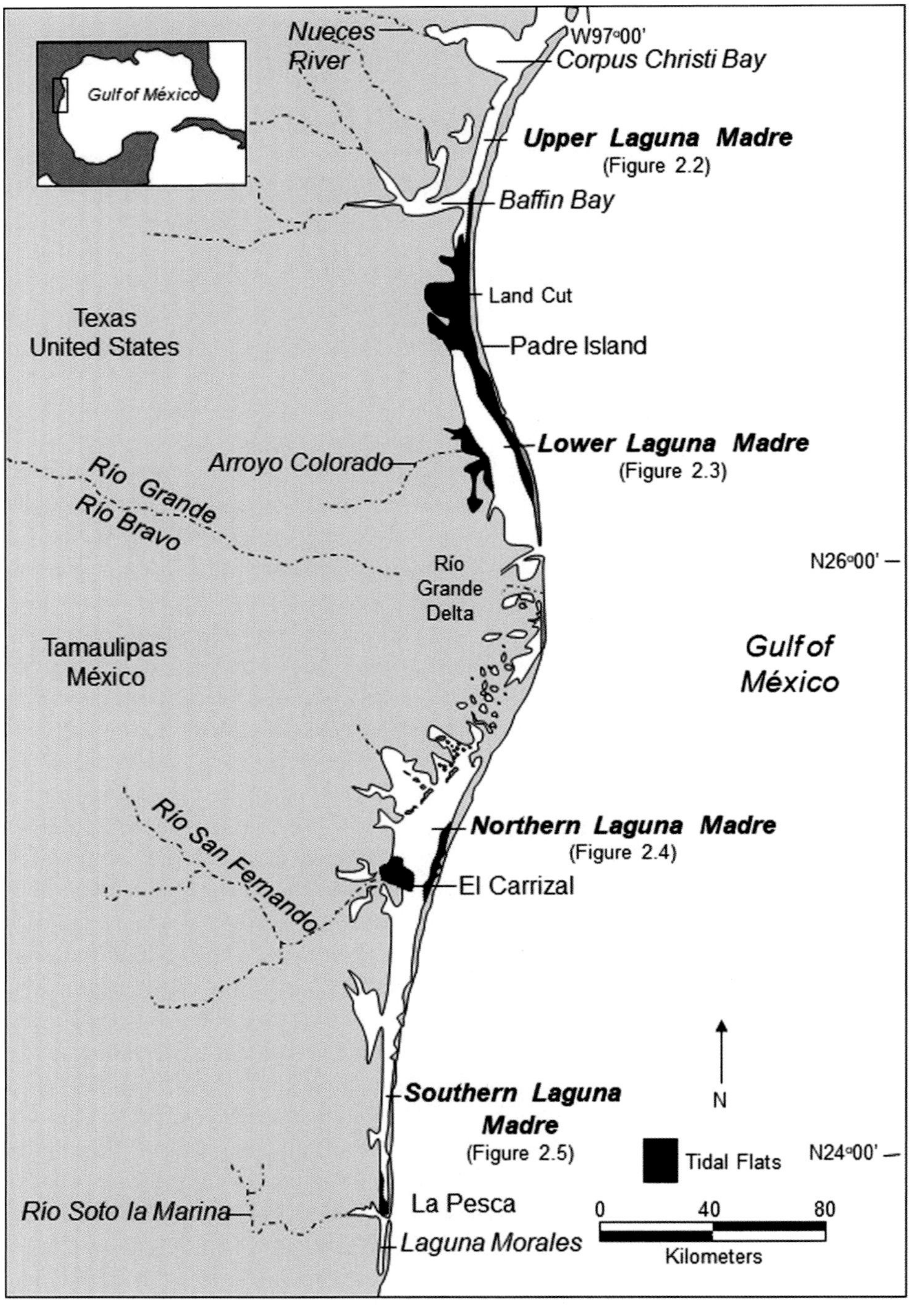

long and averages only 6 km (3.7 mi) in width (see table 2.1). Caution is advised when calculating lengths, widths, and average dimensions of both Laguna Madre systems because of extensive wind-tidal flats and shallows. Normal tides and calm winds leave hundreds of square kilometers of tidal flats exposed, but days of strong winds can push water for many kilometers across these flats, inundating vast areas (see chapter 9). Elevated tides associated with tropical storm systems also inundate these normally emergent tidal flats.

The lagoon systems in Texas and Tamaulipas are quite similar. Both are shallow, averaging about 1.0 m (3.3 ft) deep, but surface area in Texas is somewhat smaller

Table 2.1. *Physiographic data for the Laguna Madre of Texas and Tamaulipas*

	Length[a] km (mi)	Width km (mi)			Depth m (ft)	Surface Area km^2 (mi^2)		
		Max.	Min.	Ave.		MLW[b]	MSL	MHW[b]
Texas								
Upper Laguna Madre	76 (47)	7 (4.3)	3 (1.9)	6 (3.7)	0.8[b] (2.6)	191 (74)		277 (107)
Baffin Bay	23 (14)	7 (4.3)	3 (1.9)	5 (3.1)	2.3[b] (7.5)	219 (85)		247 (95)
Land-Cut	18 (11)	18 (11)	16 (10)	17.5 (11)	+0.6[c] (2.0)			
Lower Laguna Madre	91 (57)	12 (7.5)	6[d] (3.7) 11[e] (7)	8[d] (5) 14[e] (9)	1.4[b] (4.6)	727 (281)		1,364 (527)
Total Texas Laguna Madre	185 (115)	12 (7.5)	3 (1.9)	7 (4.3)	1.2 (3.9)	1,140 (440) 1,137[f] (439)	 1658[f] (640)	1,888 (729) 2,292[f] (885) 2,142[g,h] (827)
Tamaulipas								
Northern Laguna Madre	58 (36)	30 (19)	10 (6)	20 (12)	0.7[i] (2.3)		1,100[i] (425)	
Southern Laguna Madre	127 (79)	12 (7.5)	1 (0.6)	6 (3.7)	1.1[j] (3.6)		900[i] (347)	
Total Tamaulipas Laguna Madre	185 (115)	30 (19)	1 (0.6)	12 (7.5)	1 (3.3)		2,000[i] (772) 2,152[h] (831) 2,028[k] (738)	 2,728[k] (1,053)

Note: Multiple data are shown in some categories, since there is no consensus on surface area; furthermore, size calculation techniques are uncertain or unknown, so these data, except for DUMAC (1996, from database, Monterrey, Nuevo León), should be used with caution. MLW = Mean Low Water, MSL = Mean Sea Level, MHW = Mean High Water.

[a] Map calculations (approximate).
[b] Depths and surface area at MLW and MHW from Diener 1975, unless otherwise indicated.
[c] Fisk 1959.
[d] Width across water.
[e] Water width plus wind-tidal flats.
[f] G. Powell, pers. comm.; note that MHW Texas total of 2,292 km^2 represents "gross surface area" for estimating the input of direct precipitation on the Laguna and includes habitats not normally submerged.
[g] NOAA 1994; MHW Texas total of 2,142 km^2 represents "total estuary surface area" (= land and water combined?).
[h] Leija-Tristán et al. 2000.
[i] Yañez and Schlaepfer 1968.
[j] Jorge Brenner, pers. comm.
[k] DUMAC 1996 (from database, Monterrey, Nuevo León).

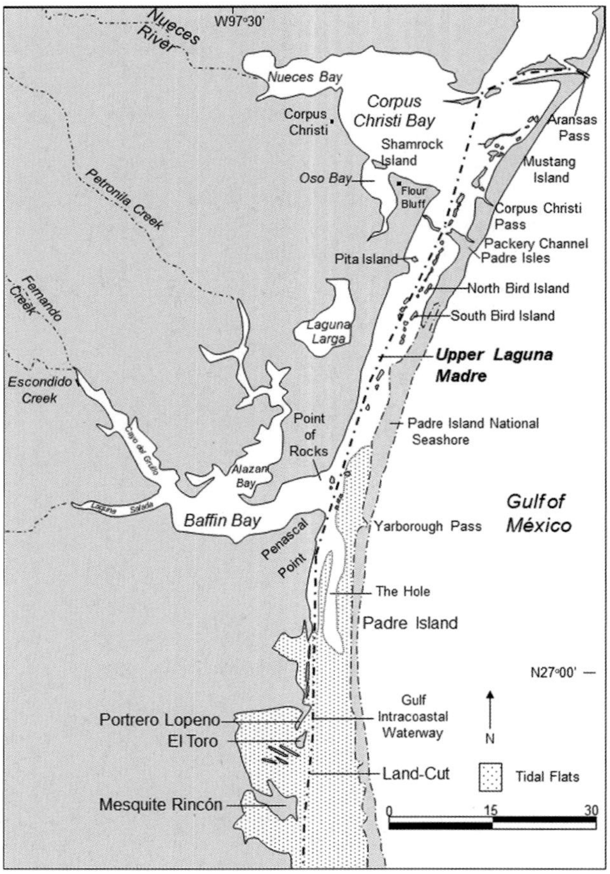

Figure 2.2. Upper Laguna Madre of Texas

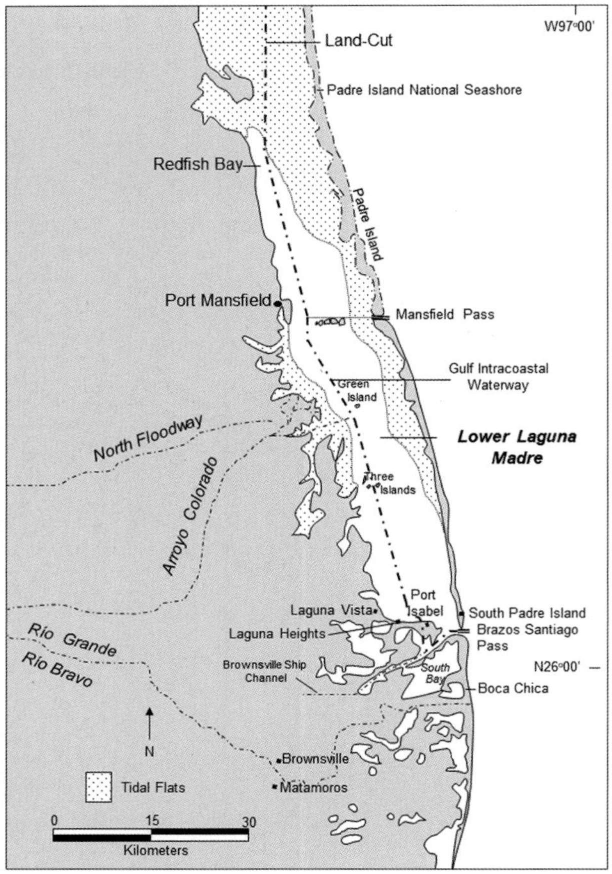

Figure 2.3. Lower Laguna Madre of Texas

(see table 2.1). Some data in table 2.1 conflict because they were not all obtained using the same method; however, similarities between the two systems are evident. Diener's (1975) work in Texas is the most detailed and frequently cited, but data are reported as Mean Low Water (MLW) and Mean High Water (MHW) and seem quite conservative when compared to data from the Texas Water Development Board (TDWR 1983) reported as Mean Sea Level (MSL). Perhaps the most recently compiled data from remotely sensed satellite data are the most accurate for Mexico (DUMAC 1996, from database, Monterrey, Nuevo León).

Islands Natural islands are more common in Laguna Madre de Tamaulipas than in Texas. Over 100 islands are found in the northern portion (fig. 2.6), and slightly fewer than 100 are located in the southern part. Many of the islands, especially in the north (e.g., Isla las Flores, Isla del Vaquero), are erosional Rio Grande Delta remnants with steep sides and are inhabited by small fishing villages. Others are low, sedimentary shoals and spits, especially behind the barrier islands and in the narrow Southern Laguna (e.g., Isla la Vaca, Isla la Pita). In Texas, only Crane Island, Pita Island, and North and South Bird Islands (fig. 2.7) exist as natural islands in the Upper Laguna, and Green Island and Three Islands are found in the Lower Laguna. Many other "islands" such as Mullet Island, Hawk Island, and Yucca Island, are located along the mainland shore in the Lower Laguna and on the inland side of Padre Island (Deer Island and Los Bancos de en Medio), and others can be found in, and just north of, the

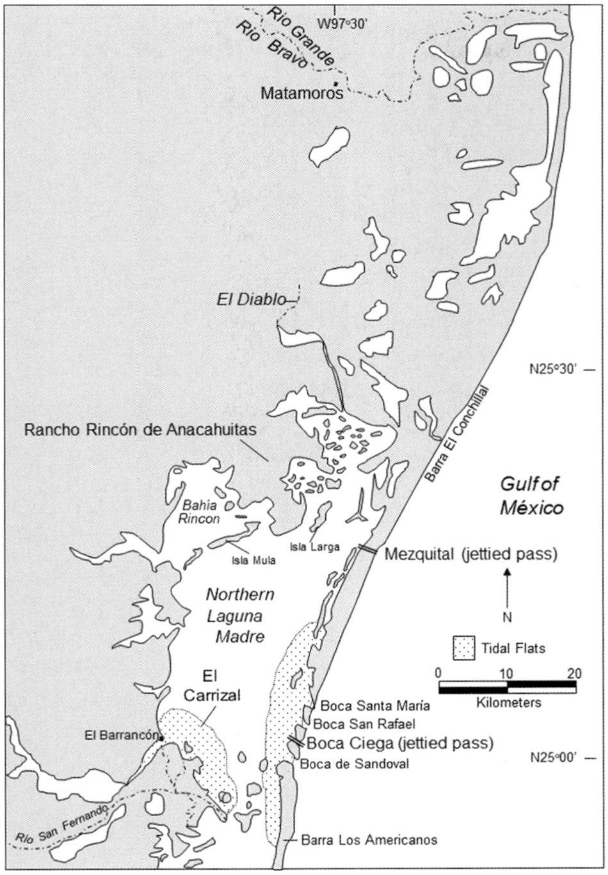

Figure 2.4. *Northern Laguna Madre de Tamaulipas*

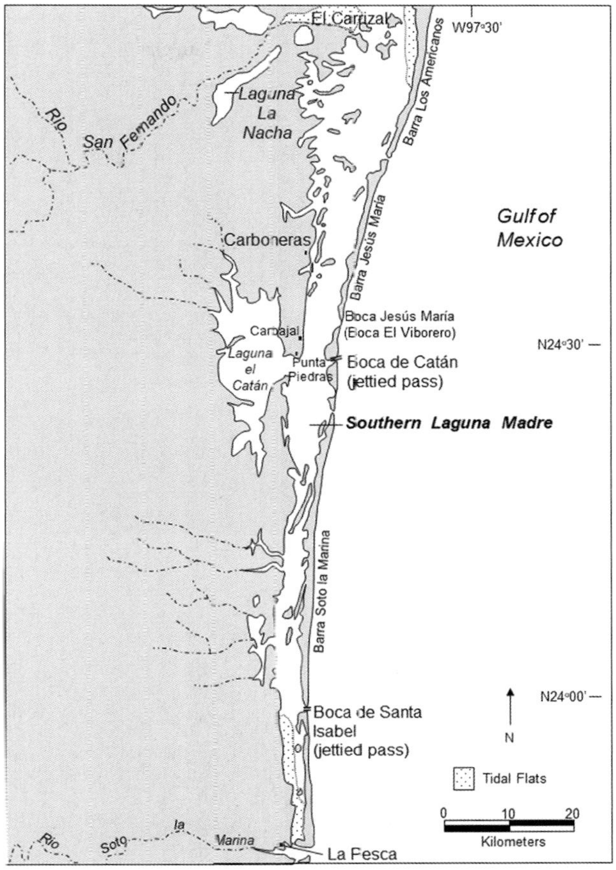

Figure 2.5. *Southern Laguna Madre de Tamaulipas*

Land-Cut, such as Trés Marías, El Toro, and Potrero Lopeno (fig. 2.8). These so-called islands are unique in that they are not surrounded by water but by wind-tidal flats except during high wind tides or storm tides. These tidal-flat islands are referred to locally as *lomas* (hills, for those that are round to irregular) or *potreros* (pastures, for those elongate and parallel to the mainland shore). All offer distinctive, island-like habitats to their flora and fauna.

Although natural islands are few in the Laguna Madre of Texas, dredged-material islands are widespread and abundant along the GIWW (fig. 2.9) and associated oil and gas field channels (fig. 2.10). Many of these islands become vegetated and stabilize over time—some are used as rookery islands for colonial nesting waterbirds or for oil and gas activities, and small fishing cabins are built on others. In open, more exposed waters, the islands tend to erode faster. Maintenance dredging of the GIWW by the US Army Corps of Engineers requires periodic placement of new materials on old or new sites. Locating new sites for placement has become increasingly difficult because of occupation or use of dredged-material islands by humans and birds and the negative impact on aquatic resources caused by open-water placement.

Counties and Population Centers Political subdivisions of the Laguna Madre and adjacent lands include five counties in Texas (from north to south)—Nueces, Kleberg, Kenedy, Willacy, and Cameron—and three in Tamaulipas—Matamoros, San Fernando, and Soto la Marina. Population centers exist at Corpus Christi in the northernmost,

Figure 2.6. *Oblique aerial view to the south over some of the many natural islands in the northernmost Laguna Madre de Tamaulipas (Photograph by J. W. Tunnell, February 1999)*

Figure 2.7. *Oblique aerial view to the north over Pita Island (right, mid-ground), one of the few natural islands in the Upper Laguna Madre of Texas (Photograph by J. W. Tunnell, January 1999)*

Figure 2.8. *Oblique aerial view to the northeast over the wind-tidal flats and El Toro "island" in the Land-Cut, Kenedy County, Texas; GIWW in background. Vegetated islands in foreground are composed of dredged material from an oilfield channel. (Photograph by J. W. Tunnell, July 1987)*

Figure 2.9. *Vertical high altitude (18,288 m [60,000 ft]) aerial photograph of Lower Laguna Madre over the Arroyo Colorado. Note numerous dredged-material islands adjacent to GIWW and oilfield channels. (USGS, 1973)*

Figure 2.10. *Aerial view of oil and gas channels and dredged-material islands just south of John F. Kennedy Causeway, Upper Laguna Madre, Texas (Photograph by J. W. Tunnell, January 1999)*

Table 2.2. Population (2010 US Census) of counties, cities, and towns along the shores of Laguna Madre of Texas

County (North to South) City/Town	Total County Population	City/Town Population
Nueces County	340,223	
Corpus Christi		305,382
Kleberg County	32,061	
No towns		
Kenedy County	416	
No towns		
Willacy County	22,134	
Port Mansfield		226
Cameron County	406,220	
Laguna Vista		3,117
Laguna Heights		3,488
Port Isabel		5,006
South Padre Island		2,816
Total	**801,054**	**320,035**

Source: https://www.txcip.org/tac/census.

Upper Laguna Madre; Port Mansfield along the south-central western shore in the Lower Laguna Madre; and Laguna Vista, Laguna Heights, Port Isabel, and South Padre Island on the southernmost Lower Laguna Madre (table 2.2). Only Padre Isles (a subdivision of Corpus Christi; fig. 2.11) and South Padre Island (fig. 2.12) are communities located on the barrier island of Padre Island. Due to the large landholdings of two federally protected areas, Padre Island National Seashore and Laguna Atascosa National Wildlife Refuge, and several large privately owned ranches, King Ranch, Kenedy Ranch, Yturria Ranch, and others, no other communities exist along the shores of the Texas Laguna Madre.

The Laguna Madre is remote and isolated in Tamaulipas, so there are only small fishing villages and no large cities along its shores (table 2.3). The city of Matamoros is the nearest large metropolitan area, and it is located over 40 km (25 mi) to the north of the northernmost point of the lagoon. Mezquital and Higuerillas are the only villages located on the outer coast, and they are situated on the northwestern bank of Mezquital Pass. On the mainland shoreline, east of the town of San Fernando, Carboneras is the largest town on the Laguna and has one of only two good paved roads reaching the Laguna's western shore. The other road, completed in the late 1990s, extends through Los Vergeles to El Barrancó. At its southernmost extent, the fishing and hunting village of La Pesca is located just west of the Laguna on the bank of Río Soto la Marina. Numerous small fishing villages exist on the islands and shores of the Laguna Madre de Tamaulipas.

Population around the Laguna Madre is 801,054 in five counties in Texas and 571,177 in three counties in Tamaulipas (see tables 2.2 and 2.3). In cities, towns, and villages immediately adjacent to the Laguna, the Texas population is 320,035 and that in Tamaulipas is 12,729. Texas has fewer but larger population centers, whereas Tamaulipas has more but smaller ones.

Figure 2.11. *Oblique aerial view to the northwest over Padre Isles development on northern Padre Island (Photograph by J. W. Tunnell, January 1999)*

Figure 2.12. *Oblique aerial photo to the north over the hotel-condominium zone on South Padre Island, Texas (Photograph by J. W. Tunnell, September 1992)*

Table 2.3. *Population (2010 Mexico census) of counties, towns, and villages along the shores of Laguna Madre de Tamaulipas*

County Town/Village	Total County Population	City/Town Population
Matamoros County	489,193	
La Bartolina		337
El Huizachal		112
El Tecolote		168
La Amistad		9
San Pedro		6
Las Yesquitas		8
Cinco de Mayo		256
El Caracol		0
Las Flores		6
La Capilla		522
Higuerillas		2,139
Mezquital		325
San Fernando County	57,220	
Gral. Francisco J. Mujica		492
La Media Luna		104
El Mezquite		16
El Barrancón del Tío Blas		1,050
Loma del Agua		1
El Carrizal		61
San Isidro Sur		2
Los Ebanos		20
El Arpa		0
Carboneras		2,693
Punta de Alambre		824
Los Legales		0
Carvajal		891
Lavaderos		74
La Florida		71
Soto la Marina County	24,764	
San José de los Leones		6
Enramadas		205
El Carrizo		315
Las Guayabas		98
Las Flores		0
Benito Juárez		125
La Pesca		1,764
Total	**571,177**	**12,700**

Source: Instituto Nacional de Estadística y Geografía (INEGI): https://www.inegi.org.mx/programas/ccpv/2010/.

Note: Four villages (El Caracol, El Arpa, Los Legales, and Las Flores) were apparently abandoned during the 20 years between the 1990 (Tunnell 2002) and 2010 censuses.

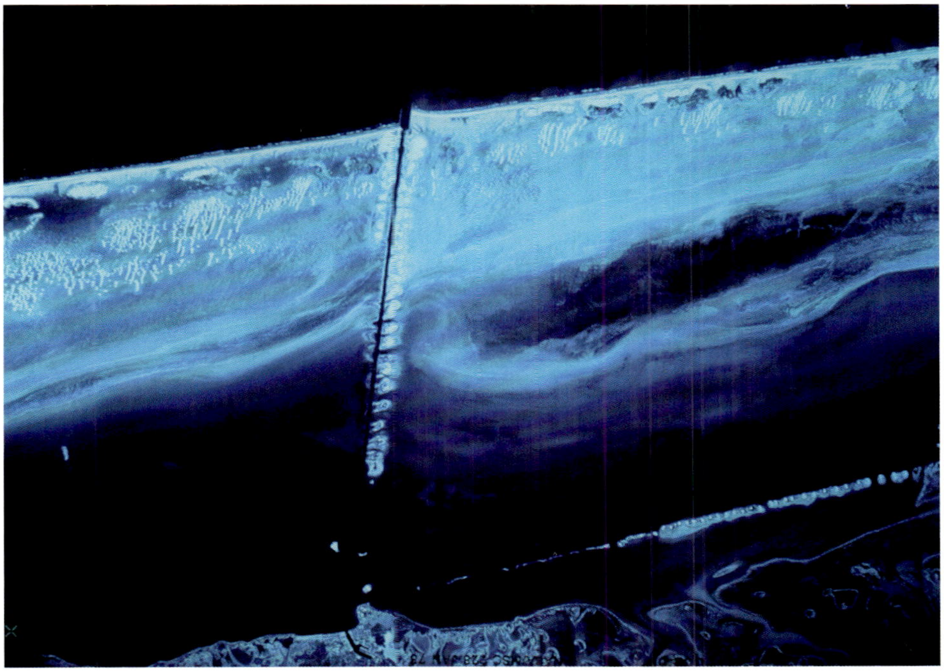

Figure 2.13. *Vertical high-altitude (18,288 m [60,000 ft]) photograph of Mansfield Pass, which was dredged through Padre Island to connect the Laguna Madre with the Gulf of Mexico (USGS, 1973)*

Barrier Islands, Peninsulas, and Passes (Tidal Inlets) Barrier islands and peninsulas protect the shallow, calm Laguna Madre from the higher-energy Gulf of Mexico. In Texas, Padre Island extends the entire length of the Laguna Madre, interrupted only by Mansfield Pass, a human-made inlet into the Lower Laguna Madre (fig. 2.13; see fig. 2.3 for map location). The northern half of Padre Island is slightly higher in elevation and more densely vegetated, thus provides more lagoonal protection from storms and has fewer washover passes. Brazos Santiago Pass, the entrance to the Brownsville Ship Channel, forms the southern coterminous of the Lower Laguna Madre and Padre Island. Brazos Island (now a peninsula, since Boca Chica, a small natural pass, closed in 1945) protects South Bay, the former southernmost part of the Lower Laguna Madre before it was cut off by the Brownsville Ship Channel in 1938 (Breuer 1962).

In Tamaulipas, the northernmost Laguna Madre is protected by a barrier peninsula, Barra el Conchillal (see fig. 2.4). To the south of this moderately wide (avg. 2.0 km [1.2 mi]) peninsula lie three narrow (500 m [1,640 ft] or less) elongate barrier islands: Barra los Americanos, Barra Jesús María, and Barra Soto la Marina (see fig. 2.5). Eight (Hildebrand 1980) to 13 (Leija-Tristán et al. 2000) or more intermittent passes (*bocas*) have been recorded or observed between the Laguna Madre de Tamaulipas and the Gulf of Mexico. These ephemeral passes were discussed at length by Hildebrand (1958, 1969, 1980) and are considered a key element governing physical hydrography and hence, productivity of the Laguna (fig. 2.14). Boom (open passes) and bust (closed passes) cycles of fisheries productivity have been linked to tropical storms and attendant, or subsequent, heavy precipitation and freshwater inflow into the system. To level out these boom and bust cycles, stop drying up of the Laguna, and allow navigation into and out of the Laguna, the Mexican government dredged and jettied four passes during the 1970s and early 1980s: El Mezquital (fig. 2.15), Boca Ciega (fig. 2.16), Boca de Catán (fig. 2.17), and Boca de Santa Isabel (fig. 2.18). El Mezquital, located on the Barra el Conchillal Peninsula, is the only pass that can be reached by car. Boca Ciega Pass is located at the southern end of the peninsula in an area of four historical,

Figure 2.14. *Oblique aerial view of narrow barrier island with two ephemeral, or intermittent, washover passes (bocas) between the Gulf of Mexico (top) and Laguna Madre de Tamaulipas (bottom)*

Figure 2.15. *Oblique aerial photo to the east toward the Gulf of Mexico over El Mezquital Pass, Tamaulipas (Photograph by J. W. Tunnell, February 1999)*

Figure 2.16. *Oblique aerial photo of Boca Ciega Pass to the west from the Gulf of Mexico across tidal flats and seagrass beds into Laguna Madre de Tamaulipas (Photograph by J. W. Tunnell, February 1999)*

Figure 2.17. *Oblique aerial photo to the east toward the Gulf of Mexico from Laguna Madre de Tamaulipas over Boca de Catán (Photograph by J. W. Tunnell, February 1999)*

Figure 2.18. *Oblique aerial photo to the east toward the Gulf of Mexico from Laguna Madre de Tamaulipas over Boca de Santa Isabel (Photograph by J. W. Tunnell, February 1999)*

ephemeral passes (north to south): Boca Santa María, Boca San Rafael, Boca Ciega, and Boca Sandoval. Southward the barrier islands of Barra los Americanos and Barra Jesús María extend to Boca de Catán opposite Laguna el Catán (aka Bahía Algodones on some maps). The barrier island of Barra Soto la Marina then extends southward to the Río Soto la Marina and is cut through by Boca de Santa Isabel about 57 km (35 mi) south of Boca de Catán.

Major Lagoon Habitats and Features Submergent seagrass meadows and emergent wind-tidal flats make up the most extensive habitat types of both lagunas. Both of these habitats are covered separately within this volume (see chapters 7 and 9, respectively). Wind-tidal flats are most common on the Padre Island side of Laguna Madre, especially south of the Land-Cut, in the Land-Cut, and along the Lower Laguna Madre mainland shore in Texas. In Tamaulipas, the broadest flats are located behind the barrier islands in the Boca Ciega Pass area, the El Carrizal or Río San Fernando Delta, and the lowermost Southern Laguna Madre near La Pesca. Seagrass beds dominate the shallows of both systems.

Mainland shorelines of both lagunas are typically of low relief, consisting of broad tidal flats or narrow beaches with vegetated berms, but erosional scarps up to 2.0 m (6.6 ft) above MSL are found in some areas of the northernmost Laguna Madre de Tamaulipas (fig. 2.19). Two other unique shoreline features are clay dunes and beach rock (coquina). Clay dunes are unique geological features that were first identified and described in South Texas along the Lower Laguna Madre mainland shores (Coffey 1909; Price and Kornicker 1961). Beach rock exists along a small segment of both the Texas (Prouty 1996) and Tamaulipan shoreline (Yañez and Schlaepfer 1968). See chapter 3 for details and photographs.

Figure 2.19. *View to the east along eroding clay bluff just off the highway leading to Mezquital in Northern Laguna Madre de Tamaulipas (Photograph by J. W. Tunnell, March 1991)*

Drainage Systems into Lagoons Drainage from inland watersheds into Laguna Madre is minimal. No rivers flow into the Laguna Madre in Texas. Within the Upper Laguna Madre several ephemeral creeks—San Fernando, Santa Gertrudis, Los Olmos, and others—flow into Baffin Bay, but only when it rains. In the Lower Laguna, Arroyo Colorado, a northern distributary channel of the Rio Grande Delta system and dredge-maintained channel, and the North Floodway (see fig. 2.3) drain much of the Rio Grande Valley agricultural lands. Although the Rio Grande at one time flowed into South Bay, it no longer directly influences any of the Laguna during normal flows and drains directly into the Gulf of Mexico.

In Tamaulipas, besides numerous dry creeks and arroyos, which flow only with substantial rains, the sources of fresh water to the Laguna are the Río San Fernando and drainage from lands irrigated from the Rio Grande in the extreme north. Flow from the Río San Fernando has decreased in recent years, as the waters have been diverted for irrigation and the sediment load has increased as a result of extensive clearing of lands for agriculture (Yañez and Schlaepfer 1968; Hildebrand 1980; Saunders and Saunders 1981). Most recently, the Río San Fernando has been diverted away from Laguna Madre into Laguna la Nacha (Brenner 1997; DUMAC 1993), then on to the Laguna Madre via a canal. Additionally, a new channel has been dug to the lagoon in the north to drain irrigation and storm waters from the Rio Grande Delta (H. Hildebrand, pers. comm.).

Upland Ecosystems and Features Landward of the Laguna Madre, terrestrial ecosystems are part of the Pleistocene Coastal Plain, including the Gulf Prairies and Marshes and Rio Grande Plains in Texas (Gould 1975; Chapman and Bolen 2018) and the Northeast Coastal Plain Province in Tamaulipas (Rzedowski 1978). The Tamaulipan Biotic Province encompasses both lagoon systems (Dice 1943; Blair 1950; see chapter 4). The topography is gently sloping to rolling. Agriculture is the main land use to the immediate west of the Laguna Madre in both countries, with cattle ranching dominating in most areas. On the Rio Grande Plains, however, cultivated lands dominate on both sides of the border. In low areas along the coast, flats dominated by Gulf Cordgrass are common. Just landward of these flats, the mesquite-grassland association is most common, except where the oak savanna dominates the Coastal Sand Plain in Kenedy County just south of Baffin Bay (Fulbright et al. 1990; Fulbright and Bryant 2002; Chapman and Bolen 2018; see chapter 11).

The Rio Grande Delta region between the two lagoonal systems forms a complex system of coastal plain uplands and wetlands, *lomas*, clay dunes, old Rio Grande (Río Bravo) distributary channels, oxbow lakes (*resacas*), playa lakes, wind-tidal flats, and small lagoons. Because of its diverse vegetative communities and migratory birds, this highly complex area of varying geomorphological features and biological habitats has been recognized as important in Mexico, where most of the delta lies—63 km (39 mi) of its 75 km (47 mi) width (DUMAC 1993). The least altered or impacted upland areas along the Laguna Madre de Tamaulipas appear to be Rancho Rincón de Anacahuitas, located on a large peninsula in the northern lagoon (J. Bergan, pers. comm.), the two peninsulas of Laguna el Catán, and the remote mainland areas south of Laguna el Catán toward La Pesca (DUMAC 1993).

Climate

The Laguna Madre region has been classified variously as semiarid (Thornthwaite 1948; Hedgpeth 1953) or subtropical steppe (Norwine et al. 1977). More recently, and precisely, the region was characterized as a subhumid-to-semiarid east-coast subtropical climate, with extreme variability in precipitation (Fulbright et al. 1990). Additional characterization includes high temperatures, along with deficiencies of moisture, and a combination of high humidity and infrequent but significant killing frosts (Fulbright et al. 1990). Interestingly, it has also been further classified as a "problem climate" (Trewartha 1961, 1968) in that it is located between two areas of substantially higher rainfall. The coasts of East Texas to the north and Veracruz to the south receive more than double the rainfall of the Laguna Madre, which makes the area unique worldwide. Significant coastal drainages are essentially lacking in the Laguna Madre region of the western Gulf of Mexico, where only two rivers, the Rio Grande and Río San Fernando, reach the coast along this 445 km (277 mi) of coastline.

Although detailed climatic data are available from long-term National Oceanic and Atmospheric Administration (NOAA) Local Climatological Data stations in Corpus Christi and Brownsville, similar data for Tamaulipas are difficult to access and of shorter duration, and stations have not been continuously occupied or located on the shores of Laguna Madre de Tamaulipas (Hildebrand 1980). Wind, rainfall, temperature, tropical storms, and hurricanes all have a significant effect on the biota of the area, and they are important influences on the hydrographic and geological processes of the region (Brown et al. 1976; Brown et al. 1980).

A growing body of literature suggests that the South Texas region (Norwine and John 2007) and the Texas coast (White et al. 2006; Montagna et al. 2007; Montagna et al. 2011) will experience significant climate change–related effects on water quality, habitats, and species. Twentieth-century climatic data trends for temperature and precipitation modeled forward for the twenty-first century predict the following global warming challenges for South Texas (Norwine et al. 2007):

- increases in high-intensity rainfall events
- increases in the average summer heat index because there are more hot days and hot nights (= increased temperatures and humidities)
- fewer instances of subfreezing temperatures (trending toward tropical)
- increased evapotranspiration
- climatic shift of easternmost South Texas from subhumid to semiarid

Along the coast, temperature, rainfall, and sea-level rise are projected to have the greatest impact. Two direct effects, which are already observable, in the past instrumental record are rapid sea-level rise and rising sea temperatures (Montagna et al. 2011). High subsidence levels in some areas exacerbate sea-level rise, as does the flat topography of the coastal area (Twilley et al. 2001; Montagna et al. 2007). Tropical Black Mangroves are expanding northward, and even cold-sensitive Red Mangroves are showing up on the Texas coast (see chapter 19). Indirectly, warmer water will cause lower solubility of oxygen and therefore hypoxic (low-oxygen) zones. These are already known to be expanding in number, extent, and duration and can be very destructive to coastal ecosystems, leading to lower biomass, productivity, and diversity and possibly altered food webs (Montagna et al. 2011).

Winds Southeasterly winds from the Gulf of Mexico prevail for most months of the year at the Laguna Madre, with northerly winds interrupting southeasterly winds during winter storms. At Corpus Christi, persistent southeasterly winds blow from March through September and north-northeasterly winds occur with winter frontal passages from October through February (Behrens and Watson 1973; Brown et al. 1976). Persistent southeasterly winds in Brownsville blow from March through November, and short-lived but strong northerly winds occur from December through February (Brown et al. 1980). Conditions are similar for the Laguna Madre de Tamaulipas region (DUMAC 1993). From a hydrographic viewpoint, wind forcing is one of the most important meteorological influences on developing coastal environments. In the inland or isolated reaches of the Laguna Madre, water levels are significantly affected by wind forcing (Ward and Armstrong 1997). Wind stress can cause water level in the lagoon to set down (lower) and set up (rise) on upwind and downwind sides of bays as much as 0.6 to 0.9 m (2.0–3.0 ft) (Fisk 1959; McGowen et al. 1977; Morton and McGowen 1980; Weise and White 1980).

Humidity At the centrally located city of Brownsville the average humidity at 6:00 a.m. ranges from 85% in December to 90% in July with an annual average of 88%. This high relative humidity causes fairly heavy dew even during drought years. Dense fogs are common at night from December through March (Lonard, Everitt, et al. 1991).

Precipitation Average annual rainfall decreases slightly along the South Texas coast from Corpus Christi (76.8 cm [30.2 in]) to Brownsville (65.2 cm [25.8 in]), then increases southward along the Tamaulipan coast to Soto la Marina (69.3 cm [27.2 in]), just inland from La Pesca (table 2.4). Rainfall in Texas is typically bimodal with higher values in May–June and August–October, peaking in September (table 2.4). Approximately one-third of the total annual rainfall is received during August–October in Corpus Christi and September–October in Brownsville. Additionally, about two-thirds of the annual rainfall falls during the six-month period from May through October. This division corresponds closely with the "rainy season" that prevails in the tropical region of eastern and central Mexico (A. Davis 1942). Most of the precipitation results from thunderstorms, and a single thunderstorm often will account for an entire month's rainfall. Flood-producing rains may occur in any season but are most frequent in September and October. Heavy rain is often associated with tropical storms and hurricanes.

Precipitation values alone are not necessarily significant, but the variability among months, years, and long-term cycles (Price 1958; Norwine et al. 1977; Hildebrand 1980; Norwine et al. 2007) coupled with high evapotranspiration (J. Carr 1967; Brown et al. 1976; Brown et al. 1980) all make for a complex and highly unpredictable water budget. For example, evaporation averages 147 cm (57.9 in) annually from a free water surface, which is more than double the average annual precipitation of 65.2 cm (25.8 in) in Brownsville. Annual rainfall in Corpus Christi since 1900 has ranged from 13.7 cm (5.4 in) in 1917 to 127.6 cm (50.3 in) in 1992. Rainfall for the Laguna Madre region should also be characterized as irregular and unpredictable for seasons, years, and decades. Behrens (1966), Price (1968), and Hildebrand (1980) have emphasized this extreme rainfall variability, as well as long-term climatic variability. The 1950s drought lasted for 80 months (Price 1968), and dry years and drought years persisted from about 1945 to 1965 (Norwine et al. 1977; Hildebrand 1980; Quammen and Onuf 1993), exhibiting multiple-decade dry periods. Recent studies indicate that South Texas experiences increased precipitation in El Niño years, especially in winter (Tolan 2007).

Table 2.4. *Comparison of average precipitation and temperature by months for Corpus Christi, Brownsville, and Soto la Marina*

Month	**Precipitation cm (in)**		
	Corpus Christi	**Brownsville**	**Soto la Marina**
January	3.8 (1.5)	3.3 (1.3)	4.1 (1.6)
February	4.3 (1.7)	3.0 (1.2)	1.0 (0.4)
March	3.8 (1.5)	2.5 (1.0)	1.9 (0.7)
April	5.1 (2.0)	3.8 (1.5)	3.6 (1.4)
May	8.1 (3.2)	5.8 (2.3)	3.8 (1.5)
June	7.9 (3.1)	7.1 (2.8)	8.9 (3.5)
July	5.6 (2.2)	4.8 (1.9)	7.1 (2.8)
August	7.9 (3.1)	6.1 (2.4)	9.7 (3.8)
September	13.5 (5.3)	13.2 (5.2)	15.5 (6.1)
October	8.4 (3.3)	8.6 (3.4)	7.6 (3.0)
November	4.3 (1.7)	4.0 (1.6)	2.8 (1.1)
December	4.1 (1.6)	3.0 (1.2)	3.3 (1.3)
Annual average	76.8 (30.2)	65.2 (25.8)	69.3 (27.2)
	Temperature °C (°F)		
January	13.7 (56.7)	16.1 (60.9)	17.3 (63.1)
February	15.7 (60.2)	16.8 (62.2)	19.6 (67.3)
March	18.9 (66.1)	20.5 (68.9)	22.4 (72.3)
April	22.6 (72.7)	23.4 (74.2)	25.3 (77.5)
May	25.7 (78.2)	26.4 (79.5)	28.3 (82.9)
June	28.1 (82.6)	28.1 (82.5)	29.6 (85.3)
July	29.1 (84.4)	28.9 (84.0)	29.1 (84.4)
August	29.3 (84.7)	29.3 (84.8)	29.5 (85.1)
September	27.4 (81.3)	27.2 (81.0)	27.9 (82.2)
October	23.6 (74.4)	24.4 (75.9)	24.9 (76.8)
November	18.7 (65.6)	19.9 (67.8)	21.4 (70.5)
December	15.0 (58.6)	16.5 (61.7)	17.9 (64.2)
Annual average	22.3 (72.2)	23.1 (73.6)	24.4 (76.0)

Source: US data are from Corpus Christi, Nueces County (period of record = 70 years), and Brownsville, Cameron County, Texas (period of record = 70 years) from NOAA Local Climatological Data Annual Summary, Almanacs for Corpus Christi and Brownsville: https://www.ncdc.noaa.gov/cdo-web/datatools/lcd. Mexico data are from WeatherBase for Soto la Marina; periods of record = 60 years for temperature and 33 years for precipitation: http://www.weatherbase.com/weather/weather.php3?s=99467.

Temperature The region is characterized by long, hot summers and short, mild winters (see table 2.4). Temperatures in Nueces County range from average winter lows of 8.3 °C (46.9 °F) to average summer highs of 33.3 °C (91.9 °F), with an average annual mean in the Corpus Christi area of 22.0 °C (71.6 °F) (Brown et al. 1976). The average winter low in Cameron County is 8.9 °C (48.0 °F), and the average summer high is 36.1 °C (97.0 °F), with an average annual mean in the Brownsville area of 23.1 °C (73.6 °F) (Brown et al. 1980). Along the Tamaulipan coast the average annual temperature ranges between 22 °C and 24 °C (71.6 °F and 39.2 °F) (INEGI 1994) with an overall regional average of 27.7 °C (81.9 °F) (DUMAC 1993).

The growing season along the South Texas coast varies from 300 days just north of Corpus Christi in San Patricio County (Everitt and Drawe 1993) to 330–365 days in Cameron County in the extreme southern end of the state (F. Judd, pers. comm.). Although freezes occur occasionally in Corpus Christi and rarely in Brownsville, two killing freezes occurred in the 1980s throughout South Texas and extended into northeastern Mexico. During 24–26 December 1983, 55 consecutive hours below freezing were reported at Brownsville, and a minimum of -6.7 °C (19.9 °F) was sustained for six hours (Lonard and Judd 1985). During the massive cold wave of 22–24 December 1989, two freezing episodes were reported at Brownsville: one with 33.75 consecutive hours below freezing and a second with 16.75 hours below freezing. A temperature at or below -8.4 °C (16.9 °F) was recorded for 11.75 hours (Lonard and Judd 1991). Freeze damage to vegetable and citrus production was estimated at $200 million in 1983 and $138 million in 1989 (Lonard and Judd 1985, 1991). The two major freezes in 1983 and 1989 also caused shallow waters to freeze in the Upper and Lower Laguna Madre that resulted in major fish kills (Martin and McEachron 1996). In 1989, surface ice formed in the uppermost Laguna Madre from the shoreline of Flour Bluff all the way across the lagoon in the JFK Causeway area to the Padre Island shoreline (J. Bronson, pers. comm.). During this freeze, mangrove communities lining the Río Soto la Marina shoreline at La Pesca were killed (J. W. Tunnell, pers. observ.).

Tropical Storms and Hurricanes Major hurricanes have had significant impacts on the natural systems of South Texas (Hayes 1967; Scott et al. 1969; McGowen et al. 1970; McGowen and Scott 1975; Judd and Sides 1983; Keim and Muller 2009) and northeastern Mexico (Hildebrand 1958, 1969, 1980). Historically, both lagoonal systems were completely (Tamaulipas) to partly (Texas) closed. It appears they were primarily driven over long-term cycles by being opened and closed due to physical processes of storm surges and torrential rainfall and runoff associated with hurricanes. Tidal inlets and washover passes are widened and deepened by hurricanes and form connections with the open sea (Gulf of Mexico). In time, after the storm, passes begin to close and return to normal conditions. The microtidal coastline in the region does not allow for sufficient water head and scour to maintain but a few passes in the absence of river outflow. The deeper, artificially maintained (dredged) passes today carry almost all flow, and storm-opened passes close rapidly. In the Laguna Madre of Texas, dredging of the GIWW and Mansfield Pass has permanently altered hydrology, circulation, and salinity. Early researchers in Tamaulipas (Copeland 1967) suggested that studies of Mexican hypersaline environments without artificial passes could allow learning by inference of how the Laguna Madre in Texas functioned before channelization. Since those early studies, however, dredged passes have been cut into the Tamaulipan lagoon.

Tropical storms and hurricanes strike the Texas coast about once every two years (McGowen et al. 1977). Hurricanes are severe tropical storms with cyclonic (counterclockwise) winds, covering thousands, sometimes hundreds of thousands, of square kilometers (Henry et al. 1980; Keim and Muller 2009). These storms are characterized by high-velocity winds, a calm area (eye) at the center, low barometric pressure, torrential rainfall, and tornadoes. Additionally, as the storm approaches land, the tidal surge causes a marked rise in water level, especially on the north side of landfall. Historical records indicate that hurricanes not only differ in intensity (Categories 1–5), but they can also differ in other characteristics. For instance, in South Texas, one hurricane (Hurricane Carla in 1961) may generate a large storm surge, another (Beulah, 1967) may be remembered for torrential rainfall, and yet another (Celia, 1970) may be characterized by extreme wind velocities (Brown et al. 1976; Brown et al. 1977; Brown et al. 1980; Keim and Muller 2009).

Depending on intensity and other characteristics, tropical storms and hurricanes may have various impacts on coastal geological processes, local biota, and ecology. Geologically, these storms accelerate coastal processes so that during only the few hours of storm passage, the degree of erosion and deposition in coastal systems might equal what normally would take months or even years to accomplish (Price 1958; Hayes 1967; McGowen et al. 1970; McGowen and Scott 1975; Brown et al. 1976; Brown et al. 1977; McGowen et al. 1977; Morton and McGowen 1980; Brown et al. 1980). Although detailed biological and ecological studies are few, opportunistic studies reveal hurricane impacts on the vegetation of South Padre Island (Judd and Sides 1983), marshes in the northern Gulf of Mexico (Harris and Chabreck 1958; Chabreck and Palmisano 1973), and southeastern coastal landscapes (Gardner et al. 1991; Gardner et al. 1992).

Within the Laguna Madre of Texas, Hurricane Beulah in September 1967 flushed the system first with marine waters from the Gulf of Mexico as a result of the storm surge, then with fresh water from rainfall and runoff (Behrens 1969). Prior to this storm, and after several consecutive dry years, average surface salinities ranged from 63 ppt in Baffin Bay to 55 ppt in the Upper Laguna Madre. Highest salinities exceeded 75 ppt during this time at the upper reaches of Baffin Bay. On 20 September 1967, Hurricane Beulah caused hurricane surges of 1.8–2.1 m (6.0–7.0 ft) from Port Aransas to Port Mansfield. Dilution of hypersaline waters with Gulf water gave estimates of 40 ppt in Baffin Bay and 35 ppt in the Upper Laguna Madre. Accompanying and following the storm's passage, some 51 cm (20 in) of rain fell on most of the area. This rainfall exceeded the all-time monthly records for the area and the total rainfall for some years. For several weeks the water in the system remained stratified, the heavier salt water being overlain by fresher water. After northerly winds promoted mixing, there was essentially no stratification remaining, and salinities in Baffin Bay and Laguna Madre were about 6 ppt and 12 ppt, respectively. Other impacts included sand deposited in the lagoon from the barrier island, mud carried into the bay by flooded streams and deposited, some carbonate solution, and a temporary change in fauna. In summary, Hurricane Beulah drastically altered the hydrology of Baffin Bay and the Laguna Madre, but it was an almost insignificant event in the depositional history of the system. Both marine and fresh waters were able to drain quickly from the Laguna Madre of Texas due to the GIWW canal system (Behrens 1969).

In 1999, Hurricane Bret made landfall on central Padre Island between Corpus Christi and Brownsville. Before going ashore, Bret had Category 4 winds of 225 kph (140 mph) traveling up from the Bay of Campeche in the southern Gulf of Mexico. After the storm, Richard Watson (marine geologist, pers. comm.) made aerial and land surveys along Padre Island, reporting 12 washover passes reopened and extending across the island and onto the wind-tidal flats as washover fans. Watson reported a maximum storm surge on the outer beach of about 1.8 m (6.0 ft). Within the Laguna Madre in the Land-Cut, the El Toro tide gauge (Blucher Institute of Surveying and Science) recorded a tide height of 1.2 m (4.0 ft) before it blew over, but the tide height probably was not much higher (R. Morton, pers. comm.). Chris Onuf (pers. comm.) flew over the Laguna Madre seagrass beds and reported "no evidence of storm damage on the lagoon side in the upper Laguna Madre." Therefore, although threatening great danger as a Category 4 hurricane, Bret had only a modest impact on Padre Island and the Laguna Madre. Since the storm weakened to Category 3 before landfall and passed over the sparsely inhabited lands of the Kenedy and King Ranches, little human impact occurred in the vicinity of the Laguna Madre.

Major hurricanes are a key force driving the hydrography and fisheries of the Laguna Madre de Tamaulipas system (Hildebrand 1958, 1969, 1980). Although the Mexican lagoon is located on a hurricane coast, major storms are rare events. They seem to be the only weather phenomenon able to reestablish the natural productivity of the lagoon after an extended drought. Hildebrand (1980) suggested that in this sparsely inhabited area, the effects of hurricanes are more positive than negative—major hurricanes open passes through the barrier islands and flush the lagoonal system with rainfall floodwaters, eliminating extreme hypersaline conditions in the lagoon. Fisheries flourish for several years afterward until passes close and the system becomes extremely hypersaline and unproductive again, completing the cycle. The hurricanes of 1909, 1933, and 1967 set off major peaks in fisheries productivity, and the *sequias* (droughts) of 1883–90 and 1960–67 were major negative events (H. Hildebrand, pers. comm.).

Hurricane season lasts from June through November, but most storms occur between August and October. In any given year, the 80 km (50 mi) segment of the southernmost Texas coast has a 9% probability of experiencing a tropical storm, an 8% probability of experiencing a hurricane, and a 2% probability of experiencing a major hurricane (Simpson and Lawrence 1971). During the twentieth century, over 50 hurricanes hit or affected the Texas coast, 24 of which were considered major, and 9 of those impacted the Laguna Madre of Texas (Henry et al. 1980; Broussard and Martin 1986; Ellis 1986).

Historical records show that 62 hurricanes affected the Mexican Laguna Madre between 1900 and 1979 (Hildebrand 1980). The maximum number of hurricanes in any single year was five in 1933, and the longest period without any hurricanes was 1904 through 1908. Three major storms completely reversed salinity regimes in the Laguna in 1909, 1933, and 1967, before establishment of dredged and jettied passes to the Gulf of Mexico during the 1970s (Hildebrand 1980). Two major hurricanes, Anita (1977) and Gilbert (1988), made landfall at or near La Pesca, but there are no documented records of impacts to natural processes or ecosystems and fisheries (table 2.5).

Table 2.5. *Hurricanes affecting the Laguna Madre of Texas and Tamaulipas, 1902–2017*

Date	Name	Category	Characteristics
Texas			
26 June 1902	Unnamed	1	Landfall on Mustang Island
18 August 1916	Unnamed	2	Landfall on Padre Island 96 km (60 mi) south of Port Aransas
15 September 1919	Corpus Christi Storm	3	Landfall on Padre Island 80 km (50 mi) south of Corpus Christi; second landfall at Corpus Christi; storm surge = 4.9 m (16 ft) in Port Aransas; destroyed most of downtown Corpus Christi
4 September 1933	Unnamed	1	Landfall just south of Brownsville; wind = 129 kph (80 mph); heavy impact from Corpus Christi into northeastern Mexico
25 July 1934	Unnamed	1	Landfall at Port Aransas
27 June 1936	Unnamed	1	Landfall at Port Aransas
25 August 1945	Unnamed	2	Landfall north of Port Aransas after following the South Texas coastline as a Cat. 4 storm
11 September 1961	Carla	4	Landfall near Matagorda; wind = 282 kph (175 mph) at landfall; wind = 12–161 kph (7–100 mph) in Laguna Madre; storm surge along entire Texas coast; 1.5–3.1 m (5–10 ft) along Padre Island
20 September 1967	Beulah	4	Landfall near Rio Grande then moved north into Laguna Madre; wind = 175 kph (109 mph); rain = 38–51 cm (15–20 in); caused major flooding
3 August 1970	Celia	3	Landfall near Corpus Christi; wind = 205 kph (125 mph) with gusts as high as 290 kph (180 mph); small but powerful storm, extensive damage to Corpus Christi
10 August 1980	Allen	3	Landfall at Port Mansfield; wind = 208 kph (129 mph); rain = 25–38 cm (10–15 in); storm surge = 3.2 m (10.5 ft) on Padre Island; extremely large, powerful storm that filled the Gulf
17 September 1988	Gilbert	4	Landfall at La Pesca 241 km (150 mi) south of Brownsville; heavy rainfall and moderate storm surge in South Texas

Table 2.5. *Continued*

Date	Name	Category	Characteristics
22 August 1999	Bret	3	Landfall between Corpus Christi and Brownsville on central Padre Island and the Land-Cut; storm surge = 1.8–2.4 m (6.0–8.0 ft) on Padre Island; 1.2 m (4.0 ft) in Laguna Madre; 10–12 washover cuts in Padre Island
13 September 2008	Ike	2	Cat. 4 through the Caribbean; landfall at Galveston 306 km (190 mi) north of Corpus Christi; very large storm; tropical storm force winds extended 342 km (212 mi) from center; wind = 177 kph (110 mph); rain = 20 cm (8 in); storm surge = 4.6 m (15 ft) and higher-than-normal water levels across the Gulf due to the storm's large size
25 August 2017	Harvey	4	Landfall in Port Aransas then Rockport; wind = 209 kph (130 mph); rain = 1.0–1.5 m (40–61 in); storm surge = 1.8 m (6 ft); extensive damage to Rockport and surrounding areas; extensive flooding in Houston
Tamaulipas			
28 August 1909	Unnamed	?	Completely flushed Laguna Madre system
5 August 1933	Unnamed	?	Completely flushed Laguna Madre system
20 September 1967	Beulah	4	Completely flushed Laguna Madre system
2 September 1977	Anita	5	Landfall near La Pesca 233 km (145 mi) south of Brownsville; wind = 281 kph (175 mph); rain = 40 cm (17 in) in Mexico; 15 cm (6 in) in South Texas; tides roughly 1.0 m (3.3 ft) above normal in South Texas
17 September 1988	Gilbert	4	Landfall 241 km (150 mi) south of Brownsville at La Pesca; no detailed information for Mexico (see Texas above)
20 July 2005	Emily	3	Landfall 112 km (70 mi) south of Brownsville in northern Tamaulipas; wind = 201 kph (125 mph); rain = 12.5 cm (5.0 in); storm surge 1.3 m (4.2 ft) in South Texas

Source: Hayes 1967; Henry et al. 1980; Hildebrand 1980; Broussard and Martin 1986; R. Watson, pers. comm., 1999, for Bret; Keim and Muller 2009.

Hydrography

The Laguna Madre of Texas has become a textbook example of a hypersaline or negative estuary (Dyer 1973; Reid and Wood 1976). The unique and characteristic hydrography of the Laguna Madre in both countries explains and justifies this long-standing, although now moderated, hallmark—little to no freshwater inflow, evaporation exceeding precipitation, shallow bathymetry, microtidal tide regime, limited circulation, and limited exchange with Gulf waters.

Freshwater Inflow Freshwater inflow is intermittent and is primarily associated with rainfall events and municipal or industrial discharges. In South Texas, the Nueces–Rio Grande Coastal Basin (27,045 km^2 [10,442 mi^2]) is the contributing drainage basin for the Laguna Madre. However, a large portion of the area is noncontributory due to flat terrain and narrow stream channels. The northern part of the coastal basin is drained by a network of intermittent streams into Baffin Bay, whereas the southern part is primarily drained by the Arroyo Colorado (TDWR 1983). There is only sheet flow, with no channelized drainage within the South Texas Eolian Sand Sheet area (Brown et al. 1977; see fig. 5.3).

Freshwater inflows into the Texas Laguna Madre varied greatly in both monthly and annual average values as a result of recurrent drought and flood conditions (TDWR 1983). Therefore, "average" must be considered with caution. On average, the total freshwater inflow (excluding direct precipitation) to the lagoon (1941–76 record) consisted of 851 million m^3 (690,000 acre-feet) annually, of which approximately 411 million m^3 (330,000 acre-feet) was contributed from gauged drainages.

Although the Rio Grande Basin is immense, it no longer contributes freshwater flow directly into the Laguna Madre; it now drains directly into the Gulf of Mexico by way of the Rio Grande. Today, the only input of fresh water is from rainfall runoff from the lowermost reach of the river, below the Anzalduas Dam at Mission (Orlando et al. 1991). In fact, the water in this lowermost reach is often estuarine because Gulf of Mexico waters flow in on flood tides. Under low-flow conditions, the Rio Grande is entirely impounded by the Amistad Dam (completed in 1968) and Falcon Dam (completed in 1953), from which waters are diverted for municipal and irrigation use. When rare Rio Grande floods do occur, the North Floodway in Texas and the South Floodway in Tamaulipas are activated, diverting floodwaters away from the normal river channel to the Laguna Madre of Texas and Tamaulipas, respectively (Orlando et al. 1991).

Old distributary channels in northeastern Mexico carry floodwaters toward the Gulf: (north to south) La Pita, Arroyo del Tigre, and Arroyo del Diablo. Only the latter, which drains through Laguna Jasso, flows into the northernmost Laguna Madre. Numerous irrigation and drainage canals interlace the rich agricultural area southwest of Matamoros (INEGI 1980). Most of the drainage influx of fresh water comes from Irrigation Districts 25 and 26 (Hildebrand 1980). Of the seven drains flowing into the Northern Laguna Madre, there was an inflow of 9 million m^3 (7,344 acre-feet) during a 13-month period between 1 September 1977 and 30 September 1978 (Sanchez 1978). To the south, the Río San Fernando has a drainage area of 13,510 km^2 (5,216 mi^2) and originates in the mountains of Nuevo León, but its runoff is highly variable (Hildebrand 1980). The average annual flow, gauged since 1931, is 665 million m^3 (539,000 acre-feet) and has ranged from 179 million m^3 (145,000 acre-feet) in 1952 to 4.678 billion m^3 (3.794 million acre-feet) in 1933 (Anonymous 1967). The Río San Fernando no longer flows directly into the Laguna Madre because its waters have been diverted into Laguna la Nacha and then to Laguna Madre via a canal (Brenner 1997). Farther south there are a number of small watersheds: Arroyo Chorreras has a drainage basin of 2,273 km^2 (878 mi^2) and average runoff of 386 million m^3 (313,000 acre-feet) per year and other small watersheds containing an area of approximately 2,432 km^2 (939 mi^2) and an estimated annual freshwater runoff of 377 million m^3 (306,000 acre-feet) (Tamayo 1949). The fresh water in all of these streams is utilized to some extent, with as much as 60%–80% of the Río San Fernando being removed, primarily for agriculture, during low-flow years before it reaches the Laguna Madre.

At the southernmost extremity, Río Soto la Marina does not contribute waters to the Southern Laguna Madre de Tamaulipas because of the extensive tidal flats and a road to the beach, which form effective barriers to exchange. Two adjacent lagunas, however, are affected by the Río Soto la Marina: Laguna Almagre, 2,720 ha (6,721 ac), to the west of La Pesca; and Laguna Morales, 3,500 ha (8,649 ac), to the south (Hildebrand 1980). The latter laguna, along with the Laguna Madre de Tamaulipas and the Mexican Rio Grande Delta area, are all considered by Mexican scientists as "la sistema Laguna Madre," extending from the mouth of the Rio Grande at 25°52' N, to the southern end of Laguna Morales at 23°32' N (Hildebrand 1980).

Evaporation Effects It is widely understood that evaporation exceeds precipitation in the semiarid region of the Laguna Madre of Texas and Tamaulipas. This key climatologic characteristic is one of the primary reasons for the hypersaline condition of both systems, usually with evaporation exceeding precipitation by two to three times. Although there is considerable year-to-year variability, a few direct examples are available for illustration. In the Texas Laguna Madre, precipitation averages only 74 cm/yr (29 in/yr); evaporation is over double precipitation, averaging 158 cm/yr (62 in/yr) (TDWR 1983). At the San Fernando weather station in Tamaulipas, where precipitation is only slightly higher than in Texas, a 34-year record (1932–65) revealed a low evaporation rate in 1941 of 158 cm/yr (62 in/yr) to a high of 230 cm/yr (91 in/yr) in 1946 (Hildebrand 1980). Between October 1955 and August 1966, precipitation exceeded evaporation during only five months: the maximum monthly evaporation rate was 26 cm/mo (10 in/mo) in July 1960 and the lowest was 6.5 cm/mo (2.6 in/mo) in February 1966. Demonstrating the high evaporation rate in Laguna Madre de Tamaulipas coupled with little or no freshwater inflow and no permanent connection to the Gulf (before dredging of passes/inlets), water levels of 1.0 m (3.3 ft) below sea level have been recorded (Anonymous 1967; Yañez and Schlaepfer 1968). Some older newspaper reports have suggested that the entire lagoon was dried up (H. Hildebrand, pers. comm.).

The information provided in these older studies regarding the importance and prevalence of evapotranspiration on the overall water budget in the Texas Laguna Madre was recently confirmed. In the Lower Laguna Madre, where evapotranspiration is greatest and far exceeds precipitation, the average freshwater inflow balance is negative (-451,882,931 m^{-3} $year^{-1}$; Schoenbaechler and Guthrie 2011a). The average inflow balance in the Upper Laguna Madre is small but positive (62,988,890 m^{-3} $year^{-1}$; Schoenbaechler and Guthrie 2011b).

Excessive evaporation over two decades of dry and drought years in both Texas and Tamaulipas focused attention on associated problems caused by this phenomenon. Concern and study in Texas focused on wind erosion and the impacts of blowing salt (fig. 2.20) on the area west of the Land-Cut (Coover and Rechenthin 1965; Price 1968). Studies in Tamaulipas focused on the evaporation and consequent drying up of the lagoon, physical hydrological characteristics, and a proposed solution to "rehabilitate" the Laguna Madre by cutting passes across the barrier island to the Gulf of Mexico (Anonymous 1967).

Figure 2.20. *Demonstration of the dramatic impact of salt blowing on vegetation. The Live Oak tree is skewed to the northwest because of strong and persistent southeast winds laden with salt crystals and clay particles. The tree is located on an upland just west of Land-Cut wind-tidal flats and clay dunes. (Photograph by J. W. Tunnell, May 1989)*

Bathymetry Shallowness of the entire lagoonal system is well documented and is an important factor ensuring a well-mixed water column. Although a recent detailed or comprehensive bathymetry map is unavailable for either lagoon, earlier workers mapped each lagoon's depths and contours (Rusnak 1960 for Texas; Anonymous 1967 and Yañez and Schlaepfer 1968 for Tamaulipas). Major bathymetric features are available on Bureau of Economic Geology maps for the Texas lagoon (Brown et al. 1976; Brown et al. 1977; Brown et al. 1980; W. White et al. 1983; White et al. 1986; White et al. 1989). The Texas Laguna Madre has a smooth, centrally sloping floor on the eastern side with a few deeper, elongate basins located on the west. Average depth is about 1.0 m (3.3 ft); the Upper Laguna Madre is slightly shallower, and the Lower Laguna Madre is slightly deeper (see table 2.1). Deeper areas of 2.4 m (8.0 ft) occur just south of the Land-Cut in Redfish Bay (here and elsewhere in this chapter, Redfish Bay refers to an area in the Lower Laguna Madre south of Port Mansfield, not the bay system between Aransas Bay and Corpus Christi Bay north of the Upper Laguna Madre) and just south of the mouth of Baffin Bay.

The shorter and wider northern lagoon in Mexico had a centrally located maximum depth of just over 3.0 m (9.8 ft) when mapped in 1963, whereas the narrower and longer Southern Laguna's maximum was 1.3 m (4.3 ft) in its northern portion and in Laguna el Catán (Anonymous 1967). In the northernmost Laguna Madre, north of Isla Larga and northwest of Isla Mula, shallow conditions, generally less than 1.0 m (3.3 ft), exist because of erosion of the islands and mainland, both remnants of the Rio Grande Delta. Elsewhere, broad washover fans with extensive tidal flats and lagoon shallows exist lagoonward of the four natural, intermittent tidal inlets around Boca Ciega.

Channelization in both Texas and Tamaulipas has since altered bathymetry, tides, hydrology, and subsequently salinity in these lagoons (see "Channelization" later in this discussion). Tidal characteristics, water circulation, and salinity are all important environmental parameters that govern the distribution of biota and habitats.

Tides Tides within the Laguna Madre of Texas, even though minimal, have generated intense interest in two divergent subject areas: law and science. In law, tide lines are used to determine ownership boundaries between the privately owned uplands and submerged state-owned lands. Much of the early knowledge about tides in the Laguna Madre of Texas was generated in regard to tidal boundary issues during the 1940s (e.g., Fisk 1949, 1959). Along with Fisk, other early workers in the Laguna Madre (Hedgpeth 1947; Collier and Hedgpeth 1950; Simmons 1957; Breuer 1957; Rusnak 1960) discovered that meteorological tides, wind tides, were far more important than astronomical tides and thus of unique scientific interest for both geological and biological processes.

In the Upper Laguna Madre, which is the most studied regarding tides and water levels, three water-level components prevail (N. P. Smith 1978):

1. Long-period, semiannual rise and fall of water level occurs in the lagoon: high water in late May and late October, semiannual lows in late February and late July; this rise and fall in water levels associated with seasons is on the order of 50 cm (20 in).
2. Superimposed on the gradual, seasonal rise and fall are water-level variations occurring over time scales of one to two weeks; this rise and fall, which is quite variable and dependent on meteorological forcing, generally is on the order of 10–20 cm (4–8 in).
3. Diurnal and perhaps semidiurnal astronomical tides appear as "noise" in the tidal signal, superimposed on the longer-period variations; this component is only on the order of 2.0–3.0 cm (0.8–1.2 in).

As one travels southward from the major Gulf of Mexico tidal inlet at Aransas Pass, the tidal amplitude decreases with ever-increasing distance and channel complexity. Tidal inlets in the Lower Laguna Madre, including Mansfield Pass and Brazos Santiago Pass, also increase tidal amplitudes locally, but the effect dissipates with distance.

During the 1990s, a widespread network of tide gauges was maintained in the Laguna Madre (Packery Channel, Bird Island, Baffin Bay, El Toro, Rincón de San José, Port Mansfield, Arroyo Colorado, Port Isabel, South Padre Island) as part of an extensive statewide system—the Texas Coastal Ocean Observation Network (TCOON). This system, operated by the Conrad Blucher Institute for Surveying and Science (CBI) at Texas A&M University–Corpus Christi, provided considerable environmental data in real time on its website and was sponsored jointly by the Texas General Land Office (TGLO) and the Texas Water Development Board (TWDB). Today, nine tide gauges (some of which have been moved slightly from their original sites) are also operated and maintained (Packery Channel, South Bird Island, Baffin Bay, Rincón de San José, Port Mansfield, Realitos Peninsula, Port Isabel, Queen Isabella Causeway, and South Padre Island Brazos Santiago Pass), primarily in the upper and lower reaches of the Laguna Madre (not the central part), and the management of these gauges now belongs to the NOAA Center for Operational Oceanographic Products and Services (CO-OPS), which maintains the National Water Level Observation Network (NWLON). All TCOON data are now available via NOAA's Tides Online website: https://tidesandcurrents.noaa.gov/gmap3/index.shtml?type=TidePredictions&region=. Partners in this network include NOAA, US Army Corps of Engineers, US Fish and Wildlife Service, Department of the Interior, TGLO, TWDB, CBI, and TCOON.

Natural tidal inlets, or *bocas*, were the avenues for tidal exchange in the Laguna Madre de Tamaulipas before the four jettied channels were dredged. Before dredging, the Laguna's tidal amplitude was only a few centimeters, and a relatively small area around the inlets was influenced by the daily tide (Hildebrand 1980). Astronomical tidal characteristics under today's conditions are not known, but wind tides are important in mixing and influencing water levels throughout the Laguna.

Wind Tides Wind tides are defined as a rise in water level on the downwind side of a body of water caused by the force of the wind on the water surface (Weise and White 1980). Numerous authors have studied and discussed the importance of wind tides and their impact on water levels and geological processes. Fisk (1959) recorded a wind tide in excess of 1.0 m (3.3 ft) in the northern part of the Lower Laguna Madre. Calculations based on wind velocity and fetch indicate that wind tides in the 0.3 to 1.2 m (1.0 to 4.0 ft) range are common within the Laguna Madre (Rusnak 1960). Copeland and coworkers (1968) noted that wind-driven water will cover "normally dry land" within the Laguna Madre. Within Kenedy and Kleberg Counties "wind driven tides may flood as much as 200 mi^2 (518 km^2) of low-lying lagoonal margin" (Brown et al. 1977). Brown and colleagues also relate that "wind-tidal flats are flooded rapidly, generally by northers or the prevailing southeasterly wind regime" and that flooding from wind tides can be localized on the flats and is a "function of the duration, intensity, and direction of the wind" (1977). McGowen and coworkers (1977) note that strong winds accompanied by spring tides and a barometric low can create tides 0.6–0.9 m (2.0–3.0 ft) higher than those produced solely by astronomical conditions. And Morton and McGowen note that "wind tides inundate areas not normally affected by astronomical tides" (1980).

An excellent example of the regional impact of wind tides can be seen in their effect on the central portion of the Laguna Madre of Texas. Winds from the south and southeast (1) raise water levels in northern Redfish Bay (Port Mansfield area), flooding a large portion of the Land-Cut; (2) drive water out of The Hole (aka Nine Mile Hole) across Middle Ground into the southern part of the Upper Laguna Madre; and (3) cause water to flow from the Laguna into Baffin Bay, thereby increasing water depth. Conversely, strong north winds produce wind tides that are low in Baffin Bay and Redfish Bay and high in Middle Ground and The Hole and adjacent northern parts of the Land-Cut (Morton and McGowen 1980).

Channelization

Channelization of the three tidal inlets (north to south: Packery Channel, Mansfield Pass, and Brazos Santiago Pass) to the Gulf of Mexico and navigation or access channels within the lagoons was the key factor in reducing salinities, causing long-term ecosystem change in both Laguna Madre systems (Collier and Hedgpeth 1950; Simmons 1957; Breuer 1962; Hedgpeth 1967; Quammen and Onuf 1993; Onuf 1994, 2007). Historically, before channelization, distance from tidal inlets in Texas and lack of continuously open tidal inlets in Tamaulipas led to excessive hypersalinity. Constrictions to water circulation into The Hole and just north of Baffin Bay in the Upper Laguna Madre of Texas caused even higher salinities within and beyond those areas, respectively (Burr 1930; Hedgpeth 1947; Collier and Hedgpeth 1950; Simmons 1957). Low seasonal tides would isolate The Hole from connection to the rest of the Laguna Madre, causing excessive salinities. Shallow-water sills between Corpus Christi Bay and the Upper Laguna Madre (see fig. 6.1; Hedgpeth 1947; Simmons 1957) and between Baffin Bay and the Upper Laguna Madre (Rusnak 1960) also impeded circulation between those water bodies.

Aside from storm events, which reduced salinities through rainfall, runoff, or opening tidal inlets, salinities prior to channelization were consistently higher in both Texas and Tamaulipas lagoons. Salinity levels two to three times higher than oceanic levels were common, if not the norm, prior to channelization (table 2.6) in both systems. In Texas, the more-restricted Upper Laguna Madre had salinities in excess of 100 ppt between 1946 and 1948, prior to dredging of the GIWW (Hedgpeth 1947; Collier and Hedgpeth 1950). In the Lower Laguna Madre, salinities over 60 ppt were recorded during this time. After dredging of the GIWW, which was completed in 1949, and before Mansfield Pass was opened in 1962, Breuer (1962) recorded Lower Laguna Madre salinities in the 30s to 50s from south to north, demonstrating elevated salinities with increasing distance from the tidal pass (Brazos Santiago Pass). After opening Mansfield Pass through Padre Island in the northern part of the Lower Laguna Madre, water circulation patterns and exchange of water between the Gulf and lagoon has kept salinities somewhat lower, usually below 40 ppt (Warshaw 1975; Brown et al. 1980; White et al. 1986). Salinities in the Upper Laguna Madre have also moderated because of exchange of water through the GIWW Land-Cut connection with the Lower Laguna Madre. Quammen and Onuf (1993) graphically illustrated these moderating conditions by plotting salinities over the 44-year period between 1945 and 1988 (see fig. 6.3).

Table 2.6. *Major channels constructed in the Laguna Madre of Texas and Tamaulipas*

Channel	Date Dredged	Width × Depth m (ft)	Reference
Texas			
Brownsville Ship Channel	1938	? × 8.5 (? × 28)	Breuer 1962
	1980	61 × 11 (200 × 36)	Orlando et al. 1991
Murdoch's Landing (= Yarborough Pass)	1941–44	24 × 2.4 (80 × 8)	Gunter 1945a
GIWW	1944–49	38 × 3.6 (125 × 12)	Warshaw 1975 Orlando et al. 1991
Channel to Harlingen	1951	38 × 3.6 (125 × 12)	Orlando et al. 1991
Mansfield Pass	1962	90 × 4.2 (295 × 14)	Kieslich 1977
Packery Channel	2005	37 × 4 (121 × 13)	Williams and Kraus 2011
Tamaulipas			
El Mezquital	1976	no data available	
Boca Ciega	?	no data available	
Boca de Catán	?	no data available	
Boca de Santa Isabel	?	no data available	

An earlier interesting but failed attempt to reduce salinities in the Upper Laguna Madre prior to the dredging of the GIWW involved cutting a pass across Padre Island opposite Baffin Bay. The Texas Game, Fish, and Oyster Commission (now Texas Parks and Wildlife Department), after recording numerous fish kills due to excessive hypersalinity and after considerable public outcry, sent its own dredge on four separate occasions to dredge a pass at Murdoch's Landing (now Yarborough Pass; Burr 1945a, 1950; Gunter 1945a). Between 1941 and 1944 the dredge cut its way through the Upper Laguna to and from the site each time from its docking place in Rockport, taking several months to do so each way. The pass silted in rapidly after each dredging, staying open only 10 months during the four-year period. When open, salinities were reduced only 0.5–1.0 ppt and only in the immediate area of the pass (Gunter 1945a).

Using the same notion that dredged passes to the Gulf would reduce excessive salinities, the Mexican government began cutting and jettying passes into the Laguna Madre de Tamaulipas in 1976. The first one was El Mezquital Pass (Hildebrand 1980). Subsequently, three other passes were dredged and jettied to allow permanent connection between the Gulf of Mexico and the Laguna: (north to south) Boca Ciega, Boca de Catán, and Boca de Santa Isabel. The first two connected into the northern basin; and the second two, into the southern basin.

The latest project to restore flow, or exchange, between the Gulf of Mexico and the Upper Laguna Madre involved the dredging and placement of jetties to stabilize Packery Channel, a historic tidal inlet (Williams et al. 2007; Reese et al. 2008; Palmer et al. 2008, 2013; Williams and Kraus 2011). Historically, this intermittent inlet, called Corpus Christi Pass, was opened after strong winter fronts or hurricanes, connecting the Gulf of Mexico to Corpus Christi Bay or the very northeastern extremity of the Upper Laguna Madre. The dredging and deepening of Aransas Pass caused the main tidal head or flow of water to funnel in and out of that larger pass, thereby closing the smaller one to the south with natural sand accumulation and longshore drift.

Studies conducted before and after the dredging of Packery Channel revealed only modest changes in the hydrography and slightly ameliorated hypersalinity in the immediate vicinity of the pass in the Laguna Madre (Palmer et al. 2008, 2013; Reese et al. 2008). Using the before-after control-impact (BACI) study design, significantly fewer nekton were found in seagrass beds after channel opening, but significantly higher densities of newly settled estuarine-dependent species (Red Drum, Atlantic Croaker, Pinfish, Blue Crab, and penaeid shrimp) were found (Reese et al. 2008). This indicates a small but positive effect in the immediate areas surrounding the Laguna Madre side of the pass on selected ecologically and economically important fishery species.

The community composition of benthic macrofauna was not altered in the area around the new inlet. There was a greater difference in community composition between deep and shallow stations than between either before-after or control-impact sites. In addition, there were no significant changes in abundance, biomass, or diversity of benthic macrofauna caused by the opening of Packery Channel (Palmer et al. 2008, 2013).

Historically in Tamaulipas, there were no continuously open tidal inlets. However, several remained in the same place and were called by their place-names in Spanish and numbers by Americans fishing the area (e.g., Boca Sandoval = 5th Pass). Several of these tidal inlets are observed on maps of the region, and they opened and closed depending on prevailing climatological conditions. The greatest number opened immediately after hurricanes, and all might be closed during prolonged droughts and remain closed for years. In the region of the current jettied Boca Ciega there was a cluster of four inlets in a 13 km (8 mi) stretch of barrier island: (north to south) Boca Santa María, Boca San Rafael, Boca Ciega, and Boca Sandoval. These passes varied in depth from a few centimeters to 1.8 m (6.0 ft), and the yearly cycle involved shoaling in summer and deepening in winter as north winds piled water in the southeastern part of the Northern Laguna Madre (Hildebrand 1980). South of this area, two other passes existed, Boca San Antonio and Boca Jesús María, passing into the Southern Laguna. Little is known of Boca San Antonio, but Boca Jesús María, also known as 8th Pass, supported a substantial fishing village and with average depths of 1.8 to 2.4 m (6.0 to 8.0 ft) was the most stable and deepest of all the passes into the Laguna. Boca el Viborero existed to the south of Boca Jesús María and became the dominant pass after Hurricane Beulah in 1967. Since there is not enough tidal prism to maintain multiple passes, it is assumed (as in Texas) that most of the time the natural passes no longer function as the result of tidal dominance by the maintained passes. However, on 14 February 1999, after a winter storm with strong north winds, Boca Sandoval was open (J. W. Tunnell, pers. observ.).

Like historical salinity levels in the Texas Laguna Madre, salinities in the Mexican lagoon no longer seem to reach the excessive hypersaline values recorded prior to channelization (dredging of barrier island inlets). Except for periods after major wet hurricanes, the Tamaulipas Laguna generally had salinities two to three times greater than oceanic salinity and frequently had recordings in the mid-100s ppt (Copeland and Jones 1965; Copeland 1967; Hildebrand 1969). Extreme salinities of 175 ppt in 1961 (Hildebrand 1969) and 295 ppt in 1965 (Copeland and Jones 1965) were recorded during the drought of the 1960s.

3 Origin, Development, and Geology

JOHN W. TUNNELL JR.

To understand distribution and ecology of present-day biota and habitats, it is important to understand both the influence of current physical processes (temperature, salinity, wind, precipitation, etc.) and the effects of origin, development, and geology on the landscape. The geomorphological structures visible today in the Laguna Madre of Texas and Tamaulipas are expressions of past historical development as well as present-day processes. The vast Rio Grande Delta separating the US and Mexican lagoons, the further division of both lagoons into smaller subunits (north and south of the Land-Cut in Texas and El Carrizal in Tamaulipas), and the lagoonal protection provided by the barrier islands and peninsulas all have profound effects on the contemporary expression of living resources. Most major present-day coastal features formed over the past several thousand years, and they are continually in the process of long-term, dynamic change (Brown et al. 1976; McGowen et al. 1977; Brown et al. 1977; Brown et al. 1980; Morton and McGowen 1980).

A wealth of geological information exists about the Texas coastline because of two factors: (1) vast oil and gas reserves and the consequent studies associated with this industry and (2) the University of Texas, Bureau of Economic Geology. Since publications on the geological aspects of the Laguna Madre de Tamaulipas are rare and those about the Texas lagoon are common, focus here is on the Texas system, with the assumption that most processes are similar for Tamaulipas. Many of the earlier ideas and concepts about South Texas geology and the Laguna Madre belong to W. Armstrong Price, pioneering coastal geologist (e.g., 1933, 1958, and many more).

Origin and Development

Laguna Madre of Texas The Laguna Madre of Texas, like the entire Texas coastline, is primarily a product of Pleistocene and Recent (Holocene and Modern) geological history. The Pleistocene ice age included over one million years of complex glacial and interglacial periods and corresponding fluctuations in sea level (Brown et al. 1976; Brown et al. 1977; Brown et al. 1980; fig. 3.1). Four major glacial episodes, during which sea level fell, were separated by warmer interglacial periods, when sea level rose.

LeBlanc and Hodgson (1959) developed a graphic model to illustrate the development of the barrier island system and the Laguna Madre (fig. 3.2A). Near the end of the final glaciation period (Wisconsin) and end of the Pleistocene—about 18,000 years before present (YBP)—sea level was some 91 to 137 m (300 to 500 ft) lower than today (Curray 1960), and the shoreline was about 80 km (50 mi) east of its present position. During this time period, streams cut deep valleys into underlying deposits in Baffin Bay (Behrens 1963), the Land-Cut area (Fisk 1959; fig. 3.3), and the Rio Grande (Fulton 1976). With the continued melting of glacial ice, these river valleys were flooded as the sea enveloped the land and sediments began filling the drowned river valleys. At the end of the Holocene—about 5,000 to 4,500 YBP—sea level

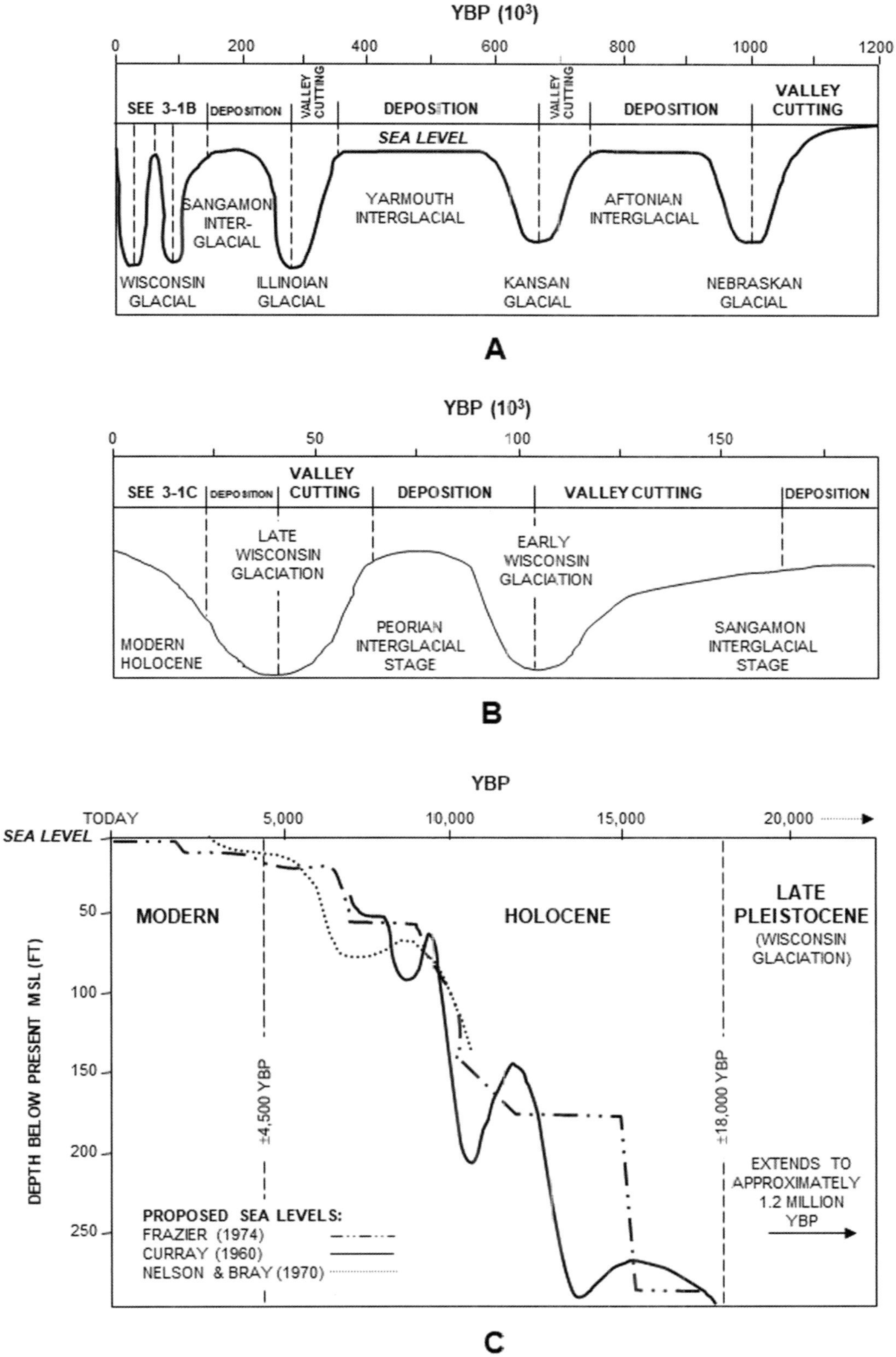

Figure 3.1. *Sea-level changes related to glacial and interglacial stages. (A) Generalized Pleistocene sea-level variations and associated erosional and depositional episodes. (B) Generalized sea-level changes during Late Wisconsin glaciation. (C) Proposed sea-level changes during last 20,000 years. (From Brown et al. 1977)*

reached about 4.6 m (15 ft) below its present level and modern geological processes became active, including the formation of sand bars and shoals parallel with the coast (fig. 3.2B). When the sea reached its present level—about 2,800 to 2,500 YBP—those shoals and bars coalesced to form Padre Island and other barrier islands (LeBlanc and Hodgson 1959; fig. 3.2C). Thus, the origin of Laguna Madre relates to the formation of Padre Island and its enclosure of the lagoonal system. Other developmental processes during the last several thousand years are responsible for certain shapes and forms of Laguna Madre features today. Baffin Bay, for instance, may have become hypersaline prior to the development of the Laguna Madre, sometime between 5,000 and 4,300 YBP (Behrens 1974).

Over the past several thousand years, hurricanes and tropical storms, in conjunction with eolian processes, have transported sediment from Padre Island landward into the

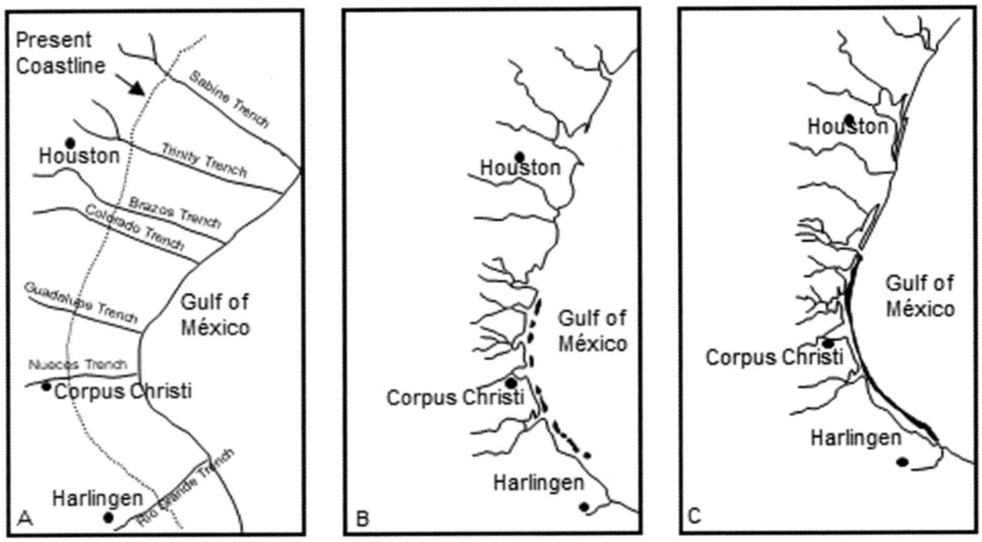

Figure 3.2. *Origin and development of the Texas shoreline. (A) Late Pleistocene falling sea-level stage about 18,000 YBP. (B) Early Modern sea-level stage with barrier islands forming about 4,500 YBP. (C) Present standing sea-level stage. (From Le Blanc and Hodgson 1959)*

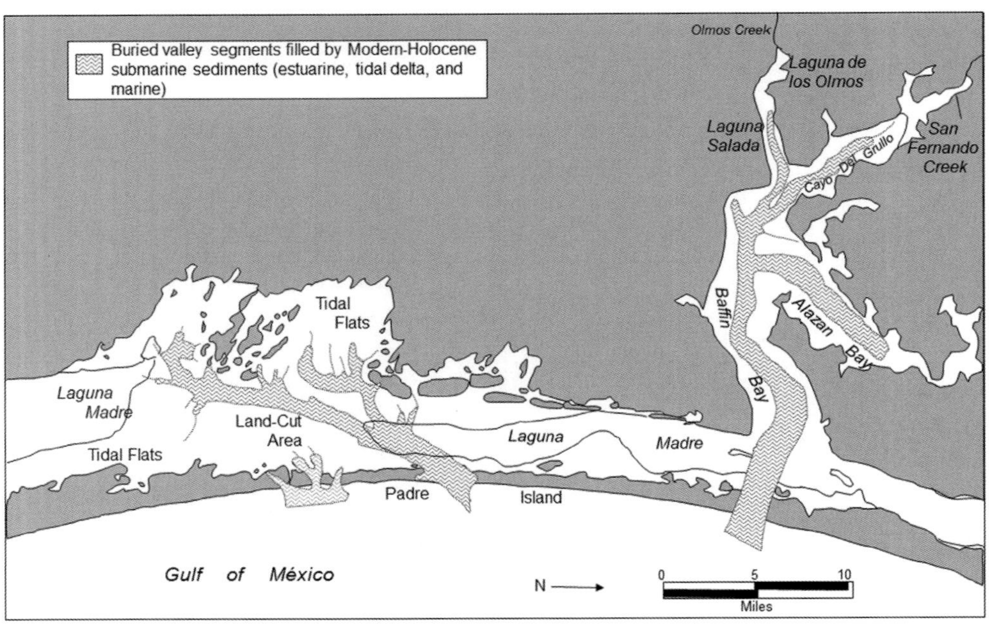

Figure 3.3. *Subsurface late Pleistocene–Holocene valleys below present-day Baffin Bay and Land-Cut. Valleys were incised during last glaciation (low sea level) and progressively filled by Holocene to Modern fluvial, estuarine, and open marine sedimentation. (Modified from Brown et al. 1977)*

Laguna Madre (Morton and McGowen 1980). These processes caused development of extensive tidal flats on the lagoon side of southern Padre Island and northward. These same processes in the Land-Cut area have caused the slow but continual infilling of the Laguna Madre until a land bridge was formed connecting Padre Island with the mainland (fig. 3.4). Scientists first proposed that this infilling occurred as a result of the massive 1919 hurricane (Gunter 1945a; Hedgpeth 1947). However, after extensive coring in the area, Fisk (1959) demonstrated that it was a much more gradual process, with final closure possibly occurring in the early nineteenth century. Most recently Watson (1989) and Morton and Garner (1993), through investigations on rates of wind-tidal-flat accumulation, indicated that closure of the Land-Cut area occurred hundreds of years ago. Even in the more vegetated area of northern Padre Island, the island has slowly migrated into the Laguna Madre, particularly during periods of drought and overgrazing, which exposed bare sand (Prouty and Prouty 1989).

Certain distinctive mainland features also relate to geological development of the area: the Rio Grande Delta, the Arroyo Colorado Delta, clay dunes, wind-tidal flats, the South Texas Eolian Sand Sheet, beach rock (coquina) at Penascal Point and Punta Piedras, and serpulid reefs in Baffin Bay.

Figure 3.4. *Satellite photo from the Space Shuttle of the Land-Cut area between the Upper and Lower Laguna Madre, Texas. Wind-tidal flats are the most common habitat type in the Land-Cut and extend southward along the backside of Padre Island. (NASA, n.d.)*

Rio Grande Delta The predominant feature of the South Texas and northeastern Mexico shoreline is the bulge caused by the Rio Grande Delta. This segment of the coast occupies more than 1,550 km^2 (600 mi^2) of deltaic plain in Texas and Mexico over what was formerly shallow waters of the Laguna Madre and inner continental shelf (Lohse 1958). This highly complex segment of coast forms a unique system of meander streams, oxbow lakes (*resacas*), playa lakes, wetlands, mudflats, *lomas* (hills), and clay dunes (figs. 3.5 and 3.6). Some of these features, such as Bahia Grande, Laguna Larga, and San Martin Lake, are remnants of the Laguna Madre (LeBlanc and Hodgson 1959). Little was known of the biology and ecology of these habitats until recently when a major ecological restoration project was initiated to restore Bahia Grande (Lichlyter 2006; Cornejo 2009; Tamez 2014; Marquez et al. 2017).

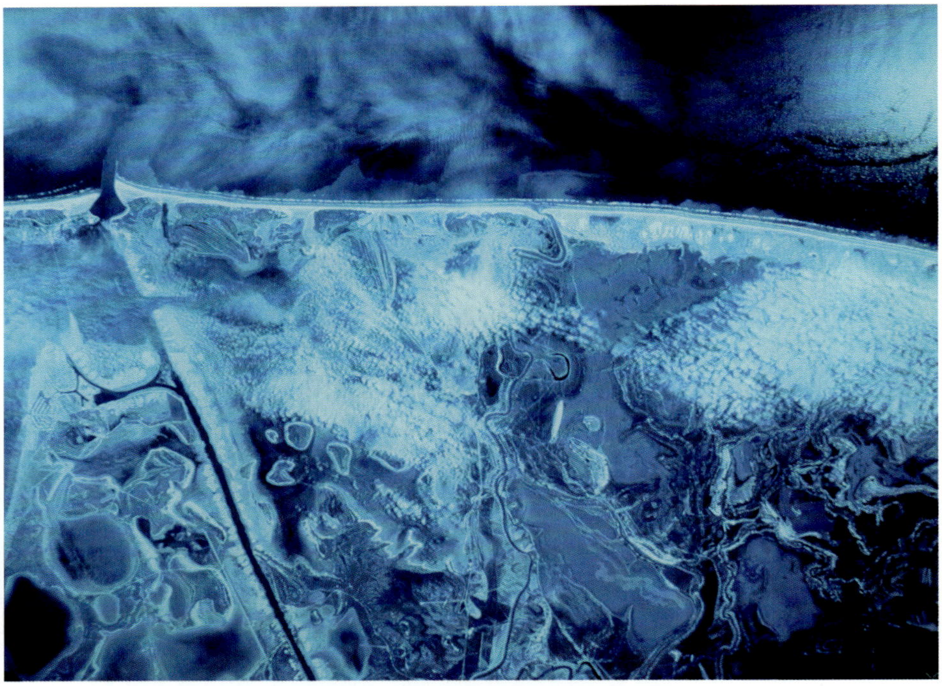

Figure 3.5. *Vertical high-altitude (18,288 m [60,000 ft]) photograph of the Rio Grande Delta, showing extreme southern Lower Laguna Madre of Texas, Brazos Santiago Pass, and Brownsville Ship Channel (left), Rio Grande (middle), and highly complex surrounding lands of playas, lakes, resacas (oxbow lakes), wetlands, mud-flats,* lomas, *and clay dunes (USGS, 1973)*

Figure 3.6. *Diagonal aerial photograph to the north over the Rio Grande Delta in Mexico. The town of South Padre Island can be seen on the horizon in the upper part of the photo. (Photograph by J. W. Tunnell, February 1999)*

Sediment discharge from the Rio Grande has continually diminished during the past 4,500 years, and during the past 50 to 60 years, discharge has almost ceased because of extensive irrigation and dam construction along the Rio Grande and its main tributaries (Brown et al. 1980). In fact, for a brief period in 2001, the mouth of the Rio Grande ceased to flow and the mouth of the river silted in, damming the connection to the Gulf of Mexico (J. W. Tunnell, pers. observ.). Consequently, the modern Rio Grande Delta is undergoing erosion and South Padre Island and Laguna Madre are retreating landward over the subsiding Rio Grande Delta. Coring, seismic studies, and carbon-14 dating conservatively indicate that the delta shoreline stood about 16 km (10 mi) gulfward of the present shoreline 3,500 YBP (Morton and Pieper 1975; McGowen et al. 1977). The current prognosis for this area is continued subsidence and gradual retreat of the Gulf and mainland shoreline (Brown et al. 1980).

Arroyo Colorado The highly irregular western Laguna Madre shoreline near the Arroyo Colorado is attributed to the earlier drainage and deposition from this system into the area, as well as current erosional and depositional processes. Although most waters of the Arroyo Colorado flow into the Laguna Madre via the dredged Arroyo Colorado Cut-Off, floodwaters from the northern and western part of the Rio Grande Valley flow via the North Floodway into the old Arroyo Colorado Delta area. A faint or remnant "bird's-foot" delta can still be observed in this area on geological environmental maps and aerial photographs (Brown et al. 1980). Additional sediments are added via Cayo Atascosa, which drains Laguna Atascosa within the Laguna Atascosa Wildlife Refuge, although flow is minimal due to small dams on the drainage system. The complexity of the shoreline here results from continued compaction and subsidence of deltaic sediments, erosion of remnant deltaic deposits, and the formation by deposition of windblown clay dunes.

Distinctive clay-sand dunes, such as Yucca Island, Horse Island, Hawk Island, Mullet Island, and numerous unnamed dunes, exist on wind-tidal flats in this area of the Laguna Madre. Furthermore, several true islands within the waters of the Laguna Madre, such as Three Islands, Rattlesnake Island, and Green Island, were formed on these western wind-tidal flats but were subsequently isolated by subsidence of the flats beneath lagoonal waters (Brown et al. 1980). "Islands" on the wind-tidal flats on the lagoon side of southern Padre Island—La Punta Larga, Los Bancos de en Medio, and Deer Island—likely had a similar origin (McGowen et al. 1977).

Clay Dunes Well-developed clay dunes are present all along the South Texas coast, from St. Charles Bay (28°13' N) southward into northeastern Mexico, along the coast of Tamaulipas to Rancho Tepehauje (23°30' N), about 24 km (15 mi) south of the Río Soto la Marina at La Pesca (Price 1933; Huffman and Price 1949; Price and Kornicker 1961). They were first described as unique depositional structures in the Port Isabel area of South Texas (Coffey 1909). Like their more common and well-known sandy counterparts (sand dunes), clay dunes are formed by windblown clay particles or small pellets, which ultimately accumulate as topographic features downwind of saline flats. Clay dunes, however, form only in dry climates (subhumid to semiarid). Although known from a few dry, inland areas in the western United States and southwestern Australia, clay dunes are most widespread and best known from coastal South Texas and Tamaulipas in the western Gulf of Mexico (Huffman and Price 1949; Price and Kornicker 1961). Like Padre Island, the longest barrier island in the world, and Laguna Madre, the largest hypersaline system in the world, clay dunes are possibly more extensive here than in any other coastal region of the world.

Clay dunes in the Laguna Madre region of Texas and Tamaulipas are even-topped, ridge-shaped eolian deposits, which form on the lee, or northwest, side of clay-floored saline playas or wind-tidal flats (Price and Kornicker 1961). Dunes grow primarily with the strong southeast winds, but only during the drier months of the year when the soil is not wet (fig. 3.7). Dune growth in South Texas extends from March through November but is most active during the drier months of the summer and during droughts. Growth is arrested during the winter when precipitation associated with northers is frequent and evaporation is reduced (Huffman and Price 1949). The highest clay dunes (10.7 m [35 ft]) are located in the Rio Grande Delta area but are lower in elevation (1.0 m [3.3 ft]) in the more humid climates to the north and south. Erosion of these structures occurs during heavy rainfall and hurricane storm surges with associated wave action.

The distribution of clay dunes in South Texas can be observed easily on the "environmental geology" maps of the Kingsville Area (Brown et al. 1977) and the Brownsville-Harlingen Area (Brown et al. 1980). The most extensive area of clay dunes exists as a transition zone, or western border, between the wind-tidal flats and uplands between Baffin Bay through the entire Land-Cut (figs. 3.8 and 3.9).

Figure 3.7. *Clay dune on western edge of Land-Cut, Kenedy County, Texas, wind-tidal flats. Note clay deposition on lower slopes and wind erosion on upper parts. Boards in foreground most likely washed in during tropical storm tides in this very remote location. (Photograph by J. W. Tunnell, May 1989)*

Figure 3.8. *View of clay dune to the south along "transition zone" between Land-Cut, Kenedy County, Texas, wind-tidal flats and uplands. Seasonal high tides and strong southeast winds pushed water into the cayos (flat-bottomed valleys between clay dunes). (Photograph by J. W. Tunnell, May 1989)*

Other concentrations are in the Rio Grande and Arroyo Colorado Deltas. Inland, throughout the coastal zone, clay-sand dunes are widely scattered on the South Texas Eolian Sand Sheet, primarily associated with saline ponds or playas. A typical clay dune profile overlying the Pleistocene Beaumont Clay is illustrated in figure 3.10 (Huffman and Price 1949).

Figure 3.9. *Clay dunes encroaching on uplands in the Land-Cut: (a) small clay dunes in the north; (b) large clay dune in the southwest (Photographs by J. W. Tunnell, May 1989)*

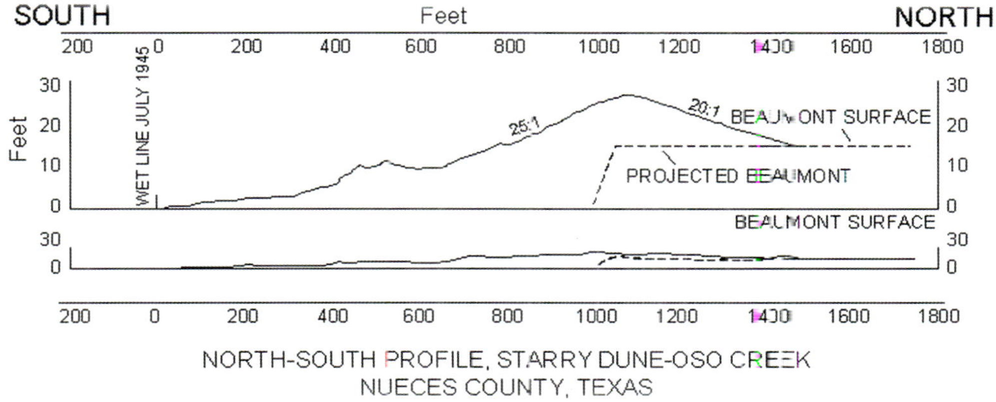

Figure 3.10. *Profile section across typical clay dune overlying Pleistocene Beaumont Clay: (top) vertically exaggerated; (bottom) to scale (Modified from Huffman and Price 1949)*

In addition to being unique depositional features, clay dunes have importance in archaeology, paleoecology, and plant distribution. Archaeological evidence (Price 1958), consisting of shell middens, indicates that early human occupants of the South Texas coastal area may have favored habitation on clay dunes at times, perhaps because of elevated viewing of the surrounding area and adjacent lagoonal location (Tunnell and Tunnell 2015). The presence of Dwarf Surf Clam, a widespread and common lagoonal clam, interbedded in clay at 2.4 to 10.0 m (8–33 ft) above sea level in clay dunes near Laguna Vista in South Texas, demonstrates the interrelationship of wind and the adjacent lagoon on the depositional history of the area (Price and Kornicker 1961). Finally, certain plants such as saltbush prefer the clayey, salty soils of clay dunes. Kleberg Saltbush, located on clay dunes in Kleberg County along Baffin Bay, is apparently found only in this specific habitat and is listed as rare because of its limited distribution in this habitat and occurrence only in Kleberg County.

Wind-Tidal Flats Like clay dunes, wind-tidal flats are unique and widespread within the Laguna Madre of Texas. These broad, low tidal flats developed over the past several thousand years and cover an area of over 900 km^2 (350 mi^2) along the

Figure 3.11. *Wind-tidal flats on the eastern side of Laguna Madre, primarily composed of sand blown or washed over the barrier islands into the lagoon: (a) sand wind-tidal flats just north and west of Mansfield Pass on Padre Island (Photograph by J. W. Tunnell, December 1982); (b) sandy wind-tidal flat and dark-colored algal mat on the western side of South Padre Island (Photograph by J. W. Tunnell, July 1980)*

margins of Laguna Madre and Baffin Bay and in the Rio Grande Delta (Brown et al. 1976; Brown et al. 1977; Brown et al. 1980). The regional semiarid climate and the hypersaline and microtidal characteristics of the waters are all necessary ingredients for producing these extensive flats.

Wind-tidal flats on the eastern side of Laguna Madre are composed primarily of sand washed or blown lagoonward from Padre Island (fig. 3.11). Western or mainland side wind-tidal flats, on the other hand, developed where wind deflation extensively eroded the shoreline. Therefore, they are mainly clay in composition and often covered with algal mats (fig. 3.12). The broad embayment or indention into the mainland along the western edge of the Land-Cut, clearly visible on both maps and satellite images of

Figure 3.12. *Wind-tidal flats on the western or mainland side of Laguna Madre are characteristically composed of clay. Photosynthetic bacteria (blue-green algae or cyanobacteria) mats, which are common in areas that are irregularly but consistently flooded: (a) the "Basin" (Fisk 1959) in north-central Land-Cut (Photograph by J. W. Tunnell, March 1993); (b) close-up of algal mat (Photograph by J. W. Tunnell, August 1982)*

the area, is a unique and graphic demonstration of the immensity of wind deflation that has taken place in this area (see figs. 1.1 and 2.2). The largest and most extensive flats are located in the Land-Cut area, the lagoon side of southern Padre Island from the Land-Cut southward, and the mainland area around the Arroyo Colorado Delta. In Tamaulipas, tidal flats are most common leeward of the barrier island near the four tidal inlets (Boca Santa Maria, Boca San Rafael, Boca Ciega, and Boca Sandoval) into the northern Laguna and also in the area of the Río San Fernando Delta, which Hildebrand (1980) termed El Carrizal Flats. Most early studies focused on the geology or geological processes of the tidal flats (Price 1933; Fisk 1959). Later interest focused on the geochemistry of the flats (Amdurer and Land 1982; Long and Gudramovics 1983). Until the 1970s, these wind-tidal flats were considered of little ecological value, but studies in recent decades have shown that they have multiple ecological values and functions, as well as linkages to surrounding habitats (Pulich and Rabalais 1986; Pulich and Scalan 1987; Withers 1994, 1996c; Withers and Tunnell 1998). Chapter 9 is therefore dedicated to the wind-tidal flats of the Laguna Madre.

South Texas Eolian Sand Sheet (Coastal Sand Plain) The predominant sand ridge that extends along the mainland coast today from Matagorda Bay to Baffin Bay has been interpreted as a former shoreline during an earlier sea-level highstand during the Sangamon Interglacial Stage. This large sand body, which averages about 4.8 km (3.0 mi) wide and occupies a major portion of the peninsulas parallel to the Texas Coastal Bend, extends along the western margin of the northernmost Upper Laguna Madre. Although interpreted differently as either a barrier island (Price 1933) or barrier strand plain (Wilkinson et al. 1975), this Ingleside Barrier (Price 1933) represents a system similar to our present-day landscape. Furthermore, several parallel water bodies and lowland flats behind the barrier may represent a former lagoonal

Figure 3.13. *Flats dominated by Gulf Cordgrass are common coastal plain habitats in South Texas and northern Tamaulipas landward from Laguna Madre. Price (1933) suggested that Gulf Cordgrass flats occupy the ancient, infilled Ingleside Lagoon that now forms the Ingleside Terrace. (Photograph by J. W. Tunnell, July 1988)*

Figure 3.14. *Diagonal aerial view of banner dune complex on South Texas Eolian Sand Sheet, Kenedy County, Texas (Photograph by J. W. Tunnell, July 1987)*

system (Ingleside Lagoon). The parallel-to-the-coast orientation of Copano Bay, Port Bay, Oso Bay, Laguna Larga (freshwater lake south of Corpus Christi on King Ranch), and Alazan Bay all seem to support this concept. The extensive coastal flats covered mainly by Gulf Cordgrass (fig. 3.13) occupy the ancient, infilled Ingleside Lagoon and now form the Ingleside Terrace (Price 1933).

During the Holocene, the South Texas climate became progressively arid, and persistent southeasterly wind systems began to develop (Brown et al. 1977). Eolian (wind) deflation or erosion of Coastal Plain sediments began and increased with intensity. The old Ingleside Barrier south of Baffin Bay was blown inland, along with sands from Padre Island, to form the South Texas Eolian Sand Sheet (Price 1933) or system (Brown et al. 1977). This area is located at a convergence of longshore drift due to winds (creating waves) and a bend in the coastline (Watson 1971; McGowen et al. 1977). The area has probably been a location of excessive sand accumulation for tens of thousands of years or more. Today this vast sand sheet covers parts of five counties, from Baffin Bay in the north to the Rio Grande Delta in the south, and extends approximately 120 km (75 mi) inland. Recently, due to the unique soil and vegetative characteristics of the sand sheet, researchers suggested that it deserves recognition as a unique area of rangeland, the Coastal Sand Plain of southern Texas (Fulbright et al. 1990; Fulbright and Bryant 2002; see chapter 11). The extensive barren and vegetated sand dunes, especially the widespread banner-dune complexes (fig. 3.14) within the region, certainly mark this area as unique (Price 1958; Brown et al. 1977; Brown et al. 1980).

Behrens (1966) suggested that an ancient beach ridge, located in Tamaulipas on the southward-projecting peninsula (south of Carbajal) enclosing the northern part of Laguna el Catán, is the same geological structure as the Ingleside Barrier deposits described by Price (1933). The orientation of Laguna el Catán parallel to this ridge and the coastline possibly represents an extension of the old Ingleside Lagoon (Price 1933) and adds support to Behrens's contention.

Beach Rock Beach rock (coquina) outcrops discontinuously along the mainland shoreline of Laguna Madre from Baffin Bay at Penascal Point southward for approximately 10 km (6 mi) (Prouty 1996). This beach rock correlates with the Ingleside Barrier (Price 1933) and represents the only lithified bedrock exposure along the South Texas coast (fig. 3.15a). It contains well-developed late Pleistocene–early Holocene karst features (solution pipes and caliche crusts) unknown elsewhere on the Texas coast (Prouty and Lovejoy 1992; Prouty 1996). The coquina appears to have accumulated in a localized zone of converging longshore Gulf of Mexico currents along

Figure 3.15. *Beach rock (coquina) outcrops: (a) Penascal Point, western shore of upper Laguna Madre, Texas, at Baffin Bay (Photograph by J. W. Tunnell, July 1987); (b) Punta Piedras, western shore of southern Laguna Madre de Tamaulipas at Carbajal (Photograph by J. W. Tunnell, December 1998)*

a Gulf beach, similar to the Little Shell and Big Shell beaches today on Padre Island (Watson 1971). Carbon dating and karst features provide evidence of regional late Pleistocene–early Holocene climate changes from more humid during the Wisconsin sea-level lowstand 18,000 YBP to more arid at the commencement of the Holocene approximately 11,000 YBP (Prouty 1996). The beach rock in this location extends as scattered rocks seaward of the shoreline (Prouty 1996) and forms part of the shallow sill across the mouth of Baffin Bay (Rusnak 1960). This sill may have restricted water circulation and elevated salinities higher than those that existed in the Upper Laguna Madre (Collier and Hedgpeth 1950) before dredging of the Gulf Intracoastal Waterway (GIWW) and a Baffin Bay channel.

The beach rock outcrop at Punta Piedras (Yañez and Schlaepfer 1968) located at the tip of the northern peninsula enclosing Laguna el Catán in the southern Laguna Madre (see fig. 3.15b), probably correlates with the Penascal Point coquina. Shelly beach sediments, similar to Little Shell on Padre Island, begin just south of Boca Jesús María on the Gulf beach opposite Punta Piedras and extend southward to the Río Soto la Marina (J. W. Tunnell, pers. observ.). Some scattered coquina deposits also occur in places along this stretch of beach.

Serpulid Reefs The lagoonal beach rock in Texas and Tamaulipas provides a natural hard-substrate habitat that is rare in both systems. Serpulid worm reefs provide another natural hard substrate. Relict serpulid worm reefs exist in a scattered distribution around Baffin Bay (Breuer 1957; Andrews 1964; Behrens 1974; Cole 1981; Hardegree 1997; Simms et al. 2010). These reef rock structures are composed of the calcareous, external tubes of serpulid polychaetes that no longer live in the area. These remnant skeletal structures indicate a past, less saline environment for colonial, suspension-feeding marine or estuarine organisms. Although some living serpulid worms are found on the reefs today (Cole 1981; Hardegree 1997), they no longer build reef structure in the hypersaline environment.

Two types of serpulid reefs are recognized within the shallow waters of Baffin Bay: patch reefs and reef fields (Andrews 1964). These are predominantly found along bay margins, but some are distributed across the mouths of Baffin and Alazan Bays. Patch reefs are small, isolated structures, whereas reef fields are comparatively large expanses of reef rock. These reef rock areas provide rare, natural rocky habitat in this typically soft-substrate region. Of the original 120 serpulid reefs located and mapped (Breuer 1957; Rusnak 1960; Andrews 1964), many are undergoing degradation from being hit by outboard motors driven by uninformed or unknowing boaters in the area, causing a reduction of this unique habitat (Hardegree 1997).

Geology

The Laguna Madre of Texas is a linear, coastal lagoon that developed on the pre-Holocene (Pleistocene) erosional surface by the buildup and enclosure of a barrier island, Padre Island (Rusnak 1960). The slowing of postglacial rise in sea level caused the sediment buildup and lagoonal enclosure. The Holocene deposits average 3 to 6 m (10 to 20 ft) in thickness, although low areas in the Pleistocene surface may have much thicker deposits. The majority of the Holocene sediments consist of fine sands (Rusnak 1960). Sands are more common in the shallow eastern lagoon near the barrier island, where storm surges and southeasterly winds are filling the lagoon, and clays and silts are more common along the mainland western shore and in the west-central basins. The Laguna Madre de Tamaulipas is characterized as a Gilbert–de Beaumont Barrier

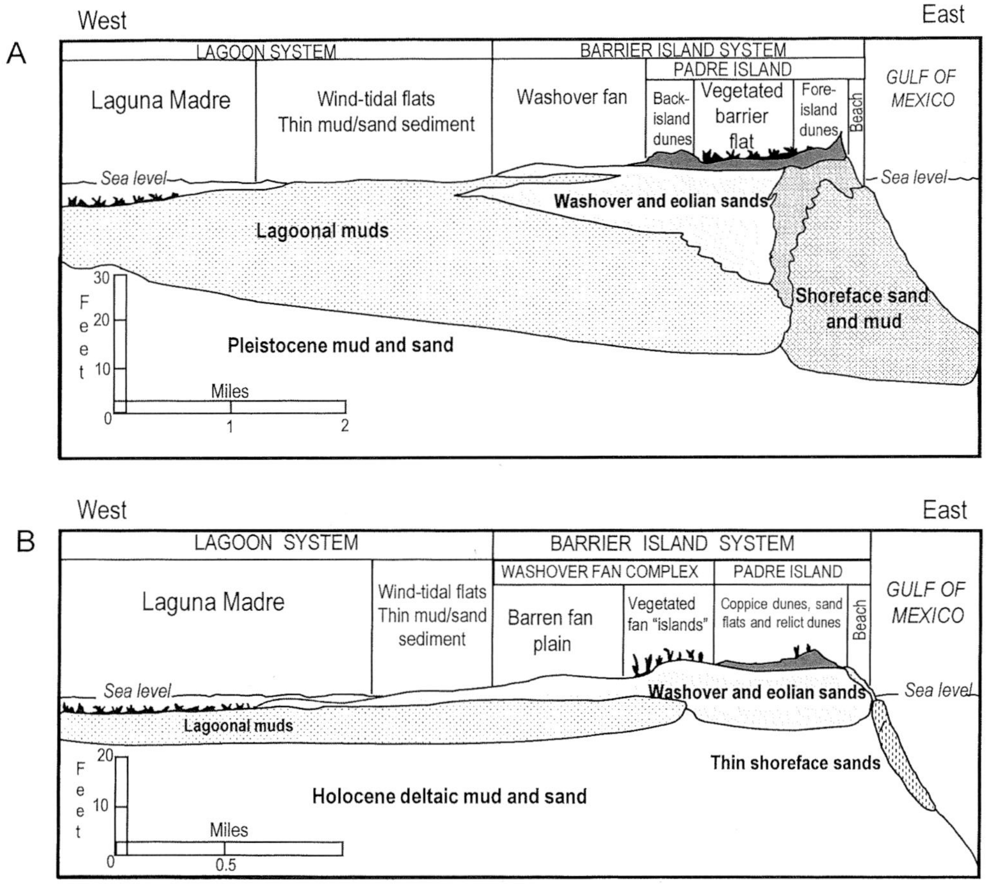

Figure 3.16. *Generalized cross sections of Padre Island barrier system: (a) Central Padre Island north of Mansfield Channel exhibits 9 m (30 ft) of Modern barrier island sand; (b) South Padre Island between La Punta Larga and Brazos Santiago Pass illustrates the thin, sand-deficient character of the island. In the south the island rests on Holocene Rio Grande deltaic sediments. The poorly developed barrier south of La Punta Larga is susceptible to extensive breaching by hurricane-tidal surge. (Modified from Brown et al. 1980)*

Lagoon (Lankford 1977), which has these same, or similar, features and processes (Yañez and Schlaepfer 1968). Figure 3.16 shows two generalized cross sections of Padre Island from the Gulf of Mexico into the Laguna Madre, demonstrating that island and lagoonal sediments are thicker north of Mansfield Pass than those to the south of La Punta Larga.

In addition to the generalized geological setting given here, as well as some of the geomorphological features and processes discussed earlier in this chapter, there are several other geochemical processes occurring in the Laguna Madre of Texas that are unique to the region. Two of these are quiet-water oolite formation and gypsum crystal formation. Oolites, which are laminated accretionary grains of calcium carbonate, are forming on the Laguna Madre shoreline in three locations: just north of the Point of Rocks at the mouth of Baffin Bay (Shepard and Rusnak 1957; Rusnak 1958, 1960; Freeman 1962); near two submerged sand bars between Baffin Bay and Padre Island (Behrens 1966); and along the northern shores of Baffin Bay, particularly around Kleberg Point (Land et al. 1979; Simms et al. 2010). The quiet-water shoal areas defied the classic theory that water agitation was necessary to form the distinctive accretionary layers of oolites. High energy does seem essential for the development of symmetrical oolites, but these rough, irregular oolites found in Baffin Bay and Laguna Madre suggest agitation is not critical for the accretionary development of carbonate coatings on grains (Dalrymple 1964; Land et al. 1979).

Bladelike crystals and rosettes—that is, clusters of crystals resembling a rose—composed of the mineral gypsum (calcium sulfate) are unique, natural structures formed in the wind-tidal flats of Laguna Madre in Texas (Weise and White 1980). They are found on both sides of the GIWW in the Land-Cut area, as well as along Mansfield Channel. Crystals apparently form when high-salinity waters periodically flood the wind-tidal flats and seep into the underlying sediments. These mineral-laden waters precipitate gypsum, forming the crystals, which grow larger with each successive

Figure 3.17. *Gypsum crystals found eroded from dredged-material islands of the Laguna Madre: (a) gray, sand-textured crystal on oil and gas channel dredged-material island in the Land-Cut, Kenedy County, Texas (Photograph by J. W. Tunnell, August 1982); (b) three types of gypsum crystals found in Laguna Madre dredge spoil (from wind-tidal flats: clear, amber colored; gray, sand-textured; and, small, light-colored, sand-textured rosettes) (Photograph by J. W. Tunnell, October 2017)*

a

b

flooding, infiltration, and precipitation (Masson 1955). When channels are dredged across the flats, the crystals, along with other dredged material, are deposited in piles. In time, wind and rain erode the sediments away, leaving the crystals and rosettes exposed on the surface. Three different types have been observed: clear, amber-colored blades/crystals; gray, sand-textured blades/crystals; and small, light-colored, sand-textured rosettes (fig. 3.17). The clear forms and smaller rosettes have been found on dredged-material deposits along oil field channels of the Land-Cut, and the larger gray forms, presumably created in a hydrogen sulfide environment that produced the grayish color, have been found in dredged material along the Mansfield Channel as well as in the Land-Cut (J. W. Tunnell, pers. observ.).

As briefly demonstrated here, but far more extensively by many publications of the Bureau of Economic Geology, it is imperative to consider geological history and processes when considering conservation or preservation of coastal natural resources. The origin and development of the coastal zone, as well as the current physical processes affecting the area, can help determine the probable success or failure of proposed efforts. There is great variety in Laguna Madre and Padre Island geomorphological features. Variability results from such factors as Pleistocene depositional and erosional history, sand availability, climatic conditions, density of vegetation, water circulation, and sediment transport (McGowen et al. 1977). With human activities superimposed on the natural landscape, the long-term dynamic balance tips toward disequilibrium. Coastal processes that have interacted for thousands of years to create the coastal features can become agents of destruction. Therefore, understanding these processes is imperative to regional coastal planning and management.

Coastal problems caused by interrupting natural geological processes are clearly elucidated in three *Environmental Geologic Atlases* covering the Laguna Madre in Texas (Brown et al. 1976; Brown et al. 1977; Brown et al. 1980). Observations caused by increased development are presented, and recommendations for orderly development compatible with the capacity and capability of natural systems are suggested. Interestingly, many of the recommendations given some 40 years ago—about how to proceed in an orderly fashion on such items as channelization, devegetation, shoreline construction, waste disposal, filling and land reclamation, artificial passes, and natural catastrophes—are still largely ignored today. Reasons for not following these guidelines are unclear but may include the lack of communication between scientists, managers, and/or decision makers or perhaps a lack of enforcement or dollars for implementation.

4 Tamaulipan Biotic Province

FRANK W. JUDD

The identification, description, and mapping of the geographic extent of biotic provinces constitute one method of organizing the biotic diversity of North America. This approach, introduced by Dice (1943), focuses on the ranges and centers of distribution of plant and animal species in physiographic regions. As defined by Dice, a biotic province is a considerable and continuous geographic area. Thus, a biotic province never occurs as discontinuous fragments, but it may include more than one major ecological association, including freshwater communities. Dice does not consider marine communities when defining a biotic province. Consequently, the focus in this chapter is on the upland and freshwater communities of the Tamaulipan Biotic Province.

Geographic Extent and Initial Characterization

Dice (1943) identified, mapped, and described 29 biotic provinces in North America. The southern boundary of the Tamaulipan Biotic Province (TBP), as mapped by Dice, begins at the Río Panuco and extends northward from a line connecting Tampico and San Luis Potosí through those portions of the Mexican states of Tamaulipas, Nuevo León, and Coahuila east of the Sierra Madre Oriental into southern Texas (fig. 4.1). Dice limited the distribution of the TBP to the southern tip of Texas, but Blair (1950) extended it northward to the Balcones Fault line on the north and the line separating pedocal from pedalfer soils on the northeast. This northeastern boundary marks the area where the vegetation changes from thorny brushlands to prairie and oak—hickory "alternes" (Blair 1950)—and is about at the San Antonio River. Throughout its length the Gulf of Mexico forms the eastern boundary of the province; the Sierra Madre Oriental forms the western boundary.

Alvarez (1963) drew the southern U-shaped boundary of the TBP in Tamaulipas beginning in the southeast at La Pesca, 180 km (112 mi) north of Tampico, and looping down to Acuña and Santa Isabel at the bottom, then northward to Ciudad Victoria in the southwest (fig. 4.2). Regardless of whether the southern boundary is drawn at Soto la Marina (Alvarez 1963) or at Tampico (Dice 1943), the entire Laguna Madre of Texas and Mexico is included in the TBP. The distribution of Tamaulipan thornscrub vegetation in Texas and Mexico delineated by Reid et al. (1990) closely matches the boundaries of the TBP (see fig. 4.1). Reid and colleagues reported that spiny shrubs and trees dominate the natural vegetation but that grasses, forbs, and succulents are also prominent.

Most of the TBP is confined to the Gulf Coastal Plain, but Dice (1943) also emphasized that TBP includes numerous rolling hills, some limestone ridges, and a few mountains. The highest are the San Carlos Mountains in Tamaulipas (see fig. 4.1), which rise to an elevation of about 1,524 m (5,029 ft).

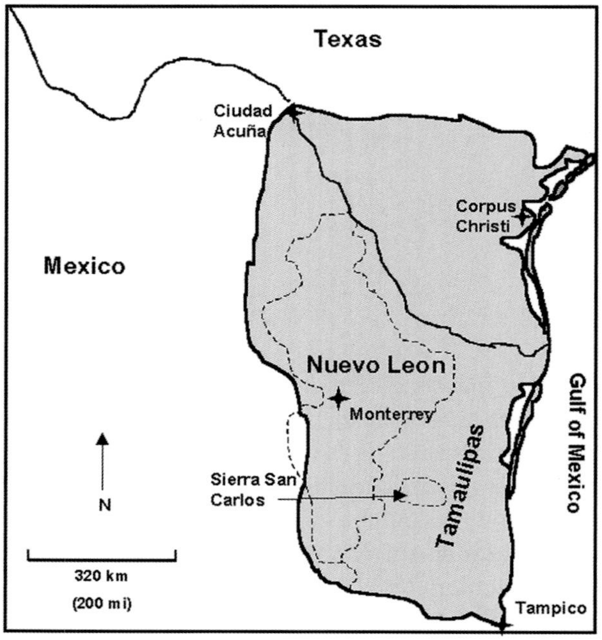

Figure 4.1. *Tamaulipan Biotic Province (shaded area) of Texas and Mexico*

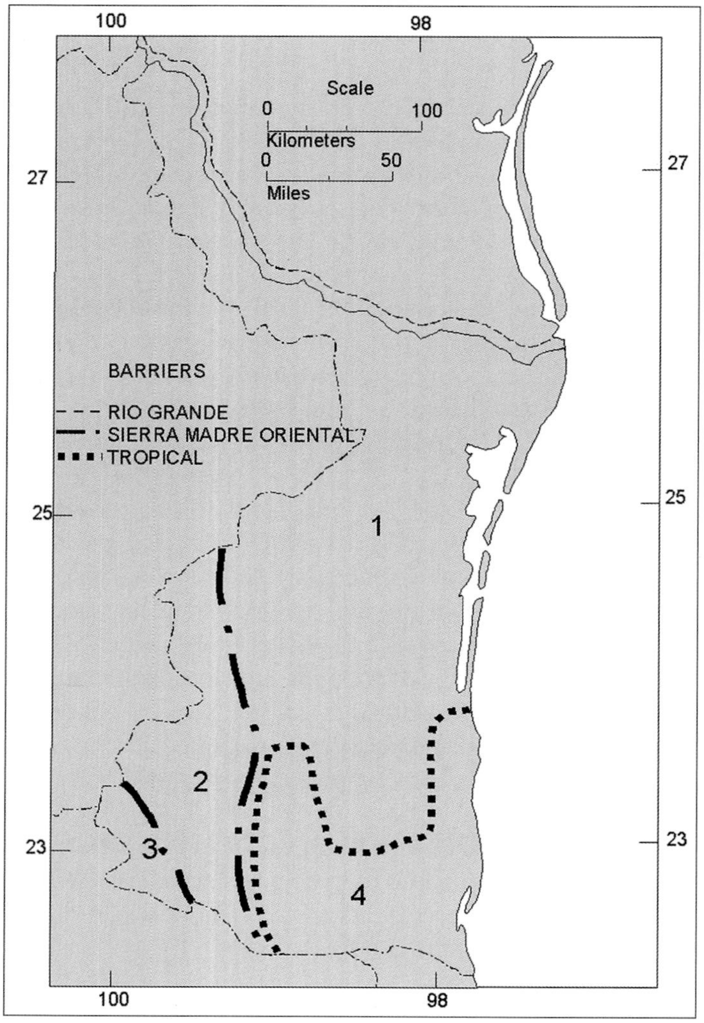

Figure 4.2. *Four biotic provinces of Tamaulipas, Mexico, recognized by Alvarez (1963): 1 = Tamaulipan; 2 = Potosian; 3 = Chihuahuan; 4 = Veracruzian*

Biotic provinces typically are characterized by peculiarities of vegetation type, ecological climax, flora, fauna, climate, physiography, and soil. In his brief description of the TBP, Dice (1943) indicated that dense thickets of shrubs and small trees are a significant vegetative characteristic. The absence of true forest, except on mountains, is attributed to the limestone substrate, which does not hold moisture well, and to high summer temperatures that amplify evaporation rates. The Gulf Coastal Plain portion of the province is reported to be characterized by a dense growth of thorny shrubs and numerous cacti. The foothills of the San Carlos Mountains also support thorny shrubs as the dominant life-form, but the species making up the vegetation are different from those of the Coastal Plain. The upper elevations of the mountains support an oak and pine forest (Dice 1937). Blair (1950) reported that thorny brush is the dominant life-form in the vegetation of the TBP in Texas, whereas Presley (2003) considered the TBP to be an ecotone community characterized by a mixture of plains, woodland, and desert-adapted elements.

Dice (1943) said little about the fauna of the TBP but cited Allen (1892, 1893), Nelson and Goldman (1926), and Dice (1937) for mammals and birds characteristic of the region. H. Smith (1939) was cited as reporting typical reptiles. Dice (1943) noted that the fauna and flora of the TBP contain many Central American and South American elements, and he mentioned parrots and palms as examples. Mammals that range a little north of the province include the White-nosed Coati, Jaguar, Jaguarundi, and Mexican Spiny Pocket Mouse. The distribution of the Nine-banded Armadillo and Collared Peccary, considered by Dice to be species characteristic of the region, now extend well beyond the TBP in Texas. Additional information on the vertebrate fauna of the TBP is provided in a later section of this chapter.

Climatic Factors Climate is a crucial aspect in the designation of a biotic province (Dice 1943; Blair 1950). The climate of this area is described in chapter 2 but is addressed briefly here to help clarify the nature of the TBP. Blair indicated that the climate of the TBP is semiarid and megathermal. This means that temperature is sufficiently high to support plant growth throughout the year, but at times there is insufficient moisture to permit growth.

Southeasterly winds from the Gulf of Mexico prevail in every month except December, when a northwest wind is common. The average relative humidity at Brownsville, Texas (about midway on a north-south axis through the province), at 6:00 a.m. ranges from 85% in December to 90% in July, with an annual average of 88%. This high relative humidity causes fairly heavy dew even during drought years. Dense fogs are common at night from December through March. Precipitation is highest near the Gulf of Mexico and decreases westward. For example, in the northeastern corner of the province, total annual precipitation is 91.0 cm (35.8 in), and in the northwestern corner it is 51.0 cm (20.1 in). The normal annual rainfall at Brownsville is 68.2 cm (26.9 in), and the peak of precipitation occurs in September and October (Leonard, Everitt, et al. 1991). About one-third of the total annual rainfall occurs during this period. Some 69% of the annual rainfall is received during the six-month period from May through October. This division corresponds closely with the "rainy season" that prevails in the Tropics of eastern and central Mexico. Most precipitation results from thunderstorms, and often a single thunderstorm accounts for an entire month's rainfall.

Temperatures in summer and early fall are fairly constant in the mid-30s °C (86 °F) during the day and mid-20s °C (68 °F) at night. Freezes occur on a regular basis in the northern portion of the province, but from the Lower Rio Grande Valley southward the length of the frost-free period is 330 to 365 days.

Subdivisons in Texas

The Tamaulipan Biotic Province in Texas is bounded on the north by the Balconian Biotic Province and to the northeast by the Texan Biotic Province. Blair (1950) subdivided the TBP into two districts, the Matamoran Biotic District and the Nuecian Biotic District (fig. 4.3). The Matamoran Biotic District was separated from the Nuecian because of the unique flora and fauna of the Lower Rio Grande Valley of Texas (Cameron, Hidalgo, Starr, and Willacy Counties). The extent of the Matamoran Biotic District in Mexico has not been delineated, and no subdivisions of the province have been named in Mexico.

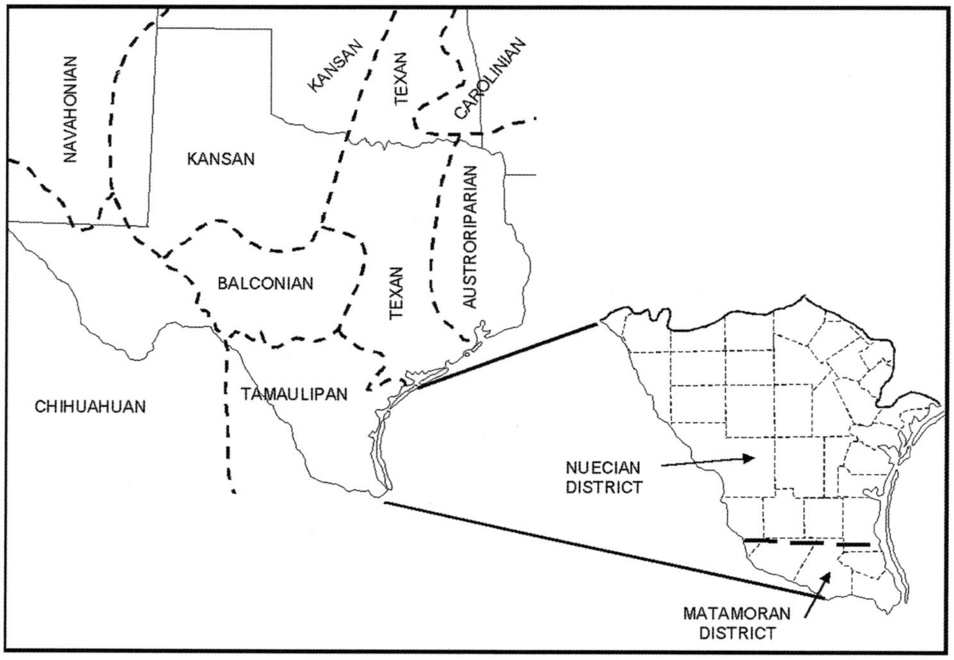

Figure 4.3. *The approximate boundaries of biotic provinces in Texas and biotic districts of the Tamaulipan Biotic Province in Texas*

Matamoran Biotic District As defined by Blair (1950), the Matamoran Biotic District corresponds to the Lower Rio Grande Valley of Texas (Lonard, Everitt, et al. 1991). Less than 5% of the Lower Rio Grande Valley of Texas remains in native shrubland (USFWS 1983; Vora 1989, 1990). The vegetation of the region has been the subject of numerous scientific reports. Clover (1937) and M. Johnston (1955, 1963) described the vegetation of the entire area in considerable detail. Blair (1950, 1952) provided an overview and identified important woody species. Lonard, Everitt, et al. (1991) focused their comprehensive treatment on the woody plants of the Lower Rio Grande Valley. Lonard and Judd (1985, 1991) reported on the effects of the severe freezes of 1983 and 1989 to the woody plants of the Lower Rio Grande Valley. Judd and Sides (1983) described the effects of Hurricane Allen on the nearshore vegetation of South Padre Island, and Judd and Lonard (1985, 1987), Judd et al. (1989); Lonard, Judd, et al. (1991); Judd et al. (1991); and Lonard and Judd (1993) examined the effects of other disturbance factors on the vegetation of South Padre Island. Judd et al. (1977), Lonard et al. (1978); Lonard and Judd (1980, 1981); and Lonard, Judd, et al. (1991) described the vegetation of South Padre Island. Lonard and Judd (1993) described the phytogeography of the Lower Rio Grande Valley, and Lonard and Judd (2002) discussed the riparian vegetation of the lower Rio Grande. Davis (1942) described and discussed the vegetation of Texas Palmetto groves in Cameron County. Based on this information, Lonard, Everitt, et al. (1991) partitioned the natural vegetation of the Lower Rio Grande Valley into four major habitats: (1) River Floodplain, (2) Coastal Prairies and Marshes, (3) Barrier Islands, and (4) Brush-Grasslands.

Blair (1952) briefly described the vegetation at four sites in the Matamoran Biotic District. About 10 km (6 mi) south of McAllen, Hidalgo County, Texas, the dominant plants were large Cedar Elms mixed with various thorny brush species including Texas Ebony, Granjeno, Honey Mesquite, Huisache, and Palo Verde. Texas Prickly Pear was scarce and confined primarily to disturbed areas.

At La Gloria Ranch, 27 km (17 mi) northwest of Edinburg, Hidalgo County, Texas, near the northern limit of the district, Blair (1952) reported the principal brush species to be Honey Mesquite, Granjeno, Palo Verde, Texas Ebony, Texas Sage, Guayacan, Allthorn, Huisache, Texas Lantana, Lime Prickly Ash, Anacahuita, Desert Yaupon, Leatherstem, and Coyotillo. The most common grass was Red Grama. He noted that the soil was sandy at this site. On another site with deep, loose, sandy soil, about 16 km (10 mi) northwest of Raymondville, dense thickets of brush and prickly pear and savanna-like clumps of mesquite were scattered over grasslands dominated by threeawn. On black clay soil just west of Port Isabel, Cameron County, Texas, the principal plant species were Gulf Cordgrass and Allthorn.

Brush-Grasslands This vegetation type extends throughout the upland areas away from the river and the coast. It is the most widespread habitat in the Matamoran Biotic District. The vegetation is characterized by a great variety of low trees, shrubs, cacti, and yuccas of variable density with small grassy areas in open sites (Lonard, Everitt, et al. 1991). Species richness is due to the overlap of western desert, northern plains, and tropical floras. Presence or absence of woody species is largely determined by soil and moisture conditions. The communities are usually stratified into an overstory of species such as Honey Mesquite, Huisache, or Texas Ebony with an almost impenetrable understory that includes Texas Sage, Coyotillo, Granjeno, Guajillo, Guayacan, Texas Lantana, Leatherstem, Lime Prickly Ash, Lotebush, Narrow-leaf Forestiera, Border Paloverde, Texas Prickly Pear, Anacahuita, and Blackbrush. On sandy to gravelly soils where drainage is good, numerous cacti are present. Prominent species include Pencil Cactus, Texas Prickly Pear, Pitaya, Lace Echinocereus, Hedgehog Cactus, and Little Nipple Cactus. In areas transected by arroyos, shrubs usually are larger near the banks where water stands after rains. Huisache and Retama are often abundant in these places. In areas of deep sands, grasses are more abundant and the vegetation is usually a savanna type with large mesquites scattered in small clumps (mottes) or occurring as single individuals. Areas such as irrigation- and drainage-canal banks, railroad rights-of-way, and highway rights-of-way typically support a dense cover of Buffelgrass. Buffelgrass, introduced from India and widely planted in "improved" pastures, establishes quickly in cleared areas and becomes dominant.

River Floodplain Vegetation of the River Floodplain includes riparian forest along the banks of the Rio Grande and along *resaca* banks. Dominant species may be Cedar Elm, Honey Mesquite, Texas Ebony, Sugar Hackberry, Anacua, or a combination of two or more of these species. Perhaps the most important community in the eastern section of the River Floodplain is the native Texas Palmetto.

Blair (1950) stated that characteristic plant species of the Matamoran Biotic District included Paloverde, Texas Ebony, Anacahuita, and Anacua. He noted that Cedar Elm was dominant in the floodplain forest of the Rio Grande at some localities, and at others there was an alternation of Cedar Elm and brush species. Blair reported that Texas Palmetto reached its northern limit at Southmost, below Brownsville on the Rio Grande, but Lockett and Read (1990) found an isolated population in Jackson County on the central coast of Texas.

Shelford (1974) included the Matamoran Biotic District in the Southern Temperate Grassland Biome and mapped the vegetation as an acacia-grassland community. Odum (1971) recognized the Broad-leaved Evergreen Forest as a distinct biome that is represented in the hammocks of Florida and in the Live Oak forests along the Gulf and South Atlantic Coasts of the United States. Benson (1979) characterized the vegetation of the southern portion of this biome as the American Subtropical Flora and stated that this vegetation is represented in the United States chiefly in the hammocks and everglades of southern Florida and to a slight extent in the swampy regions along the Gulf of Mexico, particularly near Brownsville, Texas. Odum (1971) reported that palms such as Texas Palmettos (or Cabbage Palms) are often prominent members of the community and that vines and epiphytes are characteristically present.

The Texas Palmetto community has been recognized as distinct by Clover (1937), A. Davis (1942), Odum (1971), Benson (1979), Diamond et al. (1987), Bezanson (2000), NatureServe (2002), Griffith et al. (2007), and Chapman and Bolen (2018). In 1852 stands of Texas Palmetto extended along the Rio Grande from a point near its mouth to about 130 km (81 mi) inland from the Gulf of Mexico (Clover 1937). By the late 1930s clearing for agriculture had reduced the extent of this palm forest in the United States to a small reach of the Rio Grande from a point 16 km (10 mi) below Brownsville, Cameron County, Texas, upriver 6.4 km (4.0 mi) (Clover 1937). The most extensive growth of palms was at Rabb Ranch, located approximately 16 km (10 mi) southeast of Brownsville at a bend where the river reaches its southernmost point (Clover 1937; Davis 1942). The Audubon Society purchased a 70 ha (172 ac) tract of the ranch in 1971 to establish the Texas Palmetto Grove Sanctuary (fig. 4.4). Today about 13 ha (32 ac) of palm forest are present, with the remaining land in citrus orchards and row-crop farm fields.

Figure 4.4. *The remnant forest of Texas Palmettos at the Texas Palmetto Grove Sanctuary forms a dense canopy. (Photograph by Brian R. Chapman)*

Clover (1937) included the Boscaje de la Palma as a coastal climax association of the Lower Rio Grande Valley of Texas. She pointed out that this is one of only four arborescent palm communities in the continental United States outside Florida, the other three being located in the southeastern Atlantic area, the Mississippi Delta area, and the southern California desert. Clover provided a list of 81 species associated with the Texas Palmetto community. Davis (1942) focused on the Boscaje de la Palma in Cameron County, Texas, and she also provided a description of the distribution of Texas Palmetto in the Rio Grande Delta area. Diamond et al. (1987) recognized the Texas Palmetto Series (dominated by Texas Palmetto) as a distinct late seral stage forest in Texas, and they identified it as endangered. Indeed, it was one of only three communities (of 78) in Texas to be listed as endangered. The Texas Organization for Endangered Species (Carr et al. 1993) considers Texas Palmetto a threatened species in the state.

Everitt et al. (1996) used remote sensing and spatial information technologies to map Texas Palmetto in the Lower Rio Grande Valley. Future censuses may be compared with their map (and imagery) to quantify changes in population densities. The map may also prove useful to resource managers in identifying land for acquisition for conservation and reestablishment of Texas Palmettos. For example, the map may be useful in assessing the feasibility of developing corridors connecting presently fragmented populations.

Resacas are often selected as sites for housing developments, but native vegetation remains along the banks of many such sites. Important woody species include Paloverde, Huisache, Texas Ebony, Black Willow, and Black Mimosa.

In the western section of the River Floodplain, important woody species along the banks of the Rio Grande that are subject to flooding include Huisache, Black Willow, Paloverde, Sugar Hackberry, and Mexican Ash. Floodwaters disturb or destroy the smaller vegetation along the river, leaving stretches of deep sand, which often become covered by almost pure stands of Groundsel Bush. On terraces that are rarely flooded, the dominant woody species are Huisache, Honey Mesquite, and Granjeno.

Nuecian Biotic District Blair (1952) listed the plant communities of the Nuecian Biotic District: (1) Live Oak–dominated; (2) mesquite-cactus; (3) mesquite savanna with the grasses Red Threeawn, Hairy Grama, and Hooded Windmillgrass; (4) Sugar Hackberry–Paloverde–Blackbrush; (5) Creosote Bush–Texas Prickly Pear; (6) a mixed community containing Honey Mesquite, Granjeno, Guayacan, Bluewood, Leatherstem, and Texas Mountain Laurel; (7) a mixed community containing Honey Mesquite, Granjeno, Huisache, Guayacan, Texas Sage, Beebrush, and Texas Ebony; and (8) a community in which La Coma is the characteristic tree and Red Grama the dominant grass.

Jones (1975) and Lehman et al. (2005) published on the flora (exclusive of grasses) of the Texas Coastal Bend, which includes the northeastern portion of the Nuecian Biotic District. Gould and Box (1965) published the companion volume on the grasses of the Texas Coastal Bend, Gould (1975) published *Grasses of Texas*, and Shaw (2012) produced *Guide to Texas Grasses*, a comprehensive volume describing each species of Texas grass. Jones (1975) reported that at the time of initial settlement by European peoples (about 1800) this area was largely treeless prairie that varied from flat and poorly drained to undulate or rolling. Grasses prevailed, but there were nongrassy herbs and probably a scattering of shrubs, including Honey Mesquite and Texas Prickly Pear. Trees were generally scarce but occurred wherever there was a steady supply of moisture. There were extensive woods along the Nueces and San Antonio Rivers and

Figure 4.5. *Dense riparian forests line the banks of rivers in the Gulf Prairies and Marshes region (Photograph by Brian R. Chapman)*

smaller wooded areas along the Mission and Aransas Rivers (fig. 4.5). Tree species Jones identified as dominants are consistent with extant communities identified in the South Texas Plains and Gulf Coast Prairies and Marshes vegetation areas discussed in considerable detail later in this chapter.

Plant Communities and Characteristic Plant Species

Blair (1950) identified the species characteristic of thorny brush communities throughout the TBP in Texas as Honey Mesquite; various species of *Acacia*, *Mimosa*, and *Condalia*; and Granjeno, Guayacan, Texas Sage, Beebrush, Texas Prickly Pear, Pencil Cactus, and Allthorn. Blair identified coastal marshes dominated by Gulf Cordgrass as a distinct plant community and identified vegetative patterns:

1. From the coast westward the brush cover becomes sparser as available moisture decreases.
2. The brush on sandy soils differs in species and aspect from that on clay soils. Honey Mesquite in open stands and mixed with various grasses is characteristic of sandy soils. Clay soils support a variety of species, including Honey Mesquite.
3. The brushlands of the Lower Rio Grande Valley in Cameron, Hidalgo, Starr, and Willacy Counties are more luxuriant than the brushlands farther south (in Mexico), and they are characterized by several plant species that decrease in abundance northward.
4. The most luxurious brush occurs in the immediate floodplain of the Rio Grande.

Gould (1969) recognized two major vegetational areas in the Texas portion of the TBP: the Gulf Prairies and Marshes and South Texas Plains. The South Texas Plains is called the Rio Grande Plains by some authors (Davis and Schmidly 1994; Schmidly and Bradley 2016) and the South Texas Brushland by others (Chapman and Bolen 2018). Gould's distinction is based largely on the presence or absence of trees and shrubs. The Gulf Prairies and Marshes area is primarily grassland, whereas the South Texas Plains may have originally supported a savanna grassland. Johnston (1963), who discussed the past and present grasslands of the TBP, objected to allegations that woody vegetation had encroached on grassland in southern Texas and particularly that the brush species had invaded from Mexico. He maintained that the brush species already occupied their present ranges when the first collector visited the region in 1828. Johnston emphasized that many grasslands were infested with mesquite long ago but that the growth of mesquite was stunted due to recurrent fires. He argued that the rapid takeover of the mesquite brush resulted from fire control and involved increase in height of the aerial parts of the plant and increase in density of stands rather than invasion by mesquite into previously brushless areas.

McMahan et al. (1984) followed Gould (1969) in recognizing two major ecological areas in southern Texas—the Gulf Coast Prairies and Marshes and the Coastal Texas Plains. Within these two ecological areas 12 vegetation types, excluding crops, were recognized (table 4.1). The Laguna Madre is bordered throughout its length by communities of the Gulf Coast Prairies and Marshes ecological area. Consequently, we focus attention on these barrier island and mainland communities that border on the Laguna Madre. The Marsh–Barrier Island Vegetation Type occurs on the eastern margin of the Laguna Madre, and this vegetation type and all others identified in the Gulf Coast Prairies and Marshes ecological area occur on the western margin of the Laguna Madre.

Table 4.1. *Vegetation types recognized by McMahan et al. (1984) in the Texas portion of the Tamaulipan Biotic Province*

Vegetation Type	South Texas Plains	Gulf Prairies and Marshes
Bluestem Grassland		X
Texas Sage–Blackbrush–Creosotebush Brush	X	
Mesquite-Blackbrush Brush	X	X
Mesquite-Granjeno Parks	X	X
Mesquite-Granjeno Woods	X	
Mesquite–Live Oak–Bluewood Parks	X	X
Live Oak Woods and Parks	X	X
Post Oak Woods, Forest, and Grassland Mosaic	X	
Post Oak Woods and Forest	X	
Bald Cypress–Water Tupelo Swamp	X	
Marsh–Barrier Island		X
Native or Introduced Grasses	X	X

The types are exclusive of crops.

Barrier Islands The vegetation of barrier islands bordering the eastern margin of the Laguna Madre is essentially one layered, and species richness is relatively low compared to that of barrier islands of the US Atlantic Coast. Judd et al. (1977) pointed out that the major difference between the vegetation of South Padre Island, Texas, and that of the barrier islands on the eastern coast of the United States is the absence of shrub or tree zones on South Padre Island. Virtually all species on South Padre Island are herbaceous. Honey Mesquite and Texas Prickly Pear occur on the island, but only as rare, low-growing, isolated individuals. Except for trees and shrubs planted by residents, the only conspicuous woody vegetation consists of Black Mangrove, which is abundant on the margin of the Laguna Madre near the Queen Isabella Causeway and eastward along the lagoon shorelines (see chapter 19). On North Padre Island there is a small grove of Live Oak on the Laguna Madre side of the island (Weise and White 1980).

An idealized profile of South Padre Island includes the following topographic zones and vegetation (fig. 4.6): (1) a barren foreshore zone, (2) a backshore zone with a belt of Sea Purslane nearest the Gulf and a landward zone dominated by Sea Oats, (3) a primary dune ridge supporting Beach Morning Glory and Sea Oats as dominants, (4) a zone of either secondary dunes or vegetated flats or a mosaic of both in which Seacoast Beardgrass is dominant, and (5) a tidal-flat zone that is largely bare but that supports typical halophytic vegetation in a narrow zone bordering the vegetated flats. Species of importance include Seashore Dropseed, Camphor Daisy, Sea Lavender, and Shoregrass. This profile is also typical for North Padre Island, but Weise and White (1980) used slightly different terms for the topographic zones: (1) forebeach rather than foreshore, (2) backbeach rather than backshore, (3) fore-island dunes rather than primary dunes, (4) vegetated barrier flat and back-island dunes rather than secondary dunes and vegetated flats, and (5) back-island sand flats and wind-tidal flats rather than tidal flats. Weise and White did not name dominant species in each of the topographic zones, but the lists of associated species they provided suggest that the dominants may be similar to those on South Padre Island.

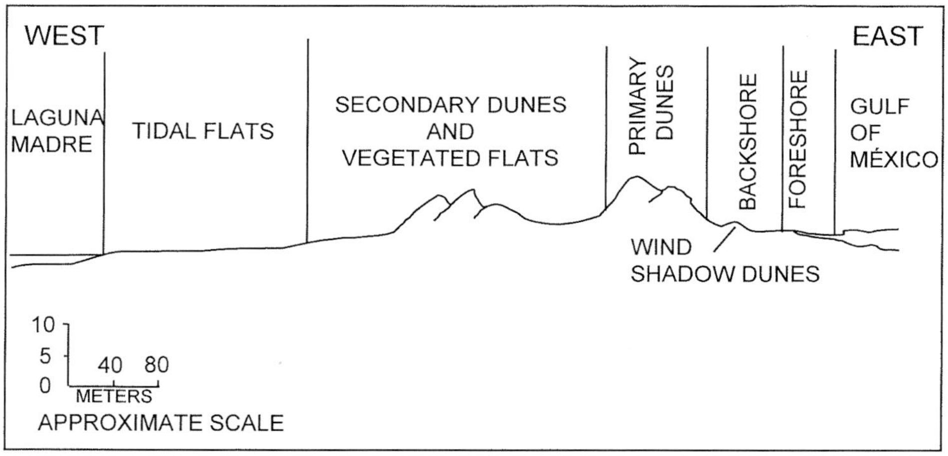

Figure 4.6. *Transverse profile of South Padre Island, Texas, showing major topographic zones*

Gulf Coast Prairies and Marshes On the Texas mainland, the vegetation of the Gulf Coast Prairies and Marshes ecological area includes three distinct habitats (Lonard, Everitt, et al. 1991): salt flats, marshes, and *lomas* (clay dunes). Salt flats are characterized by periodic inundation by salt water from storms and high tides. The soils are saline and largely barren except at the margins of outwash slopes of *lomas*. Most of the plants are halophytic herbaceous species. Dominant species are Shoregrass, Saltwort, Annual Seepweed, Sea Oxeye, and glasswort. Mainland coastal marshes are dominated by Gulf Cordgrass. A few shrubs such as Honey Mesquite, Huisache, and Paloverde may be scattered through the marsh.

Auffenberg and Weaver (1969) described the vegetation of *lomas* in eastern Cameron County as a chaparral association consisting of several species of thorny shrubs and low trees. Dominant woody species include Lotebush, Coyotillo, Honey Mesquite, and Huisache. Texas Prickly Pear, Pencil Cactus, Lime Prickly Ash, and Spanish Dagger grow along the edges of the dense shrubs and trees and in open patches. Buffalograss is dominant in open areas on the top or crest of *lomas*. At the periphery of the slope and extending in a narrow band almost around the entire *loma* is a zone composed of Seep Willow and Gutta-percha.

Bluestem Grassland Bluestem grasslands occur on the Texas mainland coast between the Mission and San Antonio Rivers (McMahan et al. 1984). This community is particularly manifest in Goliad, Victoria, and Refugio Counties (fig. 4.7). Seacoast Beardgrass is dominant. Associated species include Bushy Bluestem, Slender Bluestem, Silver Bluestem, threeawns, Buffalograss, Bermuda Grass, Brownseed Paspalum, Gulfdune Paspalum, Smut Grass, Gulf Cordgrass, Hooded Windmillgrass, and Rio Grande Dewberry. Woody species scattered in the grassland include Live Oak, Honey Mesquite, Huisache, and Groundsel Bush.

Figure 4.7. *Extensive coastal prairies dominated by bluestem grasses and scattered forbs were interrupted only by narrow riparian forests. (Photograph by Sandra S. Chapman)*

Mesquite–Blackbrush Brush This community type occurs primarily in the South Texas Plains, but it is present in the Gulf Coast Prairies and Marshes of Texas in Goliad, San Patricio, Nueces, Willacy, and Cameron Counties (McMahan et al. 1984). A small portion of one population borders Corpus Christi Bay, north of the mouth of the Nueces River. The dominant species are Honey Mesquite and Blackbrush. Associated species include Lotebush, Texas Sage, Guajillo, Narrow-leaf Forestiera, Spiny Allthorn, Beebrush, Bluewood, Granjeno, Guayacan, Leatherstem, Texas Prickly Pear, Pencil Cactus, Texas Kidneywood, yucca, Desert Yaupon, Allthorn, Purple Threeawn, Pink Pappusgrass, tridens, Slim Tridens, Hairy Grama, Matted Sandmat, coldenia, Dogweed, Knotweed Leafflower, and Twoleaf Senna.

Mesquite-Granjeno Parks Mesquite-Granjeno parklands occur primarily on sandy or loamy upland soils in the South Texas Plains (McMahan et al. 1984), but this vegetation type is also common in the Gulf Coast Prairies and Marshes south of Corpus Christi Bay. It is the principal community in Kleberg and Kenedy Counties. A small population occurs in Refugio County and reaches the shores of Copano Bay. Associated species include Bluewood, Lotebush, Coyotillo, Guayacan, Texas Colubrina, Pencil Cactus, Texas Prickly Pear, Pan American Balsamscale, Gulfdune Paspalum, Hooded Windmillgrass, Tanglehead, Purple Threeawn, Tumble Lovegrass, Lindheimer Tephrosia, Texas Bullnettle, croton, Slender Dwarf Morning-glory, Texas Lantana, Silverleaf Nightshade, and Indian blanket.

Mesquite–Live Oak–Bluewood Parks This community type comes within 25.0 km (15.5 mi) of Aransas Bay in San Patricio County, Texas. The dominant species are Honey Mesquite, Live Oak, and Bluewood. Associated species include Huisache, Twisted Acacia, Beebrush, Granjeno, Lotebush, Berlandier's Wolfberry, Blackbrush, Desert Yaupon, prickly pears, La Coma, Pencil Cactus, Agarito, Texas Persimmon, Purple Threeawn, Pink Pappusgrass, Hall's Panic Grass, White Poppymallow, Sensitive Briar, Twoleaf Senna, and Matted Sandmat (McMahan et al. 1984).

Live Oak Woods and Parks This community type occurs mainly on sandy soils and borders portions of Copano Bay in Refugio and San Patricio Counties. It is most abundant in Kenedy County, where it borders the south shore of Baffin Bay and the Lower Laguna Madre of Texas. The dominant species is Live Oak, and associated species include prickly pears, Lime Prickly Ash, greenbriar, sunflowers, Tanglehead, Crinkleawn, Gulfdune Paspalum, Fringed Signalgrass, Lindheimer Tephrosia, croton, Silverleaf Nightshade, Texas Bullnettle, Texas Lantana, dayflowers, Silverleaf Sunflower, and Shrubby Wood-sorrel.

Native or Introduced Grasses Native and introduced grasses occur in the South Texas Plains where brush has been cleared but also are represented abundantly in Willacy and Cameron Counties along the lower mainland coast of the Lower Laguna Madre. Buffelgrass, an introduced species, is commonly planted to create "improved pasture."

Community Classifications and Dominant Species

Diamond et al. (1987) provided a classification of community types in Texas. They divided the state into 11 natural regions, which were subdivided into community types (fig. 4.8). In keeping with Gould (1969) and McMahan et al. (1984) they included as natural regions in the Texas TBP the South Texas Brush Country (the South Texas Plains of Gould [1969] and McMahan et al. [1984]) and the Gulf Coast Prairies and Marshes. However, they added a third natural region called Coastal Sand Plains (see chapter 11).

The South Texas Brush Country natural area was subdivided into three vegetative types: the Brush Country, Bordas Escarpment, and Subtropical Zone. The Gulf Coast Prairies and Marshes natural area was similarly divided into three vegetative types: Dunes/Barrier, Estuarine Zone, and Upland Prairies and Woods. The Coastal Sand Plains natural area was not subdivided.

Diamond et al. (1987) recognized eight distinct forest and woodland communities in the Texas TBP (table 4.2). Two of the communities occur in both the Gulf Coast

Figure 4.8. *The natural regions of Texas recognized by Diamond et al. (1987)*

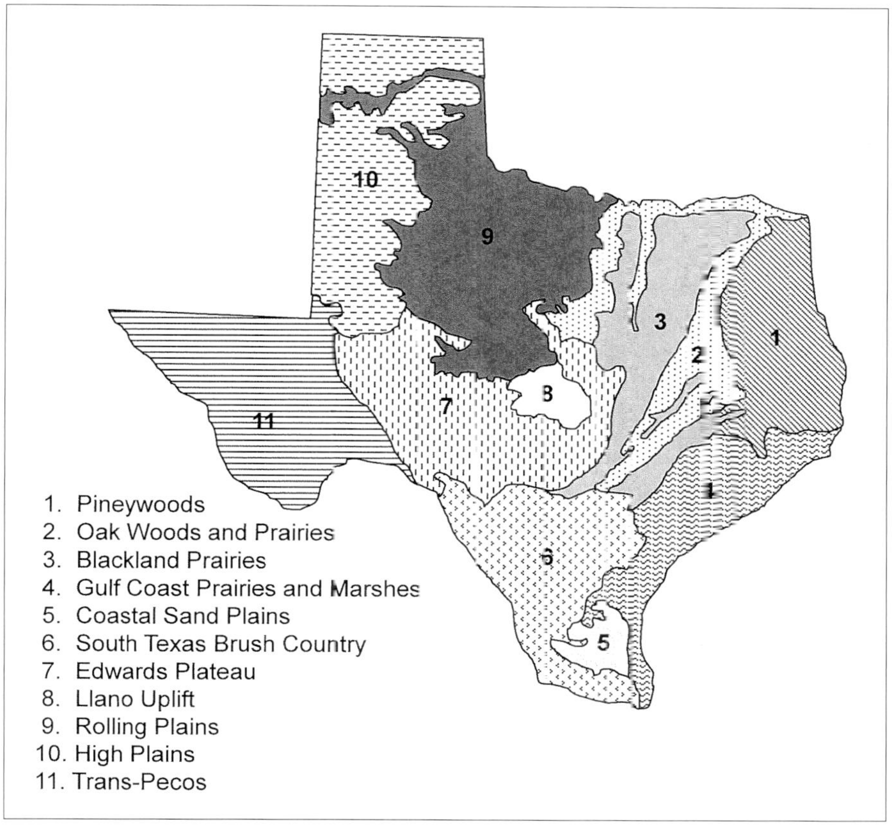

Table 4.2. *Comparison of forest and woodland communities among the natural regions of Diamond et al. (1987)*

Community	**Distribution in the Region**		
	Gulf Coast Prairies and Marshes	**Coastal Sand Plains**	**South Texas Brush Country**
Deciduous Forest			
Sugar Hackberry–Elm Series	Throughout		Throughout
Water Oak–Live Oak Series	Upland Prairies and Woods		
Deciduous Woodland			
Live Oak–Pecan Series	Upland Prairies and Woods		
Live Oak–Post Oak Series	Upland Prairies and Woods		
Honey Mesquite–Huisache Series	Upland Prairies and Woods		Brush Country
Evergreen Forest			
Texas Palmetto Series			Subtropical Zone
Evergreen Woodland			
Live Oak–Seacoast Beardgrass Series		Throughout	
Texas Ebony–Anacua Series			Subtropical Zone

Common names reflect current usage.

Prairies and Marshes and South Texas Brush Country natural areas. The Coastal Live Oak–Seacoast Bluestem community is limited to the Coastal Sand Plains.

Communities where shrubs constitute the dominant vegetation are identified only in the South Texas Brushlands (Diamond et al. 1987). In the Brush Country subdivision three communities are recognized: (1) an Evergreen Shrubland consisting of a Cenizo Series dominated by Texas Sage, (2) a Deciduous Shrubland including a Blackbrush Series where Blackbrush is dominant, and (3) a Deciduous Shrubland with a Fern Acacia Series where Guajillo is dominant. This is the only community recognized in the Bordas Escarpment region. The Subtropical Zone includes a Deciduous Shrubland comprising a Texas Ebony–Snake-eyes Series, where the dominant species are Texas Ebony and Snake-eyes.

Table 4.3 provides a comparison of herbaceous-dominated communities occurring in the natural regions of southern Texas as reported by Diamond et al. (1987). Swamps (wetlands with woody dominants) include a Black Mangrove Series dominated by Black Mangrove, which occurs in the Subtropical Zone of the South Texas Brushlands, and a Buttonbush Series dominated by Common Buttonbush, which occurs on Upland Prairies and Woods of the Gulf Coast Prairies and Marshes natural area.

Marshes (wetlands with herbaceous dominants) identified by Diamond et al. (1987) include a Rush-Sedge Series dominated by rushes, which occurs in all of the natural regions. A Gulf Cordgrass Series dominated by Gulf Cordgrass occurs in each of the subdivisions of the Gulf Coast Prairies and Marshes and the South Texas Brush Country. A Smooth Cordgrass Series dominated by Smooth Cordgrass occurs in the Dunes/Barrier and Estuarine Zone of the Gulf Coast Prairies and Marshes. A Saltgrass series dominated by Saltgrass occurs in the Coastal Sand Plains and South Texas Brush Country.

A focus of Diamond et al.'s (1987) work was the identification of the conservation status of the communities they recognized. These were ranked in four categories ranging from endangered (1) to secure (4). Communities occurring in the Texas TBP considered endangered were the Texas Palmetto Series, Texas Ebony–Anacua Series, Gamagrass-Switchgrass Series, and Seacoast Beardgrass–Indiangrass Series. The Texas Ebony–Snake-eyes Series was considered threatened. It is significant that only three forest and woodland community types were considered endangered in Texas, and two of these occur in the Rio Grande Delta region of the Texas TBP.

Diamond and Fulbright (1990) examined the grassland communities of the Coastal Sand Plains in detail and found that Seacoast Beardgrass is the prevailing dominant species. Gulfdune Paspalum is a dominant species in swales and on moderately drained flats. Camphorweed is important on dune ridges and well-drained flats. Principal component analysis showed that there were two community types. A Gulfdune Paspalum–Seacoast Beardgrass community occupied swales and moderately drained flats, while a Camphor Daisy–Seacoast Beardgrass community occurred on dune ridges and well-drained flats. They concluded that the composition of climax communities prior to disturbance by humans is problematic. However, they suggested that the prevailing dominant was likely Seacoast Beardgrass.

McLendon (1991) classified the vegetation of South Texas but omitted coastal wetlands and the Lower Rio Grande Valley. He divided the vegetation into Grasslands, Woodlands, and Shrublands and identified associations and communities within associations, recognizing 29 communities within 10 associations. Table 4.4 shows the associations and their dominant species.

Table 4.3. *Comparison of herbaceous-dominated communities among the natural regions recognized by Diamond et al. (1987)*

Community	Distribution in the Region		
	Gulf Coast Prairies and Marshes	**Coastal Sand Plains**	**South Texas Brush Country**
Tall Grassland			
Seacoast Beardgrass Series	Dunes and Barrier Islands	Throughout	
Gamagrass-Switchgrass Series	Upland Prairies and Woods		
Seacoast Beardgrass–Brownseed Paspalum Series	Upland Prairies and Woods		
Seacoast Beardgrass–Indiangrass Series	Upland Prairies and Woods		
Medium-Tall Grassland			
Cane Bluestem–Honey Mesquite Series			Brush Country
Forb-Dominated Vegetation			
Sea Purslane–Beach Morning Glory Series	Dunes and Barrier Islands		

Note: Common names reflect current usage.

Table 4.4. *Plant associations in South Texas (exclusive of the Lower Rio Grande Valley and coastal saline areas) recognized by McLendon (1991)*

Vegetation Type	Associations and Dominant Species
Grasslands	Seacoast Beardgrass–Trichloris Seacoast Beardgrass–Pan American Balsamscale
Woodlands	Sugar Hackberry–Huisache Live Oak–Post Oak
Shrublands	Mesquite–Granjeno Huisache–Prickly Pear Mesquite–Prickly Pear Guajillo–Texas Sage Blackbrush–Twisted Acacia Creosotebush–Prickly Pear

Note: Common names reflect current usage.

Communities of the Mexico Portion of the TBP

The community ecology of the Mexico portion of the TBP is less well-known than the Texas portion. However, since about 1990 a great deal of work on the Tamaulipan thornscrub of northeastern Mexico, particularly the area near Linares, Nuevo León, has been published. The bulk of the province occurs in Tamaulipas. Alvarez (1963) reported that the vegetation of the TBP in Tamaulipas is, in general, Mesquite Grassland except for the vegetation of the Sierra San Carlos and Sierra de Tamaulipas. Two plant formations, Mesquite Scrub and Gulf Bluestem Prairie, are recognizable in the Mesquite Grassland. The dominant species of the Mesquite Scrub are Honey Mesquite and associated woody species, including Anacahuita, several species of *Senegalia* and *Vachellia*, Spanish Dagger, and prickly pears. The dominant grasses are grama grasses and bluestems. In the Gulf Bluestem Prairie, bluestems are the dominant grasses on well-drained sites, whereas sloughs and depressions are occupied by cordgrasses. The Gulf Bluestem Prairie has been invaded by Honey Mesquite and other shrubs in many areas.

In the foothills of the Sierra de Tamaulipas and in the area between it and the Sierra San Carlos, the vegetation is Thorn Forest (Alvarez 1963), where Alvarez identified the dominant genera as *Senegalia, Ichthyomethia, Ipomoea, Prosopis,* and *Cassia*. At 300 m (984 ft) and 700 m (2,297 ft) elevation in the Sierra de Tamaulipas the trees reach heights of 20 m (66 ft) and the canopy averages 8 m (26 ft) high. This community is identified as Tropical Deciduous Forest. Common tree species are identified as belonging to the genera *Tabebuia, Ipomoea, Bombax*, and *Conzattia*. Species of *Bursera, Acacia,* and *Cassia* are less abundant. In low canyons *Bursera, Ceiba,* and *Psidium* are draped with lianas and epiphytes.

A Pine-Oak formation is present above 800 m (2,625 ft) in the Sierra de Tamaulipas. It is characterized by Mexican Pinyon, Nelson's Pinyon, Aztec Pine, and Arizona White Oak. Montane scrub vegetation occurs at elevations of 600–900 m (1,969–2,953 ft). This vegetation is formed by Huisache along with a few oaks and some trees of the Tropical Deciduous Forest (Alvarez 1963). Vegetation of the Sierra San Carlos is similar to that of the Sierra de Tamaulipas and has been described in some detail by Dice (1937).

Recent work near Linares, Nuevo León, Mexico, has focused on germination (Jurado et al. 2000; Jurado et al. 2001), seedling survival (J. Garcia 2011), seedling establishment (Jurado et al. 1998; Jurado et al. 2006), plant attributes with emphasis on seeds (Jurado et al. 2001), reforestation and agroforestry (Stienen 1990; Foroughbakhch et al. 2001), leaf biomass productivity and diversity (Návar et al. 2014; Foroughbakhch et al. 2009), stand biomass (Návar et al. 2002), establishment and growth potential (Foroughbakhch 1992), precipitation and throughfall chemistry in legume species (Návar et al. 2009), biomass equations for Tamaulipan thornscrub shrub species (Návar et al. 2004), plant-water relations in thornscrub species (Gonzalez-Rodriguez et al. 2004), contribution of stemflow variation to soil moisture content and aquifer recharge (Návar 2011), floristic and structural variation (Reid et al. 1990), diversity and distributional patterns of legumes (Estrada et al. 2006), and adaptability, growth, and firewood volume yield of multipurpose tree species.

Amphibians, Reptiles, and Mammals

Blair (1950) confined his treatment of the vertebrate fauna of the Texas TBP to the groups he knew best: mammals, amphibians, and reptiles. He noted that the vertebrate fauna of TBP includes a considerable element of Neotropical species, a substantial portion of primarily grassland species that range northward into the Texan and Kansan Biotic Provinces, some Austroriparian species, and some species shared with the Chihuahuan Biotic Province.

The Neotropical element is what distinguishes the TBP from other biotic provinces in Texas. Since the publication of Blair's (1950) paper three additional Neotropical mammal species have been added to the mammalian fauna of the Texas TBP (Davis and Schmidly 1994; Schmidly and Bradley 2016): Mexican Long-tongued Bat, Southern Yellow Bat, and Gulf Coast Kangaroo Rat. The Texas Pocket Gopher is endemic to the TBP. Blair (1950) noted that 15 of 61 species of mammals in the TBP have Neotropical affinities (table 4.5) and that six of these are limited in Texas to the Matamoran Biotic District: Ghost-faced Bat, Northern Yellow Bat, White-faced Hog-nosed Skunk, Jaguarundi, Mexican Spiny Pocket Mouse, and Coues's Rice Rat. The Ghost-faced Bat has now been found farther north than the TBP in western Texas, and the Northern Yellow Bat extends northeastward past the TBP.

As it is for mammals, the distinctiveness of the amphibian and reptile fauna is determined by the tropical species that reach the northward limits of distribution in the TBP and do not occur elsewhere in Texas or the United States (table 4.5). According to Blair (1950), the amphibian fauna of the Texas TBP comprised 3 salamander species and 19 frog species. The reptilian fauna included 19 lizard species and 36 snake species. Blair did not identify the number of turtle species but mentioned 2 species of terrestrial turtles: Ornate Box Turtle and Texas Tortoise.

Seven species of amphibians are found in the TBP, and all except the Black-spotted Newt and Sheep Frog do not extend north of the Matamoran Biotic District. Ten reptilian species reach their northern limit of distribution in the TBP. Six species (Blue Spiny Lizard, Rose-bellied Lizard, Regal Black-striped Snake, Speckled Racer, Mexican Hook-nosed Snake, and Cat-eyed Snake) do not range north of the Matamoran Biotic District. Two reptiles, the Texas Tortoise and Reticulate Collared Lizard, are endemic to the TBP. Indeed, the geographic range for the Texas Tortoise is a virtual match of the geographic extent of the TBP. It is the species that best characterizes the province.

Ward et al. (1990) used cluster analysis and detrended reciprocal averaging to analyze the distribution of terrestrial reptiles in Texas. They found that cluster analysis

Table 4.5. *Neotropical vertebrates (excluding birds and marine species) identified by Blair (1950) as extending into the Texas portion of the Tamaulipan Biotic Province*

Amphibians	**Reptiles**	**Mammals**
Black-spotted Newt	Texas Tortoise (endemic)	Virginia Opossum*
Mexican Burrowing Toad	Reticulate Collared Lizard (endemic)	Ghost-faced Bat*
Sheep Frog	Keeled Earless Lizard	Northern Yellow Bat
Cane Toad	Blue Spiny Lizard	White-nosed Coati
Mexican Treefrog	Rose-bellied Lizard	Hog-nosed Skunk
Rio Grande Chirping Frog	Regal Black-striped Snake	Ocelot
White-lipped Frog	Speckled Racer	Mexican Spiny Pocket Mouse
	Texas Indigo Snake	Fulvous Harvest Mouse
	Mexican Hook-nosed Snake	Coues's Rice Rat
	Cat-eyed Snake	Collared Peccary
		Nine-banded Armadillo

* Mammals that range farther north.

produced seven regions that corresponded closely to the biotic provinces proposed by Blair (1950).

Alvarez (1963) listed the mammal species characteristic of the Mexican (= Tamaulipas) portion of the TBP (table 4.6). All species except the Tamaulipan Mole and Black-eared Rice Rat have populations in Texas. The Tropical Pocket Gopher is not included because Alvarez placed it in the Veracruzian Biotic Province. All Neotropical mammals that occur in the Texas portion of the TBP also occur in Tamaulipas, Mexico.

A recent review of the status of the herpetofauna in Tamaulipas designated several physiographic regions of which the Llanura Costera Tamaulipeca (Terán-Juárez et al. 2016) most closely follows the boundaries of the TBP in Tamaulipas. The fauna consists of 18 amphibian species (15 frogs, 3 salamanders), 57 reptile species (19 lizards, 30 snakes, 8 turtles), and 1 crocodilian. With the exception of Green Iguana, species that occur only in the Llanura Costera Tamaulipeca are fairly wide ranging (e.g., Green Anole) or at the southernmost extent of their ranges (see table 4.6). The Area de Protección de Flora y Fauna, Laguna Madre y Delta de Río Bravo, which is found within the Llanura Costera Tamaulipeca physiographic area, protects 13 amphibian species; 36 snakes, lizards, and turtle species (including all 5 species of sea turtle); and 1 crocodilian.

Table 4.6. *Neotropical vertebrates characteristic of the Tamaulipan Biotic Province in Mexico*

Amphibians	**Reptiles**	**Mammals**
Spotted Chorus Frog	Green Anole (nonnative)	Tamaulipan Mole*
Rio Grande Lesser Siren	Eastern Massasauga	Eastern Cottontail
Greater Siren	Green Iguana*	Desert Cottontail
	Keeled Earless Lizard	Black-tailed Jackrabbit
	Six-lined Racerunner	Spotted Ground Squirrel
		Yellow-faced Pocket Gopher
		Merriam's Pocket Mouse
		Gulf Coast Kangaroo Rat
		Black-eared Rice Rat
		Fulvous Harvest Mouse
		Brush Deermouse
		Hispid Cotton Rat
		Coyote
		American Badger
		Striped Skunk
		Ocelot
		White-tailed Deer

Source: Mammals (Alvarez 1963); amphibians and reptiles (Terán-Juárez et al. 2016).
Note: Excluding birds and marine species.
* Species restricted to the province in Mexico that do not range into the Texas portion of the TBP.

Biogeographic Importance of the Tamaulipan Biotic Province

The TBP is an area of great biodiversity (Blair 1950, 1952; Alvarez 1963). The TBP within Texas exhibits greater faunal diversity than any of the other biotic provinces in the state. The TBP is also important as a dispersal route for tropical species moving northward and for temperate forest and grassland species moving southward. These movements have often resulted in isolation of populations and subsequent speciation (Blair 1952).

The massive movements of bird species that breed in temperate or boreal North America and winter in the Tropics have been studied for more than 50 years. At least half of the North America avifauna—approximately 338 of the roughly 650 species that regularly occur north of Mexico—migrate south of the Tropic of Cancer for the winter (Rappole 1995). Another 80 to 90 species breed in the subtropics of southern Texas, Arizona, New Mexico, and northern Mexico and migrate into the Tropics, but their movements are poorly known (Rappole 1995). In general, some bird species migrate through the length of the TBP as they travel around the Gulf of Mexico through the coastal zone, while other species travel through the width of the TBP as they migrate across the Gulf of Mexico (Rappole et al. 1979). Thus, large flocks of hawks, shorebirds, and passerine birds move through the TBP in spring and fall each year. Riparian forests, such as those along the Rio Grande, may be particularly important stopover sites because of the more abundant food, cover, and fresh water there. Beach grasslands and wooded areas along the coast are especially important during storms when birds migrating over the Gulf of Mexico attempt to make the nearest landfall. Because the TBP is the first landfall for many birds migrating northward across the Gulf of Mexico, it is crucial that this area continues to provide necessary habitats and sufficient resources within these habitats to ensure the continued existence of the migrating bird species. Degradation of habitats in the TBP could lead to extinction of many migrating species.

Lands comprising the five counties (Nueces, Kleberg, Kenedy, Willacy, Cameron) bordering the Texas Laguna Madre provide habitat for 87 species of rare, threatened, and endangered species of plants and animals (table 4.7). Federal preserves such as Laguna Atascosa National Wildlife Refuge and Padre Island National Seashore contribute significantly to the protection of these species. Many of these rare, threatened, and endangered species continue to persist because of the large ranches that occur in the area, such as the King, Kenedy, Armstrong, and Yturria Ranches. These ranches preserve native habitat and restrict entry so that there is minimal disturbance by humans and little or no development on these lands. There are few structures, mostly homes and barns. There are no towns or commercial enterprises. Clearly, the wise stewardship provided by governmental agencies and ranch owners is responsible for the continuing high species diversity in the Texas TBP.

Despite the great faunal and floral diversity of the TBP, the province has suffered serious human perturbation. For example, Jahrsdoerfer and Leslie (1988) reported that since the 1920s more than 95% of the original native brushland in the Lower Rio Grande Valley has been cleared and converted to agricultural and urban use. In the past decade, however, a variety of private and governmental organizations have recognized the value of the biodiversity of the lands along the Rio Grande and in other areas of the Matamoran Biotic District, and these entities have been purchasing lands and working to reestablish native habitat. Wise planning and stewardship are helping maintain and increase the important resource known as the Tamaulipan Biotic Province.

Table 4.7. *Endangered, threatened, and rare species found in counties adjacent to Texas Laguna Madre*

Common Name	**Status**		**County**				
	Federal	**State**	**Nueces**	**Kleberg**	**Kenedy**	**Willacy**	**Cameron**
Plants							
Texas Windmill Grass		R	X				
Giant Ball Moss		R		X	X	X	X
Elmendorf's Onion		R	X		X		
Chandler's Craiglily	E	R	X				X
Mexican Mud Plantain		R					X
Runyon's Huaco		R					X
Texas Sea Purslane		R			X		
Lundell's Nailwort		R			X		
Slender Rushpea	E	E	X	X			
South Texas Rushpea		R		X			
Vasey's Wild-lime		R					X
Texas Ayenia	E	E					X
Black-laced Cactus	E	E		X			
Runyon's Beehive Cactus		R					X
Texas Sandmint		R	X				
Runyon's Water-willow		R					X
Rio Grande Ragweed	E	E	X	X			
Plains Gumweed		R	X				X
San Patricio Tansyaster		R	X				
Insects							
Superb Grasshopper		R				X	
Los Olmos Tiger Beetle		R			X		
Smyth's Tiger Beetle		R					X
Manfreda Giant Skipper		R	X	X			
Fishes							
Opossum Pipefish		T	X	X	X	X	X
Texas Pipefish		R	X				
River Goby		T					X
Mexican Goby		T					X
Amphibians							
Rio Grande Siren, large form		T	X	X	X	X	X
Black-spotted Newt		T	X	X	X	X	X
Mexican White-lipped Frog		T					X
Mexican Treefrog		T					X
Sheep Frog		T	X	X	X	X	X

Table 4.7. Continued

Common Name	Status		County				
	Federal	State	Nueces	Kleberg	Kenedy	Willacy	Cameron
Reptiles							
Texas Diamond-backed Terrapin		R	X				
Green Sea Turtle	T	T	X	X	X	X	X
Hawksbill Sea Turtle	E	E	X	X	X	X	X
Kemp's Ridley Sea Turtle	E	E	X	X	X	X	X
Leatherback Sea Turtle	E	E	X	X	X	X	X
Loggerhead Sea Turtle	T	T	X	X	X	X	X
Texas Tortoise		T	X	X	X	X	X
Saltmarsh Watersnake		R	X	X			
Speckled Racer		T					X
Texas Indigo Snake		T	X	X	X	X	X
Scarlet Snake		T	X		X		
Regal Black-striped Snake		T			X	X	X
Cat-eyed Snake		T		X	X	X	X
Mexican Black-headed Snake		R		X			
Keeled Earless Lizard		R	X	X	X	X	X
Spot-tailed Earless Lizard		R	X	X	X		
Texas Horned Lizard		T	X	X	X	X	X
Birds							
Reddish Egret		T	X	X	X	X	X
Wood Stork		T	X	X	X	X	X
White-faced Ibis		T	X	X	X	X	X
Common Black Hawk		T				X	X
Zone-tailed Hawk		T		X			X
White-tailed Hawk		T	X	X	X	X	X
Northern Aplomado Falcon	E	E		X	X	X	X
American Peregrine Falcon		T	X	X	X	X	X
Whooping Crane	E	E					
Snowy Plover		R	X	X	X	X	X
Piping Plover	T	T	X	X	X	X	X
Eskimo Curlew	E	E	X				
Interior Least Tern	E	E		X	X	X	
Sooty Tern		T	X	X		X	X
Cactus Ferruginous Pygmy-Owl		T			X	X	X
Northern Beardless Tyrannulet		T			X	X	X
Rose-throated Becard		T					X

Table 4.7. *Continued*

Common Name	**Status**		**County**				
	Federal	**State**	**Nueces**	**Kleberg**	**Kenedy**	**Willacy**	**Cameron**
Tropical Parula		T			X		X
Brownsville Common Yellowthroat		R					X
Texas Botteri's Sparrow		T	X	X	X	X	
Henslow's Sparrow		R	X				
Sennett's Hooded Oriole		R	X		X	X	X
Audubon's Oriole		R		X	X		X
Mammals							
Yuma Myotis		T					X
Southern Yellow Bat		T	X	X			X
Texas Pocket Gopher		R	X	X			
Coues's Rice Rat		T			X		X
White-nosed Coati		T					X
Eastern Spotted Skunk		R	X	X			X
Jaguarundi	E	E	X	X	X		X
Ocelot	E	E	X	X	X		X
West Indian Manatee	E	E				X	X

Source: TPWD 2003.
Note: E = endangered; T = threatened; R = rare; X = presence in a county.

5 Ranching Heritage

NANCY L. HILBUN AND
AMY E. (KOLTERMANN) NUÑEZ

Both the eastern and western shores of the Laguna Madre have extensive ranching histories. Padre Island is no longer devoted to ranching, but the western shore primarily comprises large ranches (fig. 5.1). Ranching has affected the lands bordering the Laguna Madre and, consequently, the structure, health, and productivity of the lagoon. Ranching enterprises have limited access to the Laguna Madre and kept human perturbations at a markedly lower level than they might otherwise have been. This chapter chronicles the history of ranching on lands bordering the Laguna Madre and discusses the effects ranching has had on the ecology of the Laguna Madre.

Early History

The first European discovery of South Texas was made by Alonso Álvarez de Piñeda in 1519. While searching for a water passage to the Orient, Piñeda mapped the Gulf of Mexico coastline from Florida to Vera Cruz, Mexico. He stopped at a small river south of the Rio Grande and claimed the land for Spain (Thompson 1997).

Prior to Piñeda's voyage the coastal areas of South Texas and northern Mexico were inhabited by various groups of Indians. The Karankawas and Coahuiltecans were the predominant groups along the Texas and northern Mexico coasts. The Coahuiltecans consisted of many subgroups, such as the Katuhanno, Bobole, Carrizo, Payaya, Aranama, Tamique, Oregón, Pachal, and Kesale-Terkodam (Thompson 1997). Other noted native groups of the area included Malaguitas, Manos De Ferro, Pasnacas, Patun, Piguisas, Pimaraqui, Comecrudos, and Tanaquiapemes (M. Salinas 1990; La Vere 2004).

Significant Spanish influence on coastal southern Texas and northern Mexico and its inhabitants did not occur until the seventeenth century, when missionaries brought cattle into South Texas to supply the missions near San Antonio. These missions served as some of the first formal ranches in Texas. The livestock provided a dependable food supply that fed the priests, and they were also used to attract and hold Indians at the missions (Jackson 1986). Due to the harsh living conditions, the area between the Nueces River and Rio Grande was thought unfit for habitation. During this time Spain controlled Mexico, while the French and English were building colonial empires in other parts of North America. The threat of the expanding English and French Empires prompted Spain to send José Escandón to the Rio Grande area to initiate colonization (Weise and White 1980; G. Cruz 1988). Escandón named the area Colonia del Nuevo Santander in 1746 and sent seven units to investigate its resources the next year. Escandón reported large numbers of wild horses and cattle. In 1749 Escandón was appointed governor of the new Spanish province, and settlements were erected along both sides of the Rio Grande, several of which were successful, including Camargo,

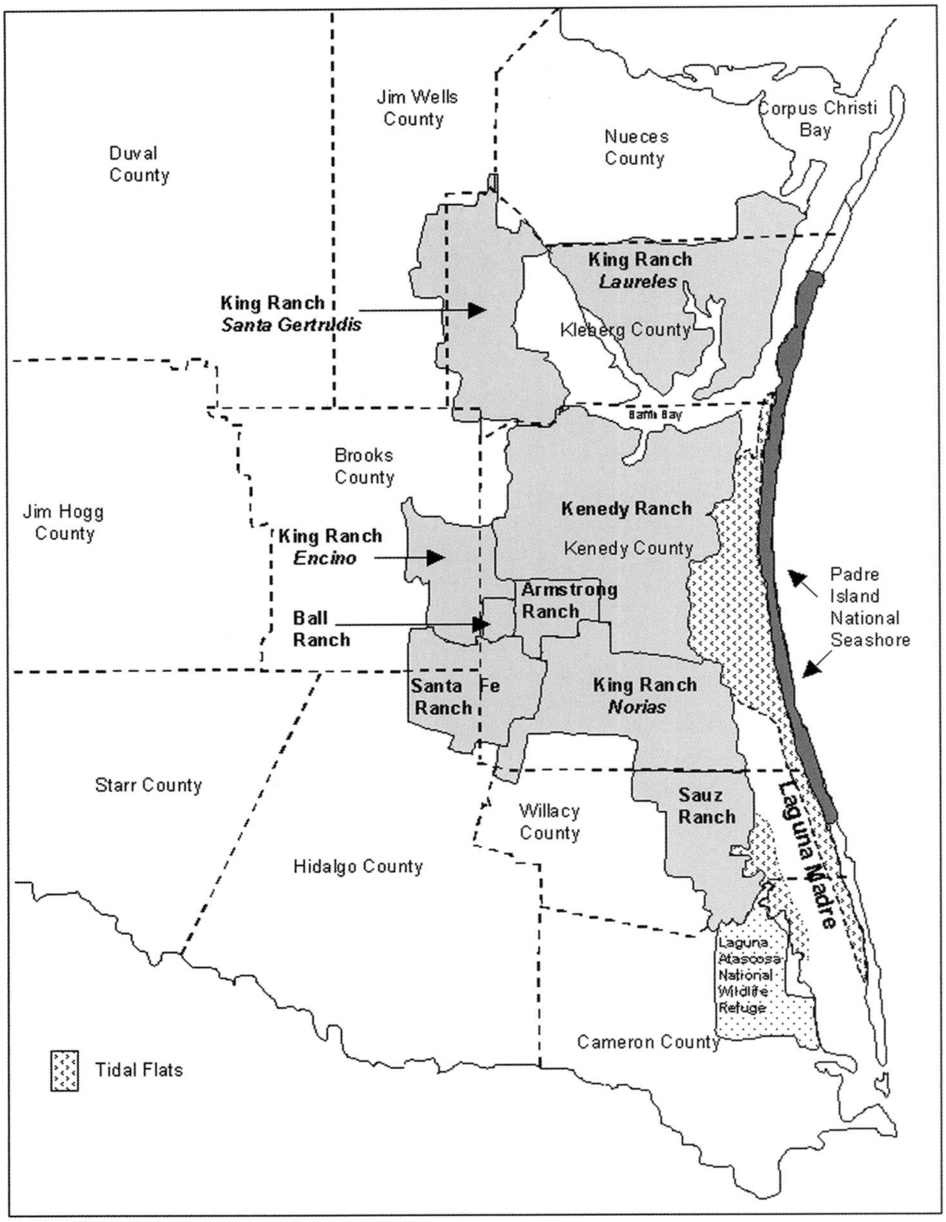

Figure 5.1. *Private ranches in South Texas, including the divisions of the King Ranch (Modified from Nixon 1986. Copyright King Ranch, Inc.)*

Reynosa, Revilla, Mier, Laredo, and Dolores. Colonization was successful from the mouth of the Rio Grande to the Nueces River by the late 1700s (Smylie 1964a).

In the 1760s Spain sent Juan Armando de Palacio and José de Ossorio y Llamas to survey towns, allocate land to individual settlers, and establish clear title to the land from the Crown of Spain. The colonists were classified as original, old, or recent, and land was divided on the basis of merit and seniority. Settlers of these early colonies began requesting individual allocations of land as early as 1753, but bureaucratic delays postponed land distribution for over a decade (TGLO 1988). Spanish land grants were divided into three distinct groups. The first group was elongated quadrangles, or *porciones* (Scott 1966; Jackson 1986). This type of grant provided settlers with a long strip of land with narrow frontage on a river so crops and livestock could have dependable water. About 170 *porciones* were located within present-day Texas, most consisting of about 2,023 ha (5,000 ac) with a small area bordering the Rio Grande.

Table 5.1. *Original Spanish and Mexican land grants bordering Laguna Madre*

Original Grantee	Name of Property	Size ha (ac)	County	Date	Country
José Narciso Cabazos	San Juan de Carricitos	243,487 (601,657)	Hidalgo Kenedy	1792	Spain
José Salvador de la Garza	Espíritu Santo	115,102 (284,416)	Cameron	178[illegible]	Spain
José Franciso Ballí	La Barreta	50,303 (124,297)	Kenedy	1804	Spain
José Perez Rey, José Maria Perez Rey, and José Manuel Garcia	Rincón de los Laureles	40,813 (100,848)	Kleberg Nueces	1807	Spain
Ramon de Ynojosa	El Rincón de Corpus Christi	32,945 (81,407)	Nueces	1832	Mexico
Alvino de la Garza and Domingo de la Garza	La Parra*	26,882 (66,426)	Kenedy	1833	Mexico
Nicolas Ballí and Juan José Ballí	Padre Island	20,609 (50,926)	Cameron Kenedy Kleberg Nueces Willacy	1829	Mexico
Enrique Villarreal	Rincón del Oso	17,337 (42, 840)	Nueces	1831	Mexico
Rafael Ramirez	Rincón del Penascal	15,941 (39,390)	Kenedy	1833	Mexico
Rafael Garcia	Santa Isabel	13,094 (32,355)	Cameron	1828	Mexico
Manuel de la Garza y Soza	El Potrero de Buena Vista	12,229 (30,218)	Cameron	1828	Mexico
Gabriel Trevino	Rincón de la Bóveda*	11,649 (28,785)	Kleberg	1834	Mexico
José Ygnacio de Trevino	San Martin	11,044 (27,290)	Cameron	1827	Mexico
Juan Antonio Ballí (Cabazos)	El Paistle*	10,437 (25,790)	Kenedy	1834	Mexico
Ygnacio Villarreal	Rincón de Mirasoles	9,647 (23,837)	Kenedy	1833	Mexico
Leonardo Salinas	La Barreta	8,960 (22,140)	Kenedy	1834	Mexico
Garcia Berrardo	El Infiernillo*	8,689 (21,471)	Kleberg	1834	Mexico
Teodora Garza	El Alazán*	5,476 (13,532)	Kenedy	1834	Mexico

Source: TGLO 1988.
* Land grants that border only Baffin Bay.

The second group of grants, involving much larger tracts of land, were issued for ranching or grazing purposes. Most of these grants were given to influential citizens of Camargo and Reynosa. This land was usually bordered by the Gulf of Mexico or the Laguna Madre and often backed up to the *porciones*. At least 35 of these grants were made, and the largest was San Juan de Carricitos (table 5.1). It comprised 243,487 ha (601,657 ac) and was granted to José Narciso Cavazos (TGLO 1988). The third group of land grants was vacant land used to prevent settlement by the French or English (Scott 1966; Jackson 1986).

The influential Hinojosa and Valli families acquired large amounts of land through the land grant process. The latter name has been written numerous ways, including Valli, Vallin, and most recently Ballí. These two families established pioneer ranches in South Texas (Scott 1966; Jackson 1986). Captain Juan José de Hinojosa was chief justice of Reynosa during this time. His influence in the government, classification as a "Primitive Settler," and personal wealth allowed him to acquire large amounts of land. Don José Maria Ballí was a local surveyor who began working with Captain Hinojosa (Robertson 1985). Many of the *porciones* and larger land grants went to the Hinojosa and Ballí family members. The bond between families was further strengthened when Don José Maria Ballí married Doña Rosa Maria Hinojosa, the daughter of Captain Hinojosa. As many as 11 large land grants and three *porciones* were acquired by the families from 1770 to 1829, making the Hinojosa and Ballí families the largest landowners in the Lower Rio Grande Valley. These influential families used the land to graze large herds of cattle and sheep (C. Garcia 1979; Jackson 1986).

In 1821 Mexico gained independence from Spain, and the land south of the Nueces River became part of the State of Tamaulipas (Calvert and DeLeón 1990). Mexico continued to grant land for grazing purposes. Conversely, by 1836 the First Congress of the Republic of Texas claimed all land north of the Rio Grande (Weise and White

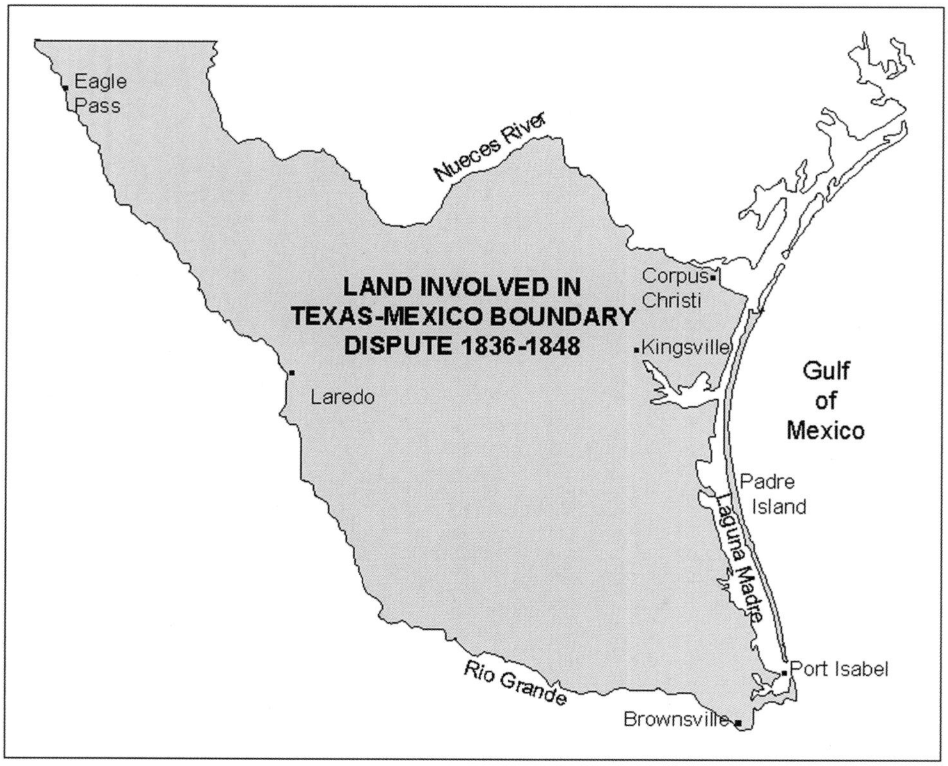

Figure 5.2. *South Texas between 1836 and 1848. Texas and Mexico both claimed the Nueces Strip, the land between the Nueces River and the Rio Grande. (Adapted from Weise and White 1980)*

1980). Therefore, the land between the Nueces River and the Rio Grande was claimed by both Texas and Mexico (fig. 5.2). This area became known as the Nueces Strip and was considered a no-man's-land where thievery, smuggling, and cattle stealing were rampant. Indians were recorded attacking the area in the first half of the nineteenth century and forcing many people to abandon their ranches, further promoting a no-man's-land (D. Smith 1986). Throughout this time Mexico continued to issue land grants in the area. Texas was annexed by the United States in 1845, and in 1846 General Zachary Taylor was dispatched to South Texas to establish the US claim on all land north of the Rio Grande (D. Smith 1986; Thompson 1997). During this time, the Mexican army, which broke diplomatic relations with Texas after the annexation, was driven south of the Rio Grande by Taylor's army. In 1848 the Treaty of Guadalupe Hidalgo was signed and the Rio Grande was established as the border between Texas and Mexico. The treaty did state that land grants made by Spain or Mexico until 1848 were valid (Robertson 1985). A committee was then established to confirm these grants. Commissioners William Bourland and James Miller were in charge of the operation, and 152 grants were confirmed in 1852 (TGLO 1988).

King and Kenedy Dynasty

During the 1840s two men arrived in South Texas who would forever change the ranching history of the world. Richard King was born to Irish-immigrant parents in New York City in 1824. He was apprenticed to a Manhattan jeweler at the age of 9 and ran away at the age of 11. King stowed away on a boat headed for Mobile, Alabama, and then worked his way from cabin boy to steamboat pilot. King then enlisted to fight in the Seminole Wars, where he met his future business partner, Mifflin Kenedy (Cheeseman 1997). Kenedy was from a Quaker family in Pennsylvania and had held various jobs, including schoolteacher. When Kenedy and King met, Kenedy already had a plan to captain a US Army boat to help move supplies for General Zachary Taylor (Douglas 1968). In 1847 King joined Kenedy, who by that time had already founded a small shipping enterprise on the Rio Grande. King and Kenedy were very successful moving supplies for the US Army (Cheeseman 1994). After the end of the Mexican War in 1848, King and Kenedy joined other residents of Matamoros and Brownsville and purchased the boats from the US Army. Kenedy and King formed a thriving business under the name M. Kenedy and Company and soon bought out the other partners. At one time Kenedy and King owned and operated 26 boats along the Rio Grande (Rowe 1953).

In 1852 King made a trip on horseback from Brownsville to the Nueces River and back. He was impressed by the vast potential of the land, and after returning to Brownsville, he inquired about the land surrounding Santa Gertrudis Creek (Rowe 1953). The land, known as Rincón de Santa Gertrudis, was part of a grant from Mexico to Juan Mendiola dated 21 December 1832 and was then owned by the Mendiola heirs (TGLO 1988). King formed a partnership with a friend, Texas Ranger Captain Gideon K. ("Legs") Lewis, to establish and maintain a livestock operation on Rincón de Santa Gertrudis (McCampbell 1952; Nixon 1986). King purchased 6,273 ha (15,500 ac) of land for less than two cents an acre and started to import Mexican cattle to stock his land (Lea 1957a). As his ranching enterprise grew, King continued to buy land and cattle. In 1855 Lewis was shot and killed, leaving no heirs (Nixon 1986).

By 1860 the King Ranch had become a major cattle enterprise, and Kenedy offered to buy an interest in the holdings. A three-way partnership was formed under the name R. King and Company (Cheeseman 1994). Richard King and Mifflin Kenedy each received three-eighths, and James Walworth, another investor, received two-eighths (Lea 1957b). Cattle, sheep, goats, and horses were the principal livestock King and Kenedy raised. Their successful steamboat operation allowed them to keep in close contact with the market demand, resulting in a thriving ranching business (Rowe 1953).

Ranching became an obsession with Captain King, leading him to hire attorneys to research old land grants of the area, locate the owners, and purchase the lands. Stephen Powers, and later James Wells, were responsible for many of King's land acquisitions (Lea 1957b). Although King was an ambitious man, he also demonstrated great compassion. During a trip to Mexico to purchase stock, King traveled to a very poor village where many people were hungry. He purchased their livestock and offered to relocate the entire village to his ranch. These people became his workforce, and their descendants have continued to work the King Ranch; they became known as *Kineños*, "King's men" (Nixon 1986; Monday and Colley 1997). In contrast, Kenedy hired skilled Mexican vaqueros from the Rio Grande area, who became known as *Kenedeños*. King and Kenedy both developed close relationships with their employees, and this helped form a loyal and dedicated workforce (Crimm 1997).

After the onset of the Civil War, the steamboat business was once again booming. Working under the Confederate flag, King and Kenedy supplied the Confederacy with meat and horses. They also imported lead from Mexico and carried cargo, mainly cotton, upriver. Later in the war, Brownsville was taken by the Union army, and since King and Kenedy were Confederates, they were forced to move across the river and operate under the Mexican flag (McCampbell 1952). They were, however, able to continue ranching during this time (Rowe 1953). In 1865 Walworth's share of the ranching enterprise was sold back to King and Kenedy for $50,000. King had accumulated 59,082 ha (146,000 ac) by the end of the Civil War and branded his cattle for the first time with the Running W (its origin remains unknown). By 1867 King and Kenedy had decided to dissolve the partnership in the interest of their heirs. King kept the Santa Gertrudis and Agua Dulce Ranches, while Kenedy maintained the Laureles Ranch (Lea 1957a, 1957b). In 1882 Kenedy sold the Laureles Ranch to a Scottish firm, the Texas Land and Cattle Company, and began to purchase La Parra Ranch (Cheeseman 1994). Henrietta King later bought the Laureles from the Scottish firm. King and Kenedy remained in the steamboat business until it was no longer profitable and sold the company in 1874 (Kelley 1986).

The two separate ranches continued to successfully operate. Captains King and Kenedy were the first in the state to fence large tracts of land. They both believed this was the only means to protect their livestock (Rowe 1953). They were the recognized leaders of the cattle industry in the West, doing more than any others to stabilize the industry in the United States. Captain Kenedy continued in many profitable business ventures, such as the building of a railroad from San Antonio to Aransas Pass. Kenedy also established the Kenedy Pasture Company with his son John, and this company owned 1,554 km^2 (600 mi^2) of rangeland in Cameron County (Daniell 1892). Captain Kenedy and Petra Vela de Vidal had six children. Their daughter Sarah J. Kenedy married Dr. E. A. Spohn of Corpus Christi, for whom Spohn Hospital in Corpus Christi was named. Today the Kenedy Ranch covers over 202,347 ha (500,000 ac) in Kenedy County (Fulbright et al. 1990) and is divided into two ownership entities, the John G. Kenedy Charitable Trust and the John G. and Marie Stella Kenedy Memorial Foundation.

In the early 1880s Robert Kleberg came to the attention of Richard King by beating him in a lawsuit. King was impressed with Kleberg's potential and hired him. Kleberg later married King's youngest daughter, Alice Gertrudis King (Rowe 1953; Nixon 1986). With the death of Richard King in 1885, Robert Kleberg was made manager of the King Ranch by King's widow, Henrietta. In 1888 the town of Alice, named after King's daughter, was settled; located at the railway 32 km (20 mi) north of the King Ranch, it became the largest cattle shipping point in the world (Nixon 1986). Kleberg was also one of the first to recognize the effects of overgrazing and encroachment of mesquite. He recognized the connection of ticks on the cattle in South Texas with Texas Fever and pressed the government into research that confirmed his belief. Kleberg designed the first dipping vat to treat for ticks and parasites on cattle (Nixon 1986). In 1899 Richard King's successor, Robert Kleberg, dug the first water well in South Texas at a depth of 162 m (532 ft). This well was an indicator of sufficient and reliable water in the area and led to more extensive settlement in South Texas. Kleberg was just as driven as Richard King and continued to increase the wealth of the King Empire, with purchase of new lands, selective breeding of cattle and horses, development of farms and orchards, and introduction of a railroad to the area.

After the death of Henrietta King in 1925 and Robert Kleberg in 1932 many felt that a chapter in South Texas ranching history was closed. Kleberg represented the last of the true cattle barons who made fortunes and history (Nixon 1986). He is recognized as a driving force in the development of South Texas, including Kingsville, Bishop, and Corpus Christi. To keep the empire alive during the depression, Bob Kleberg focused on renewing oil leases with Humble Oil and Refining Company (now Exxon), a partnership that expired in 2013.

From the beginning, Richard King had wanted to improve the cattle stock on his ranch, which needed to be able to withstand the heat and low water supply of South Texas, while maintaining quality beef for the consumers. His descendants fulfilled his dream with the development of the "Santa Gertrudis" breed, which is a cross between the Brahman and Shorthorn. Caesar Kleberg, cousin to Richard and Robert Kleberg, saw the potential of crossing a Brahman on a visit to Ed Lasiter's ranch. Dick Kleberg first saw their dream realized in a bull calf they eventually called "Monkey" (Nixon 1986). In 1940 the Department of Agriculture recognized the Santa Gertrudis as the first beef breed developed in the United States.

In addition to the dream to develop quality cattle, there was the desire to produce quality horses, and Robert Kleberg Jr. was the driving force for development of the King Ranch Quarter Horse program (Nixon 1986). Bob Kleberg's breeding endeavors eventually produced the prize racehorse "Assault," who claimed the Triple Crown in 1946 (Toland 1996). Improvements on the King Ranch continued under the management of Dick and Bob Kleberg (Nixon 1986).

By 1935 Alice Kleberg and her children had created a corporation and divided the stock equally. They decided the ranch would not split but stay together under family ownership (Hollandsworth 1998). During the 1970s and early 1980s the ranch corporation was run by a committee of various family members. In the late 1980s the King Ranch Corporation voted to hire executives outside the family to lead the company and serve on the board of directors. Today, the King Ranch Corporation is run from its new corporate headquarters in Houston, Texas, under the leadership of Jack Hunt (Grant 1998).

Today, the King Ranch covers approximately 333,873 ha (825,000 ac) of South Texas and is divided into four divisions. About one-half is leased for hunting, and 24,300 ha (60,000 ac) are used as farmland for cotton and grain sorghum. The ranch today carries only about 50,000 head of cattle on its Texas divisions. The King Ranch also owns a cattle operation in Brazil; a St. Augustine sod, citrus, and sugar cane operation in Florida; and an alfalfa, onion, and honeydew melon farm in Arizona (Ford 1998). Through these and other activities, the King Ranch Corporation continues to thrive.

Padre Island

A group of fierce Indians known as the Camones lived on Padre Island during the sixteenth century. During the seventeenth century the Negro and Malaquito Indians inhabited the island. These were thought to be closely related or subtribes of the Coahuiltecans, who occupied the adjacent mainland (WRT 1950). By the nineteenth century, the Karankawas inhabited many of the barrier islands and coastal plains of Texas, including Padre Island (Reese 1938; WRT 1950; Ricklis 1996; La Vere 2004), and lived a nomadic life along the South Texas coast. Their diet consisted mostly of fish, shellfish, birds, and bird eggs obtained from the Laguna Madre and surrounding areas (Weise and White 1980; Ricklis 1996).

Alonso de Piñeda was the first European to explore and name Isla Blanca, now Padre Island (Weddle 1985; Thompson 1997). The island was ignored until Spain initiated the land grant program in response to the expanding English and French Empires (Weise and White 1980). In 1800 Padre José Nicolas Ballí, son of Don José Maria Ballí, applied for a Spanish land grant for Padre Island, then called Corpus Christi Isla or Isla Santiago. The first application was not clear, and another was made in 1829. This time Nicolas Ballí included his nephew Juan José Ballí in the application as his business partner (Garcia 1979). Padre Ballí wanted to establish missions on the island and convert Indians, while Juan José Ballí wanted to establish a ranch. Rancho Santa Cruz, meaning Holy Cross, was built 42 km (26 mi) north of the island's southern tip. The Ballís successfully imported large herds of cattle, sheep, and horses to the island. The land was officially granted to Nicolas Ballí by Mexico in 1829 (Reese 1938). Unfortunately, Padre Ballí also died in 1829 (Smylie 1964a). The Ballí family continued to ranch on Padre Island until 1840; however, they began selling parts of the island as early as 1830. The last of the Ballí descendants had fled the island by 1844 in response to threats from the United States to annex the region (Weise and White 1980).

The Singer family came to Padre Island in 1847, after being blown off course on their way to Port Isabel, and chose to stay. Although the island was unoccupied, the northern half was still owned by José María Tovar and the southern half, by Ballí's heirs. The Singers purchased one-seventh of the southern half of Padre Island in 1851, which was the beginning of a profitable ranching business (Weise and White 1980). During this time, it is believed that John Singer funded his brother Isaac, who was the inventor of the Singer sewing machine, suggesting that Padre Island is the location for the first use of the sewing machine. John Singer's ranching success was short-lived, as he was forced to leave the island in 1861 by Confederate forces because of his sympathies toward the Union. Prior to his hasty departure, he buried $80,000 in silver in a jar near his home. After the Civil War Singer was unable to find his treasure, thus leading to one of the island's most famous treasure stories (Smylie 1964b; Weise and White 1980).

For a time, it seemed that Padre Island was unoccupied; however, it was never overlooked as an important strategic area for the military and smugglers such as Colonel H. L. Kinney. Kinney had established squatter's rights near Corpus Christi, later focusing on the development of Corpus Christi as a major port city (Smylie 1964a; Weise and White 1980).

During the confusion of the Civil War, King and Kenedy continued to purchase ranchland. As early as 1854, King acquired 4,856 ha (12,000 ac) on Padre Island from a niece of Padre Ballí. In the 1870s King and Kenedy leased additional land and established a thriving ranch operation on Padre Island. A major storm in 1880 greatly reduced King and Kenedy's ranching operations on the island, and by 1886 the operation was completely abandoned (Smylie 1964b).

In the late nineteenth century Patrick Dunn moved to the forefront of Padre Island's ranching history. Dunn was the son of Irish immigrants and had become a successful rancher by taking advantage of the open range. After the invention of barbed wire, Dunn turned his interests to Padre Island, which was bound by "natural fences," that is, the Gulf to the east and Laguna to the west (Weise and White 1980). Dunn first leased land on the island in 1879, and by 1926 he owned nearly the entire island (52,610 ha [130,000 ac]) (Weise and White 1980). Dunn efficiently ranched the island by building four major line camps placed about 24 km (15 mi) apart, the distance a herd of cattle could be driven in one day (figs. 5.3 and 5.4). Dunn's men would start rounding up cattle at the southern end of the island and drive the cattle northward and eventually across the shallow Laguna Madre, near Flour Bluff, to market (Smylie 1964b). Dunn, known as the Duke of Padre or Don Patricio, became a legend of the time. He not only ranched the island but also conducted experiments with native grasses. Furthermore, Dunn was elected to the state legislature, where he served from 1910 to 1916. In 1926 Dunn sold the surface rights of the island to real estate developer Colonel Sam Robertson; Dunn, however, retained mineral and grazing rights. Dunn ranched on the island until his death in 1937. The Dunn ranching business continued through his son, Burton Dunn, until 1971, when grazing was terminated within the Padre Island National Seashore (Rabalais 1977; Weise and White 1980).

Recreational interest in Padre Island dates back to the 1930s (NPS 2005). Between 1930 and 1950 development on Padre Island was scarce, but after World War II recreation interests on the island were revived. These interests culminated when Senator Ralph Yarborough proposed the first Padre Island Bill in 1958. In 1962 the government purchased the land for $17 million and Padre Island National Seashore was established (Sheire 1971).

Figure 5.3. *Portions of a Dunn Ranch line camp and corral remained on northern Padre Island in 1978. Some of these structures were restored by Padre Island National Seashore. (Photograph by Brian R. Chapman)*

Figure 5.4. *Patrick Dunn drilled wells and built windmills to provide fresh water to cattle in line camp corrals. This windmill still existed in 1978. (Photograph by Brian R. Chapman)*

Other Significant Ranches

While the King and Kenedy Ranches are the largest and best known in South Texas, there are other large ranches. The Armstrong Ranch, located in Kenedy County, was founded by Major John H. Durst, a close friend of Richard King and Mifflin Kenedy and an influential citizen of the time. In 1852 Durst purchased almost 37,637 ha (93,000 ac) of the La Barreta land grant held by the Ballí family for 1,600 pesos (D. Smith 1986). The Santa Fe Ranch, located in Brooks, Kenedy, and Hidalgo Counties, consists of over 60,704 ha (150,000 ac). Today, the ranch is owned by the heirs of Alice East, granddaughter of Captain Richard King (Rowe 1953). Likewise, the Sauz Ranch (45,326 ha [112,000 ac]) in Willacy County is owned by the heirs of Nettie King Atwood, the daughter of Captain Richard King (Douglas 1968).

The San Juan de Carricitos Spanish land grant, originally made in 1790, is the source of many large ranches in Willacy County. La Becerra Ranch, approximately 316 ha (781 ac), is owned by the descendants of José Narciso Cavazos (also spelled Cabazos), the original grantee of a large land grant. Other ranches that stemmed from the San Juan de Carricitos land grant were started by Francisco Yturria, who was a powerful man of the late nineteenth century from Matamoros. At the time of his death in 1912, Yturria owned over 55,038 ha (136,000 ac). These lands have since been divided among heirs into separate ranches: Santa Berta Ranch, Tres Mesquites-Thomas Ranch, and F. Yturria Ranch (TDA 1983). The El Devisadero Ranch in Kenedy County is also owned by Yturria's heirs. The Tres Norias Ranch in Willacy County can also be traced back to the San Juan de Carricitos land grant. Abundio Garcia purchased land from the wife of José Narciso Cavazos during the nineteenth century, and the land, about 177 ha (437 ac), is still owned by Garcia heirs today (TDA 1985).

Ranching and the Laguna Madre

What roles have historical and present-day ranching played on the ecology of the Laguna Madre? As early as the mid-nineteenth century King built dams across dry creeks to save fresh water after rains, thus altering the freshwater flow into the Laguna Madre (Nixon 1986). This probably did not have a major effect on the Laguna Madre, since drainage was minor and intermittent, but it was the beginning of human impact on the environment. As ranching in the area increased, so did changes in the landscape and environment. Although cattle ranching was a major enterprise at the time, it was not the only form of ranching in South Texas. Other livestock also significantly impacted the native grasslands of South Texas in the nineteenth century. In some locations, sheep populations outnumbered cattle populations 2:1 (Lehmann 1969). It has been said that when Mifflin Kenedy first saw Padre Island, it was "as green as a garden" (Price and Gunter 1942). Padre Island became denuded of vegetation shortly after 1870, and the fauna changed (Rabalais 1977). This phenomenon also occurred throughout South Texas from the Blackland Prairies to the coast with the expansion of the cattle industry from 1870 to 1880. Because of this loss of vegetation, filling in of the Laguna Madre increased as a result of sand blowing in from Padre Island carried by hurricanes and the relatively constant southeast winds (Price and Gunter 1942). Cattle ranching continued on Padre Island until 1971 (Rabalais 1977). The Laguna Madre was also suffering the effects of runoff due to the lack of vegetation. Price and Gunter (1942) concluded that the faunal and floral changes in South Texas and on Padre Island were due to introduction of farming, drainage of lands, and livestock grazing. These activities coupled with severe droughts and storms exacerbated the problems.

Studies have also been conducted on the effects of cattle on the shoreline vegetation of the natural and human-made ponds in South Texas (Whyte 1978). To provide water for cattle during summer or drought, human-made ponds (tanks) are constructed. These ponds exhibit an "edge effect," creating an ideal microecosystem for birds and waterfowl. Although constant grazing by cattle damages vegetation on the edges of watering ponds, fencing modifications can limit damage (Whyte 1978).

Despite these changes, the land management of South Texas has greatly benefited the Laguna Madre bay system. The lack of urban development in the area has had the effect of preserving numerous critical ecosystems in the Laguna Madre and surrounding uplands. These ecosystems thrive because of the lack of urban pressures. Rappole (1993) states that on the King Ranch alone at least 18 threatened or endangered species are known or suspected to be present.

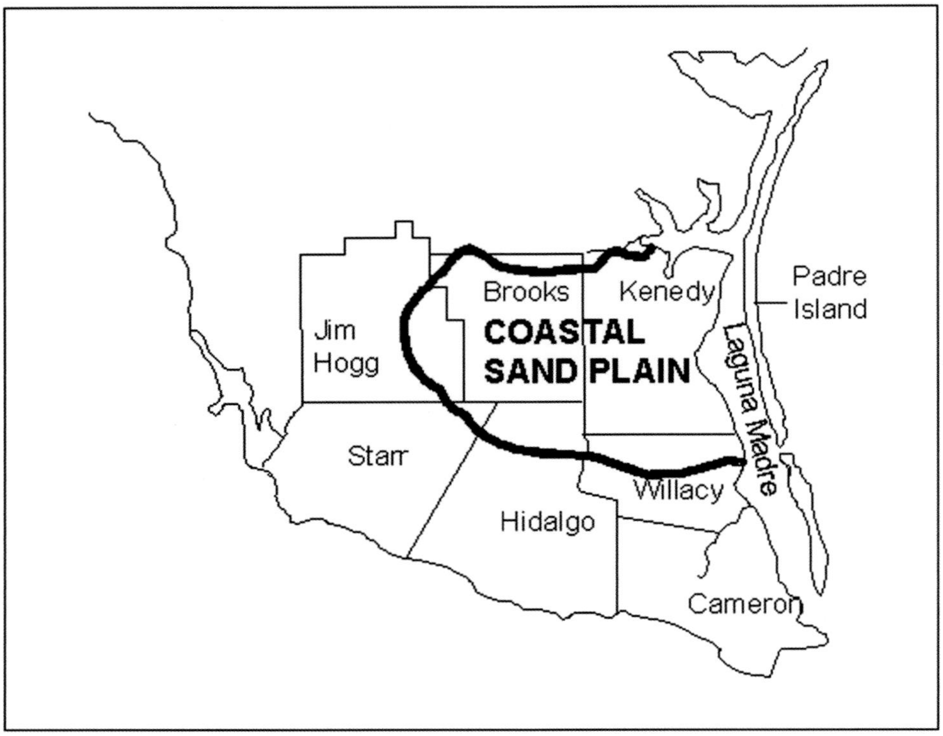

Figure 5.5. *The Coastal Sand Plain (South Texas Eolian Sand Sheet) covering parts of six counties in South Texas (Fulbright et al. 1990)*

The Coastal Sand Plain (South Texas Eolian Sand Sheet) is a unique ecosystem in the area that is protected by large ranches and lack of urban development (fig. 5.5; see chapter 4). Two of the four King Ranch divisions are located in the Coastal Sand Plain, consisting of more than 809,356 ha (2 million ac) of rangeland in South Texas. Active sand dunes make up 5% of the entire region (Fulbright et al. 1990). Several habitats exist on the Sand Plain, including extensive uplands, which are home to open grasslands with large Southern Live Oak mottes, Honey Mesquite mottes, and coastal wind-tidal flats and saline habitats. Diversity of wildlife, birds, and waterfowl found on the Coastal Sand Plain is attributed to its diverse habitats.

Mottes are especially important for Neotropical migrant songbirds because during spring migration, bird densities within mottes can be up to 20 times higher than during other times of the year (Anonymous 1993). This area is also important to endangered and threatened species such as the Ocelot and Cactus Ferruginous Pygmy-Owl (Fulbright et al. 1990; Rappole 1993). Research is ongoing at the Caesar Kleberg Wildlife Research Institute regarding the importance of habitat conservation for these and many other threatened species (Nixon 1986). Wetlands located on the King Ranch are prime breeding habitats for many species of waterfowl. These wetlands are in danger of being drained and filled for Gulf Intracoastal Waterway (GIWW) dredged-material placement by the US Army Corps of Engineers (Lonard 1993; Rappole 1993). Since wetlands serve an important ecological function, they are protected areas. The ranch wetlands provide habitats for many species of ducks and other waterfowl, as well as an important rest stop for migratory birds (Rappole 1993). Rappole has noted that any destruction of coastal habitat bordering the Laguna Madre and King and Kenedy Ranches would be detrimental to the habitats of many

Table 5.2. *Land delineation along the Laguna Madre shoreline*

Land Use	Approximate Distance along Shoreline km (mi)	%
King and Kenedy Ranches	142 (89)	34
Laguna Atascosa Wildlife Refuge	26 (16)	6
Padre Island National Seashore	112 (70)	27
Other	141 (88)	33
Total Laguna Madre Shoreline	421 (263)	100

species, including the Texas Tortoise, Brown Pelican, Black Skimmer, Least Tern, White-tailed Hawk, and White-faced Ibis. Tewes (1993) reported that the King and Kenedy Ranches are major habitats for the endangered Ocelot.

Although the overgrazing from ranch livestock has not been a positive influence on the ecology of the Laguna Madre, it cannot be ignored that the remoteness of the large ranches has contributed to the increased habitat for many plants and animals (Anonymous 1993). Between the large, privately owned ranches and federally protected lands, approximately 67% of the Texas Laguna Madre shoreline is protected from development or other human encroachment (table 5.2). Setting aside these lands has preserved many habitats that might have otherwise long ago been destroyed through commercial development (Simmons 1957). With responsible land management the Laguna Madre and its surrounding ecosystems should continue to benefit from protection provided by large ranches and federally protected lands.

Laguna Madre de Tamaulipas

Remoteness and inaccessibility, chiefly because of the lack of roads, have played an even larger role in Tamaulipas than in Texas in protecting much of the shoreline of the Laguna Madre de Tamaulipas. As a result of this lack of development, some large expanses of mainland remain intact and mostly undisturbed. Two areas of note include Rancho Rincón de Anacahuitas in the northwest and remote mainland areas south of Laguna el Catán. The property of Rancho Rincón de Anacahuitas is the largest conservation project in northeastern Mexico. Including 12,141 ha (30,000 ac) of native thornscrub brushland, 51 km (30 mi) of shoreline, considerable wetland acreage, nesting colonies of American White Pelicans and Roseate Spoonbills, 400 species of birds, and 170 species of plants, the ranch is an ideal land conservation model. The land, located 48 km (30 mi) south of Matamoros, has been owned by the Jorge Martinez family for many generations. The long-term family land management philosophy has been to leave the native brush intact. At present, with funds from the Ford Foundation and the efforts of Pronatura Noreste, the leading conservation organization in northeastern Mexico, the Texas Center of Policy Studies, and The Nature Conservancy of Texas, the ranch is moving from mostly cattle ranching to eco-tourism (camping, sailing, hiking, bird-watching) and fishing (J. Martinez, pers. comm.). This model is encouraged for other ranches along the Tamaulipan lagoon.

Part 2 Natural Resources

6 The Environment

JOHN W. TUNNELL JR.

The Laguna Madre of Texas has been characterized as a negative estuary or hypersaline lagoon (Hedgpeth 1947, 1967; Carpelan 1967; Gunter 1967). Rather than being a typical estuarine ecosystem, characterized by freshwater inflow, phytoplankton, and salt marsh productivity, it is a system characterized by lack of freshwater inflow with clear, normally phytoplankton-free waters, barren shorelines, and hypersaline conditions where seagrass meadows predominate. The food chain in the Upper Laguna Madre is based primarily on submerged benthic (bottom-dwelling) plants rather than free-floating phytoplankton, and most animals feed via an abbreviated food chain dependent on decomposition of dead seagrasses (Fry and Parker 1979). This truncated detrital food chain in an enclosed lagoon system makes the transfer of carbon (energy) from seagrasses to higher trophic levels much more efficient than the typical phytoplankton-based food chain in typical estuaries (Pulich 1980). This mechanism of high productivity is undoubtedly what explains the observations of earlier workers that the Laguna Madre supplied about 60% of the total Texas bay finfish production, even though it composed only 20% of the total Texas bay area (Hedgpeth 1947; Simmons 1957). Complementing these data during an early study of Upper Laguna Madre productivity, Hellier (1962) calculated the efficiency of conversion of plant production into fish production at 0.074% on a dry weight basis, which is higher than that calculated for offshore fishing banks or coral reefs (Pulich 1980).

The Laguna Madre of both Texas and Tamaulipas currently consists of ecosystems in a state of long-term (decadal) change due primarily to anthropogenic influences. Historically, both systems were driven more by natural climatic events, such as droughts and hurricanes, which produced salinities ranging from brackish (or mixosaline, 0.5–30 ppt) after major wet hurricanes to eusaline (30–40 ppt) or hypersaline (40–80 ppt) during "normal" climatic conditions, to brine (more than 80 ppt) during severe droughts (Hedgpeth 1947; Hildebrand 1958, 1969; Copeland 1967). In both Texas and Tamaulipas, increased water circulation and the consequent moderating of salinities appear to be responsible for recent, ongoing ecosystem changes (Hildebrand 1980; Quammen and Onuf 1993). H. Hildebrand (pers. comm.) thought that hurricane rainfall floodwaters and El Niño wet cycles, coupled with long term drought cycles, were more important regulators of change than circulation changes due to channelization in either system. He also emphasized, however, that water-storage structures, such as Falcon Lake, now greatly restrict freshwater inflow into both systems.

Prior to the hydrological changes in the Texas Laguna brought about by dredging the Gulf Intracoastal Waterway (GIWW), few studies were conducted that characterized the ecosystem completely. However, the early works by Gunter (1945a), Hedgpeth (1947), and Collier and Hedgpeth (1950) painted an interesting picture of the Upper Laguna Madre. Underwater sills existing between the Upper Laguna Madre and Corpus

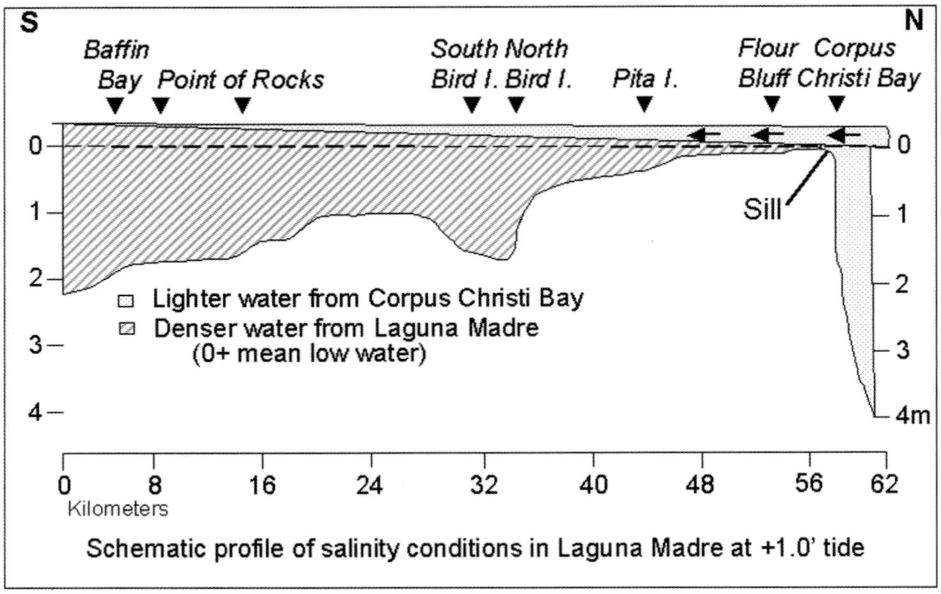

Figure 6.1. *Diagram of sill, sometimes referred to as "bulkhead," at junction of the Upper Laguna Madre with Corpus Christi Bay (Adapted from Hedgpeth 1947)*

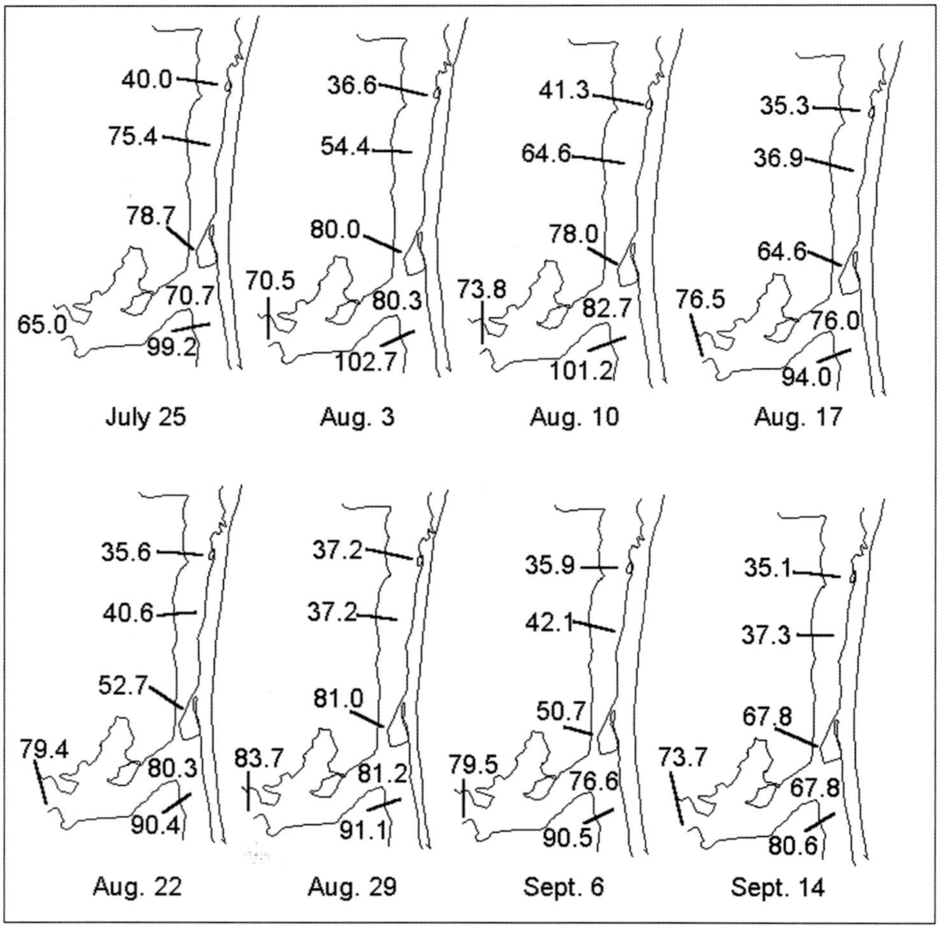

Figure 6.2. *Pattern of surface salinities in the Upper Laguna Madre during July–September 1946 before completion of the GIWW (Adapted from Hedgpeth 1947)*

Christi Bay (fig. 6.1), a similar one at the mouth of Baffin Bay, and shallow-water constrictions across the lagoon combined to result in higher salinities in Baffin Bay and southward toward the Land-Cut. Twenty-eight months of salinity samples, taken weekly by light seaplane, offer the best data set of pre-GIWW conditions (Hedgpeth 1947). Figure 6.2 gives an example of a three-month period (July–September) in 1946 when water exchange between Corpus Christi Bay and the Upper Laguna Madre reduced salinities because of the autumnal rise in tides (Hedgpeth 1947). This seasonal exchange seemed to provide the only regular moderation in salinities besides that caused by droughts and hurricanes. Frequent fish kills caused by excessive salinities and freezes are well documented prior to channelization, which now allows an escape route for fishes that would formerly have been trapped (Gunter 1945a; Hedgpeth 1947; Simmons 1957; Gunter 1967).

After completion of the GIWW in Texas and the beginning of the moderation of Texas Laguna salinities, some Texas researchers studied the Laguna Madre de Tamaulipas as a hypersaline ecosystem, providing information about fisheries and hydrography (Hildebrand 1958, 1969), community metabolism (Copeland and Jones 1965), and environmental characteristics (Copeland 1967). Hildebrand (1958, 1969) presented a model of “boom or bust” cycles of fisheries production tied to wet cycles due to hurricane rainfall and opened Gulf passes and subsequent dry, excessive hypersaline conditions, respectively. Copeland and Jones (1965) described sparse plant and animal populations and low community metabolism, similar to those of polluted environments. Five years of salinity data, ranging from 51 ppt in 196[illegible] to 295 ppt in 1965 (Copeland 1967), graphically demonstrate the relationship between increasing salinity and decreasing fish species richness. At 51 ppt, 29 species of fish were encountered; at 95 ppt, there were 8 species; and, at 110 ppt, only 5 species were present. When salinities remained between 110 and 120 ppt for almost one year, only the Sheepshead Minnow and Tidewater Silverside remained. When the salinity reached 140 ppt, both of these species disappeared, and after 150 ppt, only Brine Shrimp persisted. At 295 ppt, Brine Shrimp had disappeared and sodium chloride was precipitating, although there were large windrows of Brine Shrimp eggs on the windward shore (Copeland 1967). In late fall 1965, extensive windrows of these eggs were observed on the barrier island lagoonal shores of Boca Jesús María (J. W. Tunnell, pers. observ.).

During the major drought of the 1950s, Hildebrand (1958) was the first to present preliminary information on the biology of the Laguna Madre de Tamaulipas. The northern and southern parts of the lagoon were distinctively different. In the larger, northern area, salinities exceeded 100 ppt (108–17 ppt in March 1955) and fish or plant communities were absent. These briny waters supported only a Brine Fly community. In contrast, the southern lagoon around Punta Piedras supported a dense meadow of submerged vegetation composed primarily of Shoal Grass and Mermaid’s Wine Glass growing at depths of 0.3–1.5 m (1.0–5.0 ft). Salinities in this southern area were 41–48 ppt (March 1953 and 1954), and an intense fishery existed. A review of fishery records provided the information for Hildebrand to develop his concept of productive and nonproductive cycles for this ecosystem. Records since the first commercial fishery was established in 1911 indicate a boom and bust fishery for this northern lagoon, which during the mid-1950s was no more than a shallow brine pool. The presence of oyster shells a few inches down in the Northern Laguna Madre sediments was also evidence of a wetter past climate, trending toward a drier one (Hildebrand 1958).

In 1967 Hurricane Beulah struck the northeastern Mexico coastline with subsequent extensive rainfall and provided strong support for Hildebrand’s model of boom and bust. In November 1967, three months after the storm, he measured a salinity of only

9 ppt and dragged a minnow seine that yielded 15 species of fishes, including two freshwater species, the Common Carp and Mosquitofish. Within two years, commercial landings of Spotted Seatrout went from zero to over 453,000 kg (1 million lb) (Hildebrand 1969).

Through these and other early studies, several basic principles and concepts were developed about biota in hypersaline ecosystems. Hypersaline environments tend to have fewer species but a larger number of individuals per species (Parker 1959; Gunter 1967). Gunter suggested that the reduced number of species in hypersaline environments follows from the well-known biogeographic principle that impoverishment of species takes place as physical environmental factors vary from optimum (Sanders 1968; Briggs 1974). Additionally, fish tend to be larger in hypersaline environments (Gunter 1967), and invertebrates tend to be smaller (Parker 1959). Hypersaline "optimum" conditions for vertebrate and invertebrate species appear to be between 40 and 50 ppt, as numbers of Red Drum, for example, were severely limited in the Upper Laguna Madre in salinities greater than 50 ppt. Additionally, year-one juveniles avoided entering the Upper Laguna when salinities were greater than 54 ppt (Gunter 1967). Simmons (1957) considered a salinity of 80 ppt or greater to be a lethal concentration for most species.

In this stressed hypersaline environment, which has been described as "pioneer climax" (Pulich 1980), trophic-level interactions are simplified. Resident species of plants and animals compete not so much with one another as with the severe physical and chemical conditions of their environment. As salinities increase to the brine level, the reduction in species continues until essentially monocultures are achieved.

Organisms that thrive under hypersaline conditions are not those especially adapted to this environment but are euryhaline species able to tolerate both high and low salinities (Carpelan 1967). Tremendous osmoregulatory abilities are found within the hardy species that can survive.

Convincing evidence now indicates that anthropogenic-induced hydrological changes (dredging of channels in Texas and tidal inlets in Tamaulipas) have caused moderation of excessive hypersaline conditions and set in motion long-term ecosystem changes (Hildebrand 1980; Pulich 1980; Quammen and Onuf 1993; Onuf 1999, 2007). Although far more evidence and data exist for the Texas Laguna, information is sufficient for both systems to surmise that the lethal concentrations of salinity over 80 ppt, and certainly the briny waters over 100 ppt, are environments of the past. Unless dredge maintenance of barrier island passes and inland channels is halted, those excessive salinities and briny waters may never again be reached.

Dredging of the GIWW in Texas has had the greatest effect on moderating salinities (Simmons 1957; Breuer 1962; Hedgpeth 1967; Quammen and Onuf 1993). Although data prior to dredging the GIWW are sparse and some segments are missing, a salinity time-series graph ranging from the mid-1940s to 1990 clearly shows the ameliorating conditions caused by GIWW dredging and the subsequent increased water circulation (fig. 6.3). Continuing long-term succession of the Laguna seagrass ecosystem also seems to be in progress (Quammen and Onuf 1993; see chapter 7). Three factors working in concert appear to be responsible for moderating the Texas Laguna Madre hypersaline conditions: (1) increased water exchange with the Gulf of Mexico resulting from channel dredging; (2) increased precipitation since 1965 compared to that of the previous 20 years; and (3) increased flow into the Laguna from Arroyo Colorado and North Floodway, the two principal drainage systems of the Rio Grande Valley agriculture and municipal areas (Quammen and Onuf 1993).

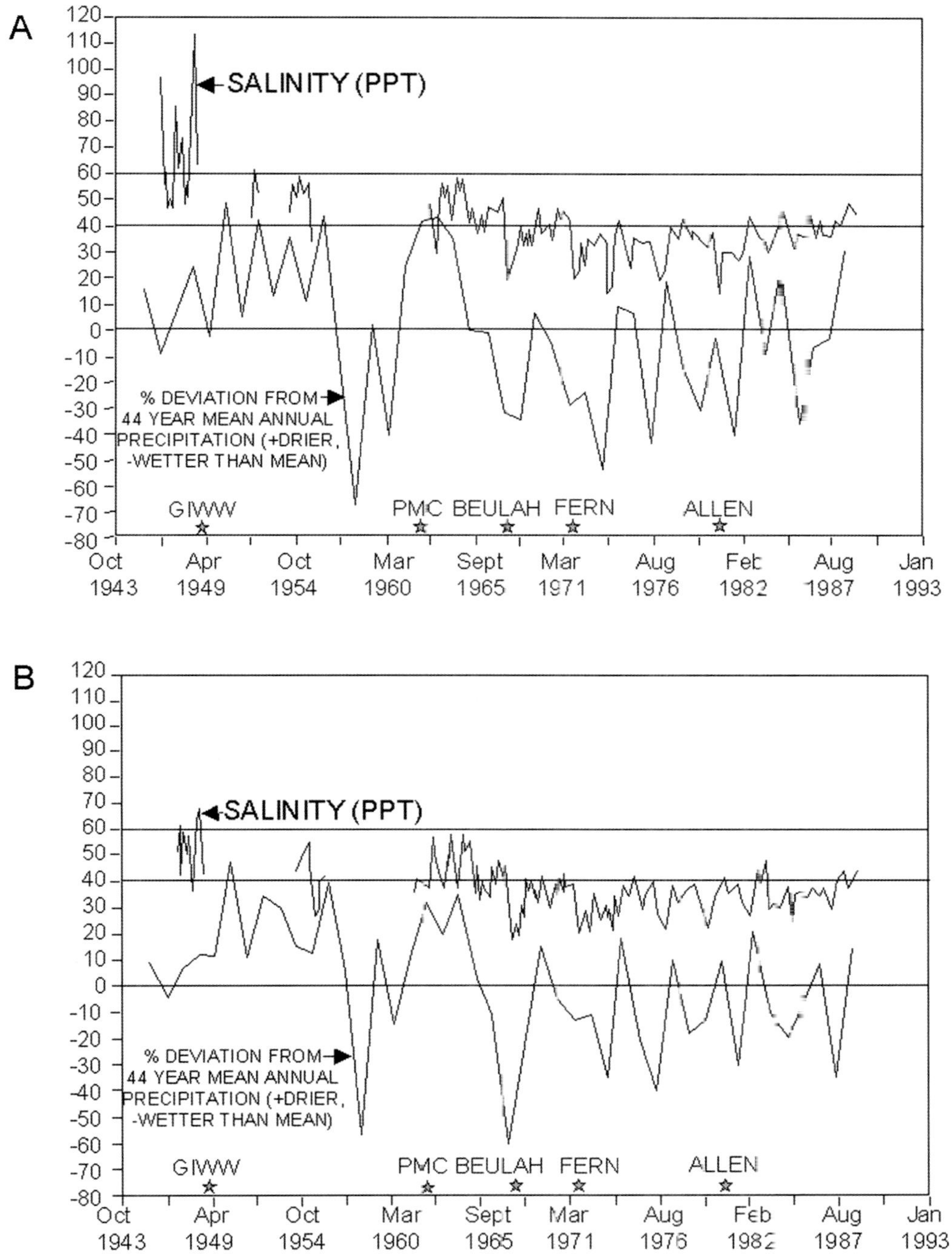

Figure 6.3. *Relationship between salinity and annual precipitation in the Texas Laguna Madre at stations with the highest salinities before completion of the GIWW: (A) Upper Laguna Madre, south of Baffin Bay; (B) Lower Laguna Madre, north end. Reference lines at 40% and 60% are included to facilitate comparisons over time and between locations. Precipitation is expressed as % deviation from mean annual precipitation during 1945–88 (44 years). Years drier than the long-term mean are above the 0 line; years wetter than the long-term mean are below the 0 line. GIWW = Gulf Intracoastal Waterway completed; PMC = Port Mansfield Channel made permanent; Hurricanes Beulah, Fern, and Allen were responsible for massive freshwater inputs over short periods of time. (From Quammen and Onuf 1993)*

Habitats

The Laguna Madre ecosystems in both Texas and Tamaulipas are similarly dominated by seagrass and bare-bottom habitats in the aquatic system, while the shoreline comprises wind-tidal flats and shallow sandy margins. The Laguna Madre of Texas has a larger areal extent of both seagrasses and tidal-flat habitat, while the Tamaulipas lagoon has more bare, open-bay bottom (table 6.1). This last condition is probably related to the fact that drastic environmental fluctuations characterized the Mexican lagoon until more recently, prior to channelization of barrier island passes (inlets), than they did in Texas. Seagrasses may currently be expanding in Tamaulipas as a result of moderated salinities and less frequent extreme salinity fluctuations between brackish and brine.

Calculation of areal extent of habitat types was accomplished in the 1990s for the Tamaulipas system by Ducks Unlimited de México, which classified all wetlands and upland habitats within and around the Laguna Madre de Tamaulipas (DUMAC, unpub. data). This digitized data set covers 39 quadrangle sheets at a scale of 1:50,000—the entire northeastern region of Mexico surrounding the Laguna Madre de Tamaulipas

Table 6.1. *Approximate areal extent of major habitat types in Laguna Madre and Baffin Bay of Texas and Laguna Madre and Laguna el Catán of Tamaulipas*

	Laguna Madre and Baffin Bay, Texas			**Laguna Madre and Laguna el Catán, Tamaulipas**		
	Areal Extent km^2 (mi^2)	**% MSL**	**% GSA**	**Areal Extent[a] km^2 (mi^2)**	**% MSL**	**% GSA**
Mean sea level (MSL)[b]	1,235 (477)			2,028 (783)		
Gross surface area (GSA)[b]	2,160 (834)			2,728 (1053)		
Tidal flats	917[c] (354)	—	42	508 (196)	—	19
Vegetated bottom	753[d] (291)	61	35	420[e] (162)	21	15
Emergent marsh	8[c] (3)	—	<1	192 (74)	—	7
Bare bottom	482[f] (186)	39	22	1,609 (621)	79	59
		100	100		100	100

Source: Texas data are compilations from C. Onuf (pers. comm.); Brown et al. (1976); Brown et al. (1977); and Brown et al. (1980) unless otherwise noted.

[a] Unpublished data provided by DUMAC.
[b] See table 2.1 for details on MSL and GSA. MSL = vegetated bottom + emergent marsh + bare bottom; GSA = tidal flats + MSL.
[c] Brown et al. (1976); Brown et al. (1977); Brown et al. (1980).
[d] Quammen and Onuf (1993), Laguna Madre seagrass = 732 km^2 (283 mi^2); Pulich (1998), Baffin Bay seagrass = 21 km^2 (8 mi^2); therefore, total Laguna Madre (732 km^2/283 mi^2) and Baffin Bay (21 km^2/8 mi^2) = 753 km^2 (291 mi^2).
[e] 338 km^2 (131 mi^2) seagrass, 82 km^2 (32 mi^2) algae = 420 km^2 (162 mi^2) total vegetated bottom (see table 6.2).
[f] Unvegetated lagoonal bottom: 284 km^2 (110 mi^2), Laguna Madre (C. Onuf, pers. comm.); 198 km^2 (76 mi^2), Baffin Bay; total = 482 km^2 (186 mi^2).

(table 6.2). A similar digitized data set is not available for the entire Laguna Madre of Texas. However, Pulich and Hinson (1996) produced a classified habitat map from November 1992 TM (Thematic Mapper) imagery for the northern portion of the Upper Laguna Madre southward to the Middle Ground region.

Submerged habitat in the Laguna Madre of Texas is predominantly vegetated bottom at 61%, whereas bare bottom occupies 39% (see table 6.1). In Tamaulipas, bare bottom is more widespread (79%) than vegetated bottom (21%). When considering the gross surface area (all submerged areas plus emergent marsh and tidal flats), the Texas lagoon has more tidal flats (42%) than the Tamaulipas lagoon (19%). Conversely, the Texas

Table 6.2. *Wetlands and uplands classification surrounding Laguna Madre de Tamaulipas*

	Class	Ha	Ac
M10W	Marine Subtidal Open Water	275,948	681,868
M2BB	Marine Intertidal Beach/Bar	1,567	3,872
E10W	Estuarine Subtidal Open Water (bare bottom)	160,850	397,460
E2ABP	Estuarine Intertidal Aquatic Bed (seagrasses)	33,776	83,460
E2AB	Estuarine Intertidal Aquatic Bed (algae)	8,199	20,260
E2FL	Estuarine Intertidal Flat	49,884	123,263
E2EM	Estuarine Intertidal Emergent	19,170	47,369
E2FLMD	Estuarine Intertidal Flat Modified	965	2,385
L10W	Lacustrine Limnetic Open Water	15,788	39,012
L20W	Lacustrine Littoral Open Water	7,188	17,762
L2AB	Lacustrine Littoral Aquatic Bed	2,770	6,845
POW	Palustrine Open Water	55	136
PAB	Palustrine Aquatic Bed	199	492
PEMWT	Palustrine Emergent Wetland	36,330	89,771
PEMMD	Palustrine Emergent Modified	5,970	14,752
RIVER	Riverine	5,384	13,304
RVEG	Riverine Vegetated	1,392	3,440
RX	Riverine Artificial	543	1,342
UPAGR	Upland Agriculture	1,082,482	2,674,813
UPBAR	Upland Barren	28,733	70,999
UPDEV	Upland Developed	5,520	13,640
UPHAL	Upland Halophyte Vegetation	87,187	215,439
UPSS	Upland Scrub/Shrub	933,290	2,306,160
CLOUDS	Clouds (cloud cover)	5,122	12,656
Total		2,768,312	6,840,500

Source: DUMAC, unpub. data. Digitized data from Earth Observation Satellite Corporation image (Landsat Thematic Mapper), 18 March 1990, covering 39 1:50,000 scale quad sheets in northeastern Mexico surrounding Laguna Madre de Tamaulipas.

system has less than 1% emergent marsh, whereas in Tamaulipas this occupies 7% of the area. Mendelssohn et al. (2017) recently compared vegetated coastal habitats in all bays and lagoons of the entire Gulf of Mexico (Cuba, Mexico, and United States).

In addition to the three major habitats (seagrass beds, bare bay bottom, and wind-tidal flats), several other habitats have small areal coverage but are important and unique to the Laguna Madre ecosystem: jettied tidal inlets, coquina (beach rock), serpulid worm reefs, and oyster reefs. Interestingly, since publication of the first book on the Laguna Madre (Tunnell and Judd 2002), Black Mangroves have continued to expand in the Texas lagoon and beyond to the north (Yañez-Arancibia et al. 2014; Armitage et al. 2015; Smee et al. 2017). Therefore, an entire chapter (chapter 19) is dedicated to this topic in this new volume. Because the major habitat types of seagrasses (see chapter 7) and wind-tidal flats (see chapter 11) are covered in detail later, the smaller habitats just mentioned are discussed only briefly here. For the most part, other artificial hard substrates besides jetties—including petroleum platforms, riprap, pier pilings, navigation buoys, floating debris, causeway bridges, and channel markers—have not been studied.

Jettied Tidal Inlets Jettied tidal inlets (passes) between the Gulf of Mexico and the Laguna Madre include (north to south) Packery Channel, Mansfield Pass, and Brazos Santiago Pass in Texas and Mezquital, Boca Ciega, Boca de Catán, and Boca Santa Isabel in Tamaulipas. There also are jetties at the southern extremity of the Mexican lagoon at the mouth of the Río Soto la Marina at La Pesca. Jetties are made of huge granite chunks in Texas (fig 6.4), but the sandstone blocks used in Tamaulipas are typically higher, wider, and shorter than their Texas counterparts. Jetties provide an artificial hard-substrate habitat lining these tidal inlets. Introduction of these structures created a stable, hard-substrate habitat within a region of predominantly unstable, soft sediments. As this new ecological environment was introduced, new hard-bottom communities were thus established where they did not previously exist in nearshore coastal waters (Hedgpeth 1954; Britton and Morton 1989; Hicks 2010).

Figure 6.4. *Texas jetties, like this one at Port Aransas, are constructed of huge granite rocks, creating an artificial rocky seashore habitat. (Photograph by Brian R. Chapman)*

No information is available on the jetty habitats in Tamaulipas, and little is available for Texas, except for Aransas Pass, located on the north end of Mustang Island just north of Padre Island and Laguna Madre. Ecology of the Aransas Pass jetties at Port Aransas has been studied since the 1940s (Whitten et al. 1950; Hedgpeth 1954; Hoese et al. 1968; Edwards and Kapraun 1973; Kapraun 1980; Rabalais 1982; Britton and Morton 1989; Hicks 2010) and continues to undergo investigation (Whorff 1992; Hicks and Tunnell 1993, 1994; Alvarado 1996).

Ecological zonation of biota is readily evident on the Texas jetties, where organisms can attach firmly to the granite surface (fig. 6.5). In Tamaulipas, however, zonation is not quite as evident or as distinct, possibly because the biota finds it more difficult to attach to the weathering, softer surface of sandstone. Moreover, the Texas jetties at Port Aransas are over 100 years old, while those in Tamaulipas are only about 40 years old. A distinctive pattern does exist, however, although it is somewhat simple in comparison to typical natural rocky-shore biotic zonation. This may be a function of the jetties' relative newness compared to natural rocky shore, and/or to the low tidal range of less than 0.5 m (1.5 ft) (Britton and Morton 1989; Alvarado 1996; Hicks 2010). The supratidal (supralittoral) zone is characterized by blue-green algae (Cyanobacteria) and the Sea Roach. The supratidal fringe is dominated by the Interrupted Periwinkle, which is sometimes present in great abundance. In the intertidal (midlittoral or eulittoral) zone, the Fragile Barnacle and the False Limpet are characteristic inhabitants (Hicks 2010; Tunnell et al. 2010). Below this, in the subtidal fringe, macroscopic and epiphytic algae predominate and clearly distinguish this zone. Over 100 species of algae have been reported from the Port Aransas jetties, and a distinctive seasonal succession has been noted (Hedgpeth 1953, 1954; Edwards and Kapraun 1973; Kapraun 1980; Britton and Morton 1989; Alvarado 1996; Lehman 1999, 2013). Temperate species dominate in the winter, and tropical ones are prevalent during the summer. Common genera of green algae (Chlorophyta) include *Enteromorpha*, *Cladophora*, and *Ulva*. Brown algae (Phaeophyta) are represented by genera such as *Padina*, *Ectocarpus*, and *Petalonia*; and red algae (Rhodophyta), by genera including *Gelidium*, *Bangia*, and *Gracilaria*.

Figure 6.5. *Jetty boulders are colonized by algae, snails, and limpets in zones corresponding to the amount of exposure and water depth, mimicking the zonation of organisms on rocky seashores elsewhere. (Photograph by Brian R. Chapman)*

Recent studies on the newly established jetties at Packery Channel, which was opened in 2005–6 (Palmer et al. 2008, 2013; Williams et al. 2007; Williams and Kraus 2011), and the older jetties at Mansfield Pass opened in 1962 (Kieslich 1977) have greatly increased our knowledge of macroalgae recruitment, succession, and zonation. Twenty species of macroalgae recruited to Packery Channel within the first month of jetty completion, and 40 species established themselves within the first year (Fikes and Lehman 2008, 2010). The 40-species total is comparable to the 48 known from Mansfield Pass, which has a well-developed algal community (Kaldy et al. 1995; Klootwyk 2006; Fikes et al. 2010). Recruitment was faster and biomass was greater in higher-wave-energy habitats at Packery, and species richness and biomass were greatest in winter months. Some species were common year-round, and others preferred select seasons (Fikes and Lehman 2010).

Lowest on the jetties, the subtidal (sublittoral) zone is more diverse than its shallower counterparts and includes numerous algae, sponges, hydrozoans, the Warty Anemone, the Ivory Bush Coral (a stony coral), numerous bivalves and gastropods, the Common Octopus, numerous crabs, the Purple Sea Urchin, and the Caribbean Rock-boring Urchin (Alvarado 1996). At least three species of soft corals (sessile gorgonians, order Alyconacea)—the Colorful Sea Whip, Regal Sea Fan, and Pallas's Sea Whip—also inhabit the jetties at this depth. Individual tiny polyps of these species form colonies that attract juvenile fishes and harbor small invertebrates. Associates of the soft corals include two snails, the One-toothed Simnia and Flamingo Tongue Snail; a bivalve, the Atlantic Winged Oyster; a shrimp, the Seawhip Shrimp; and a barnacle, the Seawhip Barnacle (Williamson et al. 2011).

Although the jettied inlets provide a migratory route between the Gulf and the bays and lagoon for many species of shellfish and finfish, certain fishes are characteristic of the jetty habitat. Jetty-associated species include herbivorous fishes, including three blennies, the Hairy Blenny, Molly Miller, and Crested Blenny; the Sergeant Major Damselfish; and the Atlantic Spadefish. Carnivores include the Goliath Grouper, Atlantic Needlefish, American Halfbeak, and Gulf Toadfish (Alvarado 1996). The most common birds using the jetties for resting and adjacent waters for feeding are gulls and terns, but herons and egrets sometimes fish from the rocks. Additionally, some sea turtles, especially the Green Sea Turtle, and the Bottle-nose Dolphin use the jettied tidal inlets for passage and feeding areas (Shaver 1990b; Schmidly and Shane 1978).

Due to their proximity to navigation lanes, jetties are vulnerable to introductions of exotic (nonindigenous) species from foreign shipping traffic. Consequently, in February 1990 the invasive, edible Brown Mussel was discovered on the Port Aransas jetties (Hicks and Tunnell 1993). Within four years, it spread a short distance north to the Matagorda jetties and expanded southward all the way to southern Veracruz, Mexico, a distance of over 1,300 km (807 mi) (Hicks and Tunnell 1994). Due to the significant macrofouling characteristics of the species in other parts of its range (Venezuela and Brazil), the potential for major disruption to the established jetty community and its ecology was great. Densities of 10,000–12,000/m^2 were commonly found on South Texas jetties, and densities of up to 27,200/m^2 of small individuals were recorded (Hicks and Tunnell 1993). Although some species were displaced by the mats of Brown Mussels, other species—some nestling, burrowing, sedentary, and clinging organisms—took advantage of the protection provided by the mussel beds and increased in diversity and abundance during the early to mid-1990s (R. Smith 1996). By 1996 the mussel beds began to decline, and by 1998 they were difficult to find, except in preferred high-energy habitats near the seaward end of jetties. The reason for this decline is not fully understood.

The jetty habitat provides an interesting ecological and biogeographic environment where bay and Gulf species as well as temperate and tropical species mingle. For example, among the oysters, Eastern Oyster, the common estuarine species, can be observed adjacent to the Crested Oyster, an offshore species. Using tree oysters (Isognominidae) as a biogeographic example, the Bicolor Purse-oyster, a species found in temperate environments, is found on Texas jetties, whereas at La Pesca, the tropical Flat Tree Oyster is present. Analogous to the seasonality of macrobenthic algae just mentioned, jetty fish populations, which are predominantly warm temperate, can attract tropical reef species during the summer months when warm tropical waters from the southern Gulf move up along the Texas shoreline (J. W. Tunnell, pers. observ.).

Thus, in addition to providing for water and migratory exchange between the lagoons and Gulf, jettied inlets provide a hard-substrate habitat for a variety of estuarine, Gulf, temperate, and tropical species that normally would not be present, or abundant, in a region dominated by soft substrates.

Beach Rock (Coquina) Natural, rocky coquina outcrops (see fig. 3.15) are located in both the Texas and Tamaulipas lagoons at the mouth of their respective bays: Penascal Point at Baffin Bay (Watson 1971; Prouty 1996) and Punta Piedras at Laguna el Catán (Yañez and Schlaepfer 1968). Origin of these unique, natural hard substrates are discussed in chapter 3, but unfortunately, there is no biological or ecological information about their role or functioning in the lagoonal ecosystem. Although nothing is known of their biology or ecology today, their presence indicates a Gulf beach environment occurred at these localities during a previous highstand of the sea during the Pleistocene and before the existence of the present Laguna Madre.

Serpulid Worm Reefs Standing today as an indicator of a past, wetter climate are the relict serpulid worm reefs of Baffin Bay. Like beach rock, they provide a rare, naturally occurring hard-substrate habitat (Breuer 1957; Andrews 1964; Behrens 1974; Cole 1981; Hardegree 1997; Simms et al. 2010). Described as patch reefs and reef fields

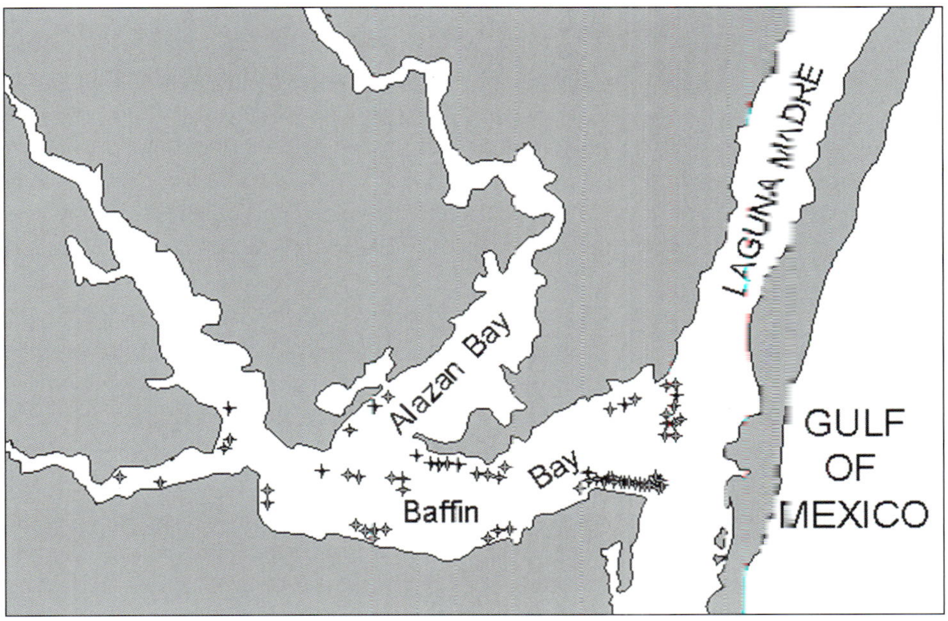

Figure 6.6. *Serpulid reef distribution in Baffin and Alazan Bays, Texas (Modified from Andrews 1964)*

(Andrews 1964) scattered in various localities around the Baffin Bay system (fig. 6.6), the reefs are composed of the calcareous tubes of serpulid polychaete worms. Although some serpulid polychaetes, such as *Hydroides dianthus*, are alive today on the reefs, serpulids no longer appear to be building reef structure.

Approximately 16 km^2 (6 mi^2) of serpulid reefs have been recorded in the Baffin Bay system (Brown et al. 1977). Patch reefs are found near the mouth of Baffin Bay and vary in shape and size from small circular structures about 8 m (26 ft) in diameter to larger ellipsoidal ones, approximately 40 m (131 ft) in length (Andrews 1964). Height ranges from 0.5 to 2.0 m (1.5 to 6.6 ft) above the sediment. Reef fields, on the other hand, are large expanses of reef rock, protruding only slightly above the bay bottom, that are scattered across the upper reaches of the bay.

Serpulid worm reefs have long been favorite fishing grounds for Baffin Bay anglers because of their generally high level of catch success. An ecological study was conducted on these reefs to determine if there were more or larger fish around the reefs and if productivity was increased because of them (Hardegree 1997). Neither ichthyofaunal diversity nor catch rate differed significantly between reef and nonreef areas. Likewise, little evidence was found to support the hypothesis that fish congregate and are larger around the reef structures, although the selective sampling technique used (trammel net) would not allow a conclusive answer. High species richness (29 species of fishes) and an abundance of invertebrates were noted on the reefs, and the biomass and photosynthetic rates of epiphytic algae were comparable to those found in Upper Laguna Madre seagrasses (Hardegree 1997).

The serpulid reefs of Baffin Bay are now apparently degrading in size and being reduced in distribution. Comparing their reduced distribution and size from earlier works (Breuer 1957; Brown et al. 1977), Hardegree (1997) suggested that reef degradation progresses by a combination of natural wave energy, increased boat traffic (prop damage), and trampling by anglers. Although the size and abundance of fishes around the reefs were not different than on other lagoon locations, the high historical catch rates for the Baffin Bay system indicates that the entire system may be more productive because of the reefs. The serpulid worm reefs, therefore, appear to be a fragile and important component of the Baffin Bay–Upper Laguna Madre system that merits protection from anthropogenic degradation.

Oyster Reefs Oyster reefs are natural accumulations of Eastern Oyster shells that accrete in the same locality for successive generations (Norris 1953). Oyster shells and living oysters provide a hard substrate for settlement of many sessile organisms and a protective habitat for many mobile species. Although the Eastern Oyster is broadly euryhaline, oyster reefs thrive best in salinities of 10–30 ppt. However, a population of oysters in South Bay in the southernmost Lower Laguna Madre has recently drawn attention because of its survivability in normal to hypersaline conditions. These oysters are of interest because they have been reported to be physiologically (Breuer 1962) and genetically (Groue and Lester 1982; Buroker 1983) divergent from other populations (King et al. 1994). Oysters in the Lower Laguna Madre are atypical in that they have adapted to hypersaline conditions (Breuer 1962). The South Bay population is probably a remnant from when the Rio Grande flowed into that area (White et al. 1986). Research in the early 1990s showed that Laguna Madre oysters are genetically divergent from oysters inhabiting other parts of the Texas coast and farther eastward (King et al. 1994). Most recently, further studies confirmed this and showed that there is a transition zone (or hybrid zone) in Corpus Christi and Aransas Bays between these southern and northern populations of Texas oysters (Hoover and Gaffney 2005;

Varney et al. 2009; Anderson et al. 2014). If the Eastern Oysters in Laguna Madre of Texas are indeed unique, they warrant protection.

In Laguna Madre de Tamaulipas, oysters are harvested commercially. The amount harvested varies greatly (see fig. 17.16) and is probably governed by lagoon salinities. These oysters may also be genetically distinct, but no research has been conducted on these populations.

Biota

No comprehensive surveys or checklists exist for the biota of both Laguna Madre systems combined. However, several studies allow for an overview of what is currently known and what information gaps exist. In Texas a comprehensive checklist of all species occurring within the estuarine and lagoon waters and islands of the Texas Coastal Bend includes the Upper Laguna Madre and Baffin Bay system (Tunnell and Alvarado 1996). There is no such compilation for the Lower Laguna Madre, although three similar compilations for major taxonomic groups are available for the Tamaulipas lagoon (DUMAC 1993; Leija-Tristán et al. 2000; CONANP 2015). Tables 6.3 and 6.4 compare the Texas Coastal Bend, the Upper Laguna Madre–Baffin Bay complex, and the Laguna Madre de Tamaulipas. Review of these tables reveals two important matters: (1) less work has been done on some taxonomic groups in Tamaulipas, and (2) species richness is dramatically lower (about one-third the number of species) in both hypersaline lagoon systems than in the more normal (positive) estuaries to the north. A third consideration is that tabulation or listing of species has not been accomplished for some groups. With further research, the Tamaulipas lagoon may prove to have greater species richness due to the presence of tropical species within that system. Greater numbers of macroalgae, molluscs, and fish species in the present list are indicators that this pattern is probable.

Table 6.3. *Number of species found in each division (plants) or phylum (animals) within the estuaries of the Texas Coastal Bend, Upper Laguna Madre, and Baffin Bay of Texas and Laguna Madre de Tamaulipas*

Taxon	Texas Coastal Bend[a]	Upper Laguna Madre and Baffin Bay	Laguna Madre de Tamaulipas
Division Cyanophyta	9	2	23
Division Bacillariophyta	341	177	57
Division Pyrrhophyta	43	19	14
Division Chlorophyta	33	5	37
Division Phaeophyta	17		18
Division Rhodophyta	68	1	65
Division Spermatophyta	6	6	6
Phylum Sarcomastigophora	31		
Phylum Ciliophora	13	6	
Phylum Porifera	11		
Phylum Cnidaria	36	7	17
Phylum Ctenophora	2	2	

Table 6.3. *Continued*

Taxon	**Texas Coastal Bend**[a]	**Upper Laguna Madre and Baffin Bay**	**Laguna Madre de Tamaulipas**
Phylum Platyhelminthes	3		
Phylum Nemertea	33	5	
Phylum Gastrotricha	2		
Phylum Nematoda	50	3	
Phylum Rotifera	8		
Phylum Kinorhyncha	1		
Phylum Gnathostomulida	1		
Phylum Annelida (Class Polychaeta)	289	73	28
Phylum Mollusca	230	72	86
Phylum Arthropoda	343	110	96
Phylum Sipunculida	3		
Phylum Echiurida	1		
Phylum Tardigrada[b]	*		
Phylum Bryozoa	19	4	
Phylum Phoronida	2	1	
Phylum Brachiopoda	1		
Phylum Echinodermata	29	9	16
Phylum Chaetognatha	4	3	
Phylum Chordata			
Subphylum Urochordata	14	1	
Subphylum Hemichordata	2	1	
Subphylum Vertebrata			
Class Elasmobranchiomorphi	23	7	18
Class Osteichthyes	211	79	122
Class Reptilia	3	3	3
Class Aves	157	130[c]	98
Class Mammalia	4	2	2
Total	**2,043**	**728**	**706**

Source: For Texas, Tunnell and Alvarado 1996; for Mexico, Leija-Tristán et al. 2000 and DUMAC 1993.

[a] Texas Coastal Bend comprises three estuaries: Mission-Aransas Estuary (Copano and Aransas Bays); Nueces Estuary (Nueces and Corpus Christi Bays); Upper Laguna Madre and Baffin Bay.

[b] Not in literature but known to exist.

[c] This is an arbitrary number and should be used with caution. The Chaney et al. (1996) checklist does not list geographic distribution for birds, so 130 of 157 aquatic birds from the Coastal Bend were estimated for the Upper Laguna Madre and Baffin Bay.

Table 6.4. *Overview of numbers of species in major groups from Texas Coastal Bend and Upper Laguna Madre, and Baffin Bay of Texas and Laguna Madre of Tamaulipas*

Group[a]	**Texas Coastal Bend**	**Upper Laguna Madre and Baffin Bay**	**Laguna Madre de Tamaulipas**
Phytoplankton and microalgae	393	198	94
Macroalgae	118	6	120
Vascular plants	6	6	6
Total plants	**517**	**210**	**220**
Invertebrates	1,112	295	243
Protochordates	16	2	0
Vertebrates	398[b]	221[b]	243
Total animals	**1,526**	**518**	**486**
Total plants and animals	**2,043**	**728**	**706**

Source: For Texas, Tunnell and Alvarado 1996; for Mexico, Leija-Tristán et al. 2000.

[a] See table 6.3 for breakdown of species within each group.

[b] Aquatic birds are listed in these numbers of vertebrates from Texas Coastal Bend and Upper Laguna Madre and Baffin Bay only for comparative purposes with aquatic bird list from the Tamaulipas lagoon. Since Chaney et al. (1996) did not separate bird distribution in their checklist, 130 of 137 aquatic birds from the Coastal Bend were estimated to occur in the Upper Laguna Madre and Baffin Bay.

7 Seagrass Meadows

KIM WITHERS AND KEN DUNTON

Seagrass meadows are among the most common coastal ecosystems and are extremely valuable because of their diverse roles within the landscape. Seagrasses play a significant role because their leaves provide complex structure in the water column and the roots and rhizomes anchor sediments (Phillips and Meñez 1988). Physically, this structure baffles waves, reduces erosion, and promotes water clarity by removing suspended sediments. Seagrass leaves increase bottom area by supporting the complex epiphytic communities that develop on their surfaces. Dense meadows may consist of more than 4,000 plants per square meter (about 4,800 per square yard) with an associated increase in bottom area of 15 to 20 times (McRoy and Helfferich 1977). Biologically, seagrass meadows provide nursery areas, refuge, and rich foraging grounds for numerous estuarine fishes and invertebrates, including many commercially and recreationally important species. Although consumed by few organisms, seagrasses are essential foods for some waterfowl and sea turtles. Live seagrasses play a major role in nutrient cycling within the water column and sediments. Dead seagrass detritus becomes an important source of organic material in coastal and nearshore ecosystems.

Seagrass habitats are among the most productive submerged habitats in marine environments. Annual production of seagrasses and associated primary producers (epiphytes, benthic algae, and phytoplankton) in subtropical areas, such as Texas, ranges from 1,000 to 4,000 grams Carbon per square meter (g C/m^2; McRoy and McMillan 1977). Daily productivity of seagrasses in the Upper Laguna Madre approaches 3 g C/m^2/day (Tomasko and Dunton 1995). At these values, seagrass communities compare favorably with the most productive terrestrial monocultural crops, such as corn, and often exceed productivity of coral reefs and oceanic upwelling areas. Only salt marshes are more productive. Entire fisheries may depend directly or indirectly on production by seagrass habitats (McRoy and Helfferich 1977). Historically, seagrass beds have supported over 50% of the Texas inshore finfish catch (Hedgepeth 1967).

Five seagrass species representing three families flourish in bay systems along the Texas and Mexico coasts (fig. 7.1): Shoal Grass and Manatee Grass (family Cymodoceaceae), Widgeon Grass (family Ruppiaceae), and Star Grass and Turtlegrass (family Hydrocharitaceae). In the Texas Laguna Madre, seagrass meadows are the most common submerged habitat type. Although permanent meadows of two perennial species, Shoal Grass and Turtlegrass, and an annual species, Widgeon Grass, occur in nearly all bay systems along the Texas Gulf Coast, the majority (94%) of the state's seagrass cover is found in the Laguna Madre (Pulich and Onuf 2007; table 7.1). The concentration of seagrasses in the Texas Laguna Madre is related primarily to geographic and climatic factors. Seagrasses tend to dominate in shallow-water systems where rainfall is low, evaporation is high, and average salinities exceed 20. Seagrasses are largely excluded from the upper, fresher portions of most Texas estuaries and bays north of Pass Cavallo,

Figure 7.1. *The five common seagrass species in Texas coastal waters: (a) Shoal Grass; (b) Manatee Grass; (c) Widgeon Grass; (d) Turtlegrass; and (e) Star Grass. Periphyton coating seagrass leaves enhances the productivity of seagrass beds. (Photographs from K. Dunton, www.texasseagrass.org)*

Table 7.1. *Comparison of the numbers of invertebrate families and species recorded in seagrass meadows in the Texas Laguna Madre and Baffin Bay and Laguna Madre de Tamaulipas*

Taxon	ULM	LLM	Baffin Bay	Tamaulipas
Phylum Bryozoa	X	X	?	?
Phylum Platyhelminthes	X	?	?	?
Phylum Rhynchocoela				
Families	2	3	0	?
Species	2	3	0	?
Phylum Annelida				
Class Oligochaeta	X	?	?	?
Class Polychaeta				
Families	20	6	3	?
Species	47	9	3	?
Phylum Mollusca				
Class Gastropoda				
Families	19	10	?	4
Species	33	13	?	4
Class Bivalvia				
Families	16	20	?	7
Species	26	30	?	11
Subphylum Crustacea				
Class Malacostraca				
Order Ostracoda	X	?	?	?
Order Cumacea				?
Families	1	1	?	?
Species	2	1	?	?
Order Amphipoda				
Families	8	2	?	?
Species	11	2	?	?
Order Isopoda				
Families	3	2	?	?
Species	4	2	?	?
Order Tanaidaceae				
Families	1	1	?	?
Species	1	1	?	?
Order Decapoda				
Families	11	?	?	?
Species	14	?	?	?
Class Cirripedia				
Order Thoracica				
Families	1	?	?	?
Species	1	?	?	?
Subphylum Chelicerata				
Class Pycnogonida	1	?	?	?
Phylum Echinodermata				
Class Holothuroidea	4	?	?	?
Class Ophiuroida	1	?	?	?
Class Asteroida	X	?	?	?

Note: ULM = Upper Laguna Madre; LLM = Lower Laguna Madre; X = present but no other information; ? = unknown.

where rainfall and inflows are high and salinities average less than 20. Thus, seagrasses in the Laguna Madre constitute a unique resource that cannot be duplicated elsewhere on the Texas coast.

General Seagrass Ecology

Factors Affecting Species Composition and Zonation The presence of seagrass species and their distribution within a community are determined by substrate composition, wave energy, water depth (related to light penetration and exposure tolerance), salinity tolerance, and successional stage. Seagrasses typically colonize low-energy, shallow-water areas with restricted circulation. Although Shoal Grass and Turtlegrass may occur on sandy substrates, both species grow better on muddier substrates. The other seagrass species are found almost exclusively on muddy substrates, and Star Grass tolerates practically liquid mud and highly polluted areas (den Hartog 1977).

Most seagrasses require a minimum of 11%–25% surface irradiance (SI, the light present at the water's surface relative to the plant surface) to maintain themselves (Duarte 1991; Kenworthy and Haunert 1991; Fonseca 1994). However, 41%–46% SI provided ideal growing conditions for Shoal Grass at an average depth of 1.25 m (4.1 ft), whereas 17%–19% SI resulted in signs of light stress (Dunton 1994; Onuf 1994). Neither Shoal Grass nor Turtlegrass was able to tolerate light levels below 14%–16% SI, particularly during spring and summer, when increased water temperatures increase the metabolic demands for growth and tissue maintenance (Czerny 1994; Czerny and Dunton 1995). In turbid waters, seagrasses are usually restricted to areas shallower than 1.0 m (3.3 ft) (Thayer et al. 1975).

Seagrasses must be submerged in water and are not tolerant of prolonged exposure. Shoal Grass and Star Grass have the widest range of depth tolerances, extending from mean highest high water (MHHW) to considerable depths (McNulty et al. 1972; den Hartog 1977). However, Star Grass usually inhabits deeper water and is tolerant of low light situations. Widgeon Grass is typically found only in shallow water between mean lowest high water (MLHW) and mean lowest low water (MLLW) (McNulty et al. 1972). Turtlegrass is found below MLHW to depths of 10.0–12.0 m (32.8–39.4 ft) (McNulty et al. 1972; den Hartog 1977). Manatee Grass is not tolerant of desiccation, so it is restricted to areas below MLLW (den Hartog 1977).

Salinity tolerances of seagrass species are important determinants of species composition and growth. Shoal Grass has the widest salinity tolerance (3.5–70) with an optimum around 45. Death occurred after 28 days at 70 in the laboratory (McMahan 1968) and dwarfing was noted at 60 (Simmons 1957), although growth continued in laboratory experiments (McMillan and Moseley 1967). Star Grass has the narrowest salinity range (13–50). Turtlegrass and Manatee Grass are intermediate in their tolerances. Turtlegrass tolerates salinities of 0–60 (optimal 20–36). Manatee Grass tolerates 3.5–50 (optimal 35) but ceased growth at 45 in laboratory experiments and died within 21 days at 52.5 (McMillan and Moseley 1967). Rates of photosynthesis declined in Turtlegrass and Manatee Grass when salinities decreased below 35 (Zieman 1975). Widgeon Grass is essentially a freshwater species with a restricted salinity tolerance (0–70) (McMillan and Moseley 1967; Kantrud 1991), although one of the authors (KHD) found this species thriving and flowering at salinities above 100 in Baffin Bay.

Succession Succession in seagrass communities is determined by species size, tolerance of environmental variability and disturbance, and competitive ability (den Hartog 1977; Zieman 1982). Shoal Grass and Star Grass are small plants with little capacity to compete with larger, more aggressive species and are typically found in

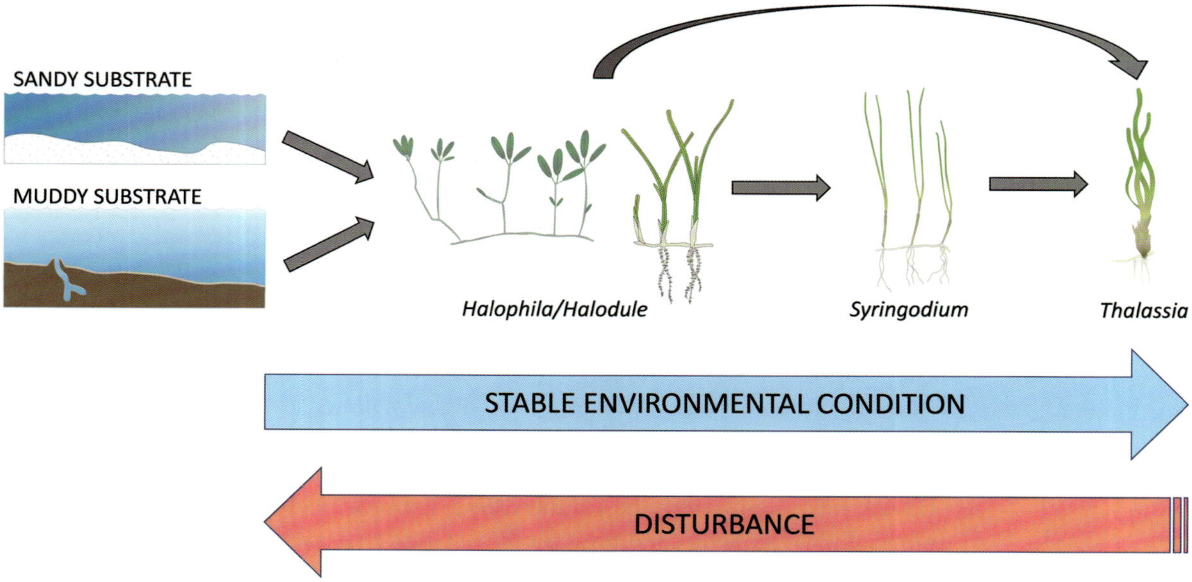

Figure 7.2. *Succession in seagrass meadows in the Laguna Madre (Modified from Zieman 1982)*

habitats unsuitable for other species. Turtlegrass and Manatee Grass are medium-sized species with fairly high tolerance to environmental variability that can outcompete most other species. These differences in ecological capacities determine how succession will proceed (fig. 7.2). Shoal Grass and Widgeon Grass are pioneer species that can rapidly colonize bare areas either by seed or vegetative branching. They stabilize sediments and help protect the integrity of sediment surfaces by buffering the effects of currents. Shoal Grass colonizes bare or algae-covered substrate, possibly with Star Grass in the understory or at the edges of meadows in deeper water. Although frequently absent, Manatee Grass may appear next, often intermixed with Shoal Grass in earlier successional phases. Ultimately, sediments become stabilized and Turtlegrass begins to colonize the area, eventually constituting the climax.

As succession in seagrass meadows proceeds toward climax, structural changes occur. The most obvious change is increased leaf area that allows epiphyte colonization (Zieman 1982). Communities characterized by Shoal Grass and/or Star Grass develop rapidly after disturbance and are generally a simple mosaic of species. Epiphytic flora is composed mainly of diatoms. On the larger leaves of Turtlegrass and Manatee Grass, epiphytic communities may contain hundreds of species of algae (den Hartog 1977). Grazing probably controls epiphytic community composition to some extent. Dominance of tightly adhering and/or encrusting species is at least, in part, the result of selective grazing (van Montfrans et al. 1984).

Distribution and Community Composition in the Laguna Madre

Upper Laguna Madre and Baffin Bay In the Upper Laguna Madre, seagrasses currently cover about 243 km^2 (94 mi^2) (Quammen and Onuf 1993; fig. 7.3). Shoal Grass is the dominant species, but all species have been reported. Three distinct species associations are common: sparse, monotypic Shoal Grass in higher-energy environments such as the mainland shoreline near the entrance of Baffin Bay; dense meadows of Shoal Grass with Star Grass as subdominant vegetation in shallow, low-energy

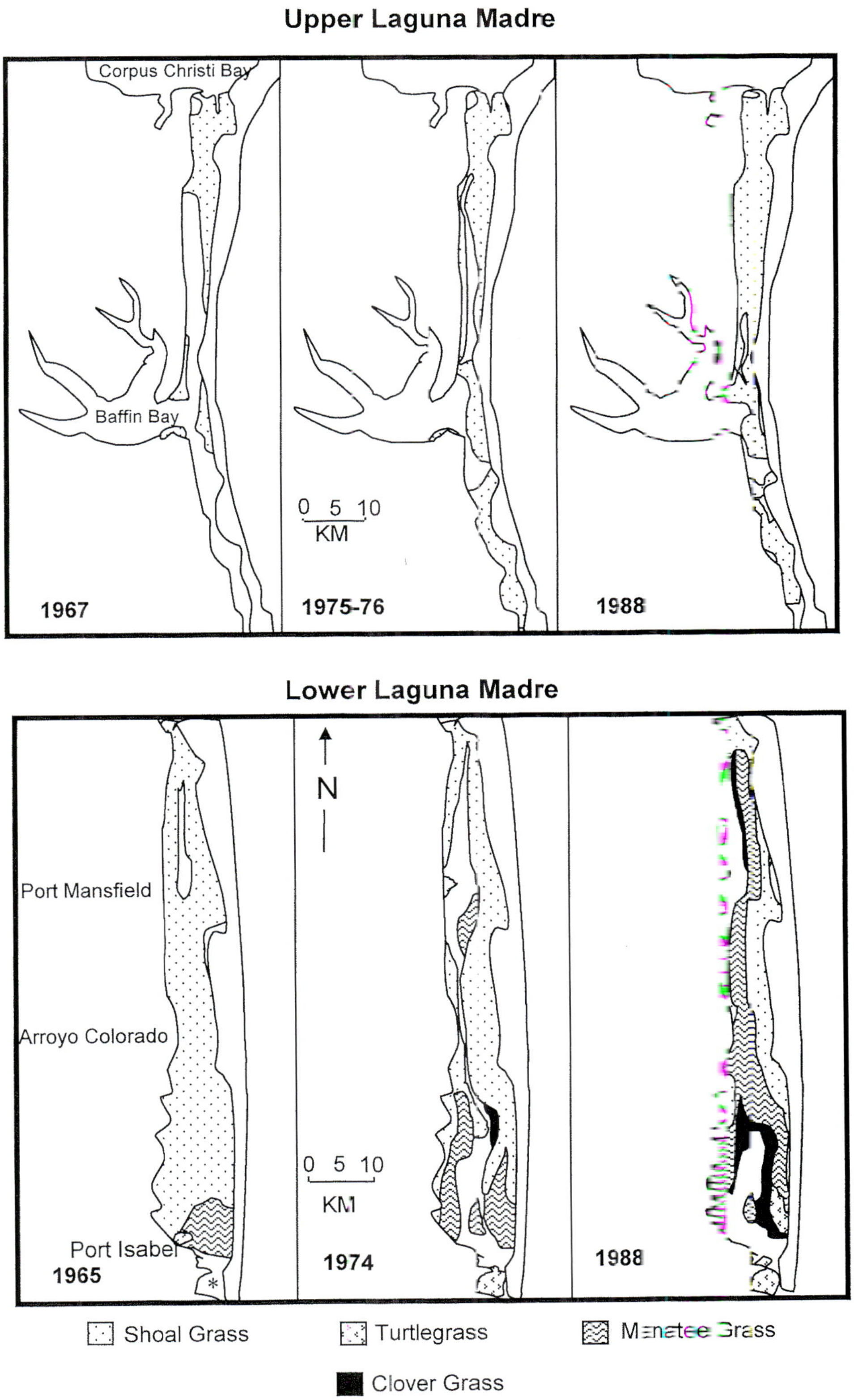

Figure 7.3. *Distribution of seagrasses in the Texas Laguna Madre between 1965 and 1988; * = no samples taken from South Bay in 1965 (Redrawn from Quammen and Onuf 1993)*

areas; and Star Grass–dominated areas with Shoal Grass as subdominant vegetation, primarily in deeper and/or more turbid areas such as along the edges of channels (Brown et al. 1976; Merkord 1978; Quammen and Onuf 1993). Turtlegrass, Manatee Grass, and Widgeon Grass occur locally (R. Phillips 1974; Merkord 1978; McMillan 1979; Williamson 1980). Widgeon Grass is abundant locally and seasonally (C. Onuf, pers. comm.) and occurs in channels paralleling the John F. Kennedy Causeway, east of Marker 21, near Pita Island, and at the northern end of The Hole (Merkord 1978). Seeds of Widgeon Grass were common in sediments, suggesting it may be more abundant than has been documented (McMillan 1985). Turtlegrass may be found in deeper water (deeper than 0.5 m [1.6 ft]) with moderate salinities (Brown et al. 1977). No Turtlegrass meadows have been mapped in the Upper Laguna Madre (Merkord 1978; Quammen and Onuf 1993), but one fairly large meadow was observed near the Humble Channel in 1989 (Withers 1996b). Manatee Grass cover has increased since the 1988 surveys and is becoming dominant in some areas (C. Onuf, pers. comm.). Seagrass density decreases south of Baffin Bay; only poorly developed meadows are found near the Land-Cut and within The Hole (Brown et al. 1977).

Seagrass meadows within Baffin Bay and its secondary bays covered only about 21 km^2 (8 mi^2) in 1992 (Pulich and Onuf 2007). Shoal Grass is the dominant species with lesser amounts of Star Grass and Widgeon Grass. Dense, marginal meadows grow on the narrow, shallow shelf along the north shore of Baffin Bay, along the southeast shore of Alazan Bay, and in Cayo del Grullo and Cayo del Infiernillo (Kreuz 1973; Dokken 1981; Cornelius 1984).

Lower Laguna Madre Seagrasses cover about 480 km^2 (185 mi^2) in the Lower Laguna Madre (Pulich et al. 1997). Although Shoal Grass once dominated the system, diversity has increased over the last 30 years. Once only a minor member of the community, Manatee Grass has come to dominate the system, and Star Grass and Turtlegrass coverage has also increased (Quammen and Onuf 1993). Widgeon Grass may be found locally, particularly along the Gulf Intracoastal Waterway (GIWW) north of Three Islands (Chin 1978; Brown et al. 1980). Turtlegrass had been largely confined to South Bay and southerly portions of the Lower Laguna Madre near passes (Brown et al. 1980; Quammen and Onuf 1993) but has become more widespread since 1988 (C. Onuf, pers. comm.). Star Grass is also found mostly in the southern portion of the bay, in deeper water where meadows transition to bare bay bottom (Quammen and Onuf 1993).

Drift Macroalgae and Phytoplankton Algal species are common components of all seagrass meadows. In the Upper Laguna Madre, attached algae are uncommon, but several species of green and red algae may be found in soft bottoms or attached to shell or other solid substrates within a seagrass meadow. Mermaid's Wine Glass was abundantly distributed in the Lower Laguna Madre, particularly east of GIWW (Chin 1978). Unattached or "drift" algae often occur in large numbers. Communities composed largely of red algae were common in Upper and Lower Laguna Madre Shoal Grass meadows (Chin 1978; Merkord 1978; Williamson 1980). Drift algae tend to accumulate in depressions, in the taller leaves of Manatee Grass associated with Shoal Grass, or wherever movement is hindered. Drift algae were found at more than 70% of stations sampled in the Upper and Lower Laguna Madre. Drift algae contributed more to baywide biomass than any other single seagrass species (Onuf 1996a). Leaves of seagrasses provide substrate for many epiphytic micro- and macroalgae in the Laguna Madre (Humm and Hildebrand 1962; Conover 1964). These algae are highly productive, sometimes contributing over 50% of the total productivity per unit area of seagrass meadow (Morgan and Kitting 1984). In addition, they are palatable, and because they are directly grazed by many species of invertebrates, they assume a more important trophic role than their seagrass hosts (Kitting et al. 1984). Diatoms dominate the phytoplankton community associated with seagrass meadows in the Laguna Madre, and dinoflagellates are common (Simmons 1957; Breuer 1962; Hildebrand and King 1978; Chaney 1988).

One of the principal causes of reduced light availability in coastal regions worldwide is increased nutrient input (Short et al. 1995; Tomasko et al. 1996), which often promotes phytoplankton blooms, epiphyte growth, and the accumulation of drift macroalgae (Pulich et al. 1997). In the Upper Laguna Madre, light was reduced by a persistent Texas Brown Tide bloom (Whitledge et al. 1999), which caused a 50% decline in the biomass of Shoal Grass over a five-year period (Dunton 1996; Onuf 1996a). In the Lower Laguna, bare patches of various sizes and shapes are clearly evident; some clearly result from drift macroalgal accumulation (K. Dunton, pers. observ.). Macroalgae can also negatively impact seagrasses by decreasing oxygen concentrations, especially if macroalgae are trapped near the sediment surface and restrict water movement.

Kopecky and Dunton (2006) examined the potential problem of excessive drift macroalgal abundance and its threat to seagrasses in Texas seagrass beds, including the Lower Laguna Madre. Drift macroalgae compete with seagrasses for light and nutrients, and in Redfish Bay, near Corpus Christi Bay, Cowper (1978) noted that a 20% reduction in SI by drift macroalgae had the potential to reduce the maximum rate of photosynthesis in seagrasses by 50%. Kopecky and Dunton (2006) found that in Redfish Bay (the bay between Aransas Bay and Corpus Christi Bay, not the region near Port Mansfield in the Lower Laguna Madre) and the Lower Laguna Madre, drift macroalgal distribution was widespread, with abundance equal to aboveground biomass of Turtlegrass, the dominant seagrass. In the Lower Laguna Madre, drift macroalgae were dominated by red algae (*Digenia simplex, Dictyota cervicornis, Jania capillacea, Agardhiella subulata*) and *Laurencia poiteaui*, a brown alga. These drifting seaweeds were positively associated with reduced seagrass cover and shoot density, which averaged 73 g m^{-2} (Kopecky and Dunton 2006). The presence of drifting seaweeds in the Lower Laguna is not unusual, but accumulations of large macroalgal mats that suffocate seagrasses over time are often associated with physical structures that interrupt water circulation or trap drifting mats. High nutrient loads in some areas are also likely responsible for large concentrations of drift seaweeds in some areas.

Historical Distributional Patterns Extreme hypersalinity prior to completion of the GIWW in 1949 probably excluded seagrasses from much of the Laguna Madre (Quammen and Onuf 1993); however, in the mid-1920s, the bottom of the Upper Laguna Madre north of Baffin Bay was described as grassy or muddy "depending on proximity to sediment-laden rivers or creeks" (J. Pearson 1929). Although this description seems incongruous because no rivers or creeks run into the Upper Laguna Madre, Pearson may have been describing water flowing across the large sandy flat that almost completely separated Corpus Christi Bay from the lagoon shown on the map in his report. No other information concerning seagrass extent or species composition was available until the Texas Parks and Wildlife Department (TPWD) began surveys in the late 1950s. It seems probable that Laguna Madre seagrass populations waxed and waned in response to changing salinities caused by drought, tropical storms, and hurricanes prior to completion of the GIWW. In this disturbance-dominated regime,

Shoal Grass and Widgeon Grass may have been the only species in the Upper Laguna Madre. Manatee Grass and Turtlegrass may have been present in the Lower Laguna Madre, particularly near Brazos Santiago Pass, where salinities would have been tolerable but Shoal Grass and Widgeon Grass likely dominated.

In 1965, Shoal Grass was the dominant seagrass in the Lower Laguna Madre and the only species present in the Upper Laguna Madre (see fig. 7.3). Manatee Grass and Turtlegrass were present near the pass at Port Isabel in the Lower Laguna Madre (McMahan 1966, 1967b). By 1974, Manatee Grass had expanded considerably in the Lower Laguna Madre, displacing Shoal Grass (Merkord 1978). Shoal Grass coverage had increased in the Upper Laguna Madre, and Star Grass existed where meadows transitioned to bare bottom. By 1988, Manatee Grass extended nearly the length of the Lower Laguna Madre at intermediate depths but had been replaced by Turtlegrass in the extreme south (Quammen and Onuf 1993).

Between 1965 and 1988, Shoal Grass in the Upper Laguna Madre covered nearly the entire basin. Manatee Grass was not found in samples taken from the Upper Laguna Madre in 1988, but patches were observed between sampling stations (Quammen and Onuf 1993) and north of the JFK Causeway (J. W. Tunnell, pers. comm.). Manatee Grass cover has increased dramatically since 1988 and currently constitutes about 17% of total seagrass cover (C. Onuf, pers. comm.).

Succession appears to be proceeding toward a Turtlegrass climax in the Lower Laguna Madre where further expansion of Manatee Grass is not expected (Quammen and Onuf 1993). Manatee Grass is already established in areas with suitable depth and water clarity, but there is ample room for Turtlegrass expansion. Small patches of Turtlegrass exist as far north as Arroyo Colorado, and these will probably expand and coalesce, displacing Manatee Grass. However, at the present rate of expansion, it is unlikely that Turtlegrass will become dominant in the Lower Laguna Madre within the next 50 years (Onuf 1996a). In the Upper Laguna Madre, the spread of Manatee Grass suggests that succession is proceeding slowly toward a Turtlegrass climax (Quammen and Onuf 1993). Within the next few decades, the composition of Upper Laguna Madre seagrass meadows will likely resemble the present composition of the Lower Laguna Madre meadows (Onuf 1996a).

Initiation of Long-Term Monitoring

A statewide long-term seagrass monitoring program was begun in 2011 by Dunton et al. (2011) based on the findings of the Texas Seagrass Conservation Plan (Pulich and Calnan 1999). The monitoring protocol followed a restricted random sampling design to ensure even coverage across the study region while still maintaining random station selection (Dunton et al. 2011; Neckles et al. 2012). To generate the sampling stations, maps from the National Oceanic and Atmospheric Administration's 2004/2007 Benthic Habitat Mapping program (https://coast.noaa.gov/digitalcoast/data/benthiccover.html) were used to create a shapefile delineating seagrass extent along the Texas coast. Grids of tessellated hexagons (each side 750 m [~2,461 ft] for a total area of ~1.46 km^2 [~0.56 mi^2]) were overlaid onto the shapefile, with hexagon side length chosen to represent the study area adequately. One sampling station was assigned within each hexagon that contained >50% seagrass cover according to the benthic habitat shapefile. A random number generator was used to assign latitude and longitude coordinates of the station within each hexagon. This design resulted in 144 sampling stations in the Upper Laguna Madre and 285 stations in the Lower Laguna (fig. 7.4).

The monitoring program was based on the hierarchical strategy for seagrass monitoring proposed by Neckles et al. (2012). For Texas coastal waters, this plan incorporated

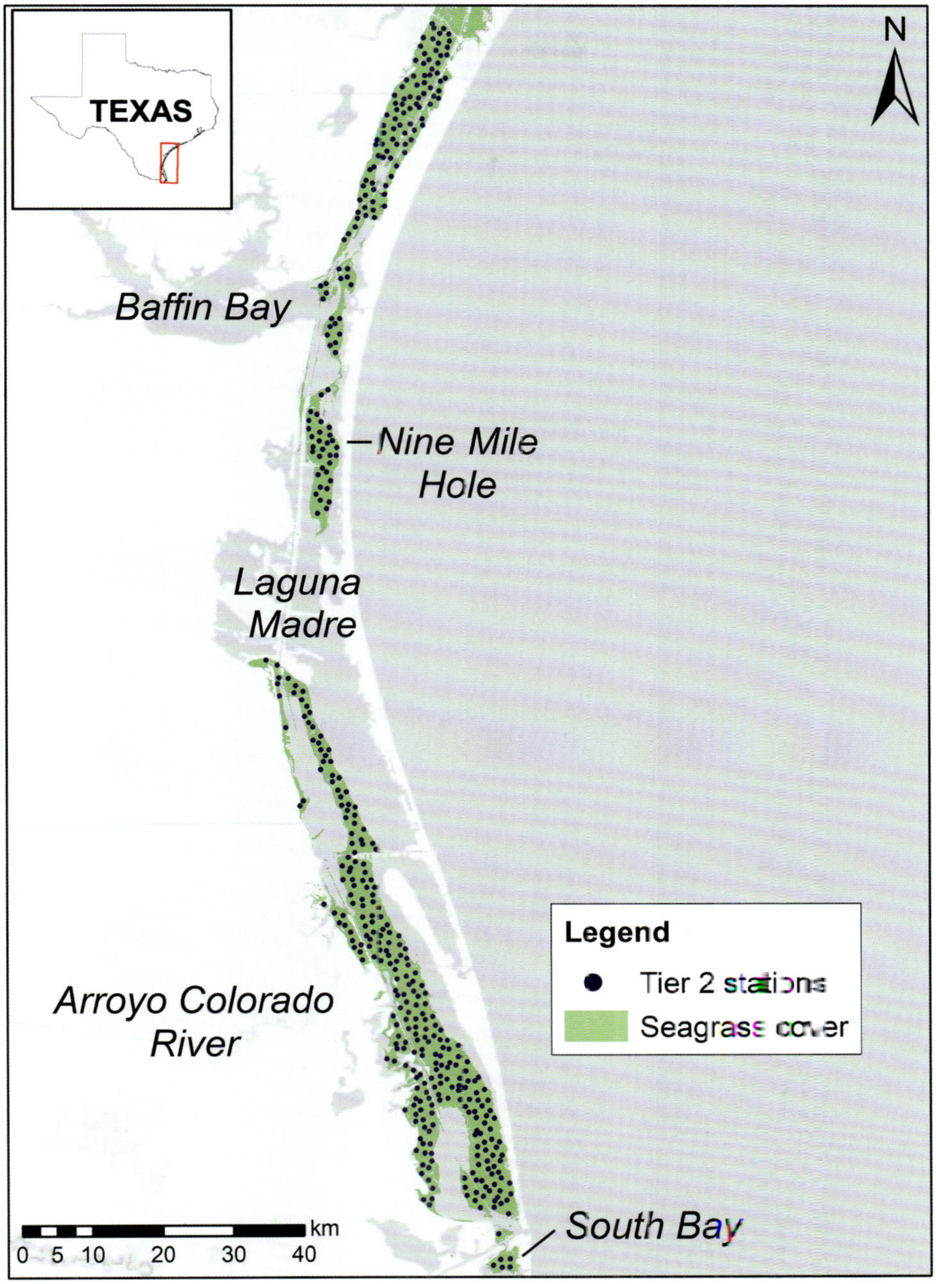

Figure 7.4. *The locations of 429 randomly assigned permanent monitoring stations established within a grid of tessellated hexagons (not shown) in the Upper and Lower Laguna Madre, Texas, occupied under the Texas Seagrass Monitoring Program (Dunton et al. 2011).*

new and historical data to establish the natural baselines of condition indicators to enable status and trends assessment for seagrass populations unique to Texas estuarine systems (Dunton et al. 2011). Measurements taken at each sampling station included those related to water quality, seagrass cover, and plant tissue condition (table 7.2). ArcGIS software was used to manage, analyze, and display spatially referenced point samples and interpolate surfaces for all measured parameters. The results of this long-term program reveal that seagrass distribution and species composition can vary

Table 7.2. *Indicators and proposed measurement frequency under a Tier 2 (annual) seagrass monitoring program*

	Indicator	Field Method
Stressor		
	*k, %SI	Underwater light sensor
	*Water transparency	Secchi
	*Depth	PDR
	*Temperature, salinity, pH, dissolved oxygen	SONDE
	*TSS	Water collection
	NH_4^+, NO_3-, PO_4^{-2}	Water collection
	*chl *a*	In situ fluorescence
	Drift algal biomass	0.25 m^2 quadrats
	Sediments (grain size/organics)	Benthic cores
	Algal epiphyte biomass	Benthic cores
Seagrass condition indicator		
	Canopy height	Benthic cores
	Shoot density	Benthic cores
	Seagrass biomass	Benthic cores
	Root:shoot ratios	Benthic cores
	*Seagrass species composition	0.25 m^2 quadrats
	C:N:P and ^{15}N:^{14}N ratios	Benthic cores
	*Percent cover	0.25 m^2 quadrats

Source: From Dunton et al. (2011).

Note: *k* = light attenuation, %SI = percent surface irradiance, PDR = Precision Depth Recorder. Asterisks denote minimum criteria for a Tier 2 sampling effort.

considerably across spatial and temporal scales in response to regional drought and land-use patterns (Wilson and Dunton 2017; Cuddy 2018).

Seasonality (Phenology) With the exception of Widgeon Grass, seagrasses are generally perennial, but populations of both Shoal Grass and Star Grass may appear or disappear quickly (Phillips and Meñez 1988) and probably did so prior to completion of the GIWW. Laguna Madre seagrasses are dormant between December and March and begin growing in April (Conover 1964). Luxuriant growth is typically observed between May and August, but biomass tends to decline in late July and early August when air and water temperatures peak. Shoal Grass biomass typically peaks in early summer and again in early fall after temperatures have moderated (Morgan and Kitting 1984).

Animal Communities

Invertebrates The invertebrate community associated with seagrass meadows is diverse and consists of epibenthic, benthic, epiphytic, and nektonic organisms. Most researchers have focused on the Upper Laguna Madre, and most studies of benthic organisms have concentrated on molluscs. Consequently, little information is available concerning the distribution or abundance of annelids and epibenthic crustaceans (crabs, amphipods), and some taxa, such as invertebrate nekton, have been almost completely ignored (see table 7.1). The low numbers for invertebrate families and species in the Lower Laguna Madre reflect a lack of studies rather than a lack of diversity. The invertebrate communities in the Lower Laguna Madre are likely just as diverse as those in the Upper Laguna Madre and may be even more so since the seagrass community is more diverse and there is a greater likelihood of finding tropical species. The potential for greater diversity in the Lower Laguna Madre may be indicated by the number of bivalve families and species collected in the few studies conducted there. Extreme salinity fluctuations, absence of extensive seagrass habitat, and low recruitment due to the lack of circulation between Baffin Bay and the Upper Laguna Madre may prevent establishment of an abundant and/or diverse benthic community in Baffin Bay.

Polychaetes may be the most abundant benthic invertebrates in Laguna Madre seagrass meadows. When polychaetes were collected and identified from the Upper Laguna Madre (Rickner 1979; White, Calnan, et al. 1983, 1989; Montagna 1992, 1993; Rocha et al. 1995) and Lower Laguna Madre (White et al. 1986), they were usually the most abundant components of the community and always had higher species richness than the molluscan component. Members of families Spionidae, Ampharetidae, Capitellidae, and Maldanidae typically were dominant. Most members of these families are deposit feeders, although some spionids are suspension feeders.

Gastropods typically outnumber bivalves in seagrass meadows but have lower species richness. Ceriths (family Cerithiidae), slippershells, and caecums (family Caecidae) were common in seagrass meadows throughout the Laguna Madre (Cornelius 1975; Rickner 1979; Williamson 1980; Powell et al. 1982; Wilhite et al. 1982; White, Calnan, et al. 1983, 1986, 1989; Rocha et al. 1995). The dominant species are deposit feeders, but a number of predators (family Pyramidellidae), scavengers (e.g., Bruised Nassa), and herbivores (e.g., Striate Bubble) also are common.

The Atlantic Papermussel and Morton Eggcockle are bivalves considered indicator species of Shoal Grass meadows (Parker 1959). Other common species in seagrass meadows throughout the Laguna Madre include tellins and macomas (family Tellinidae), the Dwarf Surf Clam, and particularly in Turtlegrass meadows, lucines (family Lucinidae) (Cornelius 1975; Rickner 1979; Williamson 1980; Powell et al. 1982; Wilhite et al. 1982; White, Calnan, et al. 1983, 1986, 1989; Rocha et al. 1995). The Bay Scallop is a commercially important member of the seagrass bivalve community. These bivalves exhibit cyclical population levels, appearing in large numbers at irregular intervals and then disappearing. All bivalves are filter feeders.

When benthic and epiphytic crustaceans (amphipods, isopods, cumaceans, and tanaids) within seagrass meadows in the Laguna Madre were studied (White, Calnan, et al. 1983, 1986, 1989; Rocha et al. 1995), they were conspicuous and often abundant members of the invertebrate community. These organisms are typically herbivores or detritivores and are trophically important because they link primary producers with higher consumers.

Shrimp and crabs are the most prominent of the epibenthic and/or nektonic crustaceans. Some of these organisms are also important commercial species, notably penaeid shrimp (Brown Shrimp, Pink Shrimp, and White Shrimp) and the Blue Crab.

Although Simmons (1957) and Breuer (1962) included invertebrate nekton in their surveys, only Chaney (1988) addressed invertebrate nekton in the Laguna Madre in detail. Shrimp dominated the assemblage, followed by Blue Crabs. Grass shrimp were the most abundant species recovered. Jellyfish (phylum Cnidaria), comb jellies (phylum Ctenophora), mantis shrimp (subphylum Crustacea, order Stomatopoda), and squid (phylum Mollusca, class Cephalopoda) were also collected. Shrimp and crabs are typically omnivorous, eating everything from invertebrate infauna to detritus and carrion. Jellyfish, comb jellies, and squid are carnivorous predators.

Echinoderms are not typically abundant but may constitute a significant proportion of animal biomass in seagrass meadows because they are large (Rocha et al. 1995). Sea cucumbers (class Holothuroidea) are frequently encountered, and brittle stars (class Ophiuroidea) may also be found (Rickner 1979; Montagna 1993; Rocha et al. 1995). Sea cucumbers are typically deposit or suspension feeders, whereas brittle stars may employ several feeding modes, including deposit, filter, and suspension feeding, as well as carnivory and scavenging. Ribbon worms (phylum Rhynchocoela) are also common deposit feeders and predators in seagrass beds (Rickner 1979; Montagna 1993; Rocha et al. 1995; Wern 1995).

Fish Fish in seagrass meadows can be categorized based on their life history and temporal use (Kikuchi 1980). Permanent residents are generally small, cryptic, less mobile species (pipefishes, gobies, blennies, and eels) that spend their entire life in the meadow. Seasonal residents spend their juvenile or subadult stages or spawning season in seagrass meadows. This group includes many important commercial or sport species, such as drums, mojarras, grunts, and porgies. Transients are present infrequently and unpredictably and include large carnivores of offshore or oceanic origin such as the Jack Crevalle (Zieman 1982). Many fish found in seagrass meadows are carnivores that prey on the abundant fish and crustacean inhabitants. Sheepshead, Black Drum, Cownose Ray (Carangelo et al. 1975), Pinfish (Darnell 1958; Carr and Adams 1973), Atlantic Needlefish (Darnell 1958), and Striped Mullet (Pullen 1960) include seagrasses in their diets.

Simmons (1957) surveyed the fish fauna of the Upper Laguna Madre without noting which species occurred within seagrass meadows. Twenty-nine species of small fish in 18 families were collected from mixed Widgeon Grass and Shoal Grass meadows between the JFK Causeway and Bird Island Basin (Hellier 1962). The most abundant species were Striped Mullet, Pinfish, Spot, and Bay Anchovy. Yearly production of fishes in the area was calculated at 156 kg/ha (139 lb/ac). In a Shoal Grass meadow near Bird Island Basin, 26 species of small fish in 16 families were collected (Chaney 1988). Tidewater Silversides and Rainwater Killifish were the most abundant species. Juvenile Spotted Seatrout were collected from midsummer to early winter, whereas juvenile Red Drum were collected from winter to early spring. Postlarval and juvenile Spotted Seatrout prefer seagrass meadows during warm weather (Lassuy 1983). Red Drum juveniles remain in estuaries for at least six months (Reagan 1985), and many remain in the bays for up to five years before moving offshore (L. McEachron, pers. comm.).

Many researchers have documented the fish fauna of Baffin Bay, particularly Alazan Bay (Breuer 1957; Fuls 1974; Tinnin 1974; Krull 1976; Dokken 1981; Cornelius 1984; Dokken et al. 1984; Hardegree 1997). However, none included seagrass meadows in their sampling protocol.

Breuer (1962) surveyed the fish fauna in the Lower Laguna Madre without differentiating which species were collected from seagrass meadows. In South Bay, Spot and Tidewater Silversides dominated in seagrass habitats (Hook 1991). Larval fish communities of Manatee Grass meadows, Shoal Grass meadows, and bare areas were compared during summer, fall, and winter spawning periods using cluster analysis (Tolan 1994; Tolan et al. 1997). Larval Bay Anchovies were the most abundant species in vegetated areas during summer and fall spawning periods. Tidewater Silversides were also important members of the Shoal Grass ichthyofaunal assemblage during fall. During winter spawning, Gulf Menhaden was especially abundant in Manatee Grass beds, whereas densities of Pinfish and Gulf Menhaden were nearly equal in Shoal Grass meadows. Overall, members of the families Engraulidae and Clupeidae dominated in vegetated areas, and juveniles generally preferred Shoal Grass meadows to Manatee Grass meadows.

Sea Turtles Turtle canneries along the Texas coast in the mid-1800s processed many tons of turtle meat before the industry collapsed around 1900 (Shaver 1990b; USFWS and NMFS 1992). Many turtles came from the Lower Laguna Madre, where Green Sea Turtles were the leading marine product by weight (Hildebrand 1981; Doughty 1984). In a three-year study of sea turtle strandings along the southern Texas coast (Rabalais and Rabalais 1980), the Loggerhead Sea Turtle was the most abundant species, followed by the Kemp's Ridley Sea Turtle, Green Sea Turtle, and Leatherback Sea Turtle. The Green Sea Turtle was the only sea turtle reported from within the Laguna Madre. More recently, more than 20 sea turtles were captured in gill nets during TPWD's fishery monitoring program in the Lower Laguna Madre (R. Blankenship, pers. comm.), whereas only one has been captured in the Upper Laguna Madre (K. Spiller, pers. comm.). Many live and dead sea turtles were stranded in the Lower Laguna Madre during a severe cold spell in February 1989, when water temperatures dropped to as low as 3 °C (37.4 °F) (Shaver 1990a). A similar cold-stunning event occurred between December 2017 and February 2018.

Leaves of seagrasses compose up to 100% of the diet of juvenile Green Sea Turtles and Hawksbill Sea Turtles (Bustard 1972; Hirth et al. 1973; Rebel 1974). Habitat preferences of one Loggerhead and four Green Sea Turtles were determined by using radio and sonic telemetry in the South Bay area of the Lower Laguna Madre (Renaud et al. 1992). The Loggerhead Sea Turtle spent ~50% of its time in seagrass bed habitats north of the Queen Isabella Causeway, whereas the Green Sea Turtles spent about 60% of their time associated with seagrasses. Additional observations were made on turtles that were not radio-tagged. Loggerhead Sea Turtles were observed only north of the Queen Isabella Causeway, which is primarily shallow and vegetated with seagrasses, whereas Green Sea Turtles were observed in vegetated, unvegetated, and jetty habitats. Turtlegrass was the dominant component of fecal samples from 12 Green Sea Turtles with carapace lengths >40 cm (15.7 in) captured near seagrass meadows. The main component of feces of Green Sea Turtles with carapace lengths <40 cm captured near jetties was algae. Smaller turtles appeared to prefer jetty habitats, while larger turtles were most often found in seagrass and channel habitats. These distributions were attributed to changes in feeding preferences of turtles as they grow.

Birds Although many species of birds use Laguna Madre habitats as feeding and resting areas (Zieman and Zieman 1989), most scientific studies have focused on the use of seagrass meadows by waterfowl (ducks), especially the Redhead. The majority of the North American population of Redheads winters in the Laguna Madre of Texas and Tamaulipas (Weller 1964). Intensive use of these lagoons is likely related to the presence of extensive and nearly homogeneous stands of Shoal Grass (Cornelius 1977).

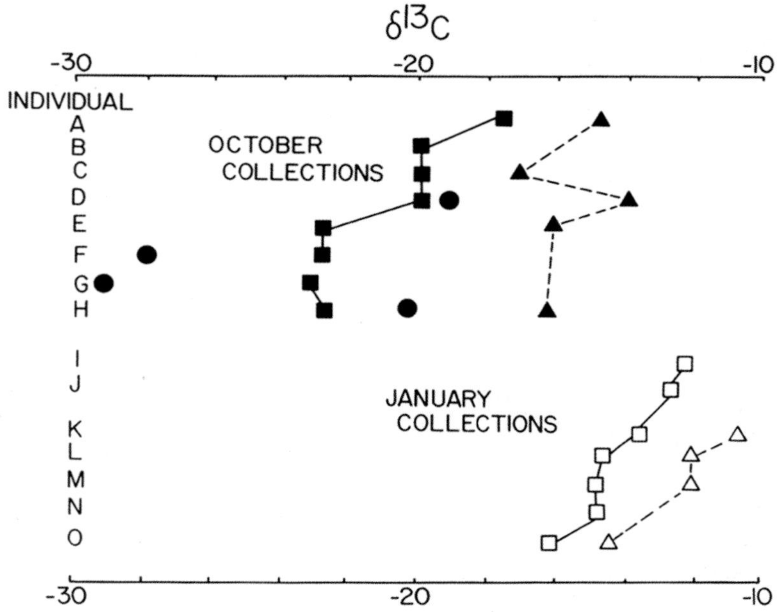

Figure 7.5. *Seasonal shifts in dependence of overwintering ducks (Redheads) on seagrass carbon based on carbon isotopic evidence. Squares indicate muscle tissue; circles indicate fat tissue; triangles indicate liver tissue. (Redrawn from Parker et al. 1992; Hammer et al. 1998)*

Redheads typically feed in shallow Shoal Grass meadows along the perimeter of the Laguna Madre (McMahan 1970; Cornelius 1975, 1977; Adair et al. 1990). Shoal Grass rhizomes were the most common food found in stomachs of Redheads wintering in the Upper Laguna Madre, as well as in those of American Wigeons and Northern Pintails (Koenig 1969; Woodin 1996). The winter dependence of these species on belowground tissues has been clearly demonstrated by studies of muscle-tissue carbon isotope ratios (fig. 7.5) and fatty acids (Anderson et al. 1992; Hammer et al. 1998).

Marine Mammals The use of Laguna Madre seagrass meadows by marine mammals remains an open subject for investigation. The Bottlenose Dolphin, which feeds over seagrass meadows in Florida (Zieman and Zieman 1989), often migrates between Texas bays and the Gulf of Mexico (Schmidly 1981; Shane and Schmidly 1978; Brager 1993). Seagrass meadows within the Laguna Madre may occasionally be used as foraging habitat.

Community Structure

Seagrass meadows provide many marine animals an important refuge from predators. The value of a seagrass meadow as a refugium is usually attributed to macrophyte structural complexity, a measure of both plant biomass and plant architecture. Most scientists view the effect of structural complexity on prey habitat choice and predator foraging efficiency as one of the key mechanisms organizing animal communities associated with seagrass meadows (Leber 1985). In the Lower Laguna Madre, seagrass habitat complexity (expressed as a function of computed leaf aggregate index) in Shoal Grass meadows decreased from east to west. Shoal Grass meadow complexity was generally higher than Manatee Grass meadows (Tolan et al. 1997). Consequently, the more complex habitats typically held more species and higher populations of fish.

Hildebrand and King (1978) noted the nursery function of seagrass beds in the Upper Laguna Madre for numerous species. However, the nursery role of seagrasses for larval and young fish has been questioned because the protective value would likely be greatest for those individuals able to orient to the blades (Olney and Boehlert 1988).

Tolan et al. (1997) found that both unvegetated, open-water areas and seagrass meadows served as primary nursery areas for early-stage larvae of many species in the Lower Laguna Madre, but some species, notably Tidewater Silversides, showed a marked preference for Shoal Grass meadows. They concluded that seagrass meadows did not serve as the exclusive nursery area for early-larval stages for either estuarine-spawned (e.g., Tidewater Silversides) or estuarine-dependent, marine-spawned fish species (e.g., Red Drum) in the Laguna Madre. However, postlarval and young juveniles of a number of species were strongly associated with seagrass habitats, and Shoal Grass meadows were the preferred nursery habitat for these life stages.

Ecosystem Processes

Energy Flow and Food Webs Traditionally, seagrass meadow food webs have been considered detritus based, but the relative roles of detrital and microalgae-epiphytic pathways are largely dependent on local conditions and consumer community composition (Zieman and Zieman 1989). Seagrasses and associated epiphytes transfer energy to higher trophic levels by direct grazing, detrital food webs, or export of material to other systems. Often, little of seagrass primary productivity is used directly by consumers, either alive or as detritus, though the importance of Shoal Grass to Redheads and other waterfowl wintering in the Laguna Madre suggests this pathway is important seasonally. Epiphytic primary productivity often contributes more carbon than the

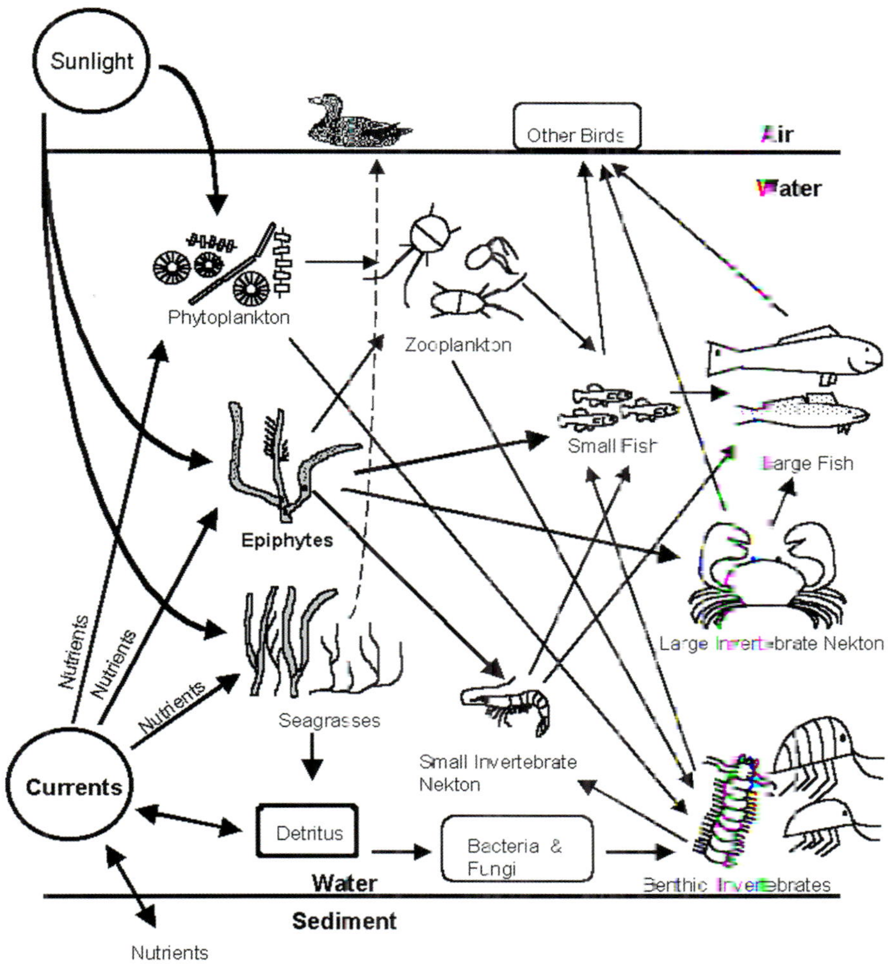

Figure 7.6. *Generalized food web for Laguna Madre seagrass meadows (Modified from Withers 1996b)*

seagrasses themselves to consumer diets (Morgan and Kitting 1984), but this pathway has not been studied in the Laguna Madre. Much of the carbon fixed by seagrasses in the Laguna Madre may be transported out of meadows by currents and storms. Large rafts of Shoal Grass are often observed along the banks of the Laguna Madre, particularly during winter, as well as following storms and high-tide events.

Although a generalized food web for Laguna Madre seagrass meadows can be developed (fig. 7.6), many consumers in seagrass meadows are opportunistic (omnivorous), so trophic structure is dynamic and fluctuates depending on changes in densities and species composition of predator and prey compartments. Predation pressure on prey species may change as relative abundance of other prey species increases or decreases. Diets of many fish change as they grow and often change diurnally in response to changes in prey availability resulting from the circadian rhythms of invertebrate prey species (Kikuchi and Peres 1977).

Nutrient Cycling Although seagrasses are biologically productive, they are often found in low-nutrient environments where they probably recycle and conserve the majority of nutrients needed for growth and maintenance (Capone and Taylor 1980). There is controversy over the relative importance of nutrient uptake out of the water column by leaves and nutrient uptake out of sediments by roots. Studies in the Upper Laguna Madre and nearby Redfish Bay suggest that nitrogen (N) and phosphorus (P) uptake by Shoal Grass and Widgeon Grass is primarily through roots (Pulich 1985). Neither species appears to derive much of its nutrients from the water column. Shoal Grass requires organic-rich sediments to obtain nutrients. Widgeon Grass is adapted to low-nutrient conditions and would be outcompeted by Shoal Grass in organic-rich sediments. Microbial recycling of organic material is much more significant in Shoal Grass meadows than Widgeon Grass meadows. No other studies of seagrass nutrient cycling have been conducted in the Laguna Madre.

Current Issues

Seagrass cover and community composition have been changing and will likely continue to change for the foreseeable future. Alterations in species composition are mostly the result of natural succession but were set in motion by moderation of salinity resulting from dredging of the GIWW and Mansfield Pass. However, decreases in cover and biomass have also been occurring. In the Upper Laguna Madre, Shoal Grass cover decreased by 3.8% (9.4 km^2 [3.6 mi^2]) between 1988 and 1994, and Shoal Grass biomass at depths >1.4 m (4.6 ft) decreased by 60% (Onuf 1996b). These decreases have been attributed to the Texas Brown Tide that flared up in 1990 and continues in some parts of the system today. Changes in species composition due to succession have been most pronounced in the Lower Laguna Madre, but a more troubling change is the increase in bare area. Overall, bare area increased to 190 km^2 (73 mi^2), up 280% between 1965 and 1988 (Quammen and Onuf 1993). Turbidity caused by maintenance dredging and other activities (fig. 7.7) has been implicated in the decline of Shoal Grass and increased bare area in the lower Laguna Madre (Onuf 1994). Light attenuation is responsible for most landscape-level losses, but scarring by boat propellers is also a concern. Changes in species composition may result in reduced resource availability for some fish and waterfowl.

Effects of Long-Term Light Reduction The amount of light reaching the lagoon bottom is a significant factor in determining seagrass meadow extent and productivity. Reduced light is linked to reductions of seagrass cover and productivity (e.g., Orth

Figure 7.7. *Turbidity over seagrass beds can be caused by maintenance dredging of channels or, as pictured here, by site preparation for placement of a gas well and associated infrastructure. (Photograph by Ken Dunton)*

and Moore 1983; Kenworthy and Haunert 1991; Dunton 1994; Czerny and Dunton 1995). In the Laguna Madre, two causes of light reduction affect changes in seagrass cover and biomass: dredging and Texas Brown Tide.

Even though dredging and dredged-material disposal have immediate and direct effects on water clarity, until recently few seagrass losses were attributed to these activities. Turbidity caused by dredging was viewed as short-lived, often because background turbidity levels are already high and limited to areas within 5 km (3 mi) of the discharge (Cronin et al. 1970; Nichols et al. 1990). However, loss of Shoal Grass in deep areas and close alignment of newly denuded areas with navigational channels between 1965 and 1974 in the Lower Laguna Madre led Merkord (1978) to suggest that turbidity resulting from maintenance dredging of GIWW caused seagrass losses.

Onuf (1994) tested the hypothesis that light reduction caused by dredging in the Lower Laguna Madre resulted in changes in seagrass cover. He postulated that seagrasses could tolerate the short periods of light reduction that occur during dredging because seagrasses survive frequent episodes of light reduction from natural causes. Therefore, he reasoned that dredge-related light reduction must persist for weeks or months to impact seagrass distribution. Intensive monitoring of the underwater light regime before and after maintenance dredging of GIWW near the Mansfield ship channel determined that light reduction attributable to dredging and dredged-material disposal persisted for 10–15 months after the dredging project in half of the areas studied. Wind-generated waves resuspended and dispersed dredged material within affected areas, and such events occurred so frequently that elevated turbidity levels were long-lasting and transported over long distances. Onuf outlined a scenario in which frequent resuspension, long-term susceptibility of deposited dredged material to resuspension, and long-distance transport of turbidity result in loss of seagrasses, particularly on the edges of meadows. Given that maintenance dredging is necessary every two to three years in this part of the lagoon (Onuf 1994), light reduction is chronic and the loss of seagrass will persist until dredge-disposal practices are modified to reduce the amount of sediment resuspension. The integrative model of seagrass distribution and productivity created by Dunton, Burd, et al. (1998) also strongly suggests that light attenuation due to dredging operations will negatively impact seagrass

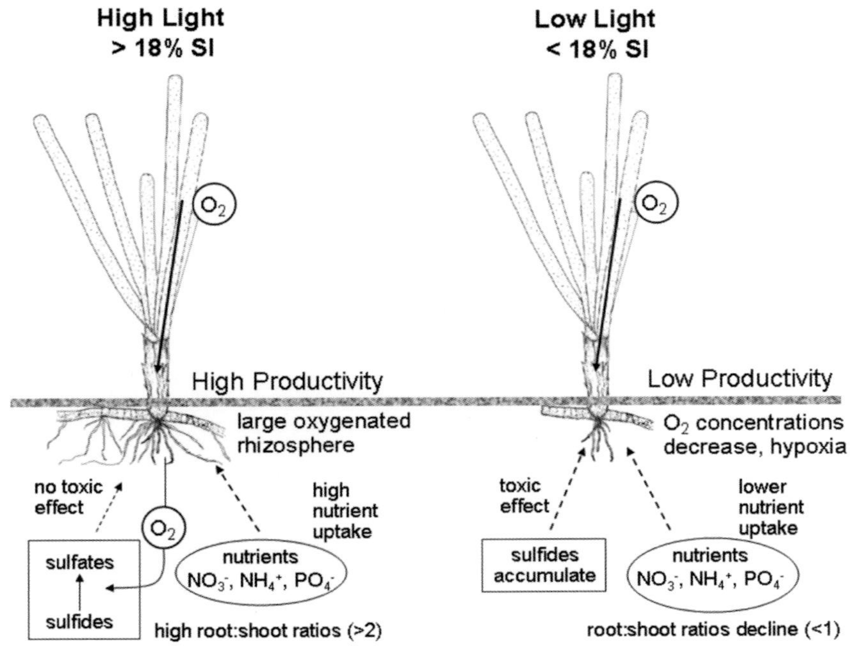

Figure 7.8. *Conceptual model for seagrass declines in response to light attenuation caused by algal blooms or dredging activity in the Laguna Madre based on studies conducted by Dunton (1994) and Onuf (1994) (Redrawn from Dunton et al. 2003)*

health based on the 18% SI minimum that Dunton (1994) and Onuf (1994) calculated independently (fig. 7.8).

The sudden appearance of Texas Brown Tide in the Upper Laguna Madre in 1990 resulted in a nearly immediate light reduction (50%; Dunton 1994) that persisted through 1997. Changes in seagrass cover were not noted during the first two years (1991–92) after initiation of the bloom, but approximately 2.6 km^2 (1.0 mi^2) were lost between 1992 and 1993, and 9.4 km^2 (3.6 mi^2) had disappeared by 1995 (Onuf 1996b). Biomass in the deepest areas (>1.4 m [4.6 ft]) decreased two years before declines in cover and increases in bare area were noted. Overall, biomass decreased by 60% in the deepest areas but did not decrease in the shallowest waters. Utilization of stored reserves allowed seagrasses to persist throughout the study at intermediate depths and for two years in deeper areas, but the trend toward loss in deeper areas was apparent. Based on light-attenuation measures, Onuf predicted that when seagrass distribution reached steady-state with the brown tide–influenced light regime, 24% of the Upper Laguna Madre would be bare, whereas only 6% had been bare before. However, in 1996, Star Grass began revegetating about half the area where Shoal Grass had been lost and Shoal Grass retreat slowed (Onuf 1999). Although water clarity improved briefly and Shoal Grass began to recover, return of the Texas Brown Tide bloom in the spring and summer of 1998 caused most regrowth to die.

Drought and Hypersalinity South Texas, an area known for increasingly frequent drought conditions, recently faced the most extreme drought on record in over a century (Cayan et al. 2010; Hernandez and Uddameri 2014; Romero-Lankao et al. 2014). This region includes the recurrently hypersaline waters of the Upper Laguna, which supports over 22,000 ha (~56,000 ac) of seagrass meadows (table 7.3). These meadows have been historically dominated by Shoal Grass, but with increased circulation and moderation of hypersaline conditions, Manatee Grass had invaded much of the northern areas of the Upper Laguna by 2012. The trend began to reverse after 2012, when Wilson and Dunton (2017) observed a massive die-off of Manatee Grass in the Upper Laguna (fig. 7.9). Manatee Grass mortality followed an extended period of extremely high salinity (50–70) during a regional drought (fig. 7.10). The prolonged period of high salinity was the main driver of Manatee Grass mortality. The effects of changes in seagrass species composition to ecosystem structure; the cycling of carbon and nitrogen; and impacts to higher trophic levels, including overwintering ducks, are significant areas of study.

Table 7.3. *Summary of total seagrass changes for Texas bay systems over four decades*

Bay System	**Late 1950s or Mid-1960s**	**Mid-1970s**	**1987 or Early 1990s**	**1998**
		Galveston Bay System		
Galveston/Christmas Bays	590[a] (1,457)	134[a] (331)	113[b] (279)	210[c] (519)
		Midcoast Region		
Matagorda Bay			1,099[b] (2,716)	
San Antonio Bay		5,000[d] (12,350)	4,305[d] (10,638)	
		Coastal Bend Region		
Aransas/Copano Bays			2,871[e] (7,094)	
Redfish Bay and Harbor Island	5,380[e] (13,293)	6,200[e] (15,320)	5,710[e] (14,109)	
Corpus Christi Bay			2,568[e] (6,346)	
		Laguna Madre System		
Upper Laguna Madre	12,321[f] (30,445)	20,255[g] (50,050)	22,903[h] (56,593)	22,44[illegible][i] (55,456)
Lower Laguna Madre	59,153[f] (146,166)	46,558[g] (115,044)	46,624[h] (115,207)	46,17[illegible][i] (114,095)
Baffin Bay			2,200[j] (5,436)	

Source: Modified from Pulich and Onuf 2007.

Note: Seagrass values are in hectares with acres in parentheses.
The late 1950s or mid-1960s data for Galveston/Christmas Bays, Redfish Bay, and Harbor Island based on 1956–58 Tobin photography. Data for Upper and Lower Laguna Madre based on field surveys during mid-1960s.
The mid-1970s data for Galveston/Christmas and Redfish Bays/Harbor Island based on 1975 (National Aeronautics and Space Administration Johnson Space Center [NASA-JSC] photography; San Antonio Bay based on 1974 NASA-JSC photography. Data for Upper and Lower Laguna Madre based on 1974–75 field surveys.
The 1987 or early 1990s data for Christmas, Matagorda, and San Antonio Bay systems from 1987 NASA-Ames Research Center photography. Data for Aransas/Copano, Redfish, and Corpus Christi Bay systems based on 1994 TPWD photography. Data for Upper and Lower Laguna Madre based on 1988 field surveys. Data for Baffin Bay based on 1992 US Fish and Wildlife Service National Wetlands Inventory photography.
The 1998 data for Christmas Bay from 1998 Galveston Bay National Estuary Program photography. Data for Upper and Lower Laguna Madre from 1998 field surveys.

a From Pulich and White 1991.
b From Adair et al. 1994.
c From Pulich 2001.
d From Pulich 1991.
e From Pulich et al. 1997.
f Areas computed for this review from McMahan (1965–67). See Laguna Madre section in Pulich and Onuf (2007).
g Areas computed for this review from Merkord (1978).
h Areas computed for this review from Quammen and Onuf (1993). See Laguna Madre section in Pulich and Onuf (2007).
i Areas computed for this review. See Laguna Madre section in Pulich and Onuf (2007).
j Areas computed for this review by Texas Parks and Wildlife Department, Coastal Studies Program, Austin (unpub. data, 2018).

Figure 7.9. *The decline in Manatee Grass was compensated by increases in Shoal Grass based on percent cover interpolations in the Upper Laguna Madre from 2011 to 2015. (Modified from Wilson and Dunton 2017)*

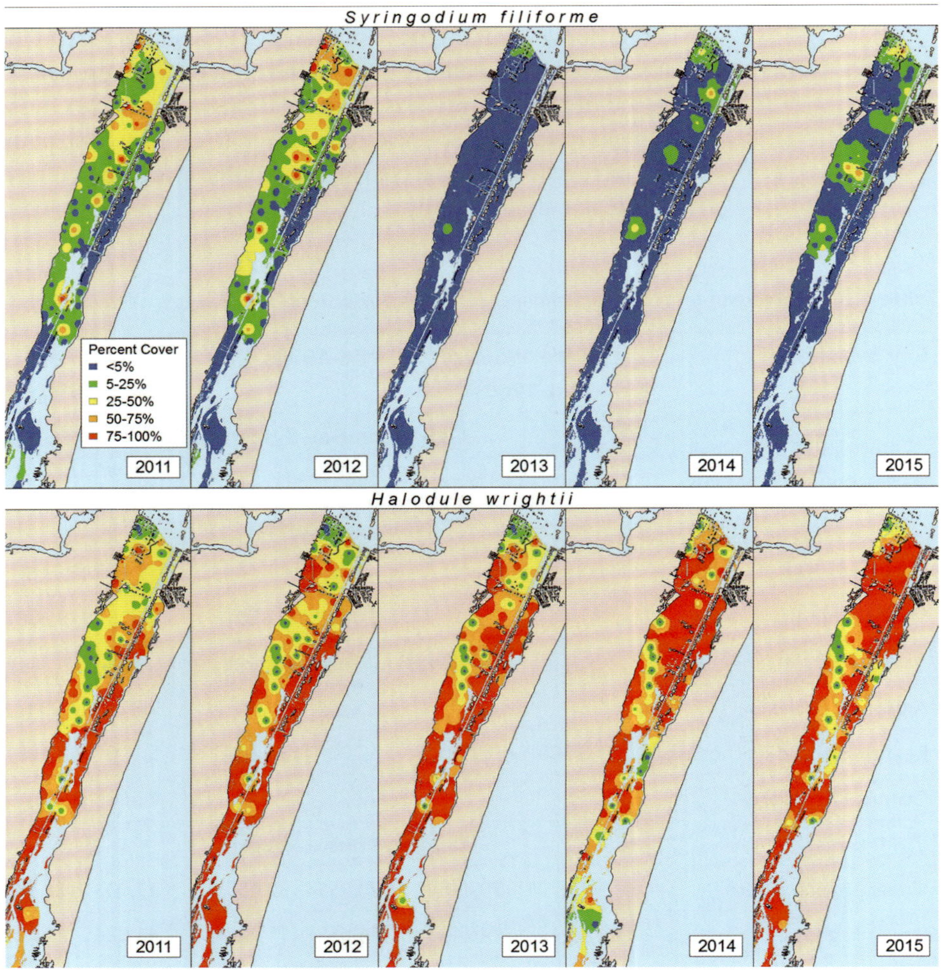

Figure 7.10. *Monthly measurements of Upper Laguna Madre salinity from National Park Service Baffin and Bird Island stations and the UTMSI LM-151 field station from June 2008 through February 2016 (boxplots), superimposed over Manatee Grass cover from 2011 to 2015. The dashed line indicates average salinity of seawater (35). Note low salinities (<20) in late summer 2010 from Hurricane Alex and extended periods of very high (>50) salinity from January to July 2013 at all stations. (Adapted from Wilson and Dunton 2017)*

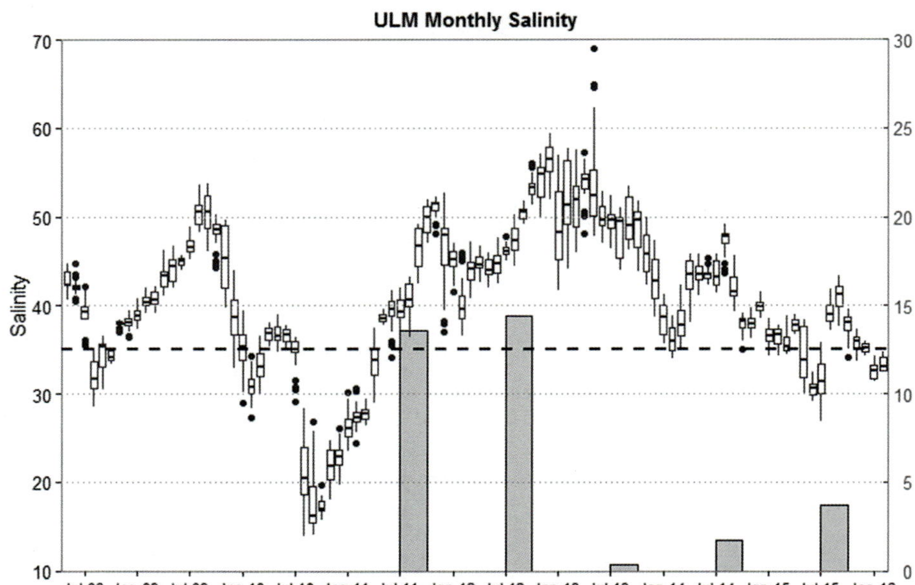

Table 7.4. *Mean and range of key tissue isotopic and stoichiometric parameters for Shoal Grass and Turtlegrass in the Upper and Lower Laguna Madre*

Region	Year	Mean $\delta^{15}N$	Std. Dev. $\delta^{15}N$	Min $\delta^{15}N$	Max $\delta^{15}N$	Mean $\delta^{13}C$	Std. Dev. $\delta^{13}C$	Min $\delta^{13}C$	Max $\delta^{13}C$	Mean NP	Std. Dev. NP	Min NP	Max NP
ULM	2011	1.3	1.4	-2.8	5.3	-11.3	1.5	-14.8	-8.1	44	9.6	19	72
	2012	2.6	1.3	-0.8	4.4	-10.4	1.3	-13.6	-8.2	34	8.9	14	59
	2013	2.1	1.2	-0.9	5.5	-11.2	1.3	-13.6	-8.0	33	8.3	14	54
	2014	1.5	2.0	-5.3	4.0	-11.9	1.4	-15.5	-9.0	39	10.3	12	62
	2015	1.7	2.1	-6.2	5.8	-11.4	1.5	-14.5	-8.4	64	18.3	34	130
LLM	2011	1.4	2.4	-3.8	7.3	-9.6	1.3	-13.1	-6.9	45	10.3	22	74
	2012	3.0	2.2	-1.6	8.1	-9.0	1.2	-12.7	-7.2	33	9.0	18	57
	2013	3.1	2.2	-2.0	8.2	-9.4	1.3	-14.5	-7.2	43	11.2	22	66
	2014	3.4	2.3	-1.8	9.8	-9.9	1.5	-14.2	-7.3	49	14.3	26	80
	2015	2.4	2.9	-4.3	9.2	-9.7	1.4	-13.6	-7.2	56	15.3	22	104

Source: From Cuddy 2018.

Nutrient Inputs and Land-Use Patterns Variations in seagrass isotopic and stoichiometric ratios can be used to compare differences among geographic regions on the Texas coast that are thought to reflect differences in light availability and nutrient loading from either point or nonpoint sources. In particular, land-use patterns can play an important role in regulating the relative proportion of anthropogenic nitrogen that enters bays and lagoons. Variations in $\delta^{15}N$ signatures and stoichiometric ratios between r-selected species (e.g., Shoal Grass) and K-selected species (e.g., Turtlegrass), also can provide short- and long-term information, respectively, on the retention of inorganic nitrogen (table 7.4).

Recently, Cuddy (2018) demonstrated that along the Texas coast $\delta^{13}C$ and $\delta^{15}N$ signatures and N:P ratios in the dominant seagrass species were related to areas of human land usage and development. Because of its responsiveness to environmental perturbations and anthropogenic impacts, Shoal Grass is a useful and effective bioindicator species, although Turtlegrass also responds to large-scale changes in environmental condition and N availability.

Seagrass N:P ratios displayed a general north-to-south gradient from P-limitation (N:P > 30) to N-limitation (N:P < 30) from the Upper to the Lower Laguna Madre (Cuddy 2018). This trend was consistent and agreed with previous work indicating that the Lower Laguna Madre is generally an N-limited system, while the Upper Laguna is P-limited and N replete (Dunton 1996; Herzka and Dunton 1997; Kaldy and Dunton 2000). One possible explanation for the north-to-south variation in N availability is the latitudinal change in sediment types along the Texas Gulf Coast. Upper Laguna Madre sediments are sandy, with porewater ammonium concentrations exceeding 100 μM (Pulich 1985; Dunton 1994). In the Lower Laguna, sediments are terrigenous and comprise fine sands and quartz, characteristics that result in sediment

N deficiencies (Brown et al. 1980; Short 1987). This is reflected in measured mean porewater ammonium concentrations of approximately 32 μM (Kowalski et al. 2009). Additionally, a strong latitudinal gradient in freshwater inflows exists on the Texas coast, with riverine inflows into the Laguna Madre greatly diminished in the south (Longley 1994). Limited opportunities for freshwater input in the Lower Laguna (with the exceptions of the Arroyo Colorado and Brownsville Ship Channel) may be partly responsible for the N-limitation found in seagrasses of this region.

Human land use may also be an important factor in seagrass isotopic and stoichiometric variation along the Texas coast. Cuddy (2018) showed that localized depletions and enrichments in $\delta^{13}C$, $\delta^{15}N$, and N:P ratios in seagrass tissues were significantly correlated with distance from outfalls draining high population areas (fig. 7.11). In particular, the influence of inputs from the Arroyo Colorado and Baffin Bay and cities such McAllen, Edinburg, Aransas Pass, and Kingsville was consistently evident. Seagrass beds growing in close proximity to these urbanized areas were typified by clusters of enriched $\delta^{15}N$ signatures, depleted $\delta^{13}C$ signatures, or elevated N:P ratios that were significantly different from the isotopic and stoichiometric characteristics of seagrass meadows located farther from areas of human influence. This was particularly notable in Shoal Grass but also reflected in Turtlegrass.

These relationships were most distinct in the Lower Laguna Madre, where enriched $\delta^{15}N$ and depleted $\delta^{13}C$ signatures in Shoal Grass were associated with a stream outlet draining the Arroyo Colorado watershed. Nitrogen loading from the Arroyo Colorado was detected by a "plume" of enriched Shoal Grass $\delta^{15}N$ signatures just north of the Arroyo Colorado, which flows through the municipalities of McAllen and Edinburg before entering the Laguna Madre. This nitrogen is then rapidly taken up by the nearshore seagrasses.

A N-inflow model developed by DeYoe et al. (2016) reported high concentrations of nitrate at flow gauges within the McAllen-Edinburg watershed in comparison with

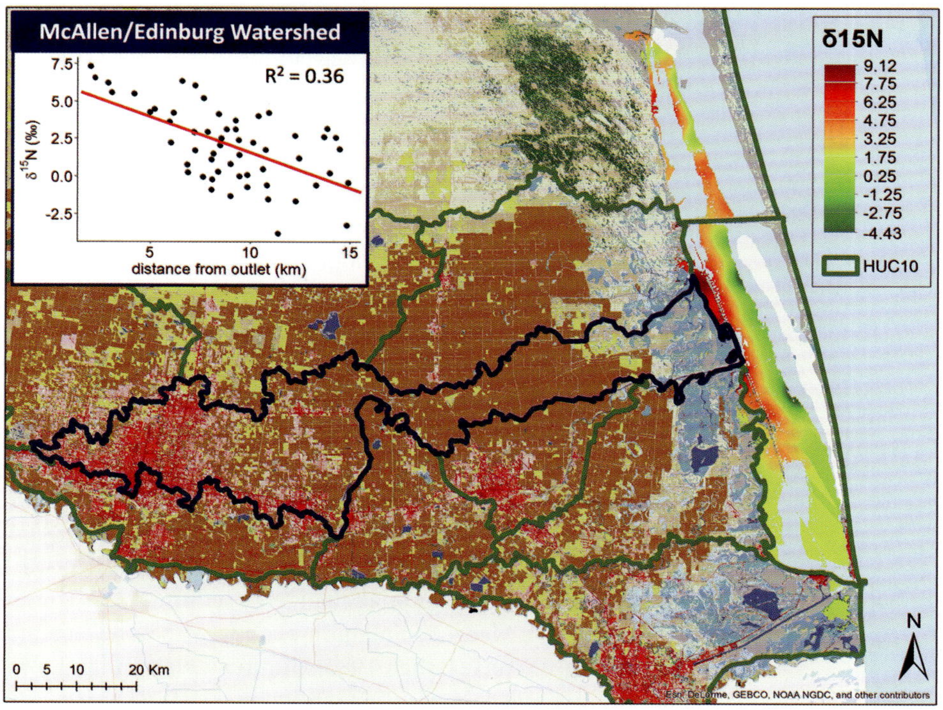

Figure 7.11. *Land use across the lower/middle Texas coast shows location of urbanized land in relation to nitrogen-enriched seagrasses in both the Upper and Lower Laguna Madre based on elevated $\delta^{15}N$ (red), depleted $\delta^{13}C$ (purple), and elevated N:P ratios (brown). (From Cuddy 2018)*

analogous data collected at other South Texas gauge stations. Signatures ($\delta^{15}N$) of the particulate N measured by DeYoe and colleagues at the McAllen-Edinburg flow gauge reflect the $\delta^{15}N$ signatures of Shoal Grass blades collected from the "plume" cluster, indicating that anthropogenic N from the arroyo watershed is likely responsible for enriched $\delta^{15}N$ signal in Shoal Grass. Seagrass N:P ratios at this location, however, while indicative of an N-replete system, were not significantly higher than the N:P ratios in seagrasses of surrounding beds. Therefore, while human-derived N appears to be entering the Laguna Madre here, the excess N delivered to the system is likely undergoing rapid microbial processing. These studies demonstrate the importance of long-term monitoring to provide a baseline for assessing change and understanding relationships between seagrass condition and anthropogenic inputs. This information can be used to develop water-quality criteria that can improve the protection of the critical seagrass habitats of the Laguna Madre.

Effects of Cover and Community Composition on Fish and Waterfowl Although seagrasses are not the only habitats used by larval and juvenile fishes, many species are strongly associated with seagrass meadows, and some exhibit preferences for Shoal Grass in the Lower Laguna Madre (Tolan et al. 1997). Widespread replacement of Shoal Grass by Manatee Grass suggests there might be a concomitant decrease in preferred nursery habitat for juvenile fishes. The effect of Manatee Grass replacement by Turtlegrass as succession proceeds in the Laguna Madre is unknown, but Turtlegrass meadows in other areas typically harbor higher densities of fish than do Manatee Grass meadows (Martin and Cooper 1981). In the Upper Laguna Madre, loss of cover due to Texas Brown Tide may have also impacted available nursery habitat for juvenile fishes. However, succession appears to be proceeding slowly and may be reversed by drought conditions, which favor Shoal Grass (Wilson and Dunton 2017). Consequently, the loss of preferred nursery habitat (Tolan et al. 1997) in the Lower Laguna Madre may not occur in the Upper Laguna Madre, especially under the projected climate-change scenarios for Southwest Texas.

The Laguna Madre of Texas and Tamaulipas is the primary wintering area for Redheads, which feed almost exclusively on Shoal Grass. About 75% of the world's population of Redheads winter in these lagoons (Weller 1964). Other traditional wintering areas such as Chesapeake Bay, Pamlico Sound, and Galveston Bay have been abandoned because seagrasses declined or disappeared, leaving the Upper Laguna Madre and Laguna Madre de Tamaulipas to support nearly the entire wintering population. Degradation and loss of Shoal Grass meadows are of particular concern in the Texas Laguna Madre (e.g., Onuf 1995), especially if it continues or escalates, because that would leave the Laguna Madre de Tamaulipas as the only area with suitable habitat. As of 1995, most seagrass losses were confined to the deepest waters (>1.4 m [4.6 ft]), which are used infrequently by Redheads because they typically dabble in shallow water. In addition, deeper water may be avoided because many of these areas are near navigational channels. Redheads choose habitats based not only on food availability but also on disturbance and proximity to the freshwater ponds they require for drinking water. However, losses of seagrasses in shallow water represent a potentially serious impact.

Scarring Recreational boating is another potential source of negative impacts to seagrass habitats (fig. 7.12). Propellers on boat motors can destroy above- and below-ground seagrass tissues (R. Phillips 1960; Zieman 1976; Eleuterius 1987; Dunton, Montagna, et al. 1998), which results in scarring. The scar itself is not the end of the

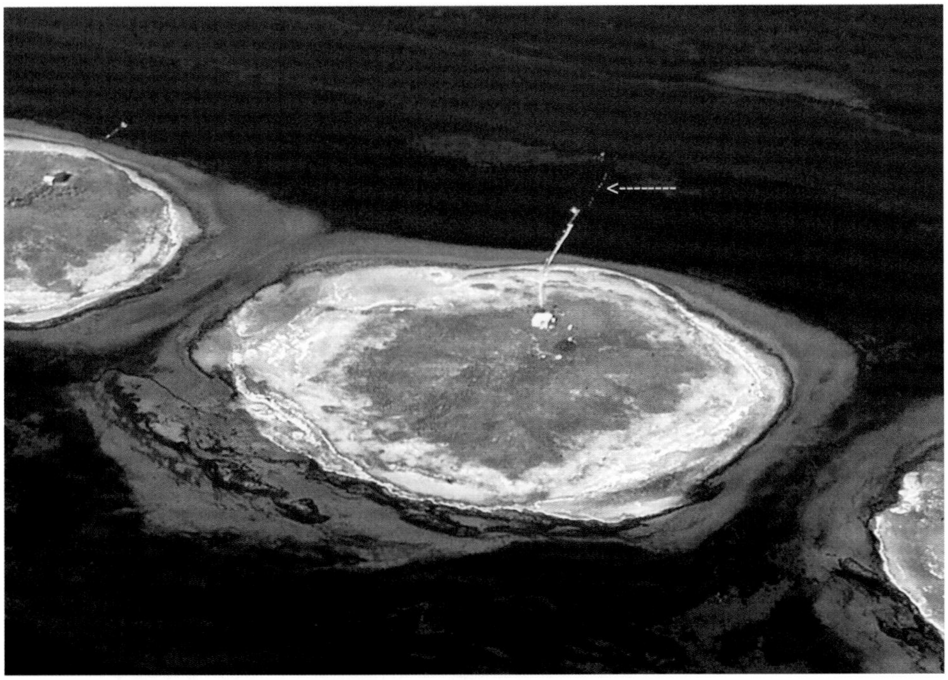

Figure 7.12. *Propeller scarring of a seagrass meadow between the GIWW and a fishing pier on a dredged-material island in the Laguna Madre is clearly visible in this aerial view (see arrow). (Photograph by Brian R. Chapman)*

damage. Erosion of the scar by wave action causes the disturbed area to deepen, and scouring may widen the scar, resulting in additional losses along the scar margins (Zieman 1982; Eleuterius 1987). Turbidity may also increase, resulting in light attenuation and inhibition of seagrass growth and recovery in the vicinity of the scar (Zieman 1982). Most scarring occurs in areas less than 1.0 m (3.3 ft) deep at low tide (Zieman 1976). Scarring lasts longer in Turtlegrass meadows and is more likely to become permanent since Turtlegrass is a slower-growing, perennial species that does not recover as quickly as Shoal Grass, a colonizing species. Scarring may be an important factor in seagrass loss in densely populated areas and is contributing to seagrass loss in shallow areas of the Redfish Bay/Harbor Island complex just north of Corpus Christi (Pulich et al. 1997).

Dunton, Burd, et al. (1998b) determined extent of seagrass scarring in the Upper Laguna Madre around the JFK Causeway and north to the mouth of Corpus Christi Bay. This area is heavily used by recreational boaters because there are few public boat launches between the causeway and Port Mansfield. Excluding the GIWW and other channels, water depth ranged from 0.3 to 0.5 m (1.0 to 1.6 ft) and seagrass cover totaled about 2,850 ha (7,042 ac). Dunton, Montagna, et al. (1998) categorized seagrass scarring based on the density of scars (% seagrasses impacted: light, <5%; moderate, 5%–20%; and heavy, >20%). Approximately 34.6%, or 493 ha (1,218 ac) of the area examined was propeller-scarred, with the majority of scarring in the moderate category. Although anglers heavily use seagrass beds, all recreational boaters bear responsibility to some degree. Four explanations accounted for the moderate to severe scarring observed in the area: (1) proximity of seagrass meadows to densely populated areas, including waterfront homes; (2) shortcuts taken at channel junctions, around shallow areas, and between islands, as well as accidental straying out of channels; (3) entry to shallow meadows from blind channels dredged for gas well or pipeline access; and (4) channels that are illegally marked and maintained through frequent and intensive boat traffic directly through meadows.

Laguna Madre de Tamaulipas

Seagrass meadows (*pastos marinos*) cover about 340 km^2 (131 mi^2) in the Laguna Madre de Tamaulipas (DUMAC 1996; fig. 7.13). Another 82 km^2 (32 mi^2) are vegetated by attached algae. The ecology of the Laguna Madre de Tamaulipas, including the seagrass meadows, is thoroughly described in chapter 12. Since the first edition of this book was published (2002), the invertebrate communities associated with seagrass meadows in Laguna Madre de Tamaulipas have been well documented (see tables 12.2–12.6). The seagrasses and invertebrate resources normally attract several hundred thousand waterfowl, primarily Redheads and Northern Pintails, to the Laguna Madre de Tamaulipas during the winter months (see chapter 13).

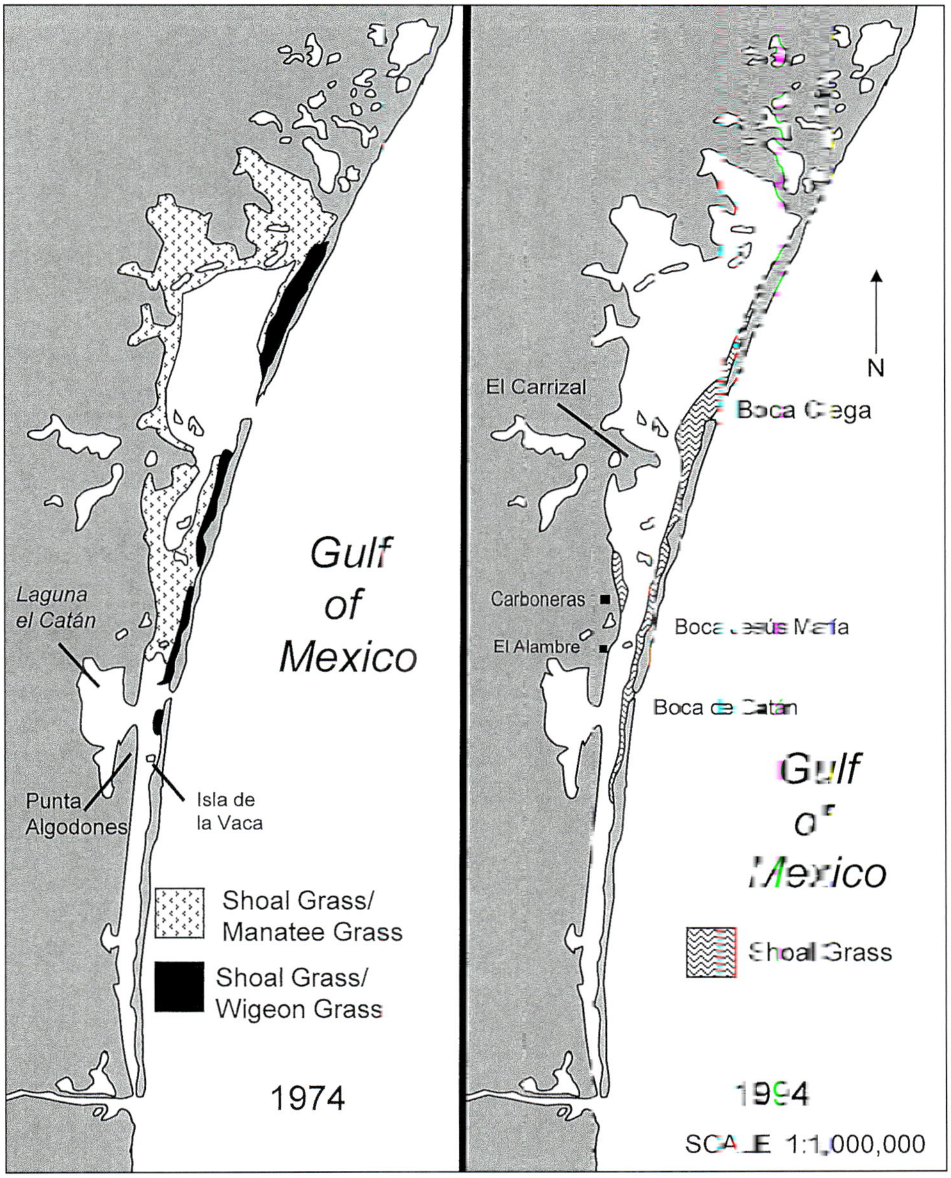

Figure 7.13. *Distribution of seagrasses in the Laguna Madre de Tamaulipas in 1973–74 (redrawn from Cornelius 1975) and 1994–95 (redrawn from DUMAC 1996). Only Shoal Grass was mapped by DUMAC in 1994–95.*

8 Open Bay

JENNIFER BESERES POLLACK,
NATASHA BREAUX, AND KIM WITHERS

Open bays are one of the most extensive and productive habitats in estuarine systems (Kritzer et al. 2016). These unvegetated subtidal areas are often undervalued in coastal assessments because their productivity is primarily driven by cryptic infauna and microbial processes, making them difficult to quantify (Thrush et al. 2001). However, productivity in open bays influences the overlying waters and adjacent habitats such as seagrasses, tidal flats, marshes, tidal inlets, and riverine systems (Armstrong 1987) and plays a key role in maintaining marine biodiversity (Thrush et al. 2001; Kritzer et al. 2016).

Laguna Madre of Texas

Extent, Distribution, and Trends The extent of unvegetated bay bottom in the Laguna Madre varies relative to the extent of seagrass cover. The majority of open-bay habitat in the Upper Laguna Madre is found in Baffin Bay (≈198 km^2 [76 mi^2]), the deeper waters of the Gulf Intracoastal Waterway (GIWW), and the fairly turbid waters of The Hole (see fig. 7.3). Water clarity and salinity are primary factors controlling seagrass distribution in South Texas (Pulich 1998), and these fluctuate with freshwater inflow and other dynamic processes. Open-bay habitats likely dominated the Upper Laguna Madre prior to completion of the GIWW, which resulted in long-term salinity decline through increased connectivity (Quammen and Onuf 1993). Open-bay habitat decreased dramatically from 1967 to 1988 as seagrass cover, primarily Shoal Grass, increased by 110% after dredging of the GIWW and the resultant freshening of the system (Quammen and Onuf 1993). Between 1988 and 1994, unvegetated bottom increased, likely due to light attenuation from a prolonged brown tide bloom (Onuf 1996b).

In 1998, unvegetated habitats in the Laguna Madre covered approximately 357 km^2 (138 mi^2), with approximately 130 km^2 (50 mi^2) in the Upper Laguna Madre and 227 km^2 (88 mi^2) in the Lower Laguna Madre (Onuf 2007). Finkbeiner et al. (2009) utilized 1-meter resolution digital multispectral imagery to map Laguna Madre benthic habitats in 2009. Unconsolidated sediments composed approximately 140 km^2 (54 mi^2) in the Upper Laguna Madre, 867 km^2 (335 mi^2) in the Lower Laguna Madre, and 51 km^2 (20 mi^2) in adjacent Baffin Bay. In 2015, bare substrate covered 27.3% of the Upper Laguna Madre (Congdon and Dunton 2016).

Open-bay habitats likely also dominated the Lower Laguna Madre before the GIWW was completed. Salinities were less extreme than in the Upper Laguna Madre due to tidal exchange with the Gulf of Mexico via Brazos Santiago Pass and freshwater inflow via Arroyo Colorado, making it likely that seagrass cover was more widespread and persistent. In 1965, seagrasses, primarily Shoal Grass, dominated the entire lower lagoon, but by 1988, unvegetated areas had expanded by 95% (Quammen and Onuf

1993; see fig. 7.3). Unvegetated areas were found primarily in the deepest portions of the Lower Laguna Madre and were aligned with the GIWW during 1974–75 surveys. Unvegetated habitat accounted for 47.9% of the lower lagoon in a 2015 survey (Congdon and Dunton 2016). Expansion of unvegetated areas in the Lower Laguna Madre has been attributed to turbidity caused by maintenance dredging of the GIWW, periodic hypersalinity, and siltation/water-quality degradation from inflow sources (Merkord 1978; Onuf 1994; Dunton, Burd, et al. 1998; Sheridan 2004; Congdon and Dunton 2016).

Plant Communities

Brush (2005) described the shallow-water and coastal plant communities of the Laguna Madre. Seagrass was reported to cover a large portion of the shallow Laguna Madre, particularly Turtlegrass, Shoal Grass, and Manatee Grass. Coastal salt marshes were uncommon but, where present, are dominated by Smooth Cordgrass and Saltmeadow Cordgrass. Temporary wetlands occur as a function of rainfall, wind, and tidal patterns and contain salt-tolerant species such as Saltwort, Sea Oxeye, and Saltgrass. Trees were uncommon except where planted and watered.

Phytoplankton are the dominant plants in the open bay (fig. 8.1). Phytoplankton biomass in Baffin Bay frequently exceeds state "screening levels" for impairment, has

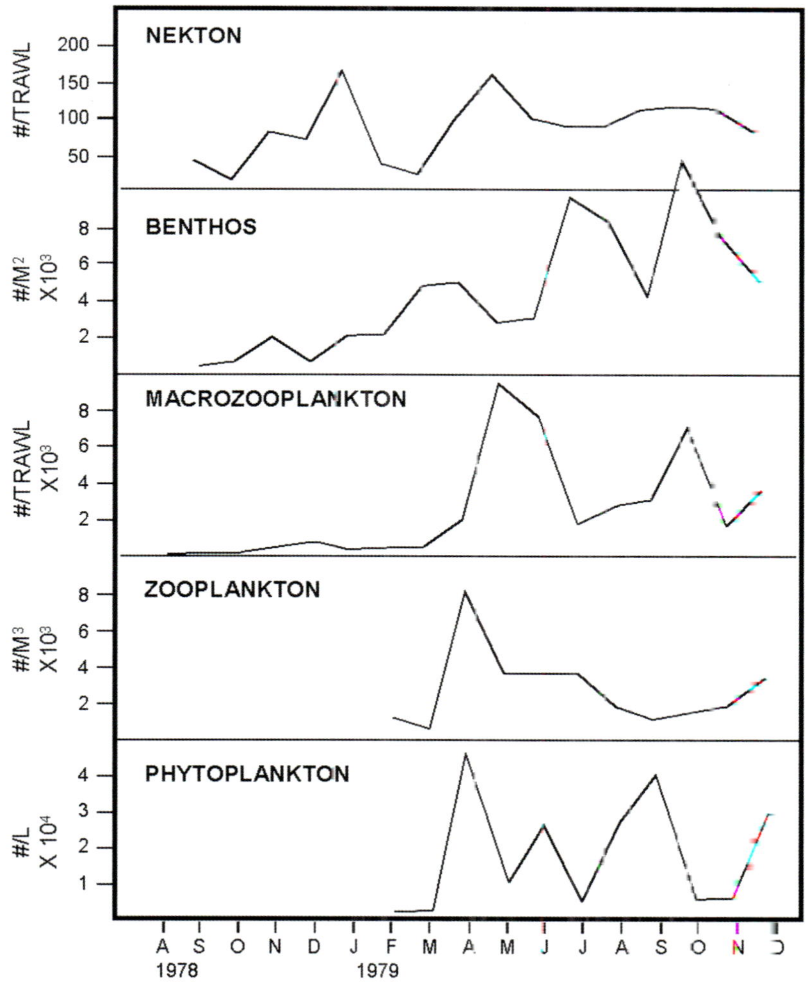

Figure 8.1. *Monthly mean abundance of organisms collected in Alazan Bay (Cornelius 1984)*

increased since the 1970s, and is now generating concern about the potential role of nutrient inputs from the surrounding watershed (Montagna and Palmer 2012; Wetz et al. 2017). Perhaps most recognized is the phenomenon of prolonged, dense blooms of the Texas Brown Tide phytoplankter that have occurred since 1990 (including a continuous >7-year bloom), likely facilitated by the typically hypersaline conditions in the range of 40–60 ppt (Buskey et al. 1998; Buskey et al. 2001). Much of the increase in chlorophyll in Baffin Bay since the 1990s has coincided with blooms of Texas Brown Tide (Wetz et al. 2017). Dominance of the brown tide organism over other phytoplankton species may be facilitated by prevalence of high concentrations of reduced nitrogen such as ammonium and dissolved organic nitrogen (Gobler et al. 2013). The ability of the Texas Brown Tide phytoplankter to use forms of phosphorus other than phosphate under severe phosphate deficiency may facilitate its success in phosphorus-limited environments, such as the Laguna Madre (Liu et al. 2001).

The brown tide organism is one of several phytoplankton species of concern in the Laguna Madre ecosystem. Texas Brown Tide can have negative effects on aquatic life, including fishes and shellfish (Thronson and Quigg 2008). Meroplanktonic larvae may also be affected. Larvae of *Streblospio benedicti*, a polychaete, experienced reduced growth rates and swimming speeds when raised in cultures with Texas Brown Tide, indicating harmful sublethal effects on this dominant species (Ward et al. 2000). A bloom of Red Tide caused by a toxic dinoflagellate led to a fish kill in early 2000 (Denton and Contreras 2004). In late 2010, another large fish kill occurred coincident with a bloom of the dinoflagellate *Pyrodinium bahamense* and the diatom *Thalassiothrix* sp. and low dissolved oxygen (Texas Parks and Wildlife, unpub. reports). Additional research is warranted to determine whether these algal blooms are symptoms of large-scale water-quality degradation or are representative of natural variability.

Phytoplankton communities in the Laguna Madre system are usually dominated by diatoms with secondary contributions by dinoflagellates such as *Ceratium furca*. Abundant diatom genera included *Chaetoceros*, *Nitzschia*, *Rhizosolenia*, *Thalassionema*, and *Thalassiothrix* in the years prior to the Texas Brown Tide (Breuer 1962; EHA 1977; Hildebrand and King 1978; Cornelius 1984). These species are still present, but since the late 1990s there have been more frequent brown tide blooms, which reduce abundance and diversity of other phytoplankton species.

Phytoplankton dynamics in the Laguna Madre system are strongly influenced by salinity. Observations during 2013–16 in Baffin Bay indicate that during droughts, densities of the Texas Brown Tide organism were high and the phytoplankter dominated the phytoplankton community; during wet periods, however, phytoplankton diversity increased and the Texas Brown Tide organism no longer dominated (fig. 8.2; Cira and Wetz, unpub. data). Blooms of diatoms (*Rhizosolenia* spp.) and *Synechococcus*-like cyanobacteria have also been observed in the Laguna Madre during reduced salinity periods with low brown tide densities (Buskey et al. 2001). Prior to the advent of frequent brown tide events in the Laguna Madre system, phytoplankton densities were generally highest at low-moderate salinities (fig. 8.3; Simmons 1957; Breuer 1962; EHA 1977; Hildebrand and King 1978; Cornelius 1984).

Ferguson-Wood (1963) described benthic diatoms in the Laguna Madre and Baffin Bay. Some 102 species were collected in the Laguna Madre, where *Amphora* spp. were frequently dominant; *Rhodpalodia* spp. were abundant; and species of *Achnanthes*, *Actinoptychus*, and *Grammatophora* were common. Twenty-nine species and a few unidentified individuals were collected in Baffin Bay. Near the mouth of the bay the community resembled that of the Upper Laguna Madre. Eighty-two species were collected in the Lower Laguna Madre, where *Cocconeis* spp. were dominant, *Rhodpalodia*

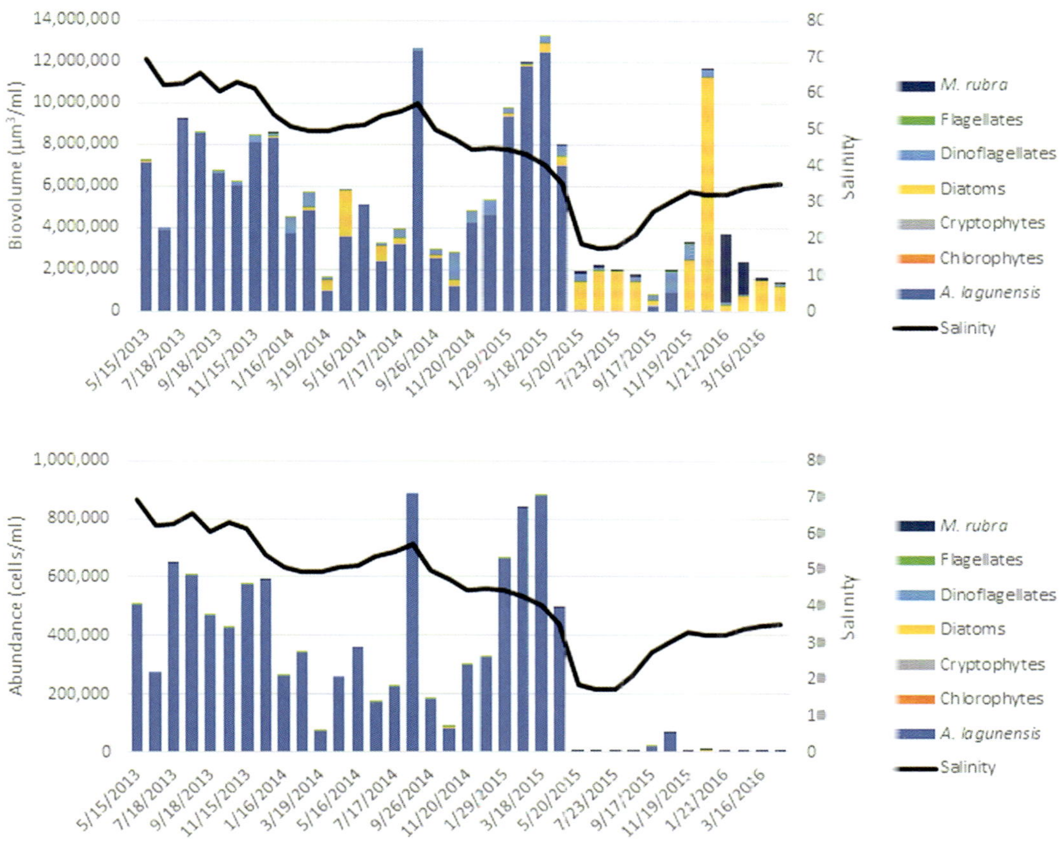

Figure 8.2. *Abundance/biovolume of the Texas Brown Tide phytoplankter, along with other phytoplankton groups (flagellates, dinoflagellates, diatoms, cryptophytes) and the phototrophic ciliate* Myrionecta rubra, *throughout a three-year sampling period (May 2013–April 2016) at a site in the main channel of Baffin Bay. Salinity is also shown. (E. Cira and M. Wetz, unpub. data 2016)*

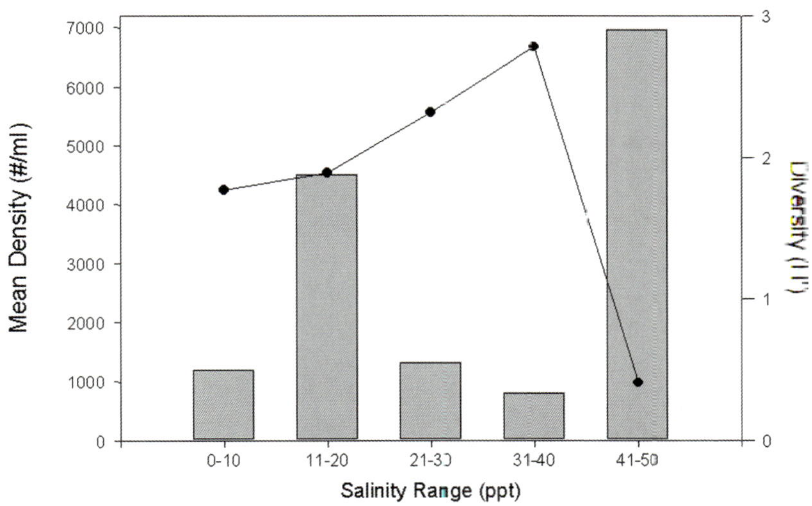

Figure 8.3. *Mean density (bars) and diversity (line) of phytoplankton in Alazan Bay in relation to salinity (Data from Cornelius 1984)*

spp. were abundant, and *Grammatophora* spp. were common. Cell counts were generally highest in sediments where the water depth was <1.0 m (3.3 ft) and lowest in water depths of 2.0 m (6.6 ft) or more (Oppenheimer and Wood 1965).

Drift macroalgae in the Upper Laguna Madre are dominated by red algae, primarily *Gracilaria* spp. and *Laurencia* spp. (Hildebrand and King 1978). In the Lower Laguna Madre, red algae are also common as well as large fragments of the brown macroalgae *Sargassum* spp., which enters the lagoon through passes during spring and drifts throughout the system (Breuer 1962). Benthic microalgae are found on and within bottom sediments.

Animal Communities

Zooplankton The dominant mesozooplankter in the Laguna Madre was the translucent marine copepod *Acartia tonsa* (Simmons 1957; Hedgepeth 1967; Hildebrand and King 1978; Cornelius 1984; Buskey et al. 2001). Other important calanoid copepods included *Pseudodiaptomus* spp. and *Centropages* spp. (Hildebrand and King 1978); cyclopoid copepods, primarily *Oithona* spp. and *Saphirella* spp., were much less abundant. Microzooplankton populations were dominated by ciliates, followed by dinoflagellates (Buskey et al. 2001). Neither meso- nor microzooplankton have demonstrated efficient grazer control of the Texas Brown Tide phytoplankter (Jakobsen et al. 2001; Buskey et al. 2003), but protozoan grazers not affected by phytoplankton toxins may potentially control bloom initiation (Buskey 2008). Meroplankton communities were dominated by trochophore larvae, bivalve larvae, gastropod veligers, and barnacle nauplii (Hildebrand and King 1978; Cornelius 1984). The macrozooplankton/micronekton community was dominated by larval and juvenile fish and mysids, primarily *Americamysis* spp. (Cornelius 1984). Mean zooplankton abundance and diversity were highest at low-moderate salinities (fig. 8.4; Cornelius 1984).

Benthic and Epibenthic Invertebrates Benthic communities can serve as biological indicators in estuaries because the organisms are relatively immobile and typically spend their entire life cycles within estuarine systems. These characteristics, along

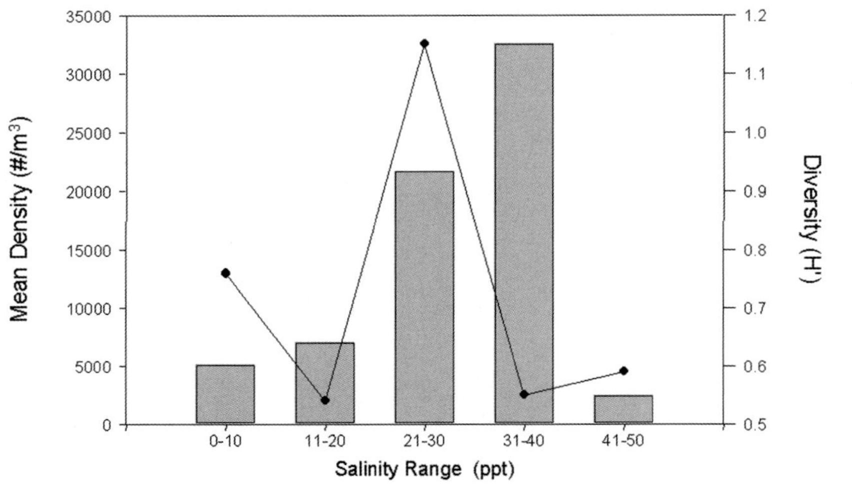

Figure 8.4. *Density (bars) and diversity (line) of zooplankton in relation to salinity in Alazan Bay (Data from Cornelius 1984)*

with longer lifespans than those of planktonic species of similar size, make benthic macrofauna useful in determining spatial and temporal trends within their habitats. Because these communities make up the basis of estuarine food webs, disturbance events that hinder the benthos may limit food resources for higher-trophic-level organisms, leaving the ecosystem vulnerable to future disturbance (Flint et al. 1982; Montagna et al. 2013).

Benthic invertebrate communities of open-bay habitats play an important role in recycling and regenerating nutrients from the sediments back into the water column for use by phytoplankton (Zeitzschel 1980). They serve as an important trophic link, converting primary productivity into available biomass for use by higher consumers. Benthic invertebrates are important primary consumers in the open-bay habitat and regulate phytoplankton standing crop in shallow water. The activities of benthic organisms appear to regulate flow of nutrients from sediments into overlying waters (Rhoads 1974; Aller 1978) and may also modify the biological, physical, and chemical attributes of sediments.

In the Laguna Madre system, benthic communities of unvegetated areas are typically dominated by molluscs and/or polychaetes in relatively high abundance but low diversity (Parker 1959; White et al. 1989; Montagna 1993). Abundant molluscs include the Dwarf Surf Clam (fig. 8.5), tellins (e.g., the Tampa Tellin and the Texas Tellin), and the Pointed Venus (fig. 8.6; Parker 1959; White et al. 1989; Montagna 1993). Dwarf Surf Clams in particular tend to have a disproportionate effect on biomass in Baffin Bay and other systems along the Texas coast (Breuer 1957; Dalrymple 1964; Flint and Younk 1983; Cornelius 1984; White et al. 1989; Montagna 1993; Montagna and Kalke 1995; Rubio et al. 2018). Because of their ability to colonize and grow rapidly and survive a range of salinity conditions (Breuer 1957; Calabrese 1969; Montagna et al. 1993), Dwarf Surf Clams have been identified as an indicator species for inflow effects (Montagna and Kalke 1995; Pollack et al. 2009). Macrobenthic biomass displays year-to-year variability in Baffin Bay but is typically far below that of the Laguna Madre (fig. 8.7; Kim and Montagna 2012).

Figure 8.5. *Dwarf Surf Clam, the dominant bivalve species in open-bay habitat of the Upper Laguna Madre and Baffin Bay (Photographs from Tunnell et al. 2010)*

Figure 8.6. *Pointed Venus, a common bivalve species in open-bay habitat of the Upper Laguna Madre and Baffin Bay (Photographs by J. W. Tunnell Jr. from Tunnell et al. 2010)*

Figure 8.7. *Comparison of mean density (A), biomass (B), and N1 diversity (C) of benthic macrofauna in open-bay habitat in Baffin Bay and seagrass habitat in the Upper Laguna Madre. Baffin Bay was surveyed from March 2014 to July 2017. The Laguna Madre was added to the study in June 2015. (Adapted from Mendenhall 2015 and Breaux 2017)*

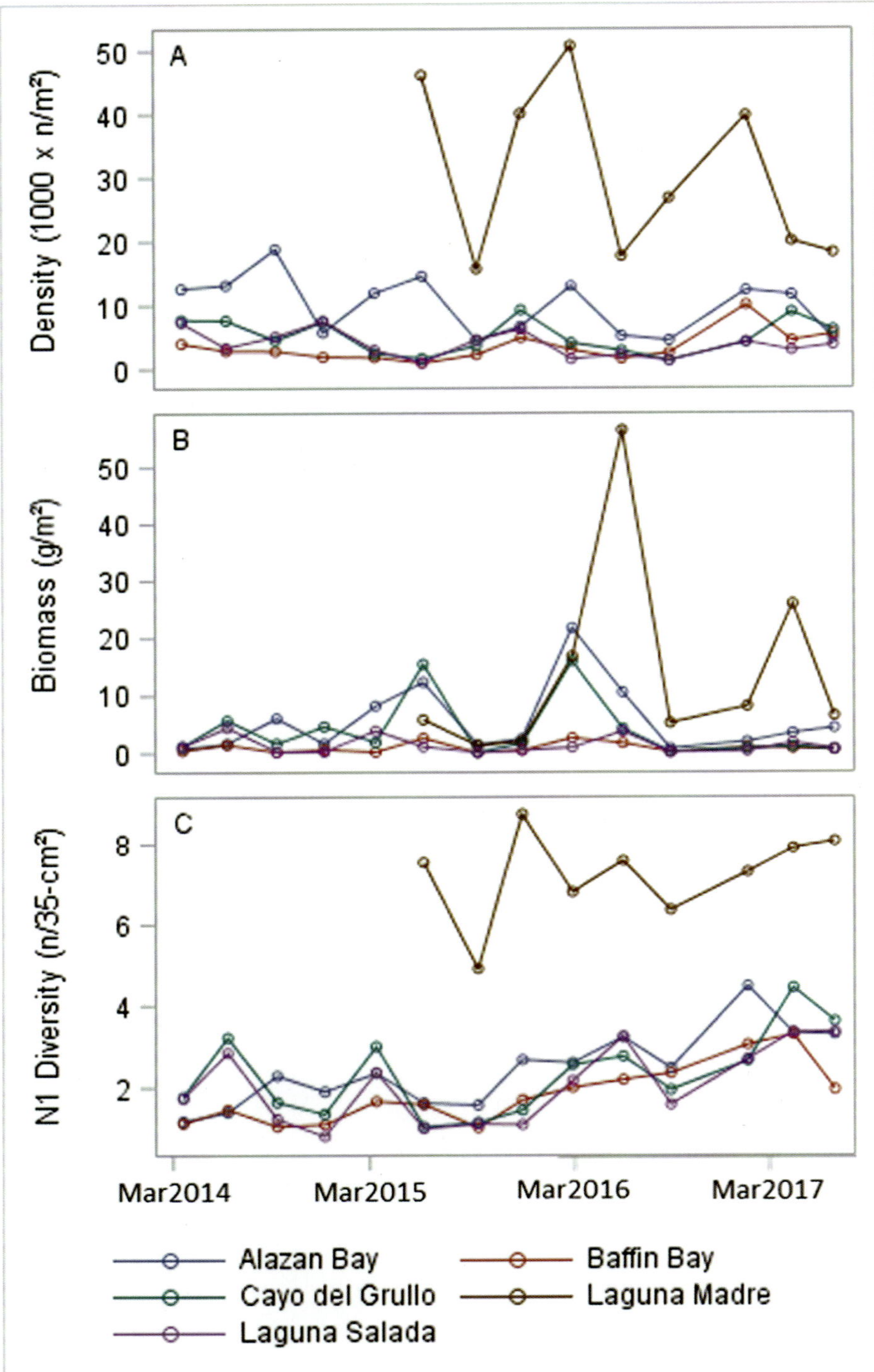

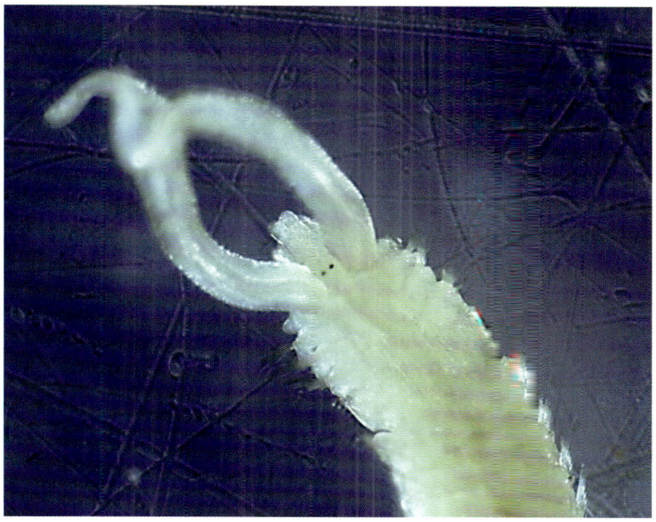

Figure 8.8. *Polychaete from the family Spionidae, which is often indicative of disturbance and found in high abundances, particularly in soft sediments of Baffin Bay*

White et al. (1986, 1989) recognized different assemblages in unvegetated areas of the Laguna Madre: the depauperate bay margin, the more diverse bay-center assemblage, and the inlet-influenced areas (Port Mansfield) in the Lower Laguna Madre. The bay margin is dominated by polychaetes, primarily capitellids and the Tampa Tellin and the Pointed Venus. In deeper, muddier areas of the lagoon, the bay-center assemblage was dominated by spionid (e.g., *Streblospio benedicti*; fig. 8.8) and capitellid polychaetes (e.g., *Mediomastus* spp.); bivalve species including the Dwarf Surf Clam, Pointed Nutclam, Texas Tellin, Glassy Lyonsia, and Minor Jackknife Clam; gastropods, such as the Beautiful Caecum and Channel Barrel-bubble; and the amphipods *Ampelisca abdita* and *Grandidierella bonnieroides*. Inlet-influenced areas are dominated by bivalves (e.g., the Common Atlantic Abra and Florida Cross-barred Venus); a gastropod, the Greedy Dovesnail; and polychaetes, such as *Armandia maculata*, *Cossura delta*, *Sigambra tentaculata*, and *M. californiensis*.

A unique, unvegetated, hard substrate also exists in Baffin Bay. Serpulid reefs, composed of the calcareous tubes of serpulid polychaetes, are found scattered throughout the bay (fig. 8.9; see chapters 3 and 6). However, these reefs currently contain few reef-building serpulid worms. Crustaceans, polychaetes, and molluscs are the most abundant organisms associated with this habitat (Mackin 1971; Martin 1979; Cole 1981; White et al. 1989; Hardegree 1997). Both Mackin (1971) and Hardegree (1997) described polychaete/amphipod assemblages dominated by amphipods in the genus *Corophium*. Organisms found at all sites in Cole's (1981) study of Black Drum feeding habits were nemerteans, Scorched Mussels, Acorn Barnacles, and the polychaetes *Diopatra cuprea* (fig. 8.10), *Polydora* sp., *Mediomastus californiensis*, *Pectinaria gouldii* (fig. 8.11), *Marphysa sanguinea*, and *Hydroides dianthus* (fig. 8.12). Hardegree (1997) also found the polychaete species *Dodecaceria concharum*, *Dorvillea rubra*, and *Naineris laevigata* and members of the families Terebellidae (fig. 8.13) and Syllidae. The serpulid reef assemblage described by White et al. (1989) was dominated by Scorched Mussels and the polychaete *Nereis succinea*. Other abundant species in this assemblage were the Dwarf Surf Clam, Glassy Lyonsia, and an isopod, *Edotea montosa*.

Living oyster reefs are uncommon in the Texas Laguna Madre ecosystem and are generally limited to small populations in South Bay, in Arroyo Colorado, and near Port Isabel. Reefs are composed of Eastern Oysters with a genetically distinct population structure that may reflect adaptation to high temperatures and salinities (King et al.

Figure 8.9. *Close-up of a piece of a serpulid reef benthic community from Baffin Bay showing species of amphipods, polychaetes, and bivalves*

Figure 8.10. Diopatra cuprea, *a polychaete species found on a serpulid reef in Baffin Bay*

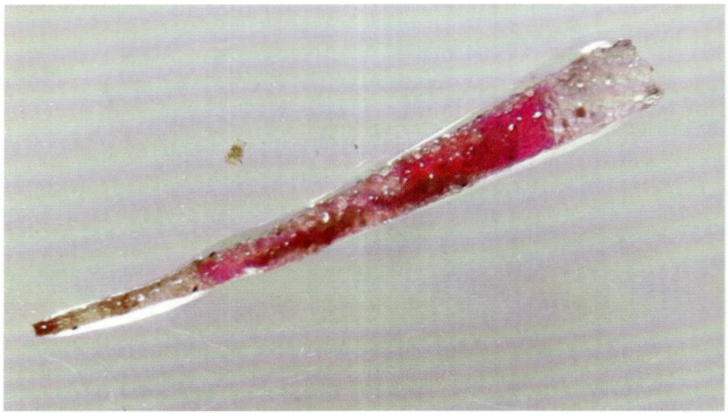

Figure 8.11. Pectinaria gouldii, *sometimes called the "ice cream cone worm," lives inside a tube built of sand grains. This polychaete species is found in Baffin Bay.*

Figure 8.12. *A serpulid worm,* Hydroides dianthus, *collected from a serpulid reef in Baffin Bay*

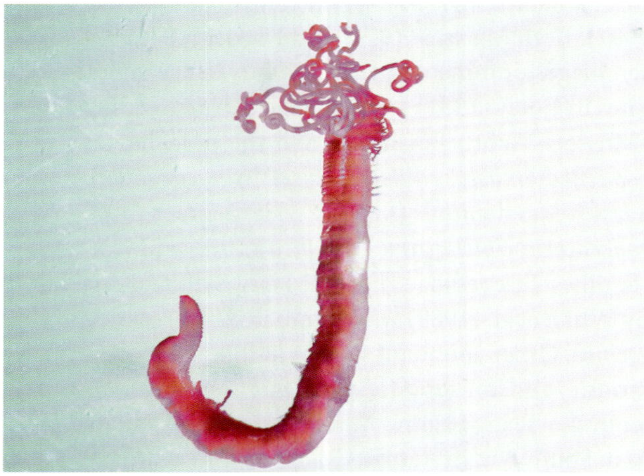

Figure 8.13. *A polychaete worm in the family Terebellidae collected from a serpulid reef in Baffin Bay*

1994; Varney et al. 2009). Other molluscs associated with this habitat are the Scorched Mussel and the Impressed Odostome (White et al. 1986). Characteristic polychaetes include *Tharynx marioni*, *Hydroides dianthus*, and *Prionospio heterobranchia*.

Nekton Since an entire chapter is devoted to fish (chapter 17), only invertebrate nekton are discussed here. Generally speaking, the most abundant invertebrate nekton in the Laguna Madre are shrimp and crabs. Sheridan and Minello (2003) examined nekton use of seagrass beds in the Lower Laguna Madre. Two of the most numerous shrimp species were the economically important Brown Shrimp and Pink Shrimp, with the former more numerous in spring and the latter more numerous during fall. Other conspicuous shrimp species were the seagrass canopy residents, the Zostera Shrimp and Arrow Shrimp. The Lesser Blue Crab and the commercially valuable Blue Crab were periodically abundant, and other dominant species included the White Spotwrist Hermit Crab and the Texas Mud Crab.

In 2005, the US Army Corps of Engineers permanently reopened Packery Channel, a historic tidal inlet, in an effort to restore flow between the Gulf of Mexico and the Upper Laguna Madre. Reese et al. (2008) found significantly higher densities and smaller sizes of estuarine-dependent Blue Crabs and penaeid shrimp post-opening. Because these species have similar life histories—adults spawn in the Gulf of Mexico near tidal inlets, and their eggs, larvae, and juveniles travel through tidal inlets into estuarine nursery habitats—increased access to Laguna Madre via Packery Channel may lead to higher fisheries productivity for these species.

There is little information concerning invertebrate nekton use of unvegetated areas of the Lower Laguna Madre. So much bottom vegetation is present that it is difficult to determine which species are associated with unvegetated areas. Cornelius (1984) identified 14 nektonic invertebrate taxa in Alazan Bay, adjacent to Baffin Bay and the Upper Laguna Madre. Three invertebrate species, Brown Shrimp, Texas Mud Crab, and Atlantic Brief Squid, constituted about 1% of the total nekton collected. Blue Crabs constituted 38% of biomass and Brown Shrimp and Bay Squid together made up <1%. Abundance of all nekton peaked in January and May (see fig. 8.11). Abundance and diversity of invertebrate nekton were greatest at low-moderate salinities (fig. 8.14).

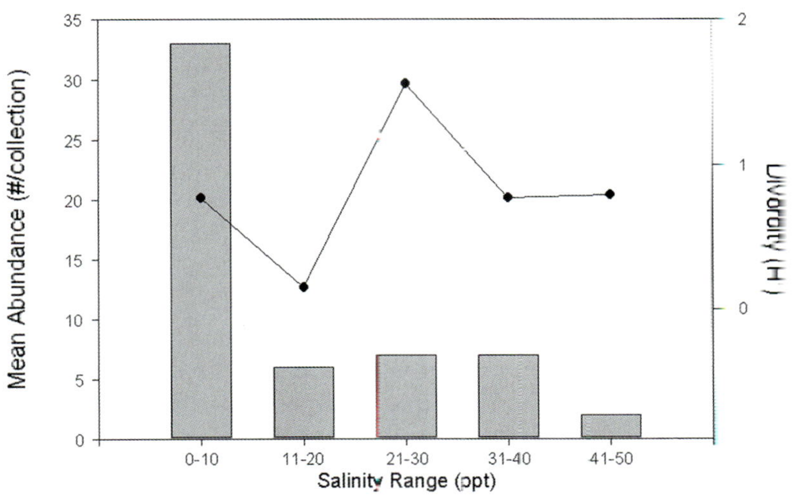

Figure 8.14. *Abundance (bars) and diversity (line) of invertebrate nekton in relation to salinity in Alazan Bay (Data from Cornelius 1984)*

Stokes (1974) reported that in the Lower Laguna Madre White Shrimp were most commonly collected in areas with little or no vegetation; Brown Shrimp abundance was greatest in areas devoid of vegetation; and Pink Shrimp were collected from channels, but otherwise only in vegetated areas.

Birds The Lower Rio Grande Valley, including the Laguna Madre, is one of the foremost birding areas in the United States. Brush (2005) described a number of bird phenomena that occur along the Laguna Madre coastline. High-density, mixed-species feeding aggregations occur along seasonally wind-exposed intertidal flats in wintertime, including Piping Plovers, Snowy Plovers, Long-billed Dowitchers, Black-bellied Plovers, and Western Sandpipers. The highest densities of wintering Redheads in North America are found in the Laguna Madre (Woodin 1996), likely due in part to the relatively stable food supply provided by Shoal Grass rhizomes (Michot et al. 2006; Ballard et al. 2010). Large flocks of Northern Pintails also winter in the Lower Laguna Madre and feed on the ample seagrasses (Brush 2005). Double-crested Cormorants are common winter residents in the Upper Laguna Madre, consuming fish prey, including Striped Mullet and Gulf Toadfish (Withers and Brooks 2004). American White Pelicans and at least six tern species also winter in large numbers (Brush 2005), but a small, resident population of American White Pelicans is present in the Texas and Tamaulipas lagoons throughout the year (B. Chapman 1988). Other fish-eating birds such as Reddish Egrets and Brown Pelicans occur year-round (Brush 2005). Greater and Lesser Scaups, in particular, make use of benthic molluscs such as the Dwarf Surf Clam. Wading birds, such as Great Blue Herons, often forage for fish and crustaceans along bay margins. Aerial hunters, such as gulls and the Osprey, also find prey in the deeper waters of the bay center.

Community Structure

Salinity Effects Salinity is the chief factor affecting biotic communities in open-bay habitats. Salinity plays an especially crucial role in the Laguna Madre, where hypersalinity is the rule rather than the exception. High-salinity conditions there favor growth of Texas Brown Tide organisms (Buskey et al. 1998; Buskey et al. 2001; Wetz et al. 2017), which can dominate plankton communities and persist for years. Salinity variance, in addition to measured salinity, can be an important determinant of community composition (Van Diggelen and Montagna 2016). Depending on the temporal scale of changes in salinity (sudden, drastic vs. gradual, long-term), communities, particularly the benthos, may respond in different ways (Armstrong 1987). Salinity changes may act as a disturbance, affecting both abundance and diversity of organisms, or it may stimulate benthic production because of increased nutrient inputs associated with inflow (Alber 2002). Across Texas estuaries, increasing salinity tends to be detrimental to suspension feeders (such as the Dwarf Surf Clam) but favorable for deposit feeders (Kim and Montagna 2012; Rubio et al. 2018).

Salinity effects on benthos and phytoplankton are likely to have cascading effects on higher-trophic-level organisms in Baffin Bay. Following prolonged drought conditions, Black Drum in the bay experienced a widespread emaciation event in 2012, prompting concern from multiple public and private entities. Affected fish appeared unusually thin, with gelatinous flesh, empty stomachs, and atrophied internal organs. Reports of emaciated fish were mostly confined to the Baffin Bay complex, with a few reported in the Laguna Madre. Limited prey availability in Baffin Bay likely was the cause of this emaciation event (Grubbs et al. 2013; Olsen 2017) because the distribution of prey communities is largely linked to salinity regimes (Montagna and Kalke 1992). Ajemian et al. (2018) found that Black Drum movements were limited during high-salinity periods, possibly due to higher energy costs associated with osmoregulation, which could further limit access to prey.

Spatial Distribution of Benthic Invertebrates In open-bay habitats within the Laguna Madre, benthic invertebrates are horizontally distributed, primarily as a function of the distance from inlets and water depth (Parker 1959; White et al. 1986, 1989). They also are distributed vertically within the sediments, with the majority occurring within the top 0–3.0 cm (0–1.2 in) (Montagna 1993). Polychaetes dominate regardless of depth but were virtually the only organisms found in sediments 3–10 cm (1.2–3.9 in) below the surface. Biomass was typically greater in the 3–10 cm than in the 0–3.0 cm section due to presence of larger individuals.

Sediment grain size varies throughout the Laguna Madre system and plays a large role in determining macrobenthic spatial distribution (fig. 8.15; Breaux 2017). Because benthic invertebrates are relatively immobile, sediment characteristics greatly affect their community structure (Rhoads 1974), with fine-grained sediments often acting as a stressor (R. Johnson 1970; Warwick and Uncles 1980; Montagna 1999; Mannino and Montagna 1997). High proportions of silt and clay can make conditions challenging for macrofauna, particularly suspension feeders (Rhoads and Young 1970). The proportion of fine-grained sediments may increase with large freshwater inputs (Bolam et al. 2004; Hermand et al. 2008).

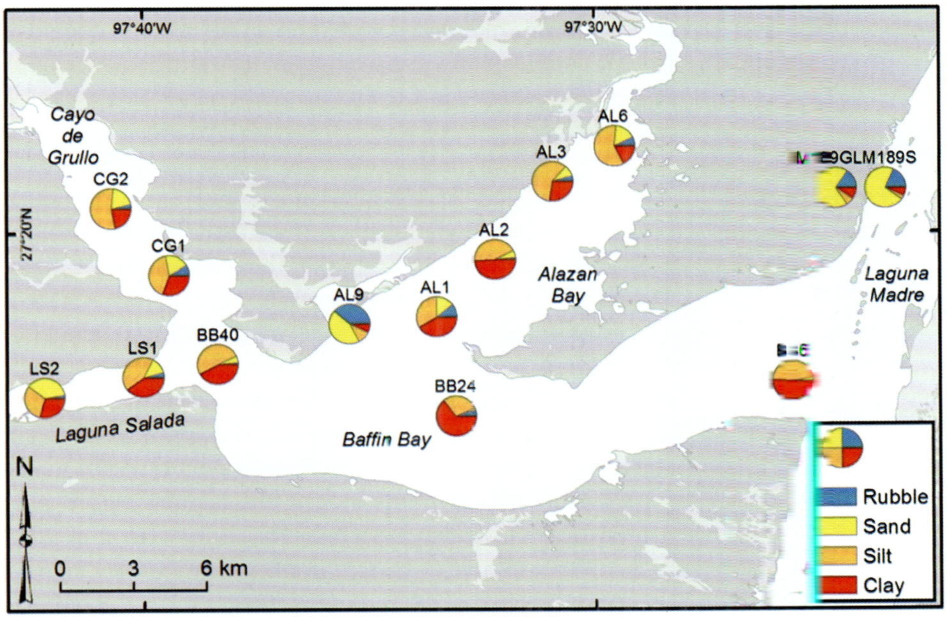

Figure 8.15. Map of the Baffin Bay complex and Laguna Madre sampling sites with pie charts depicting sediment composition (mean sediment grain size) from 2015 to 2016 (from Breaux 2017)

Energy Flow and Food Web

Despite negative inflow and resultant low terrestrial inputs, productivity in the Laguna Madre is high, and robust commercial and recreational fisheries flourish. Along the Texas coast, the Upper Laguna Madre ecosystem (Martinez-Andrade et al. 2005) supports the largest commercial Black Drum fishery in Texas (Olsen 2017). Recycled nutrients are likely important for maintaining the productivity in the system, but they also contribute to the recurring prevalence of Texas Brown Tide, which has affected the Laguna Madre for many years. In a study of adjacent Baffin Bay, Wetz (2014) reported elevated ammonium and orthophosphate concentrations in near-bottom waters during summer (despite low surface-water concentrations), indicating that enough nutrients are continually recycling to support algal growth. The Laguna Madre has among the highest levels of eutrophication in the Gulf of Mexico (Bricker et al. 1999; Bugica et al. 2020).

Stable isotope analyses are commonly used in ecological studies to help explain trophic interactions and ecosystem functioning (Phillips et al. 2014). Natural variation in isotopic compositions can be used to trace sources of organic matter and prey/consumer positions within the food web (Fry 2006). This distinction carries through to higher trophic levels, allowing determination of the origin of food resources (Fry and Sherr 1984; Peterson and Fry 1987) and inference of movement (Bowen et al. 2005; Cunjak et al. 2005; Breaux 2017). Due to the large areal coverage of seagrasses and lack of significant allochthonous detrital inputs in the Laguna Madre, the primary carbon source within its open-bay habitats is seagrass detritus. Within Baffin Bay, where seagrasses are not prominent, phytoplankton serves as a primary carbon source (Street et al. 1997). Street and colleagues and Fry and Parker (1979) found that benthic macrofauna from the Laguna Madre could be distinguished from those from Baffin Bay based on differences in carbon stable isotope composition. Within Baffin Bay, isotope composition was used to demonstrate that Black Drum consume benthic food resources produced in the estuary year-round (Rubio et al. 2018).

Current Issues in Texas

The Laguna Madre ecosystem and Baffin Bay in particular have been the subject of much recent research due to frequent algal blooms, hypoxia, and eutrophication. Climate models indicate that the southwestern United States will experience a warmer, drier climate in the twenty-first century as a result of increasing temperature and evapotranspiration (Seager et al. 2007; Montagna et al. 2007; Montagna et al. 2011). Increases in upstream water withdrawals (Montagna et al. 2013) coupled with predicted increases in aridity will lead to stronger and longer periods of hypersaline conditions and more frequent salinity disturbances (IPCC 2014). These changes will likely have negative effects on the open-bay biota of a Laguna Madre ecosystem already experiencing hypersalinity. Indeed, the Laguna Madre and Baffin Bay have already demonstrated long-term increases in salinity (Hu et al. 2015). Concurrent increases in acidification and decreases in alkalinity in the Laguna Madre (Hu et al. 2015) compound stress on calcifying organisms, such as bivalves, that provide important food resources to higher trophic levels (Rubio et al. 2018). A better understanding of how changes in water-quality parameters, especially salinity, affect estuarine resources would support resource management and conservation efforts in the Laguna Madre ecosystem.

Laguna Madre de Tamaulipas

Extent, Distribution, and Trends Unvegetated habitats in the Laguna Madre de Tamaulipas cover approximately 1,509 km² (621 mi²) (79%) of nonemergent estuarine habitats (see table 6.1). Open bay habitat has expanded considerably since 1974 (see fig. 7.13). Before inlets were dredged open and stabilized with jetties, the boom and bust character of the Laguna Madre de Tamaulipas resulted in highly variable coverage of seagrass and unvegetated bottom. During wet years and following hurricanes or tropical storms, seagrass meadows generally expanded, but during dry years and as the length of time since a hurricane or tropical storm increased, seagrass coverage typically declined. As overall salinity in the Laguna Madre de Tamaulipas declines due to greater exchange with Gulf waters, unvegetated habitat will likely decrease in favor of expanded seagrass coverage (see chapter 12 for a more detailed discussion). Additional research is warranted to describe the distribution and long-term trends of unvegetated habitats in the Laguna Madre de Tamaulipas.

The sediment of the Laguna Madre de Tamaulipas has most recently been described as mixed clay with patches of fine sand and loamy-clay (Leija-Tristán et al. 2000). Much of the Laguna in the mid-1960s was unvegetated and nearly dry (Anonymous 1967). In 1968, sediments were classified as (1) well-sorted fine sands along the backside of barrier islands, (2) clay-mud sediments with abundant shell and shell fragments parallel to the mainland, and (3) poorly sorted fine sediments (clay-mud) primarily in the center of the lagoon and Laguna el Catán (Yañez and Schlaepfer 1968).

Benthic Invertebrates Leija-Tristán et al. (2000) conducted a comprehensive taxonomic study of biota within the Laguna Madre de Tamaulipas and described 61 genera of molluscs, including 46 species of gastropods, 26 species of bivalves, 3 species of cephalopods, and 1 species of chiton. The highest biodiversity was associated with Eastern Oyster reefs, including gastropods, the Impressed Odostome, Spotted Slipper Snail, Virgin Nerite, Convex Slippersnail, and Semiplicate Dovesnail. Correa-Sandoval and Rodríguez-Castro (2013) provided an update of known bivalves along the Tamaulipas coast and included many new records. The Eastern Oyster and Bay Scallop fisheries in the Laguna Madre de Tamaulipas are described in chapter 12.

Of the 28 polychaete species associated with hard substrates identified from 21 genera and 12 families, species representing Nereididae and Eunicidae were the most abundant. Barba and Sánchez (2005) assessed peracarid crustaceans in submerged aquatic vegetation (SAV) and unvegetated soft sediments. Seventy-two percent of amphipods and isopods were observed in SAV and 28% in unvegetated soft sediments. In contrast, 81% of mysids were observed in unvegetated soft sediments and 19% in SAV. Additional sampling effort is encouraged to increase the knowledge of polychaete communities associated with soft sediments.

García-Cubas (1968) surveyed the micromolluscan fauna across the Laguna Madre de Tamaulipas during a period when little or no submerged vegetation was likely present. Seventeen species of bivalves and 16 species of gastropods were identified at abundances ranging from 0 to 675 per 10 cm³ (0.61 in³) of sediment. Molluscs tended to be more numerous in the southern lagoon, especially in and near Laguna el Catán and Punta Piedras. Bivalves, principally the Pointed Venus and Dwarf Surf Clam, dominated the collections, supporting Hildebrand's (1958) previous findings. Other important bivalves were the Broad-ribbed Carditid, Cross-barred Venus, Yellow Egg Cockle, and Tampa Tellin (see chapter 12 for updates to previously published bivalve species lists). Gastropod numbers were typically low, but the Pitted Baby-bubble and Grooved Odostome had the most widespread distributions and highest relative abundances.

Nekton The taxonomic study of biota within the Laguna Madre de Tamaulipas conducted by Leija-Tristán et al. (2000) extended to nektonic species as well. They identified 61 genera from 22 families of decapod crustaceans, with Ocypodidae (18.8%), Palaemonidae (16.0%), and Portunidae (13.4%) demonstrating the highest relative abundances. They identified 92 genera from 50 families of fish. Most species (85.2%) are carnivorous, feeding on fish (56.6%), invertebrates, (71.3%), and arthropods (33.3%). A large proportion of species, about 81%, are commercially valued for human consumption, forage, bait, or as artisanal or aquarium species. Shrimp of the genera *Farfantepenaeus* and *Litopenaeus* and crabs of the genus *Callinectes* are the most important commercial species; additional detail on commercial harvests is discussed in chapter 12.

Pérez-Castañeda et al. (2010) assessed the spatial distribution of *Farfantepenaeus* shrimp in the Laguna Madre de Tamaulipas, targeting SAV habitats. Brown Shrimp was the dominant species (39.5%), followed by Pink Shrimp (36.8%); the remaining 23.7% corresponded to small unidentifiable recruits. Shrimp abundance was positively correlated to seagrass biomass, indicating the importance of seagrass for enhancing growth and survival of juvenile penaeids in this hypersaline system (Pérez-Castañeda and Defeo 2005). Chapter 12 contains additional detail about penaeid shrimp abundance and distribution in the Laguna Madre de Tamaulipas.

9 *Wind-Tidal Flats*

KIM WITHERS AND PAUL V. ZIMBA

Tidal flats are nearly featureless sand and/or mud habitats found in many estuarine areas. These harsh, unpredictable environments are neither terrestrial nor aquatic but fall somewhere in between. The tidal flats bordering the Laguna Madre are unique. Since their submersion and emersion are affected primarily by wind and storm tides rather than lunar tides, they are termed "wind-tidal flats," which cover vast expanses of land adjacent to the Laguna Madre and Laguna Madre de Tamaulipas. Moving southward along the Texas coast, wind-tidal flats replace salt marshes as both freshwater inflow and annual precipitation decrease and eolian erosion, especially from barrier islands, increases (W. White et al. 1983). Extensive wind-tidal flats are found on lagoon sides of barrier islands in both Texas and Tamaulipas, with smaller expanses on mainland shorelines and at the mouths of creeks or other drainages (fig. 9.1). The largest flats in Texas, often known as the Laguna Madre Flats, are found in the Land-Cut, a land bridge between the mainland and central Padre Island and the separation between the Upper and Lower Laguna Madre. In Tamaulipas, large flats are found at the delta of the Río San Fernando (El Carrizal) and between the Río Soto la Marina and La Pesca (Withers 2002c). The El Carrizal Flats separate the northern part of the lagoon from the southern. The most recent estimate from 2002–4 of areal extent of wind-tidal flats in Texas is 631 km^2 (244 m^2) (White et al. 2005; White et al. 2006, 2007). In the late 1990s, areal extent in Tamaulipas was estimated at 508 km^2 (196 mi^2) (Withers 2002c).

To the casual observer, wind-tidal flats in the Laguna Madre appear unproductive and devoid of life. Irregular flooding, prolonged exposure, extreme summer temperatures, low freshwater inflow, and high soil salinities do not allow marsh plant communities to develop (Pulich et al. 1982). Algal mats, composed of photosynthetic bacteria (cyanobacteria or blue-green algae) and benthic microalgae such as diatoms, may form over and intermix with sediments. Despite lack of readily recognizable plant communities, primary productivity by algal mats may be nearly as high as that of seagrass meadows and is about 20%–40% of a typical cordgrass marsh (Pulich and Rabalais 1982, 1986; Pulich et al. 1982). On Laguna Madre flats that are frequently flooded, persistent benthic invertebrate communities develop (Withers 1994, 1996c, 1998; Withers and Tunnell 1998). These invertebrates convert primary productivity to animal biomass that is used by higher consumers such as shorebirds when flats are exposed and by wading birds, crabs, and demersal fish when they are flooded. Laguna Madre wind-tidal flats are one of the most significant feeding areas for aquatic birds on the Texas coast (Senner and Howe 1984; Haig and Plissner 1993; Withers 1994; Brush 1995). They are essential foraging habitats for wintering and migrating shorebirds and wading birds and are important to several state- and federally listed endangered or threatened species or "species of concern": Piping Plovers, Snowy Plovers, Reddish Egrets, White-tailed Hawks, and Peregrine Falcons.

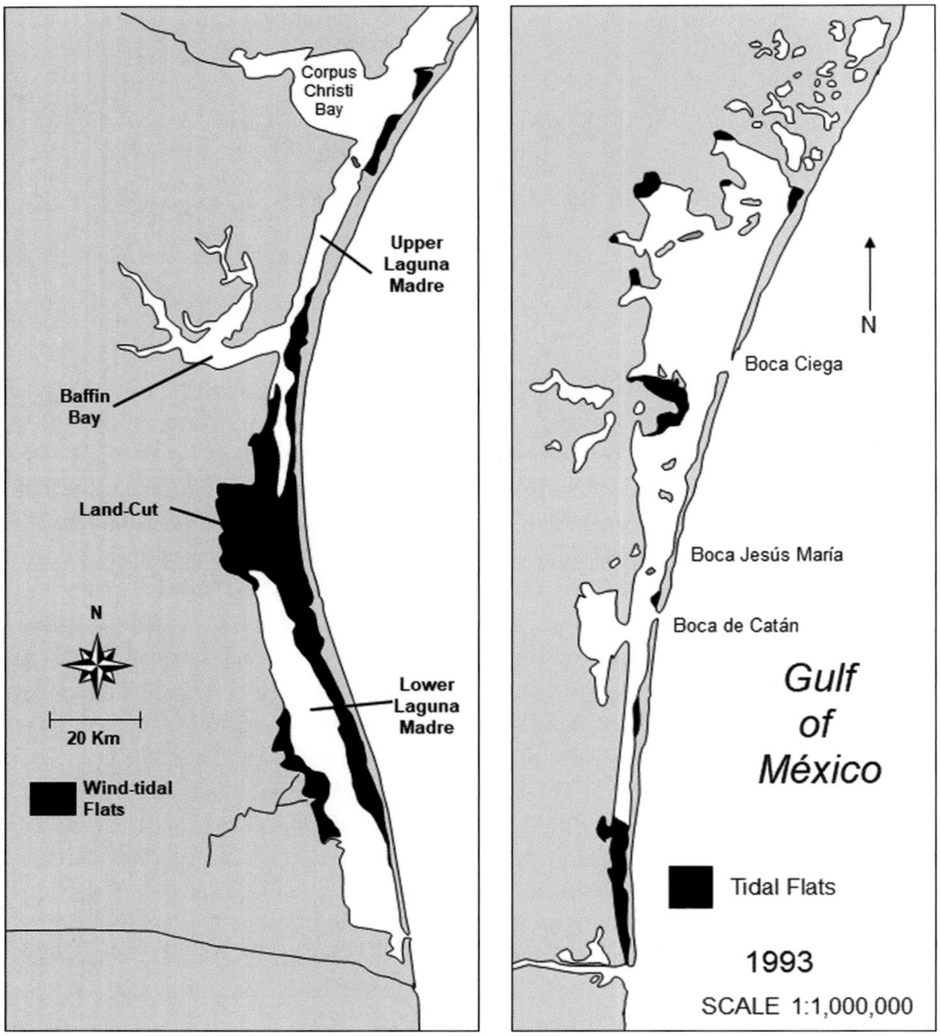

Figure 9.1. *Map of (a) wind-tidal flats adjacent to the Laguna Madre in Texas (modified from Pulich and Rabalais 1986) and (b) Tamaulipas (redrawn from Farmer and Carrera Gonzalez 1993)*

The Physical Environment

Ecosystem function of Laguna Madre wind-tidal flats is largely driven by abiotic characteristics and processes. The physical environment of wind-tidal flats can change dramatically in only hours if winds shift, rainfall occurs, or temperatures increase or decrease. Wind typically plays an outsized role in hydrology of wind-tidal flats, although rising sea levels appear to be causing more frequent and lengthier periods of inundation and less protracted periods of exposure. Elevation, orientation, flooding/exposure, sediment type, microrelief, physiography, and changing sea levels determine extent and persistence of algal mat development. In turn, mat development influences microrelief, sediment type and moisture content, and sediment stability via exopolysaccharides that bind inorganic material.

Hydrology Astronomical tides play a minor role in day-to-day flooding and exposure of Laguna Madre wind-tidal flats. Seasonal increases in systemwide water levels (September–October; April–May) result in fairly long periods when many flats are completely inundated and have little or no exposure. Seasonal decreases in water levels between July and August usually completely expose flats and there is little or no flooding, but during the winter low (January–February), flooding is rare to frequent depending on winds. For the most part, flooding or exposure of flats occurs when winds are aligned to blow directly along the axis of the estuary (Brown et al. 1977). Occurrence and persistence of flooding and exposure are irregular, and either may occur very rapidly (within a few hours) depending on speed and direction of wind (Fisk 1959; Hayes and Scott 1964; Hayes 1965; Brown et al. 1977; White and Galloway 1977; Weise and White 1980; Withers 1996c). For example, southeasterly winds of 13–17 kph (8–11 mph) forced water over flats in the Land-Cut at 0.5–1.0 km/day (0.3–0.6 mi/day) (Fisk 1959). When wind speeds increased to 30–50 kph (19–31 mph), the wind-tidal surge moved 4.2–6.1 km/day (2.6–3.8 mi/day). Strong northerly winds pushed water completely across the Land-Cut in 36 hours at rates of 12 km/day (7.5 mi/day). During a 24-hour period with north winds of 8–20 kph (5–12 mph), Amdurer (1978) noted a 6 cm (15 in) rise in water level; and when wind direction changed, floodwaters receded rapidly, with a drop of 3 cm (5 in) in four hours

Physiography Wind-tidal flats in the Laguna Madre are found between mean sea level (MSL) and 1.0 m (3.3 ft; at the outside) above MSL (fig. 9.2). They slope very gently lagoonward (≈10 cm/km [16 in/mi]) (Herber 1981); on the ground the change in elevation is essentially imperceptible until the upland is reached. The interaction of microelevational change, microrelief/microtopography, and inundation/exposure results in a mosaic of three essentially emergent microhabitats defined by water depth

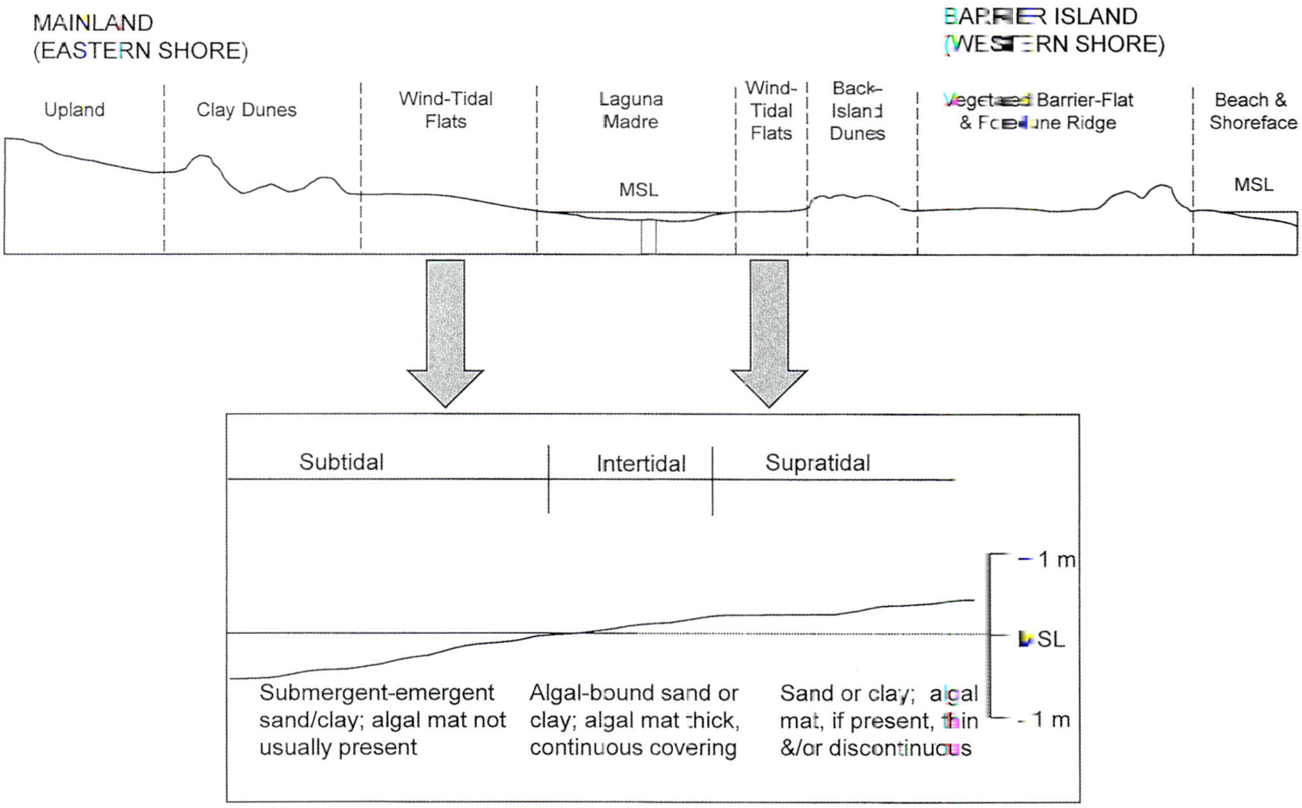

Figure 9.2. *Horizontal schematic depicting positions of wind-tidal flats in relation to other barrier island and mainland habitat types. A close-up view of the flat at mean sea level shows gentle slope, sediment type, and position of algal mat, if present.*

and substrate saturation that reflect shorebird foraging ability (fig. 9.3; Baker and Baker 1973; Withers and Chapman 1993; Withers 1994): (1) areas along the edge where the water is 2.0–5.0 cm (0.8–2.0 in) deep; (2) usually saturated but always wet areas, often covered with water ranging in depth from a "film" to about 2.0 cm (0.8 in) deep; and (3) areas that may appear dry on the surface but are damp to wet below the surface, occasionally with standing water (up to 2 cm deep) in depressions. Areal coverage of microhabitats differs with water levels, and sometimes microhabitats are nonexistent. A typical back-island flat is generally dry and sandy at the highest elevations near back-island dunes where flooding is rare. As elevations decrease, bare sand and/or mud

Figure 9.3. *Tidal flat showing microhabitats. Note the depth of the water on the legs of the foraging shorebirds in the background and the middle of the photo, with the dry, higher-elevation areas between both groups of shorebirds that are foraging in deeper water. Note in the foreground the intermixing of areas of shallow standing water, wet substrate with no standing water, and at the bottom of the photo, what appears to be dry substrate. (Photograph by Kim Withers)*

Figure 9.4. *The sharp textural boundary ("step-up") between the rarely exposed sandy/muddy areas below mean sea level and the algal flat (Photograph by Kim Withers)*

grade into sand and/or mud covered with an algal mat, particularly at and just above MSL, where flooding and exposure normally occur. There is often a sharp textural boundary between the rarely exposed sandy or muddy areas below MSL and the algal flat, with a "step up" onto the algal-bound areas (fig. 9.4; Withers 1994).

Geology and Geomorphology Wind-tidal flats in the Laguna Madre are geologically young features that began forming about 2,500 years ago when sea level approximated its present level. Tidal flats form on the flood-tidal deltas and washover fans on the baysides of barrier islands (Brown et al. 1976) and are composed primarily of medium and fine sands (Withers 1994). On the mainland, wind-tidal flats typically develop on fluvial deltaic deposits where erosion by streams diminishes as the water flows into upper reaches of bays or where winds have deflated older eolian, barrier-strandplain or deltaic deposits. Most mainland flats are small and generally composed of clays (Brown et al. 1976; Brown et al. 1977). The largest mainland flat, the Land-Cut, is a land bridge composed of infilling sand blown from Padre Island to the mainland (see chapter 11) that divides the Upper Laguna Madre from the Lower. Tidal-flat areas on the Kenedy Ranch adjacent to and associated with the Land-Cut are covered by a thin algal turf in many areas (P. V. Zimba, pers. observ.). The delta of the Arroyo Colorado in the Lower Laguna Madre is another large mainland flat.

Laguna Madre tidal flats became emergent about 200 years ago (Long and Gudramovics 1983). Flats increase in size primarily through accretion of wind-transported sand from barrier islands. Windblown sand adheres to wet or damp algal mats when they are exposed. Algal filaments extend into the water column when flooded, trapping clays (Fisk 1959) and reducing erosion. The process of accretion is usually incredibly slow. Vertical accretion in the Land-Cut gradually decreased from 0.5 to 0.25 mm/yr (0.02 to 0.01 in/yr) or less over the last 2,500 years (Miller 1975). However, drought may increase accretion rates. Rapid progradation of wind-tidal flats near Bird Island Basin between 1941 and 1969 was attributed to larger amounts of windblown sand resulting from drought-induced devegetation of dunes (Prouty and Prouty 1989). Floodwaters carry suspended sediments, especially to mainland flats. Prevailing southeasterly winds force sediment-laden waters away from Padre Island and into Baffin Bay, and the waves generated may erode tidal-flat margins. Seasonally, "northers" carry sediments back to barrier islands, but these events are infrequent and relatively unimportant (Fisk 1959; Herber 1981).

Biotic Communities

Primary Producers Algal mats are prominent features of most tidal flats in the Laguna Madre, but "algal" is used as an umbrella term for the complex mat-forming community composed of photosynthetic and nonphotosynthetic bacteria and benthic microalgae. Bacterial communities in both wet and dry areas are dominated by two gram-negative orders: Proteobacteria, which includes many bacteria responsible for nitrogen fixation; and Bacteroidetes, an order that is widespread in soils and water (Huang et al. 2020). Algae and photosynthetic bacteria growing on wind-tidal flats are extremophiles. When mats are dry, they crack, become black, and appear dead (fig. 9.5), but some species continue to actively grow and photosynthesize. Likewise, when flats are flooded with shallow water, temperatures may exceed 40 °C (104 °F), salinities may exceed 100 ppt, and nutrients are limiting, but many photosynthetic bacteria persist even though salinities of ≥50 ppt severely limit diversity of mat-forming groups. Photosynthetic bacteria in mats are able to survive desiccation and other extreme conditions by altering the thickness of their cell wall (fig. 9.6).

Figure 9.5. *Black, cracked, "dead" algal mats (Photograph by J. W. Tunnell)*

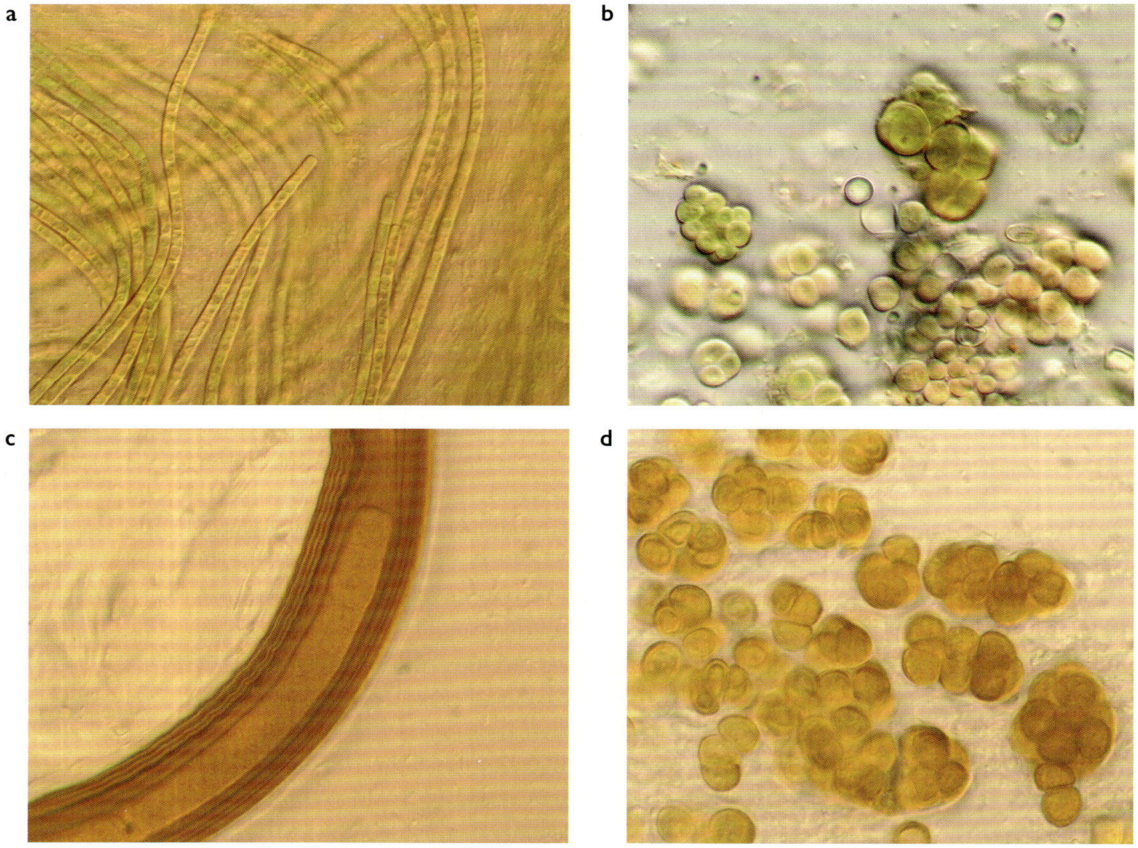

Figure 9.6. *Photosynthetic bacteria (blue-green algae or cyanobacteria) isolated from algal mats on wind-tidal flats in the Laguna Madre showing the effects of desiccation. Both filamentous forms (e.g.,* Lyngbya*) and coccoid forms show a similar thickening of the cell wall as they dry. Wet conditions: (a) filamentous, (b) coccoid. Dry conditions: (c) filamentous, (d) coccoid. (Photographs by Paul V. Zimba)*

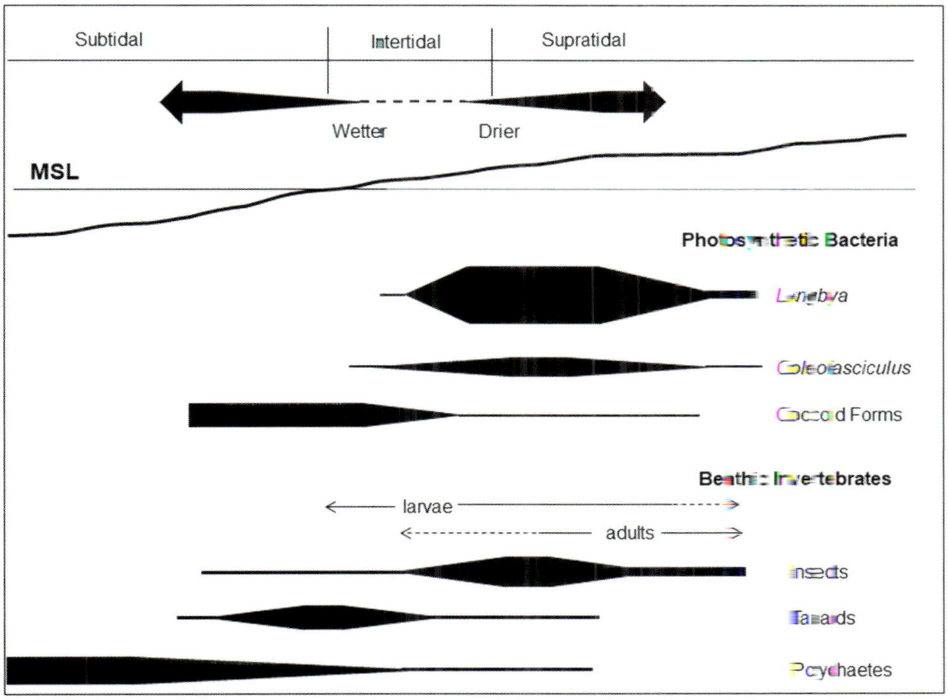

Figure 9.7. *Abundance and distribution for the major taxa of photosynthetic bacteria and benthic and epibenthic invertebrates. Weight of lines reflects relative abundances of different groups. (Photosynthetic distribution from P. V. Zimba, pers. comm., 2019; invertebrate distribution from Withers 1994)*

Photosynthetic bacteria are responsible for most of the primary productivity associated with wind-tidal flats. Filamentous forms are the most abundant and are the primary components of the felt-like or leathery mats that may be up to 3.0 cm (1.2 in) thick (fig. 9.7). *Lyngbya confervoides* (= *Microcoleus lyngbyaceus*) was originally thought to account for ~70% of the living community (Fisk 1959; Sorenson and Conover 1962; Armstrong and Odum 1964; Dykstra 1966; Zupan 1971; Gotto et al. 1981; Herber 1981; Pulich et al. 1982; Pulich and Rabalais 1986). However, recent genetic and morphological analyses suggest that there are at least four species of *Lyngbya* present in mats (P. V. Zimba, unpub data). The diversity of mat photosynthetic bacterial species is just beginning to be understood; at least 40 unique photosynthetic bacterial operational taxonomic units (OTUs) have been detected. Historically, pennate diatoms were common when mats were submerged, especially at salinities less than 60 ppt, but they were usually less than 10% of total algal biomass (Simmons 1957; Pulich et al. 1982). When mats were submerged for several weeks in hypersaline water, accumulation of fine silt shaded photosynthetic bacteria and allowed pennate diatoms *Licmophora normanii*, *Amphora* spp., and tiny *Fragilaria* spp. to become dominant (P. V. Zimba, unpub. data).

Seasonal succession of mat-forming photosynthetic bacteria is driven by water availability and salinity. *Lyngbya* spp. survive and grow at salinities from 20 to >70 ppt, whereas other forms require lower salinities. In drier mat areas subject to infrequent flooding, *Lyngbya* spp. dominate during dry periods but are replaced by *Coleofasciculus* spp. and an understory of filamentous and coccoid photosynthetic bacteria during wet periods (rains and/or submersion). If wetting/submersion in lower-salinity water is prolonged, diatom abundance increases. As the mats dry, attached diatoms are replaced by photosynthetic bacteria that tolerate higher salinities, community diversity decreases, and bacterial photosynthetic activity declines. Even when mats are wet and look photosynthetically active (green), if salinities exceed 100 ppt, photosynthetic potential is zero (P. V. Zimba, unpub. data).

Vascular plants are rare to nonexistent on wind-tidal flats in the Laguna Madre because sediment salinities are typically extremely high and there is little freshwater inflow. Scattered salt marsh vegetation may be found, primarily glasswort (*Salicornia* spp.) and Saltwort, particularly in or around depressions within the flat and during wet years. Occasionally, sparse clumps of Smooth Cordgrass are found along lagoonal edges of flats, and Black Mangroves occur along tidal channels (Herber 1981; W. White et al. 1983; Pulich and Scalan 1987).

Succession from unvegetated salt flat to a Saltwort-glasswort marsh may occur under conditions of adequate freshwater (Brodgen et al. 1977). Under optimal conditions and at suitable elevations (1) algal mats develop and stabilize sediments, promoting deposition, accumulation of organic material, and nitrogen fixation; (2) seeds of high-marsh halophytes germinate at higher elevations in depressions where water accumulates; (3) fiddler crab bioturbation allows rainwater to penetrate sediments, reducing soil/porewater salinities; and then (4) other salt marsh plants such as Saltwort, Shoregrass, and Saltgrass are able to colonize. Extended periods of drought or extremely high tides increase salinities, which halts or reverses the process. In the late 1990s and early 2000s, succession on most flats in the Laguna Madre was arrested in the algal mat stage, but there were observations of encroaching marsh vegetation on some flats north of the JFK Causeway and in the Horse Island area of the Lower Laguna Madre (Withers 2002c). The expansion of Black Mangroves occurring as a result of the "tropicalization" of the Gulf of Mexico coast has the potential to significantly alter function of tidal flats in the Laguna Madre (see chapter 19). If reductions in salinities continue (see chapter 7), mangroves will almost certainly expand onto flats, contributing to continued declines in tidal-flat areal coverage. More frequent and prolonged submersion as a result of relative sea-level rise washes salts out of sediments, and as porewater salinities decline, marsh vegetation and mangroves gain a foothold. When vegetation increases, sedimentation also increases, which increases the flat's elevation, altering hydrology and other physical processes that maintain the lack of vegetation in the habitat.

Consumers

Invertebrates Flooding is necessary for abundant benthic and epibenthic invertebrate communities to develop and persist (Withers 1994; Withers and Tunnell 1998). Invertebrates recruit to wind-tidal flats during seasonal high tides. Persistence and stability of the community are largely dependent on flooding frequency, and communities are usually more ephemeral in some areas than in others. Abundance and community composition may also be modified locally as shorebirds deplete patches of organisms. On flats that are frequently flooded, invertebrate communities can exist between late fall and midsummer. Organisms are usually most abundant during winter and spring, although a brief summer flush may be observed during June. Summer low tides extirpate communities on most flats, which become very hot, dry, and hypersaline.

The most common and abundant members of the invertebrate community associated with algal flats in the Upper Laguna Madre are dipteran larvae, polychaetes, and tanaids (Withers 1994, 1996c, 1998; Withers and Tunnell 1998). Of these three

Figure 9.8. The Laguna Madre Fiddler Crab digs deep burrows in soft sediment and aerates the soil. The tailings are deposited on the surface. (Photograph by Brian R. Chapman)

common taxa, polychaetes are typically the most diverse, with as many as 65 species. The assemblage of invertebrates associated with the previously described microhabitats as well as submerged (5–10 cm [2–4 in] deep) areas adjacent to emergent microhabitats, constitutes a subset of the assemblage (see fig. 9.7). Semiterrestrial organisms (insect larvae, insects) dominate in drier areas; abundance and dominance of aquatic-adapted organisms (polychaetes, tanaids) increase as sediment saturation and/or water depth increases (Withers 1994, 1998).

The Laguna Madre Fiddler Crab is endemic to the supralittoral zone of South Texas south to Tampico (Thurman 1984). Common along the upland margins of wind-tidal flats in the Laguna Madre and upper reaches of Baffin Bay, it is typically the only fiddler crab found in these environments (Rabalais 1983; Rabalais and Cameron 1985a, 1985b). The fiddler crab is an important bioturbator (fig. 9.8) and food source for tidal-flat transients such as Coyotes.

Vertebrates Fish may use wind-tidal flats when inundated, with counts of >5 fish/m^2 on the Kenedy flat when flooded (P. V. Zimba, pers. observ.). Sheepshead Minnows were the only fish consistently observed on flooded flats in the Upper Laguna Madre (Withers 1996c) and are probably the only fish that could be considered a regular resident of flooded flats. Sheepshead Minnows feed on algae (diatoms) and detritus when they are small and on crustaceans, insects, and small fish when they are larger (Harrington and Harrington 1972; Pfeiffer and Wiegert 1981). As flats dry, Sheepshead Minnows may become stranded and concentrated in depressions, where they are easy prey for birds, or move into deeper water, where they are preyed on by a variety of fish (Pulich et al. 1982). Occasionally, Southern Flounder have been disturbed in the shallow water adjoining flats in the Upper Laguna Madre, and "finger" Striped Mullet have also been observed sporadically either on flooded flats or in shallow adjacent waters (Withers 1996c). Large, bottom-feeding fish such as Black Drum, Red Drum, and Striped Mullet may also be found if waters are deep enough (Fisk 1959; K. Withers, pers. observ.).

Figure 9.9. *Mixed flock of migrating Willets and Hudsonian Godwits resting in the relative safety of a flooded Laguna Madre wind-tidal flat (Photograph by Brian R. Chapman)*

The wetlands and barrier islands of the Gulf of Mexico coast contain some of the most important migrating and wintering shorebird habitat in North America (Withers 2002b), and at least 32 species of shorebirds occur in the Laguna Madre (Withers 1994, 1996a, 1998; Ecoservices 1993b; Brush 1995; Withers and Tunnell 1998; see also chapter 15). Overall abundance of shorebirds in the Laguna Madre system (Texas and Tamaulipas) is at least an order of magnitude higher than anywhere else in the Gulf of Mexico (Withers 2002b). Most shorebirds use frequently flooded wind-tidal flats, although adjacent habitats, particularly washover areas, may be important, especially to breeding and roosting birds. In the Upper Laguna Madre, shorebird numbers on tidal flats generally increase in late fall (October–November) as seasonal high tides ebb and expose the flats and the invertebrates that have recruited. Shorebird numbers continue to build through the winter and early spring (December–March), then decline as migrants return to northern breeding grounds. Tidal flats are most heavily used by wintering and northbound migrants. Southbound (late July–October) migrants are common on beaches in the area, but tidal flats are not as heavily used because summer exposure extirpates prey and then seasonal high tides flood the flats, leaving no exposed area for foraging (Withers and Chapman 1993; Withers 1994). The most common and abundant species migrating through or wintering are sandpipers, but banded plovers, Black-bellied Plovers, dowitchers, and Willets are also common (fig. 9.9). During late spring and early summer, the only species present on tidal flats are those that breed in the area, such as Wilson's Plover and Snowy Plover, or nonbreeding, often immature individuals that did not complete the migration to northern breeding grounds.

Ecosystem Processes

Energy Flow and Food Web Lack of readily recognizable vegetation on wind-tidal flats conveys a false impression of insignificant primary productivity. However, primary production on Laguna Madre wind-tidal flats by benthic photosynthetic bacteria and algae can be nearly as high as that of seagrasses and has been estimated at 222 g C/m^2/yr to over 500 g C/m^2/yr (Odum and Wilson 1962; Brodgen et al. 1977; Pulich and Rabalais 1982, 1986) with an algal turnover rate of about 10.8 times/yr (Pulich and Rabalais 1982, 1986). Systemwide, this represents a substantial amount of production because of the vast areas occupied by wind-tidal flats and lack of salt marshes.

Both grazing and detrital food chains are found on tidal flats in the Laguna Madre, but the grazing pathway is likely the principal means by which primary productivity associated with photosynthetic bacteria and benthic algae moves to higher trophic levels (Pulich and Scalan 1987). Benthic microalgae are extremely important food resources in other estuarine systems including salt marsh and seagrass (Haines and Montague 1979; Sullivan and Moncreiff 1990; Moncreiff and Sullivan 2001). Seagrass detritus and moribund algae likely form the basis of detrital food chains. Significant amounts of seagrass wrack are incorporated into mats (P. V. Zimba, pers. observ.); bacterial and fungal breakdown of this carbon source should release dissolved nitrogen and phosphorus, enhancing photosynthetic bacterial growth.

Seasonal, systemwide water-level fluctuations and daily flooding and exposure by wind tides determine whether or not predators are present on tidal flats. In higher, drier areas, an abbreviated food chain based on microalgae and saltwater-adapted insects has been suggested (Pulich and Rabalais 1982; Pulich and Scalan 1987). On algal flats that are frequently flooded and exposed, polychaetes, crustaceans, and insect larvae are the primary consumers (e.g., Withers 1994). Figure 9.10 depicts a generalized food web. The presence of vertebrate predators (shorebirds, wading birds, or fish) is dependent on water depth. Shorebirds are the only predators that can use flats when they are either exposed or covered with very shallow water (<2.0 cm [0.8 in]). When the flats are flooded with more than a few centimeters of water, small fish such as Sheepshead Minnows come up onto the flats to forage and provide wading birds with prey, especially shallow-water foragers such as the Reddish Egret (Farmer and Carrera Gonzalez 1993). Waters must be fairly deep (>20 cm [~8 in]) before predacious fish can forage.

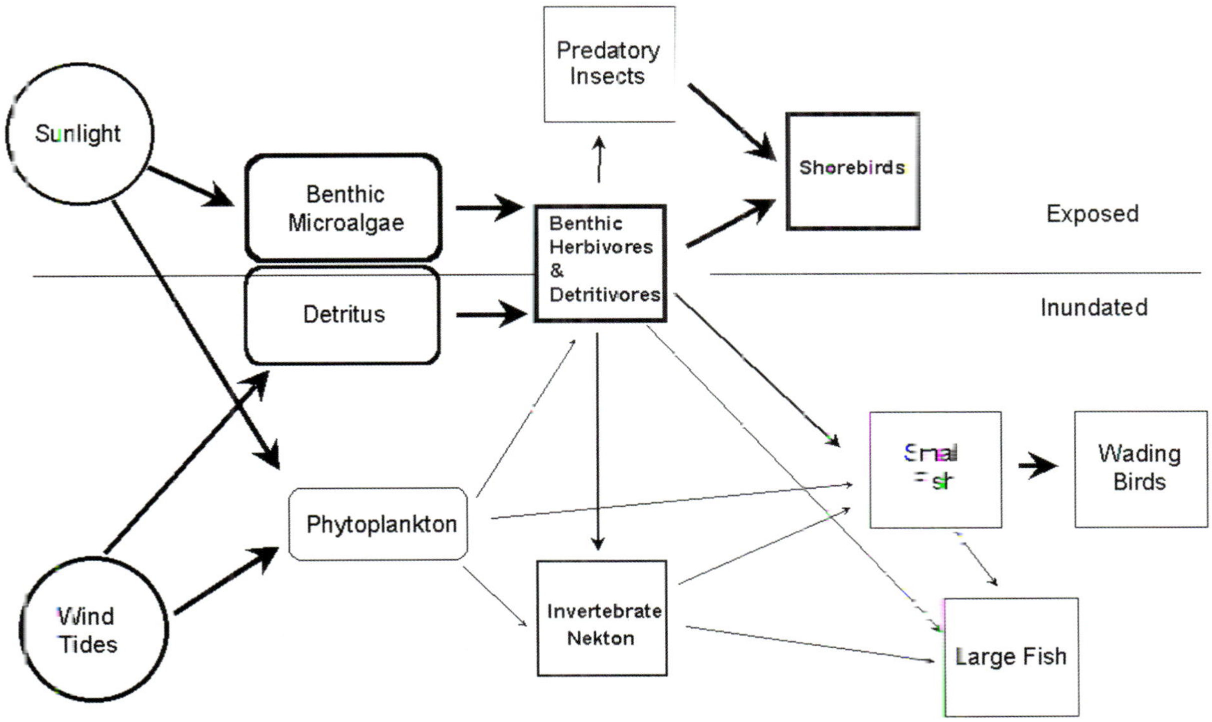

Figure 9.10. *Generalized food chain for tidal flats. Weights of arrows and boxes reflect the relative importance of each compartment and pathway. (From Withers 1996c)*

Nutrient Cycling Nitrogen-limitations are typical of marine systems, but phosphorus can also be limiting in hypersaline systems such as the Laguna Madre (Villareal et al. 1998). Nitrogen is generally available, if not abundant, on wind-tidal flats in the Laguna Madre. Photosynthetic bacteria fix the nitrogen they need; nitrogen is also fixed by anaerobic bacteria in sediments below the algal mats. Average annual nitrogen inputs from mats in the Laguna Madre to adjacent shallow waters ranged from 4–89 kg N/ha/yr (Pulich and Rabalais 1982, 1986), a considerable contribution to the system. In addition, substantial amounts of dissolved ammonia leach out of flooded algal mats, representing an important source of "new" nitrogen to adjacent waters. There is also evidence that dissolved ammonia percolates down through sediments to the water table (R. Hay, pers. comm.).

Phosphate may accumulate in algal mats in the Laguna Madre. After algal mats had been exposed to the point of desiccation and dormancy, reflooding resulted in leaching of accumulated phosphate from the mat (Pulich and Rabalais 1982, 1986). There was also evidence that phosphate was lost through downward percolation. When mats were wet or flooded and actively growing, net uptake of phosphorus was observed.

Status and Trends

Worldwide, extent of tidal flats declined by at least 16% between 1984 and 2016 (Murray et al. 2019). There are no data with which to evaluate the current status and historical trends of tidal-flat areas in the Laguna Madre de Tamaulipas. Data presented in table 9.1 are the most recent available and are from mapping of Texas estuarine wetlands using aerial photos (1950s, 1979, 2002–4) and the National Wetlands Inventory (NWI) classification system (Cowardin et al. 1979). The position of tidal flats, their low elevation, and the large effects of wind in moving water on and off the flats means that the extent of flats in aerial photos can be substantially affected by seasonal high and low systemwide water levels as well as the specific conditions at the time that aerial photos were taken. Thus, absolute areal extent of flats may be more or less than what is mapped (White et al. 2005; White et al. 2006, 2007). Greatest losses occurred between the 1950s and 1979, although some losses in this time period may be due to photointerpretation and conditions at the time the photos were taken. Slight gains were registered between 1979 and the early 2000s except in North Padre Island.

Table 9.1. *Tidal-flat area (km^2) in Laguna Madre over the last 50 years*

Region	1950s	1979	2002–4	% Net Change
North Padre Island	9.6	2.8	2.6	-73
Padre Island National Seashore[a]	372.4	309.3	353.6	-5
Lower Laguna Madre	238.0	207.0	216.7	-8
Total	620.0	519.1	572.9	-8

Source: White et al. 2005; White et al. 2006, 2007.

Note: Tidal flats were mapped using the NWI classification system (Cowardin et al. 1979) and included the following codes and descriptions with modifiers for hydrologic regime (i.e., regularly flooded, irregularly flooded, irregularly exposed): system/subsystem—E2 (Estuarine Intertidal); class—Unconsolidated (US) or Aquatic Bed (i.e., cyanobacterial/algal mats; AB).

[a] Table 7 in White et al. (2007) lists the area of tidal flat in the 1950s as 305.9 km^2. However, in the text (p. 76) a mapping error is discussed, and 66.5 km^2 are added to the 1950s areal extent, resulting in a corrected area of 372.4 km^2 and a net overall loss of 5%.

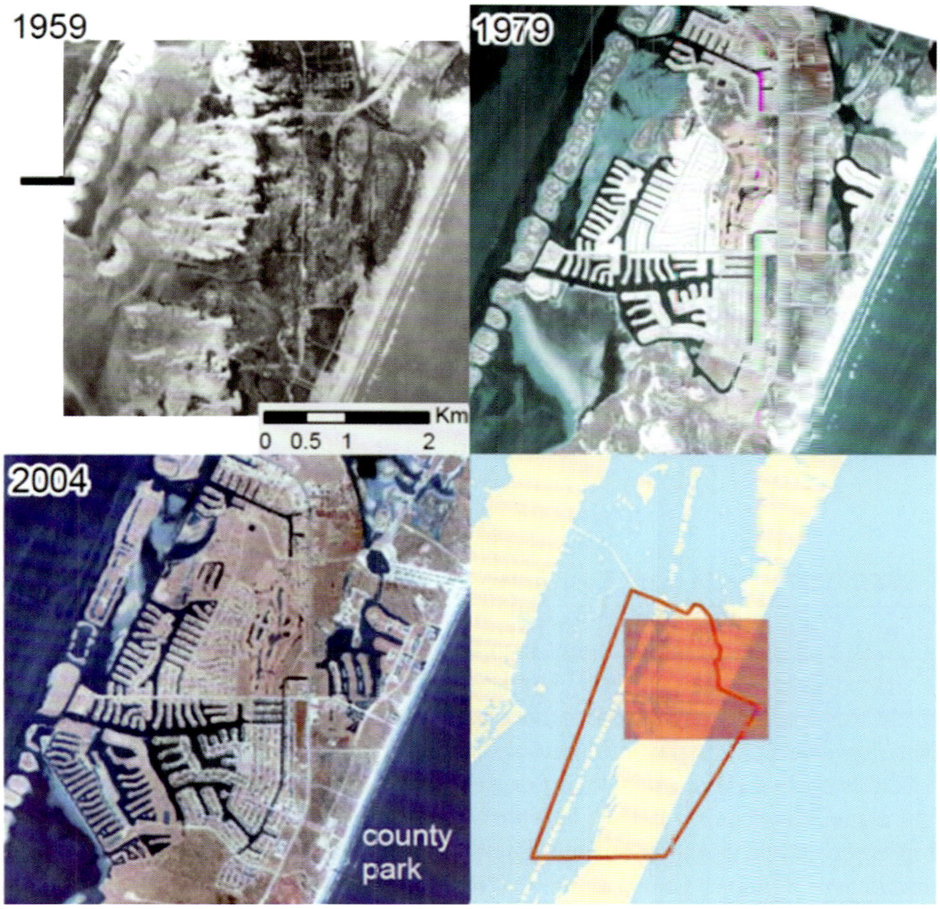

Figure 9.11. *Padre Isles residential development in the North Padre Island region showing destruction of wind-tidal flat between the 1950s and 2004 (From White et al. 2006).*

The approximate extent of the North Padre Island region was from the JFK Causeway to the northern boundary of Padre Island National Seashore (White et al. 2006). Tidal and algal flats were 82% of all intertidal habitats in North Padre Island. Compared to the other regions, the area of wind-tidal flat was small, but the mapped area was also quite small in comparison. Most of the tidal flats in this region were lost, and most of that loss was due to development of a canal subdivision (Padre Isles) on a large flat in the northern part of the region (fig. 9.11).

The Padre Island National Seashore region extends from its northern boundary south of the JFK Causeway to its southern boundary at Mansfield Pass (White et al. 2007). Of all intertidal habitats within the seashore, 96% were tidal and algal flats. There was a 5% overall loss in tidal-flat area. Most of the losses (72%) between the 1950s and 1979 were due to conversion to upland as back-island dunes migrated lagoonward. Increases between 1979 and the early 2000s were primarily (62%) due to replacement of seagrasses with tidal flats in Nine Mile Hole. Overall, gross gains were offset by gross losses, resulting in a small net loss of flat area through time.

The Lower Laguna Madre region extends from Mansfield Pass to the Brownsville Ship Channel (White et al. 2005). Tidal and algal flats were 87% of intertidal habitats in the region. Overall, there was about an 8% decrease between the 1950s and the early 2000s. From the 1950s to 1979, about 75% of the decrease was attributable to conversion to uplands, in large part as a result of dune migration. Greater areal coverage in the 1950s may have been due to drought, which apparently reduced water levels and may have exposed more flats during that time period. Development of a marina, dredged-material disposal on tidal flats, mangrove encroachment, and conversion to open water as a result of relative sea-level rise are implicated in the remaining losses. Although the 1950s tidal-flat area may be somewhat overestimated, small gains between 1979 and the early 2000s resulted in a small loss overall.

Current Issues

In the Laguna Madre, tidal flats occupy the space marshes fill in other areas. Salt marshes convert plant biomass to animal biomass at the place where the seawater meets the land; tidal flats do the same. Wind-tidal flats in the Laguna Madre systems of Texas and Mexico represent the most important habitat for wintering shorebirds at the northernmost extent of most species' wintering range, and this must be a focal point in discussing conservation of the habitat. "Wetlands" mean marshes to most people, not tidal flats. Despite the critical role tidal flats play in supporting migrating and wintering shorebirds, it is hard to convince people that tidal flats, which appear barren and unproductive, are worthy of conservation. Most people just do not understand why they should care about a habitat that has no readily seen and understood component such as cordgrass and may completely overlook the foraging shorebirds. Losses of wind-tidal flats or their degradation carry potential for both long- and short-term reductions in shorebird carrying capacities. Competition may already limit the numbers of shorebirds in microhabitats they prefer (Withers 1994); reductions in current carrying capacity could result in negative impacts on shorebird populations.

Tidal flats are being destroyed or degraded by a variety of causes, both natural and human-made. One bit of good news is that excepting the northern area of Padre Island, most losses in the Laguna Madre in Texas were relatively small through the early 2000s. Tidal flats are lost when they are converted to other habitats by sea-level rise or dune migration, when they are covered with dredged material, or when residential or commercial development occurs. Even if they are not completely lost, dredging, dredge disposal, development on adjacent lands, driving, and contaminants may degrade tidal flats by altering ecosystem function and/or physical processes. Tidal-flat habitat has been completely destroyed by commercial or residential development outside Padre Island National Seashore where extensive tracts of wind-tidal flats have been protected from all but oil and gas development (Withers and Tunnell 1998; White et al. 2005; White et al. 2006).

At the landscape level, losses of tidal-flat habitat due to conversion associated with sea-level rise or plant community succession might be permanent, but they also could be relatively short-lived, perhaps decades. Relict tidal flats in nearby Copano Bay provide evidence that the habitat may re-form at higher elevations when sea level increases (Ricklis 1995; Ricklis and Blum 1997). Rapid progradation of Padre Island tidal flats during periods of drought (Prouty and Prouty 1989) indicates that increased accretion rates associated with drought might be enough to offset some losses, but drought also contributes to accelerated dune migration and conversion to upland. In addition, sea-level rise combined with normal or increased rainfall may decrease sediment salinities enough to allow plant community succession to proceed. Expansion of high-marsh halophytes, cordgrasses, and Black Mangroves onto flats may occur, but return to drought and/or cold weather may arrest or reverse expansion. Sea-level rise also appears to be causing changes in the community structure of the algal mats. Submersion times seem to be increasing, and mats appear to be thinning as the community shifts from dominance by sheath-forming photosynthetic bacteria (= mat formers) to coccoid photosynthetic bacteria and bacteria and/or microalgae other than photosynthetic bacteria.

Vegetating tidal flats with cordgrasses, Black Mangroves, or seagrasses in mitigation projects is often viewed as enhancing existing habitat (Withers and Tunnell 1998). Tidal flats may be targeted for planting to mitigate losses of vegetated wetlands, and impacts to tidal flats are routinely mitigated by vegetating unvegetated habitats. Mitigation projects in the Lower Laguna Madre have converted several tidal flats to Black Mangrove and/or cordgrass habitats. With few exceptions, there has been no "in-kind" mitigation for tidal flats lost to development in the Laguna Madre. In fact, there is no method for attempting in-kind mitigation for tidal-flat loss.

Scars of tracking, probably due to cattle, are evident on tidal flats of Mustang Island in 1938 aerial photos. Vehicles and propeller scars from boats are evident on nearly all tidal flats in the Laguna Madre. Impacts from vehicles are much more serious when flats are wet or saturated than when they are dry. Studies in other areas concluded that maximum damage to benthic invertebrate communities occurred after only a few passes by a vehicle (Leatherman and Godfrey 1979). However, there do not appear to be any significant long-term effects on benthic communities once tracking stops as long as track scars are not very deep (Withers 1996a).

Dredging impacts tidal flats when channels are cut through them and when they are used as disposal areas (Withers and Tunnell 1998). Channelization destroys benthic fauna, reduces area for birds, and changes hydrology by blocking floodwaters from inundating flats or by causing flats to drain more quickly. Dredged-material disposal buries benthic organisms, reduces foraging area for birds (although it may increase roosting area), and changes hydrology by blocking wind tides and isolating remaining flats from inundation. At a minimum, either practice may change hydrology to the point that most function ceases. Proliferation of navigational channels and marinas with associated dredged-material disposal on tidal flats has been implicated in rapid and more extensive flooding and erosion of vegetated barrier flats, subaerial spoil, and made land (Morton et al. 1977; White and Brodgen 1977).

10 *Barrier Islands*

ELIZABETH H. SMITH AND FRANK W. JUDD

Barrier islands are long, narrow offshore deposits of sand or sediment that extend parallel to the mainland coastline and protect it from storm winds and waves that cause erosion. Barrier islands are separated from the mainland by a shallow bay, lagoon, or sound and often occur in chains with the islands separated by narrow tidal inlets. The geomorphic position of barrier islands between the open Gulf of Mexico and adjacent mainland bays and lagoons of Texas and Mexico delineates the protective importance of these island systems to sustainability and dynamics of tidal flats, lagoons, and bays. River sediments discharged into the Gulf of Mexico are the primary source of building materials for barrier islands, and the sand and shells deposited on Gulf beaches provide the material necessary for coppice dune and foredune ridge formation and maintenance (fig. 10.1). Establishment of vegetation stabilizes the dunes and provides protection of leeward habitats during storm surges. Salt spray from the Gulf of Mexico blowing across the barrier island affects the salinity of soils and freshwater ponds, availability of nutrients, and composition of the vegetative community.

A characteristic plant assemblage is associated with each topographic zone across a barrier island (Yzaguirre 1974; Negrete et al. 1999; Nelson et al. 2000; Nelson et al. 2001). Extreme high tides from storm surges may flood vegetated flats, influencing plant and animal community dynamics for several years afterward. Sand deposition on tidal flats is increased during droughts, increasing the areal coverage of unvegetated, active dunes. Seagrass meadows benefit from the physical protection of barrier islands

Figure 10.1. *When trapped by an obstacle, sand blown landward from the foreshore accumulates to form small coppice dunes that may be gradually stabilized by Sea Oats, Goatfoot Morning Glory, and other plants. (Photograph by Brian R. Chapman)*

but may become shallower due to windblown sediment deposition from adjacent island dune fields. Open-bay habitats are indirectly affected by barrier islands as a protection from the Gulf of Mexico and directly affected by transport of marine waters and organisms through island passes. Barrier islands along the Gulf of Mexico also occur off the coast of Mexico from just south of the Rio Grande to Altamira, Tamaulipas, 402 km (250 mi) south of Brownsville, Texas. Thus, the islands provide an avenue of northward dispersal for numerous organisms.

Physical Setting and Processes

Barrier islands are found only on coasts that have shallow slopes. The elongated islands are positioned parallel to the coastline and are typically isolated from the mainland by bays and lagoons. Longshore currents, low tidal range, and wave energy work together to form and maintain barrier island systems (Morton et al. 1977; Britton and Morton 1989). These islands function as protective barriers to the adjacent mainland and shallow bays and lagoons. Padre Island, part of a barrier island chain extending from the central Texas coast to the southernmost extent of Texas, is 182 km (113 mi) long—the longest barrier island in the United States (see fig. 2.1). A similar barrier island chain is located along the northeastern coastline of Tamaulipas. Unique flora and fauna inhabiting these islands contribute to Tamaulipan Biotic Province diversity. Various combinations of physical factors and degree of isolation from the mainland and adjacent barrier islands contribute to slight differences in topography, vegetation, and animal composition on each island. Constant redistribution of sands, driven primarily by prevailing southeasterly winds, contributes to formation of many of the distinct topographic features across an island. These dynamic, eolian forces continually erode some areas and build up others, causing active dune migration across islands in some areas. Abrupt changes in topography often occur as a result of hurricanes and storm surges. These events may reopen passes through an island, denude or flatten dunes, create new unvegetated dune fields, and erode beachfronts (Price 1933; Weise and White 1980). Other factors producing shoreline changes on Gulf Coast barrier islands include sea-level rise, compactional subsidence, deficits in sediment supply, and human activities (Morton and Pieper 1975). Thus, the habitats on barrier islands and the shape of the islands themselves are in a constant state of change.

Natural passes originally separated barrier islands from each other, but artificial passes have been dredged across some to connect the lagoons on their landward side to the Gulf of Mexico (see figs. 2.2–2.5). Tidal surges associated with hurricanes and strong tropical storms may scour enough sediment to create a "washover pass" across a barrier island at a low point (fig. 10.2), but such passes usually fill with sand within weeks or months (Brown et al. 1976). Padre Island is separated from Mustang Island to the north by an intermittently open washover pass, Packery Channel (historically known as Corpus Christi Pass). Other washover passes, Newport and Corpus Christi Passes, also occur in this area. Padre Island extends southward to Brazos Santiago Pass, but an artificial pass, Mansfield Pass, approximately halfway between the end points of the island, connects the Lower Laguna Madre to the Gulf. Several shallow passes were located along the Tamaulipas barrier island chain, but most were closed historically. Even so, the low areas where these passes existed can be opened temporarily by hurricane storm surges. During the past two decades, four passes through the barrier island to the Laguna Madre de Tamaulipas have been dredged and jettied (see chapters 2 and 3).

Figure 10.2. *Hurricanes and strong tropical storms generate tidal surges and large waves that can break through the foredune ridge and create washover passes that later close as sediments deposited by longshore currents refill the passes. (Photograph by Brian R. Chapman)*

North Padre Island is quite wide (up to 4.8 km [3.0 mi]; average width is about 3.0 km [1.8 mi]). Extensive fresh to brackish marshes occur within the vegetated flats in the island interior. The middle region of Padre Island becomes narrower but has similar marsh complexes within the Padre Island National Seashore boundaries. South of Mansfield Pass the island becomes a comparatively thin ribbon of sand where washover passes and wind-tidal flats are abundant (Brown et al. 1980). Expansive wind-tidal flats occur along the lagoon side of Padre Island, the Land-Cut, and the South Bay–Laguna Larga–San Martin Lake area of Texas (Brown et al. 1976; Brown et al. 1977) and border much of the Laguna Madre de Tamaulipas shorelines (Saunders and Saunders 1981; see chapters 3 and 9). Tamaulipas barrier islands are much narrower than Padre Island and have fewer marshes in the island interior (Britton and Morton 1989).

Barrier islands along the southern Texas and northern Tamaulipas coasts are typically wider at their northern ends (Selander et al. 1962; Brown et al. 1976; Brown et al. 1977; Brown et al. 1980). Several habitat types occur across the islands, and each is related to elevation, physical forces, and geomorphology of the adjacent Gulf of Mexico and lagoons (Nelson et al. 2000; Nelson et al. 2001). Physiographic zones extending the entire length of the islands include foreshore (swash zone), backshore (from high-tide line to dunes), foredunes (primary dunes), vegetated flats, secondary dunes, ponds and marshes, back-island dunes, and wind-tidal flats and coastal marshes in some places (fig. 10.3). Foreshore habitat, wind-tidal flats, and coastal marshes are affected more by hydrological and eolian forces than are primary dunes or secondary dunes and vegetated flats. Ponds and marshes in vegetated flats are a significant component of the barrier island ecosystems.

Longshore currents carry sediment deposited in the Gulf of Mexico by rivers and gradually release the suspended sediments onto the foreshore. South Padre Island is oriented on a north-south axis and the prevailing winds are onshore from the southeast, so the longshore currents in that region flow northward. The Rio Grande

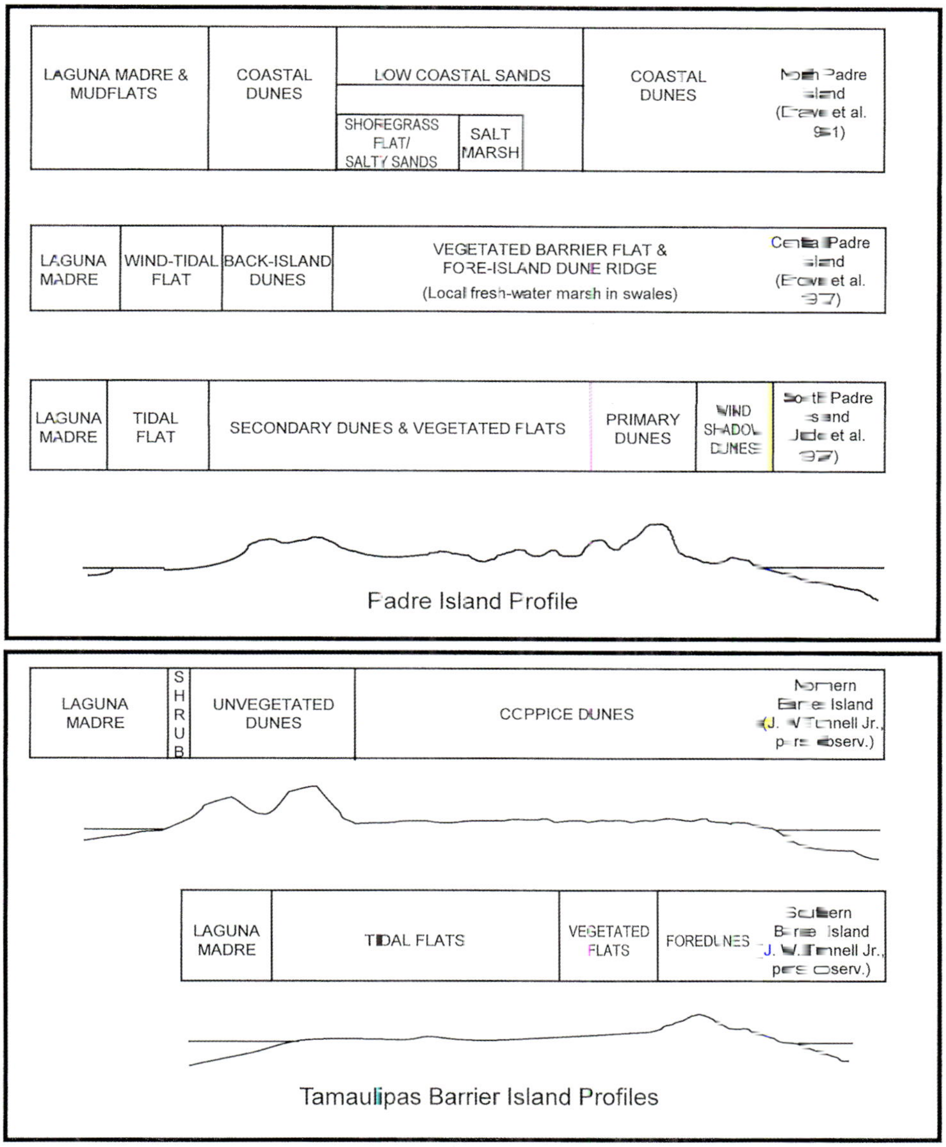

Figure 10.3. *Cross-section profiles of barrier island vegetation zonation for Padre Island, Texas and in Tamaulipas, Barra el Conchillal near El Mezquital (northern) and Barra Jesús María near Boca de Catán (southern)*

is the only major river supplying sediments to the longshore current streaming along South Padre Island. Construction of Falcon Dam on the Rio Grande in 1955 reduced the sediment transported from more than 12,336,000 m^3 (10,000 acre-feet) per year to less than 1,234,000 m^3 (1,000 acre-feet) per year. Furthermore, the jetties at Brazos Santiago Pass intercept much of the sediment transported by the longshore current. Consequently, except for a small area adjacent to the Brazos Santiago Pass that has accreted, South Padre Island has been undergoing erosion since 1854 (Morton and Pieper 1975).

The highest elevations on barrier islands in Texas occur along foredune (primary dune) ridges landward from Gulf beaches. Padre Island has the tallest dunes along the Texas coast—some are as high as 15.2 m (50 ft) above sea level (Weise and White 1980). Average dune heights range from 6.1 to 7.6 m (20 to 25 ft) on North Padre Island (Brown et al. 1976). On South Padre Island, the heights of the tallest dunes range from 0.0 to 9.8 m (0.0 to 32.1 ft).

Most dunes on Tamaulipas barrier islands are lower than those on Padre Island (Britton and Morton 1989), but a few dunes were estimated to be up to 10.7 m (35 ft; J. W. Tunnell, pers. observ.). A unique island profile begins north of Mezquital and extends southward to Boca Ciega Pass, where coppice dunes vegetated with Sea Purslane prevail across most of the island, extending 1.0 to almost 3.0 km (0.6–1.8 mi) wide. Large (13–15 m [43–49 ft]) unvegetated dunes are located near the lee side of the barrier island, covering shrubs (primarily mesquite) and grasses (Gulf Cordgrass) (J. W. Tunnell, pers. comm.). Foredunes are less vegetated from north to south in Texas due to lower average precipitation rates (Weise and White 1980).

In many areas immediately seaward of foredunes, low coppice dunes may occur. These dunes typically have low, spreading vegetation that collects sand migrating up the beach face. Wind-shadow dunes may form behind coppice dunes as sand accumulates in wind-protected areas (Weise and White 1980). Coppice and wind-shadow dune elevations typically average 1.0 m (3.3 ft) tall, but taller ones may exist where sand accumulates in vegetation over extended periods. If these dunes are not eroded by storm surges or high tides, they may coalesce with the foredune ridge or form a new ridge.

Elevations within central parts of barrier islands are not uniform due to the formation of ridges and swales resulting from east-to-west dune migration across the island (Prouty and Prouty 1989). Ridge and swale topography forms during alternating wet and dry periods (fig. 10.4). During wet periods, vegetation growth stabilizes the active dune; during droughts, the active dune migrates ahead of the stabilized ridge and produces a long, linear depression or deflation trough (Hunter et al. 1972). These troughs typically hold standing water during years of high precipitation and support a diversity of wetland flora and fauna (Prouty and Prouty 1989). Increases in vegetative cover generally stabilize the deflation-flat area, but migrating dune fields may locally cover flats, increase elevation, and bury vegetation (Longley et al. 1989).

In low areas, where wind scours the surface down to the water table, sediment moisture eliminates further erosion. Average water table depth depends on precipitation, so erosion rates vary from year to year. During wet years, these troughs hold water and are important to barrier island ecology. The number and size of ponds and marshes vary from island to island, from north to south, and from year to year, primarily as

Figure 10.4. *Grassy flats blanket most Texas barrier islands, but dune migration during alternating wet and dry periods creates a landscape mosaic of hillocks and open areas in some places. (Photograph by Brian R. Chapman)*

Figure 10.5. *Ridge and swale topography is evident beyond the trough containing a temporary pond. (Photograph by Brian R. Chapman)*

a result of rainfall variability. Consequently, areal extent of these wetland habitats is difficult to quantify. Topographic ridges and swales within vegetated flats are oriented longitudinally along northern Padre Island, so ponds and marshes often exhibit a linear configuration (fig. 10.5). Ponds and marshes located immediately leeward of foredune ridges exhibit more elliptical shapes. These low areas are formed by erosion of fore-island dunes (Weise and White 1980).

Foredune ridges are oriented parallel to the Gulf beach, but dune migration occurs from southeast to northwest. Back-island dunes on northern Padre Island are aligned east to west (Hunter et al. 1972), while back-island dune fields are more obliquely oriented in relation to the prevailing southeast wind. Back-island dunes are generally <3.0 m (10 ft) high—typically not as tall as foredunes. However, some may reach about 7.5 m (25 ft) in protected areas on North Padre Island (Weise and White 1980) and about 9.0 m (30 ft) on South Padre Island (Judd et al. 1977; Brown et al. 1980). During droughts, unvegetated dunes migrate across Padre Island in a near west-northwest direction (Prouty and Prouty 1989). When rainfall is more abundant, vegetation coverage tends to expand and stabilize dunes, thus reducing the extent of migrating dunes. Back-island dunes averaging about 5.0 m (~15 ft) on Tamaulipan barrier islands were stabilized with shrub vegetation bordering Laguna Madre de Tamaulipas flats in some places (Selander et al. 1962).

Sediment composition on the barrier islands is predominantly sand with high permeability and low water-holding capacity (McGowen et al. 1979). Approximately 95% of the sand grains on South Padre Island are in the 0.125 to 0.25 mm range (Judd et al. 1977). Shell fragments are intermixed with sand during periods of high wave and wind energy. Sediment composition changes from north to south as a result of various sources of river sediment and from east to west as a result of Gulf and bay processes. Offshore bars and older submerged sand bodies deposited by ancient river systems serve as sediment reserves. Longshore currents and waves resuspend and deposit sediments on beaches. As sediments dry, the finer particles are transported by wind to backshore and dune ridges. Southeasterly winds transfer sand from beach to foredune ridges throughout most of the year, but much larger quantities of sediment are transported duneward during hurricanes, tropical storms, and spring tides (Weise and White 1980).

The longshore current along the Texas coast generally runs north to south along the eastern Texas coast until it converges with the northward longshore current of eastern Mexico and southernmost Texas around the Big and Little Shell beaches on Padre Island (Watson 1971). The beach face in these areas is more steeply angled, and the sands of the foreshore and backshore are comingled with many loosely sorted shell fragments (Britton and Morton 1989).

Foredune ridges are typically composed of fine, well-sorted sands transported up the beach face. Migrating dunes in vegetated flats leeward of the foredune ridge and back-island dunes have similar composition (Weise and White 1980), although some finer sediments may be transported from unvegetated deflation-flat areas (Longley et al. 1989). Sediments in the vegetated flats are generally composed of sand and shell deposits, whereas some finer sediments and organic matter may accumulate and overlay sand and shell in ponds and marshes (Weise and White 1980).

Average annual rainfall decreases and average temperature and evaporation increase from north to south along Padre Island. Freshwater ponds may dry during droughts, but some fresh water collects temporarily in low depressions following rainfall. These water sources usually become brackish or evaporate completely. The lens of a shallow, perched aquifer is located about 2.4–3.0 m (8–10 ft) below the surface of the island interior (Brown et al. 1976). Rainwater percolating downward through the sand replenishes the aquifer, which is situated above the saltwater table. Differences in density between the two bodies of underground water inhibit mixing.

Water levels and pond size vary throughout the year. For example, Sissom (1990) reported that water levels in three ponds on Padre Island National Seashore were lowest during September 1989 after summer drydown. Pond levels fluctuated during Sissom's study, increasing in depth to about 0.6 m (2.0 ft), as a result of heavy rains in March 1990, increasing volumes and surface areas. Water levels, pond surface areas, and water volumes eventually diminished to pre-flood conditions as water percolated through surrounding sands. Water levels measured during the study were used to assess potential changes from wetland to upland vegetation composition. High levels of relative humidity provide an additional water source when dew collects on plants at night. Relative humidity is an important aspect of the barrier island climate regime and typically remains high throughout the year because moist Gulf air mediates air temperatures.

Trophic Levels within Barrier Island Systems

Vascular plants are the predominant primary producers in barrier island habitats. Vascular plant species richness (number of species) is high, comparable to grassland-dominated mainland systems. Species diversity (a measure of community composition) changes within habitats across the barrier island and among barrier islands (Drawe et al. 1981). The flora of South Padre Island includes at least 204 species of angiosperms representing 47 families (Judd et al. 1977). Five families, Asteraceae (26 species on Padre Island), Cyperaceae (10 species), Poaceae (46 species), Fabaceae (24 species), and Scrophulariaceae (8 species), account for 55.9% of the species diversity. Twenty-four additional families (51% of the total) are each represented by a single species. Many of these are ruderal weeds found only in the disturbed habitats along the main roadways and around buildings. Some may be represented by a single plant, clone, or colony (Judd et al. 1977; Nelson et al. 2001).

Fresh water and nutrient availability influence primary production. Prevailing southeasterly winds transport sand, silt, small organic particles, and salt up the beach face (Whelan 1975) and may also affect coppice and foredune vegetation productivity (Britton and Morton 1989). Due to high soil porosity, the available nutrients are quickly leached out, particularly in the dunes (Drawe et al. 1981). Fires may temporarily reduce standing biomass, but often productivity increases the following growing season (Drawe and Kattner 1978).

Dune plants are specially adapted to tolerate high salinities, low water availability, and sediment accretion. Conditions (lower salinities, water availability, nutrients) are less stressful in the barrier flat zone, so plant species richness often is higher there. Plants adapted to periodic flooding and drying are primary producers in ponds and marshes located within barrier vegetated flats (Britton and Morton 1989)

Little information is available concerning decomposition processes in South Texas barrier islands. Litter decomposition has been described as a gradual process (Longley et al. 1989), accelerated only by fire (Kattner 1973; Drawe and Kattner 1978). No information is available on roles of bacteria and fungi in barrier island vegetation dynamics.

Invertebrates, from herbivores to primary and secondary carnivores, scavengers, and detritivores, perform every type of consumer role in a barrier island system. Importance of invertebrates as part of the food web has not been quantified for Texas and Tamaulipas barrier island chains, but extensive species lists and habitat preferences have been compiled for Padre Island (Ortiz 1976).

Fish communities in three earthen ponds within Padre Island National Seashore were composed of only three species (Caudle 1992). The Sheepshead Minnow was the most abundant fish species in two ponds, and the Mosquitofish was the dominant species in a third pond. Low numbers of Gulf Killifish were present in all three ponds.

Ten species of frogs and toads inhabit Padre Island (table 10.1), and most tend to remain close to the ponds and other moist habitats (Duran 2004). The Green Treefrog, Hurter's Spadefoot, and Rio Grande Leopard Frog were listed as uncommon in Padre

Figure 10.6. *A Woodhouse's Toad burying itself in loose sand (Photograph courtesy of the National Park Service)*

Table 10.1. *Amphibians and reptiles documented on Padre Island*

Frogs and Toads
Green Treefrog
Rio Grande Leopard Frog
Spotted Chorus Frog
Couch's Spadefoot
Great Plains Narrow-mouthed Toad
Gulf Coast Toad
Hurter's Spadefoot
Eastern Narrow-mouthed Toad
Texas Toad
Woodhouse's Toad

Terrestrial, Estuarine, and Pond Turtles
Eastern Snapping Turtle
Ornate Box Turtle
Red-eared Slider
Texas Diamond-backed Terrapin
Texas Tortoise
Yellow Mud Turtle

Sea Turtles
Green Sea Turtle
Hawksbill Sea Turtle
Kemp's Ridley Sea Turtle
Leatherback Sea Turtle
Loggerhead Sea Turtle

Lizards
Great Plains Skink
Green Anole
Keeled Earless Lizard
Little Brown Skink
Mediterranean Gecko
Rose-bellied Lizard
Six-lined Racerunner
Slender Glass Lizard
Texas Horned Lizard
Texas Spotted Whiptail

Snakes
Bullsnake
Checkered Gartersnake
Desert Kingsnake
Desert Massasaugua
Diamond-backed Watersnake
Eastern Hog-nosed Snake
Eastern Yellow-bellied Racer
Flat-headed Snake
Great Plains Ratsnake
Gulf Coast Ribbonsnake
Marsh Brownsnake
Mexican Milksnake
Rough Earthsnake
Saltmarsh Watersnake
Southern Texas Groundsnake
Texas Coralsnake
Texas Glossy Snake
Texas Indigo Snake
Texas Patch-nosed Snake
Texas Ratsnake
Texas Threadsnake

Alligator
American Alligator

Source: Duran 2004; PINS 2018.

Island National Seashore during an inventory in the early 1980s (PINS 1984), but Duran (2004) found them to be more abundant than previously believed. Toads, such as Couch's Spadefoot, Hurter's Spadefoot, and Woodhouse's Toad, may bury themselves in the loose sand on the back side of a dune (fig. 10.6) and emerge to breed only after a warm rain. Toads can often be observed at night on damp paved roads during and after rainy periods. Rainfall also stimulated breeding of the Gulf Coast Toad and Texas Toad on northern Mustang Island (R. Moore 1976). A population of the Eastern Narrow-mouthed Toad was recently confirmed on Padre Island Seashore when the breeding vocalizations of the tiny secretive toads led to their discovery (PINS 2018). All 10 amphibian species on the island prey on small insects (Garrett and Barker 1987). The amphibian fauna of the Tamaulipas barrier islands have not been surveyed since Selander et al. (1962) and remains poorly known.

On several occasions during the past 50 years, an American Alligator has crossed the Laguna Madre to take up residence in one of the larger, more permanent ponds on northern Padre Island (Duran 2004). The ponds also provide habitat for four species of turtles (see table 10.1). Though rare, two terrestrial turtles, the Ornate Box Turtle and the Texas Tortoise, inhabit the sandy interior of Padre Island. Ornate Box Turtles, which require drinking water, are prone to dehydration and tend to remain near ponds, but they do not spend much time in standing water. During droughts and periods of extreme heat, Ornate Box Turtles bury themselves in sand. The Texas Tortoise, which is less dependent on standing water, obtains most of its water from moisture in its food. It feeds on grass, succulent vegetation, and the pads, flowers, and fruits of the Texas Prickly Pear (Auffenberg and Weaver 1969; Conant 1975; Rose and Judd 2014). However, young Texas Tortoises drink water when available by taking it in through their nose rather than by mouth. This species was also documented in the stabilized dunes on a Mexican barrier island (Selander et al. 1962).

Ten species of lizards inhabit Padre Island (see table 10.1). The Keeled Earless Lizard, which feeds primarily on insects (Judd 1976; Garrett and Barker 1987), is abundant in partly vegetated dune habitats on Padre Island (Judd 1973; PINS 1984) and Tamaulipas barrier islands (Selander et al. 1962). The Six-lined Racerunner in Texas (Duran 2004) and the Texas Spotted Whiptail in Tamaulipas (Selander et al. 1962) are also found in dune habitats, where they prey on an assortment of invertebrates (Garrett and Barker 1987). Twenty-one snake species are documented on Padre Island (see table 10.1).

Gulf beaches are essential resting and feeding habitats for many resident and migratory birds, especially shorebirds (see chapter 15), and the grasslands are especially important for declining wintering grassland birds (Lawson et al. 2011). Padre Island is located on the Central Flyway, a major avian migratory route, and more than 380 species of birds have been recorded on the island. Seed-eating birds probably rely on the seasonal abundance of several plant species within the foredune zone as well as those within the vegetated barrier flat. However, no study has been conducted identifying and quantifying the importance of dunes to resident and migrant herbivorous birds. Northern Mockingbirds, Northern Cardinals, and Northern Bobwhites were documented within clay-dune habitat at the margins of the Laguna Madre de Tamaulipas (Selander et al. 1962).

Barrier islands along the Texas and Tamaulipas coast are important stopovers for migrating Peregrine Falcons (Hunt and Ward 1988). During the fall and spring, large concentrations of migrating Peregrine Falcons use Padre Island for hunting and foraging (Hunt et al. 1975).

Table 10.2. *Mammals confirmed to be present on Padre Island during the twentieth century*

Order Didelphimorpha	**Order Lagomorpha**
Virginia Opossum	Black-tailed Jackrabbit [TAM]
	Eastern Cottontail
Order Cingulata	**Order Soricomorpha**
Nine-banded Armadillo [TAM]	Least Shrew
	Crawford's Desert Shrew
	Eastern Mole
Order Rodentia	**Order Chiroptera**
Southern Plains Woodrat [TAM]	Brazilian Free-tailed Bat
Common Muskrat	American Perimyotis
Northern Grasshopper Mouse	Northern Yellow Bat
Coues's Rice Rat	**Order Carnivora**
Marsh Rice Rat	Coyote [TAM]
White-footed Deermouse	Common Gray Fox
North American Deermouse	Bobcat
Fulvous Harvest Mouse	Hog-nosed Skunk
Northern Pygmy Mouse	Raccoon [TAM]
Hispid Cotton Rat [TAM]	American Badger [TAM]
Texas Pocket Gopher [TAM]	**Order Artiodactyla**
Hispid Pocket Mouse	Nilgai*
Merriam's Pocket Mouse [TAM]	White-tailed Deer [TAM]
Gulf Coast Kangaroo Rat [TAM]	Collared Peccary
Eastern Fox Squirrel	Feral Pig*
Spotted Ground Squirrel [TAM]	
Mexican Ground Squirrel	
Nutria*	
House Mouse*	
Brown Rat*	
Black Rat*	

Source: Padre Island: Baker and Lay 1938; Hall 1951; Blair 1952; Raun 1959; Dalquest 1968; Koepke 1969; Kennedy et al. 1973; Yzaguirre 1974; Rabalais 1975; Baccus et al. 1977; Baker and Rabalais 1978; Segers and Chapman 1984; Zehner 1985; Harris 1988; Sissom 1990; Goetze et al. 1999; Hice and Schmidly 2002; Jones and Frey 2008, 2013. Tamaulipas: Selander et al. 1962.

Note: An asterisk (*) denotes an exotic (nonnative) species; a species confirmed to inhabit the Tamaulipas barrier islands is signified by [TAM].

For more than a century, biologists have visited the barrier islands of Texas and Tamaulipas to survey the mammalian fauna (True 1889; Lloyd 1891; V. Bailey 1905; Howell 1914; Baker and Lay 1938; Hall 1951; Blair 1952; Raun 1959; Selander et al. 1962; Dalquest 1968; Koepke 1969; Thomas 1972; Kennedy et al. 1973; Yzaguirre 1974; Rabalais 1975; Baccus et al. 1977; Baker and Rabalais 1978; Yates and Schmidly 1977; Segers and Chapman 1984; Zehner 1985; Harris 1988; Sissom 1990; Goetze et al. 1999; Hice and Schmidly 2002; Jones and Frey 2008, 2013). As a result of these efforts, a total of 43 species of terrestrial mammals has been recently documented from Padre Island (table 10.2). Six of the species were exotic, and 37 were native. One exotic species, the Nilgai, occasionally wanders over to the island from ranches on the mainland but has not established a permanent population. Records of bats foraging in the air above Padre Island likely are incomplete because of the proximity to the mainland and the difficulty of documenting their presence.

The Gulf Coast Kangaroo Rat, which inhabits the sparsely vegetated dunes on Padre Island (E. Smith 1986; Jones and Frey 2013), was collected in a similar habitat on the Tamaulipas barrier island (Selander et al. 1962). The low abundance of kangaroo rats on the Mexican barrier island was attributed to predation of young rats by land crabs (Hall 1951). Kangaroo rats are primarily herbivorous but occasionally feed on insects. The Spotted Ground Squirrel, an opportunistic feeder, inhabits Padre Island and Tamaulipas barrier islands (Selander et al. 1962). Its primary food sources include tender vegetation, seeds, and roots, but insects are occasionally taken, and the remains of a Keeled Earless Lizard was found in a seed cache within a ground squirrel burrow (fig. 10.7). Spotted Ground Squirrels most commonly frequent active dunes and vegetated flats on Padre Island (Segers and Chapman 1984). In dunes, the

Figure 10.7. *Spotted Ground Squirrels are active during the day and feed on tender vegetation. (Photograph courtesy of the National Park Service)*

Spotted Ground Squirrel forages in barren areas where it can readily identify and escape from terrestrial predators. To minimize capture by aerial predators, Spotted Ground Squirrels construct several refuge burrows throughout their home ranges in both dune and vegetated areas. Home burrows, typically located high in dunes, are constructed with varying complexity, but most contain a nest chamber and one or more food cache chambers. The ground squirrel travels using runways between and underneath vegetation to forage in vegetated flats. The Texas Pocket Gopher is abundant on northern Padre Island and Tamaulipas barrier islands (Selander et al. 1962; Jones and Frey 2013), but if this species occurs on South Padre Island, it is rare (F. W. Judd, pers. comm.).

Black-tailed Jackrabbits, Coyotes, White-tailed Deer, and American Badgers frequent dunes to some extent but primarily inhabit vegetated flats on Padre Island (PINS 1984; Goetze et al. 1999; Jones and Frey 2013) and Tamaulipas barrier islands (Selander et al. 1962). Moist grassy areas near ponds and other wetlands in some areas often harbor populations of the Marsh Rice Rat, Coues's Rice Rat, and Common Muskrat. Racoons regularly wander pond margins. The grasslands carpeting the middle of the islands support abundant populations of the Hispid Cotton Rat, White-footed Deermouse, Fulvous Harvest Mouse, and Eastern Cottontail, but Merriam's Pocket Mice and Hispid Pocket Mice may be extirpated from these habitats where they were once abundant (Jones and Frey 2013). Fringe zones, where grasses are encroaching on dunes, provide habitat for the Mexican Ground Squirrel, Northern Grasshopper Mouse, and Eastern Mole.

Most mammals probably use freshwater sources on islands. Sissom (1990) documented the presence of the Bobcat, Coyote, American Badger, Raccoon, and White-tailed Deer on Padre Island by identifying their tracks around ponds. In addition, he observed several Black-tailed Jackrabbits in habitats adjacent to pond margins.

Community Structure and Zonation

Carls et al. (1990) reviewed the literature regarding the relationships of Padre Island plant communities to published designations of ecological regions and suggested that several affinities exist. Parts of the island contain vegetative communities described as Gulf Prairies and Marshes of Texas (G. Thomas 1975), Marsh/Barrier Island (McMahan et al. 1984), and the Dune/Barrier Island subregion of the Gulf Coast Prairies and Marshes natural region (Diamond et al. 1987). Griffith et al. (2007) provided a detailed account of the ecoregions of Texas and included Padre Island as ecoregion 34i, Laguna Madre Barrier Islands and Coastal Marshes, within a larger ecological unit, the Western Gulf Coastal Plain. Setting aside these overviews, it is more practical to describe the distinct habitat zones on the barrier islands.

The foreshore is the barren beach area where waves come to die. Through daily inundation and scouring by tides, this region, also called the swash zone, is devoid of plant life, but many species of invertebrates live just beneath the sand surface. Sanderlings, Willets, and other shorebirds frequent the swash zone to feed on tiny creatures exposed by receding waves. A line of shell remains, organic debris, and trash high on the backshore marks the highest reach of the waves. The wrack line is visited during the day by scavenging birds, including the Ruddy Turnstone, and at night by other opportunistic animals, such as the Atlantic Ghost Crab, Coyote, and Raccoon. Ghost crabs begin their evening activities by crawling into the swash zone to wet their gills and drink.

The backshore zone is the relatively flat area between the normal high-tide line (berm) and the seaward base of the primary dunes (Judd et al. 1977). The backshore is often dotted with low coppice dunes stabilized by driftwood, debris, or salt-tolerant plants. Three species, Sea Purslane, Gulf Croton, and Sea Oats, most commonly support coppice dunes. The mat-forming ground cover of Sea Purslane is especially important in formation of embryonic dunes. Its vegetation structure impedes movement of sand across the backshore, capturing windblown sand to form the coppice dunes (Lonard and Judd 1980). These small clumps of sand ameliorate wave forces for leeward dunes during extreme high tides. Isolated patches of Seashore Dropseed also occur in the backshore zone, but infrequent inundation by seasonal spring and storm tides precludes establishment of many other species in this zone.

Primary dunes are located upslope from the backshore zone. The dunes in this zone unite to form a distinct foredune ridge that extends the length of the island unless broken by storm surges or human activities. Foredunes are rarely inundated by tides (Judd et al. 1977), but high seasonal tides and winter storm surges may reach the bases of primary dunes, uproot vegetation, and remove sand. Prevailing winds and air-blown salt influence vegetation type on the primary dune zone. Most primary dunes bear a crown of Sea Oats, but other species also cloak the slopes. Dominant among these are Gulfdune Paspalum, Fringeleaf Paspalum, Beach Tea, Beach Evening Primrose, Goatfoot Morning Glory, Fiddleleaf Morning Glory, and Partridge Pea.

The sandy soils of vegetated flats are dominated by a dense carpet of grasses. Low vegetated dunes are interspersed with ponds and marshes and active dune fields in many areas. Topography of this zone is a result of various environmental and geological factors, and the vegetation can be quite diverse. The foredune ridge ameliorates prevailing winds most of the year, thus protecting vegetation and promoting growth. A small grove of stunted Live Oaks persisted for many years on the vegetated flats near the north end of North Padre Island, but the oaks eventually died. A copse of small Live Oaks, likely the offspring of the preceding generation, now occupy the site. Live Oaks are absent on South Padre Island. Judd et al. (1977) suggested that the absence of trees and shrubs is largely due to unsuccessful colonization rather than a lack of dispersal.

Drawe et al. (1981) analyzed low vegetated dunes within vegetated flats on North Padre Island. These slightly elevated coastal sands tend to have greater plant diversity and vegetative cover than the surrounding flats, and the grass and forb compositions are slightly different from those of nearby primary dunes. Gulfdune Paspalum, Red Lovegrass, Seashore Dropseed, Sea Oats, Fringeleaf Paspalum, and Marsh Fimbry are the predominant grasses and sedges. The most common dominant forbs included Woolly Stemodia, Seaside Heliotrope, Frogfruit, Largeleaf Marsh Pennywort, and Corpus Christi Fleabane.

The climax vegetation on barrier islands is a mid-grass prairie, but the dynamic nature of barrier islands may limit the existence of potential climax communities (Drawe et al. 1981). Inhibiting factors include continual coastal dune migration, variable soil salinities in relation to storm surge, limited nutrient availability, and high evapotranspiration rates in the region. Extensive vegetation studies in the 1970s and 1980s were valuable in determining successional dynamics. The vegetation of South Padre Island is relatively well known. Lonard and Judd published 58 papers on various aspects of the vegetation of South Padre Island between 1974 and 2016. Climax vegetation slowly progresses toward a mid-grass or even a tallgrass prairie climax dominated by Seacoast Beardgrass, Bushy Bluestem, and Bitter Panicum.

Depressed wetland areas and the margins of ponds support grasses, sedges, Bulrush, Rattlebush, and in some places thick stands of Narrow-leaved Cattail (Kattner 1973; Drawe et al. 1981). Clumps of Black Willow mark locations where water is more permanent. Shallow, small depressions that retain water for a day or two after rainstorms are generally surrounded by a fringe of low sedges.

Shoregrass is the most common species on the vegetated margins of the wind-tidal flat on the lagoon side of Padre Island (Nelson et al. 2000). Glassworts, Beach Carpet, Sea Oxeye, Carolina Wolfberry, and Camphor Daisy are often interspersed with Shoregrass and usually constitute the mixed plant community just beyond the Shoregrass fringe.

The invertebrate community within South Texas and Tamaulipas barrier islands has received inadequate study even though these taxa form an important component of grassland structure and function. Atlantic Ghost Crabs, often seen scuttling to the golf ball–sized entrances to their burrows in the backshore, represent just one of 41 crustacean species on Padre Island (fig. 10.8). Numerous species of insects—especially grasshoppers, planthoppers, beetles, flies, and wasps—are integral members of complex invertebrate communities in vegetated areas. Ortiz (1976) investigated the insect fauna on Padre Island through four seasons and counted 245 species (nine orders) in backshore, primary dunes, coppice dunes, foredunes, and vegetated flat zones. Vertical zonation of the species was generally related to wind speed and direction, as most species were found within or beneath the plant canopy and in the soil on the leeward side of dune ridges.

Spatial zonation of the vertebrate community is related primarily to prey distribution, abundance, and degree of feeding specialization. Barrier islands are dynamic systems influenced by physical factors. Species with a generalist strategy have an advantage over those with a selective feeding mode. Most resident vertebrates are either generally herbivorous or omnivorous, and their home ranges typically correspond to body size. Small rodents usually inhabit vegetated areas of islands and feed on seeds, leaves, and stems characteristic in a particular zone. Mode of locomotion through vegetation and ability to burrow in sand further define each group's range.

Figure 10.8. *Atlantic Ghost Crabs dig burrows in the backshore and are common denizens of Padre Island. (Photograph by Hans Hillewaert, Creative Commons)*

The Hispid Cotton Rat, White-footed Deermouse, and Fulvous Harvest Mouse use runways underneath vegetation for foraging and protection from predators. They feed on herbaceous plants, seeds, and small invertebrates. The primary food of the Gulf Coast Kangaroo Rats is seeds, but the rodent also ingests herbaceous vegetation and small invertebrates as available in the dune zone and along edges of dune-vegetated flats (Davis and Schmidly 1994). Larger mammals forage in all vegetated zones to search for prey items. Most of the common species feed opportunistically on both vegetation and animals, but American Badgers are specialists on ground squirrels and burrowing rodents.

Migratory birds use island resources seasonally. Peregrine Falcons visit the barrier islands of Texas and Tamaulipas during the winter, where they perch on the apex of dunes. The falcons ambush low-flying birds over the foreshore, vegetated flats, wind-tidal flats, and lagoons. Neotropical migrants use all zones for food and protection during spring and fall passages. Some resident birds, such as the Eastern Meadowlark and Loggerhead Shrike, nest in vegetated flats yet feed in more open dune areas, particularly on the leeward, less windy sides (Rappole and Blacklock 1985, 1994).

The ephemeral nature of inland ponds within vegetated flats may limit community structure complexity. Most amphibian species are opportunistic breeders, taking advantage of wetland habitat and food following heavy rains (R. Moore 1976). Several vertebrate predators, including adult frogs, snakes, birds, and mammals, capitalize on increases in adult frog and toad abundance surrounding ponds during breeding and the subsequent population boom of tadpoles. Fish population dynamics and community structure have not been investigated in natural ponds, but environmental conditions (salinity, water temperature, dissolved oxygen, and depth) affected fish species diversity and habitat complexity in three engineered ponds. Physical parameters varied widely among the ponds and may have been the cause of low species richness and the dominance of a single fish species in each pond (Caudle 1992).

The barrier island food web is similar to that in a grassland-dominated system and probably follows a grazing pathway. Most resident and migratory species are dependent on roots, rhizomes, nectar, and seed production for survival. The numerous herbivorous invertebrate and vertebrate species are utilized as prey items for primary and secondary consumers, ranging in size from spiders, lizards, and toads to snakes, raptors, and larger mammals. Productivity of the barrier island system is influenced by physical factors, including variable rainfall, high evaporation and temperatures, and tropical storm surge events. Therefore, population dynamics of most species are closely correlated with environmental variability. Omnivory may give a consumer the advantage to survive in this stressful habitat.

Effects of Environmental Disturbance

Recovery of Vegetation after Fire Almost every year, some portion of the grasslands on Padre Island is affected by fires caused by lightning or human activities (Drawe and Kattner 1978; Drawe and Ortega 1996). For example, a grassfire on Padre Island caused by celebratory fireworks scorched a large area on 3 December 1999 (Hodnett 1999). More recently, a grass fire ignited by careless handling of fireworks on 3 July 2017 eventually consumed more than 130 ha (320 ac) near Padre Island National Seashore (Churchwell 2017). Although Drawe and Kattner (1978) examined the effect of fire on litter and green plant biomass in the Padre Island grassland, secondary succession following fire in the vegetated flats and the margins of tidal flats has received little scientific study until recently (Lonard et al. 2003; Lonard et al. 2004)

Lonard et al. (2003) reported on the recovery of vegetation on the margins of tidal flats at Padre Island National Seashore after a fire. Species richness, diversity, and importance, as well as evenness and vegetation abundance, were evaluated over 26 months. Seashore Dropseed was the dominant species on both burned and nonburned transects. Species richness and diversity were restored within 108 days after the fire, but vegetative cover was slow to recover. It took 19 months for the cover of burned areas to equal or exceed the cover of nonburned areas. Biomass also was slow to recover and did not equal the biomass of nonburned transects in 26 months. The principal effect of fire in the tidal-flat community was in the reduction of standing dead plants and surface litter.

Species richness, diversity, and evenness were restored quickly in vegetated flats (51 days) and secondary dunes (71 days) following a wildfire that denuded these habitats on Padre Island National Seashore. The burned secondary dune site retained the same dominant species as its nonburned counterpart from the beginning to the end of the study, but it took 106 days for the dominant species to be restored in the burned vegetated flats. One year after the fire, live plant ground cover was virtually identical on the burned and nonburned secondary dunes. Two years post-fire, the plant communities in burned and nonburned sites in both the vegetated flats and secondary dunes were similar. However, on the vegetated flats, live plant cover on the burned site exceeded that on the nonburned site. The dead plant biomass and standing dead plant cover remained markedly lower on the burned sites in both the vegetated flats and secondary dunes after two years.

Recovery of Vegetation in Washover Zones Barrier islands are prime examples of disturbance-dominated ecosystems. They are formed, shaped, and moved by hurricanes. During storms, waves and tidal surges often overwash the foredune ridge and uproot large expanses of vegetation while covering other vegetated areas with sand. Lonard and Judd (1993) studied the recovery of vegetation along simulated washover margins created to imitate hurricane effects. The research took place on South Padre Island, which is 55 km (34.2 mi) long and has a total area of about 16,200 ha (40,000 ac). This relatively low half of Padre Island experienced 23 washovers with an average width of 434 m (1,424 ft) on South Padre Island during a hurricane in 1986 (Judd and Lonard 1987).

To simulate the uprooting and scouring of vegetation by hurricanes, Lonard and Judd (1993) removed vegetation from the north and south margins of a washover adjacent to secondary dunes and vegetated flats. To imitate coverage of vegetation by sand during a hurricane, experimental plots containing vegetation were covered by sand at depths of 10, 20, and 30 cm (4, 8, and 12 in) on the north margin of the same washover area. Two control and two sand coverage plots were established for each depth of sand.

Following vegetation removal, cover increased at a rate of 5.8% per year on the north margin and 10% per year on the south margin of the washover. At the same time, cover decreased on control plots. Cover increased much more slowly in the 10 cm (2.7% per year) and 20 cm (2.0% per year) sand coverage plots. Conversely, the average annual increase in cover was 12% in 30 cm sand coverage plots. Marsh Fimbry, a sedge, was the major colonizing species in the sand coverage plots. Species richness was usually restored within a year in both vegetation removal and sand coverage experiments, but there was a significant change in species composition and shifts in dominance. These data suggest that the vegetation of washover margins is in a continuous state of change.

Blow-Out Gaps The natural dike formed by the foredune ridge can also be breached by high winds, forming a gap. In many cases, a blow-out gap is initiated by human disturbance. Excessive foot traffic across a dune or tire tracks from off-road vehicles can destroy the vegetation, leaving a barren area open to the ravages of wind (Leatherman and Godfrey 1979). Gradually, the path can widen, creating a large gap as the wind carries the exposed sand toward the grasslands in the middle of the island. Windblown sand from a blow-out buries the mid-island grasses and creates blow-out dunes, which can continue to be moved by the wind at rates of up to 10.7 m (35 ft) in one year. If conditions permit, blow-out dunes may eventually be stabilized by vegetation and appear as low, grass-covered hills. Others develop into dune fields that slowly migrate toward the Laguna Madre. Walking in dunes is discouraged because a path can widen into a large gap.

Effects of Oil and Gas Exploration Oil and gas development is widespread on Padre Island, even within Padre Island National Seashore. During a 30-year interval from 1951 to 1981, a total of 70 oil and gas extraction procedures occurred within the boundaries of the national seashore (NPS 2005). The development of a drilling site involves several processes that destroy vegetation: construction of a road to the site, clearing of a well-pad area, coverage of the well pad with solid surface materials, mowing the fringe around the pad, and construction of pipelines.

Since much of native prairie along the Texas coast has been converted to cropland or consumed by urban development (Smeins et al. 1991), the grasslands on Padre Island have become important stopover sites for migratory birds. Most grassland species are migratory and spend about half the year in migration and on grasslands in their southern wintering grounds (Igl and Ballard 1999). Populations of birds that breed and winter in grasslands have declined noticeably in North America primarily due to habitat loss or alteration (Brennan and Kuvlesky 2005). With these concerns in mind, Lawson et al. (2011) examined how the presence of roads and active and abandoned well sites affected avian abundance. The lower height of mowed grassy areas with reduced litter surrounding active well sites seemed conducive to grassland bird use, but there was little difference in avian abundance among the three site types. Birds were attracted to well pads that had nearby perches, such as stakes, posts, other artificial structures, or low shrubs.

However, a potential major impact from oil and gas activities involves the contamination of nearby soils. Soil samples from 18 oil and gas production sites located within Padre Island National Seashore contained elevated levels of four heavy metals (barium, chromium, lead, and zinc), sodium, salinity, pH, and petroleum hydrocarbons (Carls et al. 1995). The long-term effects of substrate alteration and toxicants on the flora and fauna of the barrier island are largely unknown.

11 Ecology of the South Texas Coastal Sand Plain

TIMOTHY E. FULBRIGHT AND FRED C. BRYANT

Extending from the Laguna Madre inland to the Rio Grande Plains of South Texas, the South Texas Coastal Sand Plain is a biologically diverse landscape with a remarkable history and unique ecology. Other names sometimes applied to the region include the Coastal Sand Sheet, Eolian Sand Sheet, and Wild Horse Desert. Spanish explorers and early travelers, such as Lieutenant Colonel Ramon Valera, referred to the Sand Plain in 1842 with the foreboding name "El Desierto de los Muertos" (the Desert of the Dead; Brune 2002; Leach 2017). The Sand Plain is, however, far from dead. It is an ecologically distinct region with a rich diversity of plants and animals that we referred to elsewhere as "The Last Great Habitat" (Fulbright and Bryant 2002).

A region of prairies, Plateau Live Oak and Honey Mesquite forests, savannas, and undulating sand dunes, the Coastal Sand Plain is distinctive ecologically. The unique region is in part a product of the fine sands that somewhat abruptly replace the clay soils of the Texas Coastal Prairie south of Los Olmos Creek, a tributary of Baffin Bay. South toward Mexico, the undulating sands give way to the clays and loams of the Lower Rio Grande Valley about 80 km (50 mi) to the south of Los Olmos Creek. The

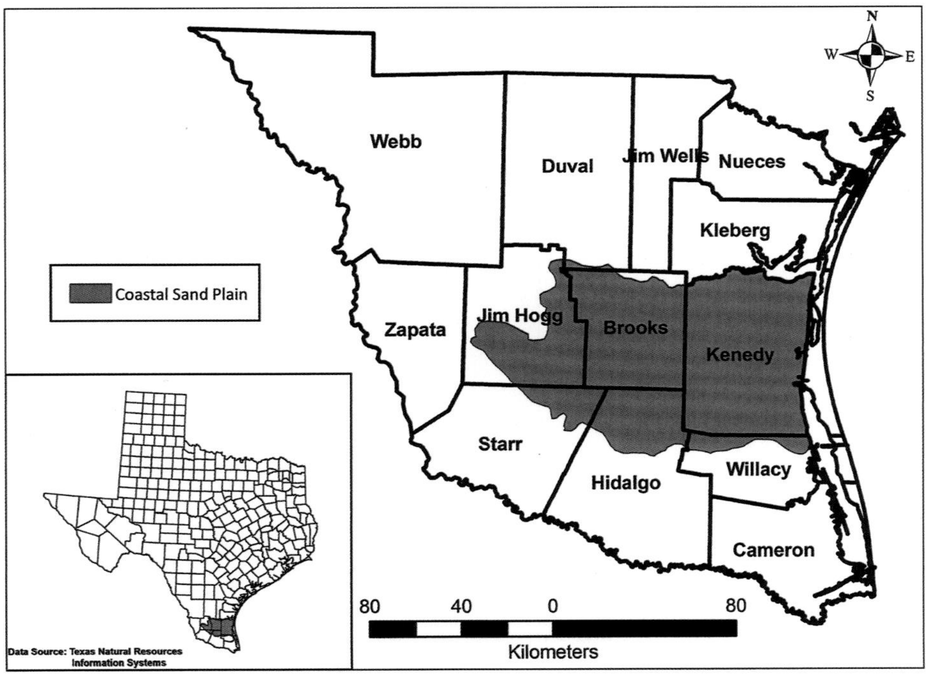

Figure 11.1. *Map of the Texas Coastal Sand Plain (Adapted from Carr, n.d.; map prepared by Humberto Perotto-Baldiviseo, Texas A&M University–Kingsville)*

Sand Plain stretches 137 km (85 mi) or more from the Laguna Madre on the east to the Bordas Escarpment, a caliche plateau formed by the edge of the Goliad Formation (Poole et al. 2007) on the west. Altogether, the Sand Plain covers >800,000 ha (2 million ac), a vast expanse the size of the states of Rhode Island and Delaware combined (fig. 11.1). Less well-known than many other similarly distinctive areas, such as the Nebraska sandhills, the Sand Plain deserves recognition as a unique and important ecological region.

Soils and Topography

Undulating sand dunes and swales create topographic relief and are distinctive features of the Coastal Sand Plain (fig. 11.2). Sands were blown inland during the Holocene and have accumulated for tens of thousands of years or more (see chapter 3). Eolian-deposited sands range from less than 3 m (<10 ft) thick to more than 8 m (>26 ft; Forman et al. 2009). The eolian-deposited sands within the upper 1.0 m (3.3 ft) of the Sand Plain were deposited <3,000 years ago. Late Pleistocene fluvial-deltaic sands and strandplain sands occur below the eolian-deposited sands. The Goliad Formation of Pliocene origin underlies the Late Pleistocene deposits and consists of sand and sandstone mixed with silt, clay, and caliche layers (McGraw 1984).

Barren, active dunes are present on about 5% of the Sand Plain (fig. 11.3). Most of the active dunes occur in the eastern part of the Coastal Sand Plain near the Laguna Madre, although individual dunes occur almost 129 km (80 mi) inland. Active dunes migrate in the direction of the prevailing wind, moving from southeast to northwest most of the time. In the eastern portion of the region, dune fields may cover several square kilometers and individual dunes may exceed 102 ha (250 ac). Maximum topographic relief in these areas is usually about 9 m (30 ft), although individual dunes may reach 15 m (50 ft) tall (W. Carr, n.d.). Active dunes often grade into vegetated dunes to form the dune and swale landscapes that characterize the Sand Plain.

Figure 11.2. *Undulating sand dunes and swales that create topographical relief are a distinctive feature of the Coastal Sand Plain. (Photograph by Timothy E. Fulbright)*

Figure 11.3. *Active sand dunes are most abundant in the eastern half of the Texas Coastal Sand Plain. (Photograph by Timothy E. Fulbright)*

Most of the contemporary landscape supports a cover of vegetation, but the undulating nature of the landscape provides visual evidence of dune activation and movement in the past. Periods of extensive dune reactivation and movement across the landscape occurred periodically during the past 3,000 years and were coincident with severe droughts (Forman et al. 2009). Droughts reduce the cover of vegetation, and when vegetation cover is below about 30%, dune systems may reactivate with prevailing winds (Pye and Tsoar 1990; Kuriyama et al. 2005). The most recent period of dune migration occurred about 1790 CE following extreme drought during 1789 and 1790 CE (Forman et al. 2009).

Well-defined drainage systems are rare in the Coastal Sand Plain (W. Carr, n.d.). Numerous internally drained, ephemeral ponds derived from blow-outs or subsidence of the underlying strata dot the landscape during periods of adequate rainfall. A fine-textured hard pan of caliche with a high water-holding capacity underlies the ponds (Martin and Hehnke 1981). The degree of salinity of these ponds depends on the underlying substrate.

Climate

Climate of the Coastal Sand Plain is also an important factor in the unique ecology of the region. The climate is classified as subhumid, subtropical (Kottek et al. 2006). Rainfall is bimodal with peaks in May and June (Fulbright et al. 1990). Mean annual precipitation ranges from about 55 cm (22 in) in the western portion to 74 cm (29 in) in the eastern portion (Fulbright 1994; US Climate Data 2017). The high coefficient of variation in annual precipitation, about 35%, is comparable to that of tropical deserts (Norwine and Bingham 1986) and results in years of severe drought followed by periods of abundant rainfall. The normal precipitation deficiency, a measure of moisture available for plant growth, is essentially the same in the Sand Plain as in the deserts of western Texas (Fulbright and Bryant 2002).

When the first Europeans arrived, the Sand Plain experienced cooler winters than the present. For example, Father Espinosa, a minister of Mission San Juan Bautista located on the Rio Grande between Laredo and Eagle Pass (west of the Sand Plain) described north winds during winter in the early eighteenth century as being so cold that they “froze water underground and brought snow that lasted several days” (Weddle 1968). The present climate is characterized by dry, mild winters with infrequent frosts and hot, humid summers—daily maximum summer temperatures frequently exceed 38 °C (100 °F). Summer heat has a profound effect on many organisms that inhabit the region, influencing survival, reproduction, daily movements, and use of habitat.

Human History

Archaeological investigation of the Coastal Sand Plain has been limited (Hester et al. 1989). Evidence of early occupation during Paleoindian and Early Archaic periods includes mammoth and bison bones found in the proximity of dart points and a fluted projectile point at a site in Kenedy County (Gonzalez et al. 2015). Middle Archaic artifacts such as Bulverde and Tortuga dart points also have been found in the region (Hester et al. 1989). Late Archaic period sites are more common than earlier sites, and Late Prehistoric sites are widespread in the Coastal Sand Plain, where small triangular arrow points are often present (Gonzalez et al. 2015). Archaeologists have speculated that the lack of permanent streams and surface water kept occupation of the Sand Plain limited to small, mobile groups of people that visited the region during wet periods (Hester et al. 1989).

Historic Native American occupants of the Sand Plain included transient Comanches and Lipans and probably nomadic bands of Coahuiltecans. Jean Louis Berlandier, a French botanist who crossed the Sand Plain traveling from Matamoros to Goliad during the early nineteenth century, commented that in 1830 Comanches came to Rancho de las Animas, 15 leagues (83 km [52 mi]) north of the Arroyo Colorado and there killed a herdsman (Ohlendorf et al. 1980). He also noted that Lipans frequented the region between Matamoros and Goliad.

During the late eighteenth and early nineteenth centuries, the Sand Plain was divided into several Spanish and Mexican land grants. These included the San Juan de Carricitos grant, which comprised 243,482 ha (601,657 ac) in present-day Hidalgo, Kenedy, and Willacy Counties, granted by Spain to Don Jose Narcisco Cavazos in 1792 (TGLO 1983). This was the largest grant made in South Texas. Heirs of Cavazos remained in possession of the land after his death until Indians drove off most of them in 1811.

Major Spanish settlements were established during the eighteenth century along the San Antonio River, to the north of the Sand Plain, and to the southwest along the Rio Grande. Lack of permanent water kept the Sand Plain sparsely populated by Spanish colonists. Berlandier crossed the Sand Plain from Matamoros to Santa Gertrudis Creek, near present-day Kingsville, in 1834 and recorded that he had seen no dwellings after leaving the Rancho de las Animas (Ohlendorf et al. 1980). He referred to the Coastal Sand Plain as a “wilderness of plains almost 25 leagues [139 km; 85 mi] wide, covered with small forests of oaks.”

Many descendants of the original Spanish and Mexican land grantees in the region moved south of the Rio Grande following the Battle of San Jacinto in 1836 in obedience to an order from the Mexican government. A dispute occurred between Mexico and Texas during 1836–46 about whether the northern boundary of Mexico was the Nueces River or the Rio Grande (McGraw 1984). Abandonment of the region by land grantees and the boundary dispute led to the Coastal Sand Plain becoming a haven for bandits, smugglers, and abandoned livestock. Wild horses were so abundant in the region during the 1840s and 1850s that early travelers referred to it as the “Wild Horse Desert” (Inglis 1964).

Large herds of wild horses and cattle on the Sand Plain demonstrated that the region had good potential for ranching (D. Smith 1986). In recognition of this potential, several large ranches were established there during the mid-nineteenth century, some of which are still in operation. In 1852, Major James Durst paid the Ballí family 1,600 pesos for the 37,634 ha (92,996 ac) La Barreta grant (D. Smith 1986). This purchase and the marriage of Durst's daughter, Mary Helena, to John Barkley Armstrong in 1878 marked the beginning of the Armstrong Ranch. Captain Richard King established the King Ranch in 1853; two of the four divisions of the King Ranch lie in the Sand Plain. A few years later, King's partner, Captain Mifflin Kenedy, established the Kenedy Ranch, which sprawls over more than 162,000 ha (400,000 ac) in the Sand Plain. Other large ranches were to follow.

The major economic activities of the contemporary Sand Plain are ranching, energy production, and hunting. The region maintains extensive areas of unfragmented wildlife habitat because many large ranches remain intact. Average size of "farms" in Kenedy County, for example, is 13,245 ha (32,728 ac), whereas the statewide average is 212 ha (523 ac; USDA 2012). Another factor in the lack of fragmentation is the sparse human population of the region. Kenedy County, which lies entirely in the Sand Plain, had a population of 0.2 persons/mi^2 in 2020; the state average was 111 persons/mi^2 (US Census Bureau 2020b, 2020c). Of the 254 counties in Texas, only Loving County in West Texas has a lower population density.

Vegetation and Wildlife

We referred to the Coastal Sand Plain as "The Last Great Habitat" because it contains extensive tracts of contiguous wildlife habitat, unique vegetation communities, high biotic diversity, and remains of keystone importance for migratory birds (Fulbright and Bryant 2002). Terrestrial plant communities range from grasslands to Live Oak forests. The diversity of plant communities and subtropical climate of the region result in a diversity and abundance of wildlife. Fourteen plant species are endemic to the Sand Sheet (W. Carr, n.d.), and distinctive birds such as the Green Jay and Northern Beardless Tyrannulet reach the northern limit of their range in the region (fig. 11.4).

Figure 11.4. *The Green Jay is a ubiquitous resident of the Plateau Live Oak and Honey Mesquite forests of the Texas Coastal Sand Plain. (Photograph by Timothy E. Fulbright)*

Figure 11.5. *Seacoast Bluestem is the prevailing dominant grass of the Coastal Sand Plain. (Photograph by Timothy E. Fulbright)*

Plant Communities A unique combination of soils and climate promotes a mosaic of different plant communities in the Coastal Sand Plain. Coastal environments on the eastern edge of the Sand Plain include extensive wind-tidal flats dominated by forbs such as glassworts and Sea Oxeye (Fulbright et al. 1990). Saltgrass occurs in salt marshes, with Saltmeadow Cordgrass important in intermediate marshlands.

Plant communities farther inland include grasslands, Plateau Live Oak forests, savannas dotted by mottes (small, isolated groves) of Plateau Live Oaks or Honey Mesquites, and mesquite–mixed-brush shrublands. The vegetation complex has been classified as the Mesquite–Live Oak–Seacoast Bluestem cover type (Fulbright 1994). Seacoast Bluestem is the prevailing dominant of the Coastal Sand Plain (fig. 11.5; Diamond and Fulbright 1990). Associated grasses include Crinkleawn, Brownseed Paspalum, and Thin Paspalum. Gulfdune Paspalum dominates with Seacoast Bluestem in swales and moderately drained flats where important associates include Big Bluestem and Gulf Muhly. Hooded Windmillgrass is another important component of Coastal Sand Plain grasslands, particularly westward from the coast.

Pan American Balsamscale often is codominant with Seacoast Bluestem westward from the coast, but Arrowfeather Threeawn may dominate with Seacoast Bluestem and Pan American Balsamscale on dry midslopes (McLendon 1991). Heavy grazing by livestock results in a decline in Seacoast Bluestem and other dominant mid-grasses and an increase of short grasses. Dominant short grasses in heavily grazed areas of the Coastal Sand Plain include Arrowfeather Threeawn, Fringed Signalgrass, Hairy Grama, Red Lovegrass, and Coast Sandbur. Gulf Cordgrass communities are scattered throughout the eastern and central Coastal Sand Plain in poorly drained areas (Fulbright et al. 1990). Bushy Bluestem and Switchgrass are sometimes important associates of Gulf Cordgrass.

Camphorweed may dominate along with Seacoast Bluestem on well-drained flats (Diamond and Fulbright 1990). Silverleaf Sunflower may be locally abundant in wet years. Other important forbs include Hoary Milkpea, American Snoutbean, Cardinal Feather, and Erect Dayflower. Vegetation of the Coastal Sand Plain also includes at least 54 species endemic to Texas (W. Carr, n.d.). Rare plants in the region include Jones' Nailwort, Lundell's Nailwort, Sand Brazoria, Short-crown Milkvine, and Texas Peachbush (W. Carr, n.d.; Poole et al. 2007).

Tanglehead was historically a dominant with Seacoast Bluestem in the western portion of the Coastal Sand Plain where the fine sands transition to sandy loams (McLendon 1991). Tanglehead began to increase dramatically during the late 1990s and has become one of the most dominant grasses in the western portion of the region (Bielfelt and Litt 2016; Wester et al. 2018). Although Tanglehead is a native grass, ecologically it has behaved as an invasive species that reduces native plant species richness (Bielfelt and Litt 2016). Reasons for the dramatic increase in Tanglehead are unclear, but shifts in rainfall patterns coupled with a regional decline in livestock stocking rates are possible enablers of the invasion (Wester et al. 2018).

Nonnative grasses of subtropical origin also have invaded portions of the Coastal Sand Plain, including Bermudagrass, Buffelgrass, Guinea Grass, Natal Grass, and several Old World bluestems. Bermudagrass largely occurs in moist areas, and Buffelgrass is most abundant in areas with sandy loam soils on the edges of the Sand Plain. Natal Grass grows in the fine sands of the Sand Plain but is generally scattered in stands of native grasses and seldom forms dense, monotypic stands. Guinea Grass was virtually nonexistent in the Coastal Sand Plain during the 1980s but has become increasingly abundant and now forms monotypic stands in parts of the region. Dominance by Guinea Grass reduces species richness of native grasses and forbs (S. Moore 2010), but a combination of burning and cattle grazing may reduce Guinea Grass dominance and increase native plant species richness (Ramirez-Yanez et al. 2007). Old World bluestems typically grow in clay-textured soils and may invade depressional areas where clays are present. Seeds of Old World bluestems tend to cling to vehicles, and the plants often become established along the edges of caliche roads in the region.

Native grasslands in the Coastal Sand Plain will continue to be degraded in the future by woody plant encroachment and increased abundance and dominance of nonnative grasses. Construction of caliche roads associated with energy development in the fine sands provides avenues for establishment of woody plants such as mesquite and exotic grasses such as Old World bluestems because the road material alters soil properties, making conditions more favorable for establishment of these species. In addition, runoff water is concentrated at the edges of the roads, making roadsides effectively more mesic than the surrounding landscape. Increased abundance of pipeline and highline rights-of-way also provide the opportunity for the ingress of exotic grasses because these are periodically mowed and the mowing equipment is frequently contaminated with seeds of exotic grasses.

Woody Plants Woody vegetation has continually increased in the Coastal Sand Plain since settlement by Europeans (Inglis 1964). During 1995–2004, for example, mesquite increased 7% on a private ranch in the Coastal Sand Plain (Rangeland Consultants, Inc., unpub. data, 2006). Live Oak mottes increased by 3% during the same period. Reasons for the increase in woody vegetation since pre-Columbian times are unclear but probably include lack of periodic fire (Bogusch 1952). Fire-return interval was about three years in pre-Columbian times, which was probably adequate

Figure 11.6. *Honey Mesquite savannas cover extensive areas of the Texas Coastal Sand Plain. (Photograph by Timothy E. Fulbright)*

Figure 11.7. *Plateau Live Oak forests occupy much of the Texas Coastal Sand Plain. (Photograph by Timothy E. Fulbright)*

to suppress woody vegetation (Frost 1998). Changes in climate, increasing atmospheric carbon dioxide concentrations, consumption and dissemination of seeds by cattle, and heavy grazing that reduced fine-fuel loads to inadequate levels to support fire are additional factors that may be contributing to this increase.

A Honey Mesquite savanna consisting of woody plant mottes scattered within a matrix of grassland occupies much of the Sand Plain where shallower sands prevail (fig. 11.6). The existing savannas developed through encroachment of mesquite into grassland, followed by development of an understory of smaller shrubs, including Spiny Hackberry and Lime Prickly Ash (Franco-Pizana et al. 1995; Franco-Pizana et al. 1996). Individual mesquites in these mottes may exceed 9 m (30 ft) in height (Fulbright et al. 1990).

Establishment of Spiny Hackberry beneath mesquite canopies during motte development in Coastal Sand Plain grassland provides an example of how extreme temperatures influence structuring of vegetation communities. Mesquite canopies provide shade so that soil-surface temperatures are lower than those in adjacent open grassland. After birds eat Spiny Hackberry fruits, perch in the canopies of mesquite, and defecate the seeds, Spiny Hackberry becomes established beneath mesquites. Whereas the extreme soil temperatures in open grasslands (>60 ° C [140 °F]) inhibit germination during early spring and summer when Spiny Hackberry fruits are produced (Fulbright et al. 1995), the cooler temperatures (about 40 °C [104 °F]) of soil surfaces shaded by mesquite canopies reduce dormancy of the seeds. Concentration of seeds from defecation by birds and reduction of seed dormancy result in a flush of seed germination when moisture conditions are favorable. Density and species richness of understory shrubs increase as the mesquites age. In some areas of the Coastal Sand Plain these mottes have expanded over time and coalesced, forming continuous mesquite-dominated woodland.

Deeper soils (>1.8 m [>6.0 ft]) of the Coastal Sand Plain are occupied by savanna with scattered Live Oak mottes (Hanselka, n.d.). Plateau Live Oak density in these mottes ranges from 308 to 1,240 trees/ha (125 to 502 trees/ac; Beasom and Haucke 1975). In some areas, Live Oak mottes have expanded to form dense, continuous Live Oak forests (fig. 11.7), which are distributed in a discontinuous fashion. Altogether, areas of Live Oak cover a quarter to a third of the Coastal Sand Plain (Fulbright et al. 1990). Included in these forests are stands of a low-growing form of Live Oak known locally as "Running Live Oak," which consists of dense trees <1.5 m (5.0 ft) tall.

Animal Communities The diverse soils, plant communities, and topography of the Coastal Sand Plain support a diversity of animals, some of which do not occur elsewhere in the United States. The region provides critical stopover habitat for Neotropical migrants. The Sand Plain hosts at least five reptile and amphibian species, seven bird species, and three mammal species that are considered threatened according to the Texas Parks and Wildlife Department (TPWD 1998a, 1998b, 1999a, 1999b, 1999c, 2003).

Herpetofauna The Texas Indigo Snake is among the many unique amphibians and reptiles that inhabit the Sand Plain (fig. 11.8). The Indigo is one of Texas' largest snakes, reaching 2.4 m (8.0 ft) or more in length (Werler and Dixon 2000). Indigos eat woodrats, birds, frogs, toads, and other snakes, including rattlesnakes, and have home ranges as large as 110 ha (273 ac) during summer. The Bullsnake and Western Diamondback Rattlesnake also are abundant in the Sand Plain.

Figure 11.8. *The Texas Indigo Snake is among the unique amphibians and reptiles that inhabit the Texas Coastal Sand Plain. (Photograph by Timothy E. Fulbright)*

Texas Tortoises, Keeled Earless Lizards, Black-spotted Newts, Lesser Sirens, and an isolated population of Woodhouse's Toad (Dixon 2000) are also found in the Coastal Sand Plain. Choruses of the Western Narrow-mouthed Toad, Plains Spadefoot, and Texas Toad enliven the night air following spring and summer precipitation events, and the populations of these amphibians may fluctuate dramatically with the sporadic rainfall on the Sand Plain (Fulbright et al. 2013). Several reptile species, including the Common Spotted Whiptail, Six-lined Racerunner, Great Plains Skink, and Texas Horned Lizard, use rodent burrows as temporary daytime or nocturnal shelters (Galan and Light 2017).

Birds A large proportion of long-distance migratory bird species in North America funnel along the Texas coast between the Gulf of Mexico to the east and the semiarid South Texas Plains to the west during autumn and spring migrations (Fig. 11.9; Rappole and Blacklock 1985). Igl and Ballard (1999) documented 190 species of birds wintering in or migrating through the Coastal Sand Plain. Migrating birds from northern and eastern North America concentrate along the coast during passage because it is energetically challenging to cross the open water of the Gulf of Mexico where stopover sites are unavailable (Walsh et al. 2017). The Texas Coastal Sand Plain, consequently, is a crucial area for resting and refueling (Rappole and Blacklock 1985). Concentration of migratory birds crossing the Texas Coastal Sand Plain is greater than is found in other migration routes in North America (Walsh et al. 2017). Based on estimates made using radar, migratory traffic rates of birds crossing the Texas Coastal Sand Plain near the

Figure 11.9. *Bird species such as this American Redstart funnel along the Texas Coastal Sand Plain during autumn and spring migrations. (Photograph by Timothy E. Fulbright)*

coast were 600% greater than rates documented elsewhere in North America (Walsh et al. 2017). Live Oak mottes of the Coastal Sand Plain are an important staging habitat where migrating birds can rebuild fat reserves before continuing migratory flight. At least 39 species of migrants, including 18 warbler species, have been recorded in Live Oak mottes (Williges 1989).

At least 88 species of terrestrial birds breed in the Coastal Sand Plain (Lipschutz 2016). Unique brushland birds that breed there include the Northern Beardless Tyrannulet, Tropical Parula, Green Jay, and Groove-billed Ani (Brush 2005). Although populations of grassland birds are declining at an alarming rate in the United States (Brennan and Kuvlesky 2005), the Texas Coastal Sand Plain provides important breeding habitat for several grassland species, including Horned Larks, Grasshopper Sparrows, Dickcissels, and Eastern Meadowlarks (Fall 1973; Walsh et al. 2017). Red-billed Pigeons also are occasionally observed in the region (Fall 1973).

Many migrating raptors funnel through South Texas. More than 300,000 Broad-winged Hawks migrate each year from their breeding grounds in the eastern United States and Canada south through coastal South Texas on their way to wintering areas in central and southern Mexico (Bildstein 1999). At least a dozen species of hawks occur in the Coastal Sand Plain as either migrants or residents (Turner et al. 2014), and the nation's largest population of Ferruginous Pygmy-Owls (including the "cactus" subspecies) occurs in Kenedy and Brooks Counties within the Coastal Sand Plain (Woodin et al. 2000). The estimated population of Ferruginous Pygmy-Owls on a 29,000 ha (71,661 ac) area dominated by Live Oak and mesquite was 745 to 1,823 (Mays 1996).

Populations of the Northern Bobwhite have declined during the past 50 years in North America as a result of habitat loss, fragmentation, and land-use changes that have reduced habitat suitability in many areas. However, the decline has been smaller in South Texas, where the Coastal Sand Plain, especially the western portion in Jim

Hogg and Brooks Counties, remains a stronghold for the species. The presence of large, unfragmented grassland-shrubland habitat dominated by native grasses on private ranches is critical to conservation and the future abundance of bobwhites in the Coastal Sand Plain. The importance of the Coastal Sand Plain to bobwhite conservation was underscored by designation of the area by the National Bobwhite Conservation Initiative as a Quail Conservation Region of National Significance.

The Rio Grande Wild Turkey is abundant on the Sand Plain, and its population density is closely tied to the availability of roosting habitat provided by the Live Oak forests and savannas (Caveny et al. 2011). The Sand Plain has served as a source of turkeys for repopulating areas where they have been extirpated or starting new populations where turkeys are absent (Forgason and Fulbright 2003).

Ephemeral freshwater playas scattered throughout the Coastal Sand Plain provide valuable habitat for wintering waterfowl and shorebirds (see chapters 13 and 15). The Laguna Madre hosts the largest concentrations of wintering Redheads in North America (Ballard et al. 2010). Ponds in the Sand Plain provide dietary fresh water for Redheads and Lesser Scaup.

Mammals Texas Pocket Gopher mounds are a conspicuous feature of the region (fig. 11.10; Galan and Light 2017). These subterranean rodents, endemic to southern Texas and northeastern Mexico, create deep, elongated tunnels in the sandy soil. The Gulf Coast Kangaroo Rat, Hispid Pocket Mouse, Merriam's Pocket Mouse, and Hispid Cotton Rat are also common in the Sand Plain (Chapman and Packard 1974; Mason 2014; Adcock 2016). Thirteen species of bats have been documented from the region (Chapman and Chapman 1990).

The Mountain Lion is the largest mammalian carnivore in the Texas Coastal Sand Plain, and though periodically seen in the region, its population status is unclear. Small populations of the endangered Ocelot occupy dense brush thickets at the western edge of the Sand Plain, but the Bobcat is a common inhabitant of the entire

Figure 11.10. *Texas Pocket Gopher mounds dot the landscape of the Texas Coastal Sand Plain. (Photograph by Timothy E. Fulbright)*

region. Two other cats, the Jaguar and Jaguarundi, once roamed the region but are now extirpated.

American Bison roamed the Coastal Sand Plain during the Pleistocene along with the Woolly Mammoth, ground sloths, and horses (Gonzalez et al. 2015), and a remnant population of the Pronghorn persisted in the western portion of the Sand Plain until the latter part of the twentieth century (Weniger 1997). The White-tailed Deer is now the most abundant large native herbivore in the Sand Plain. Originally found primarily along drainages, these deer occur throughout the region in part because of the increase in woody vegetation. The Collared Peccary is the only other large native herbivore in the region.

Several species of exotic ungulates have been introduced on various private landholdings in the Sand Plain. These animals are largely confined by high fences, but two species are free ranging—the Feral Pig and Nilgai. Feral Pigs are abundant in the Sand Plain. Although hunting Feral Pigs is a popular pursuit, the animals damage native vegetation and may contribute to the decline of native animals (Seward et al. 2004). Feral swine also harbor several diseases, including brucellosis, leptospirosis, and toxoplasmosis.

Nilgai, an antelope introduced from India, were released on the Norias Division of the King Ranch in 1929 and 1930 (fig. 11.11; Lehmann 1948). From an initial introduction of 12 animals, the population increased to 200 by the mid-1950s. The Nilgai population has recently been estimated to be approximately 38,000 animals, the majority of which are in the Sand Plain (Fulbright et al. 2017). The high degree of diet overlap between Nilgai and White-tailed Deer suggests that the two species potentially compete for food, particularly during the drought episodes that are common to the region (Hines 2016; Fulbright et al. 2017). Some Nilgai carry Cattle Fever Ticks, which cause concern among landowners in the Sand Plain and among government agencies (Foley et al. 2017).

Figure 11.11. *The Nilgai, an antelope introduced from India, is abundant in the Texas Coastal Sand Plain. Several Nilgai are seen here in the foreground foraging alongside White-tailed Deer. (Photograph by Timothy E. Fulbright)*

Future of the Texas Coastal Sand Plain

The vast, pristine nature of the Coastal Sand Plain has been under increasingly intensive assault in recent years. Many people see the large tracts of land that are unoccupied by humans as the perfect place to put things that cause environmental damage, regardless of their biological and historic value. Past proposals that would forever fragment and degrade the vegetation and wildlife communities of the Coastal Sand Plain included development of a bombing range for the US Navy, a spaceport, wider highways, and upland disposal of spoil from dredging on the Gulf Coast Intracoastal Waterway. Fortunately, most of these proposals did not happen, but energy development to meet the needs of a growing human population is expanding. Continued oil, natural gas, and wind-farm energy developments with associated roads, pipelines, and other infrastructure will inevitably result in degradation of wildlife habitat and introduction of exotic grasses across the Coastal Sand Plain. The truly unique and valuable natural landscapes of the region are worth making economic sacrifices to conserve. The forces of "progress" are formidable, but large tracts of the landscape are in the hands of farsighted, conservation-minded stewards that will help the Coastal Sand Plain remain "The Last Great Habitat."

12 *Laguna Madre de Tamaulipas*

ALEJANDRO FIERRO-CABO, CARLOS ZAMORA-TOVAR, LEONARDO ARELLANO-MÉNDEZ, AND CARLOS CINTRA-BUENROSTRO

The Laguna Madre de Tamaulipas (LMT) is separated from its Texas counterpart by the deltaic plain of the Rio Grande (Río Bravo), where layers of river sediment accumulated over time. The LMT is formed by two contiguous basins, the Septentrional, located from Las Flores to El Carrizal Island (figs. 12.1a), and Meridional from El Carrizal Island to Río Soto la Marina (figs. 12.1b, c). The Septentrional basin is larger (1,100 km^2 [425 mi^2]) and deeper (0.7 m [2.3 ft] average depth) than the Meridional (900 km^2 [348 mi^2]), which is so shallow (0.35 m [1.12 ft] average depth) that its southern end tends to dry out during the dry season (March–May).

The lagoon extends across two hydrological regions, the watersheds of the Río Bravo and the Río San Fernando–Río Soto la Marina systems. The hydrological regime of the LMT results from the balance between seawater inputs through 13 passes (4 permanent and 9 temporary passes), freshwater from storms, runoff, irrigation water surplus (in the north), and rivers—mainly from Río San Fernando but with limited inputs from Río Soto la Marina (Bello et al. 2009; SEMARNAT 2015). About 200 natural islands in the LMT are protected by three long barrier islands and a peninsula separating the lagoon from the ocean (Tunnell and Judd 2002; also see Cooper et al. 2007). The majority of the islands in the lagoons are fetch-limited barrier islands formed with much less energy than the ocean-facing barriers and are thus much smaller (Cooper et al. 2007).

Because of the LMT's location and wide latitudinal range, three climate types, according to the Köppen system as modified by García (2004), influence portions of it: temperate, semi-warm, and subhumid with scarce rains all year in the north; dry, semi-dry, and hot with summer rains in the central zone; and temperate, semi-warm with summer rains in the south. Consequently, regional climatic conditions in three zones interact with surrounding coastal dynamics and their genesis influences the distribution of flora and fauna (Villarreal 2000). These zones are coincident with the municipalities (counties) to which they belong (Tunnell and Judd 2002; Zamora-Tovar 2007; Bello et al. 2009; SEMARNAT 2015). The north zone (Matamoros municipality) is characterized by periodically flooded alluvial plains and numerous small lagoons that often interconnect with the LMT. The predominant soils are Arenosol, Solonchak, and Vertisol types, which are saline, flooded, and exposed to both wind and marine erosion. These are covered by thorn scrub, halophytes, and coastal dune vegetation. Scarce rainfall during the year (average 600 mm [23.6 in]) and eolian sand transport maintain the numerous active dunes in the area. The central zone (San Fernando municipality) includes a large bay, Laguna el Catán, which is shaped by the delta deposits of the San Fernando River. The river and summer rains averaging 650 mm (25.6 in) annually provide fresh water that decreases salinity and supports ample

Figure 12.1. *Views of (a) the southern part of the Northern Laguna Madre de Tamaulipas at the Rio San Fernando Delta and tidal flat (El Carrizal); (b) the Southern Laguna Madre de Tamaulipas north of Carboneras; and (c) the seagrass beds behind the narrow barrier island near Soto la Marina (Photographs by J. W. Tunnell, February 1999)*

a

b

c

development of seagrasses. The lagoon in this zone has a highly erosive mainland margin, and the barrier islands separating it from the Gulf of Mexico have few permanent (dredged) and temporal passes. In addition to the three soil types present in the north zone, eroded and saline Chernozem soils are present. The soils support halophytes, thorn scrub, and scattered mangroves. A gradation of landscapes exists along the shorelines of the south zone (Soto la Marina municipality), ranging from hills to low coastal plains formed by the Río Soto la Marina. The barrier islands lack passes, which results in intermittent partial drying of the lagoon and its marshes. The soils are Arenosol, Solonchak, Gleysol, and Vertisol types, which are saline, frequently flooded, and depending on location, covered by thorn scrub, dry deciduous forest, or mangroves. The Soto la Marina zone experiences summer rains averaging about 900 mm (35.4 in) annually.

A human population of more than 9,700 inhabits the LMT shores, and of these about 1,400 live on islands (Mendoza et al. 2011) and depend mainly on the fisheries of a few highly productive species (fig. 12.2). Commercial fisheries of Striped Mullet, Brown Shrimp, Blue Crab, and Eastern Oyster represent about 40% of the volume and value for all State of Tamaulipas fisheries (Leija-Tristán et al. 2000; Zamora-Tovar 2004, 2007; SEMARNAT 2015). Small-scale cattle ranching and bird-watching tourism complement the economic activities of its residents.

a

b

Figure 12.2. *Fishing boats at Carboneras, Southern Laguna Madre de Tamaulipas: (a) modern fiberglass outboard boats; (b) a restored old sailing skiff, once used to pull* chinchorros *(large bag seines) (Photographs by J. W. Tunnell, December 1998)*

Its location in a transition zone between the Nearctic and Neotropical biogeographic regions fosters the confluence of species with tropical and temperate affinities and results in remarkable biodiversity (Rzedowski 2006). Many endangered, threatened, and endemic species, as well as a multitude of migratory birds (Phillips 1911; Leopold 1959), are components of the LMT biological richness. For example, about 36% of the world's population of Redheads (ducks) spend the winter months in the LMT (SEMARNAT 2015). Overall, the area supports 458 vascular plants, 106 macroalgae, 73 mammals, 284 birds, 44 reptiles, 17 amphibians, 161 fishes, 140 crustaceans, 143 molluscs, 48 echinoderms, 28 polychaetes, and 17 cnidarians (Leija-Tristán et al. 2000; Raz-Guzmán and Huidobro 2002; Zamora-Tovar 2002, 2004; Barba and Sánchez 2005; Celis et al. 2007; Zamora-Tovar 2007; Fredericq et al. 2009; Contreras-Lozano et al. 2013; Correa-Sandoval and Rodríguez-Castro 2013; CONANP 2014; SEMARNAT 2015; Raz-Guzmán and Soto 2017).

Plant Associations

Vegetation of the LMT is typical of the Tamaulipan Biotic Province. Freshwater aquatic vegetation (submerged and emergent) is present in ephemeral and semipermanent wetlands. Common wetland species include Southern Cattail, Olney's Three-square Bulrush, White Spikerush, Waternymph, and Tropical Royalblue Waterlily. Matorral Espinoso Tamaulipeco (Tamaulipan Thorn Scrub) is an endemic regional association distributed in patches in the area and strongly threatened by land-use change. Characteristic thorn scrub species include Texas Ebony, Spanish Dagger, Texas Sage, Soapbush, and Coyotillo. The deciduous tropical dry forest is fragmented but represents the northernmost distribution along the Gulf of Mexico coast. Some characteristic species are Spiny Hackberry, Tenaza, Huisache, and Gumbo-limbo. Dune vegetation on barrier islands includes many herbaceous species, such as Sea Oats, and a diverse assemblage of woody species, including Buttonwood and Railroad Vine.

Salt marsh vegetation includes various halophytic grasses and succulents, such as Shoregrass, Saltwort, Sea Purslane, a seepweed, and Perennial Glasswort. Extensive mangrove associations in the south zone are composed of four species: Black Mangrove, Red Mangrove, White Mangrove, and Buttonwood. Only Black Mangrove occurs in the north zone. Coastal prairie habitats are dominated by grasses, such as Gulf Cordgrass, Dune Sandbur, and Fringeleaf Paspalum (Leija-Tristán et al. 2000; Zamora-Tovar 2007; CONANP 2014; SEMARNAT 2015).

Submerged Aquatic Vegetation Seagrass meadows and associated macroalgae are widely distributed within the LMT. Of the approximately 272,800 ha (674,103 ac) of submerged surface area (Tunnell and Judd 2002), about 5.3% (14,459 ha [35,729 ac]) is covered by seagrasses and other aquatic macrophytes. According to our observations (Zamora-Tovar, unpub. data), the attached macroalgae species with highest biomass in seagrass meadows are (from highest to lowest) *Penicillus capitatus, Udotea flabellum, Gracilaria tikvahiae, Chondrophycus poiteaui, Caulerpa mexicana,* and *Acetabularia crenulata*. Common drift species are also present and include *Digenea simplex, Dictyota menstrualis, Hypnea musciformis, Hydroclathrus clathratus, Jania adhaerens, Chondria littoralis,* and *Polysiphonia subtilissima*. In addition, the epiphytes *Blennothrix lyngbyacea, Pneophyllum fragile, Acrochaetium flexuosum, Neosiphonia ferulacea, Polysiphonia denudate,* and *Hincksia mitchelliae* are common. Macroalgal species richness varies markedly throughout the year: 54 species in the summer, 40 in the summer–autumn, but only 12 in the winter–spring (table 12.1). This flora bears a

Table 12.1. *Species of photosynthetic bacteria (cyanobacteria and blue-green algae), macrophytic algae, seagrasses, and mangroves documented in Laguna Madre de Tamaulipas*

Species	Zone	Season
Photosynthetic bacteria		
Anacystis aeruginosa	C	YR
Blennothrix lyngbyacea	C, S	YR
Lyngbya lutea	C	YR
Phormidium nigroviride	S	YR
Porphyrosiphon notarisii	C	YR
Schizothrix arenaria	C	YR
Spirulina subsalsa	C	YR
Xanthophyceae (yellow-green algae)		
Vaucheria sp.	C	WS
Phaeophyceae (brown algae)		
Dictyopteris delictatula	S	SA
Dictyota menstrualis	N, C	YR
Feldmannia indica	C, S	YR
Hincksia mitchelliae	N, C, S	YR
Hummia onusta	C	WS
Hydroclathrus clathratus	C	SA
Lobophora variegata	S	YR
Padina gymnospora	S	SA
Petalonia fascia	S	WS
Rosenvingea intricata	N, C	WS
Sargassum filipendula	C, S	YR
Sargassum fluitans	C, S	YR
Sargassum natans	S	YR
Spatoglossum schroederi	C, S	SA
Sphacelaria rigidula	S	YR
Sphacelaria tribuloides	C	SA
Rhodophyceae (red algae)		
Acanthophora spicifera	C	SA
Acrochaetium flexuosum	C	YR
Acrochaetium hypneae	C	SA
Acrochaetium microscopicum	C	YR
Acrochaetium secundatum	C	SA
Agardhiella ramosissima	C, S	SA

Species	Zone	Season
Agardhiella subulata	C	SA
Aglaothamnion boergesenii	C	YR
Botryocladia occidentalis	C	SA
Bryocladia cuspidata	N, C, S	SA
Bryocladia thyrsigera	N, S	YR
Catenella caespitosa	C	SA
Centroceras clavulatum	N, C, S	YR
Ceramium cimbricum	C, S	SA
Ceramium deslongchampsii	S	SA
Ceramium diaphanum	S	SA
Champia parvula	C	SA
Chondria dasyphylla	C	WS
Chondria littoralis	N, C	SA
Chondrophycus poiteaui	N, C	YR
Digenea simplex	N, C	YR
Erythrotrichia carnea	C, S	SA
Eucheuma isiforme	C	SA
Gelidium americanum	C	YR
Gelidium crinale	N, S	YR
Gracilaria blodgettii	S	YR
Gracilaria cornea	C	YR
Gracilaria gracilis	N, C	YR
Gracilaria mammillaris	N, S	YR
Gracilaria tikvahiae	C, S	YR
Grateloupia filicina	S	YR
Grateloupia pterocladina	C	SA
Haliptilon cubense	C, S	SA
Haliptilon subulatum	S	SA
Halymenia floresia	S	SA
Herposiphonia secunda	N	SA
Hypnea musciformis	N, S, C	YR
Hypnea spinella	N, C	YR
Hypnea valentiae	C	SA
Jania adhaerens	C	YR

Table 12.1. *Continued*

Species	Zone	Season
Liagoropsis schrammii	S	SA
Neosiphonia ferulacea	C	YR
Neosiphonia gorgoniae	C	SA
Pneophyllum fragile	C, S	YR
Polysiphonia boldii	N, C	WS
Polysiphonia denudata	C	YR
Polysiphonia echinata	C	YR
Polysiphonia havanensis	C	YR
Polysiphonia subtilissima	C	YR
Pterocladiella bartlettii	S	YR
Rhodymenia pseudopalmata	S	YR
Sahlingia subintegra	S	YR
Spyridia filamentosa	C	YR
Spyridia hypnoides	N, S	SA
Stylonema alsidii	S	YR
Titanoderma pustulatum	C	YR
Chlorophyceae (green algae)		
Acetabularia crenulata	C	YR
Boodleopsis pusilla	C	SA
Bryopsis hypnoides	N, C, S	SA
Caulerpa ashmeadii	C	SA
Caulerpa mexicana	C	SA
Chaetomorpha aerea	N, S	YR
Chaetomorpha brachygona	C, S	YR
Chaetomorpha linum	S	SA
Cladophora albida	C, S	WS
Cladophora coelothrix	C	SA
Cladophora dalmatica	C, S	SA
Cladophora laetevirens	N, C, S	SA
Cladophora montagneana	C	WS
Cladophora vagabunda	N, C, S	YR
Codium isthmocladum	N, S	SA
Halimeda incrassata	C	YR
Penicillus capitatus	C	SA
Rhizoclonium africanum	C, S	YR
Rhizoclonium riparium	N, C, S	WS
Udotea flabellum	C	SA
Ulva clathrata	C, S	WS
Ulva fasciata	N, C, S	YR
Ulva flexuosa	N, C, S	YR
Ulva lactuca	C, S	WS
Ulva linza	C	WS
Ulva prolifera	C, S	YR
Magnoliopsida (seagrass)		
Shoal Grass	N, C, S	YR
Star Grass	C	YR
Widgeon Grass	C	YR
Manatee Grass	N, C, S	YR
Turtlegrass	C	YR
Magnoliopsida (mangroves)		
Black Mangrove	N, C, S	YR
Buttonwood	N, C, S	YR
White Mangrove	N, C, S	YR
Red Mangrove	N, C, S	YR

Source: Zamora-Tovar, unpub. data.
Note: Zones (see text for geographic extent): N = north, C = central, and S = south. Season: YR = year-round, SA = summer–autumn, and WS = winter–spring.

floristic relationship with the Caribbean tropical flora, but there are also species with a Carolinean temperate affinity (Zamora-Tovar, unpub. data).

The five seagrass species occurring in Texas seagrass meadows are also present in the LMT and have consistently been reported for the lagoon: Turtlegrass, Star Grass, Shoal Grass, Manatee Grass, and Widgeon Grass (Hildebrand 1958, 1969; Cornelius 1975; DUMAC 1993, 1996; Garcia-Gil et al. 1993; Martínez and Novelo 1993; Mora-Olivo 1996; Withers 2002a; López-Martínez 2013). Although they are continually present, the distribution, density, and percentage cover of each species are undergoing change (Arellano-Mendez, unpub. data). All five species currently occur in the central zone (Carboneras to Punta Piedras); Shoal Grass and Manatee Grass constitute most of the beds in the northern zone (El Mezquital), while Shoal Grass, Manatee Grass, and Widgeon Grass are present in the southern zone (San José de los Leones to La Yegua).

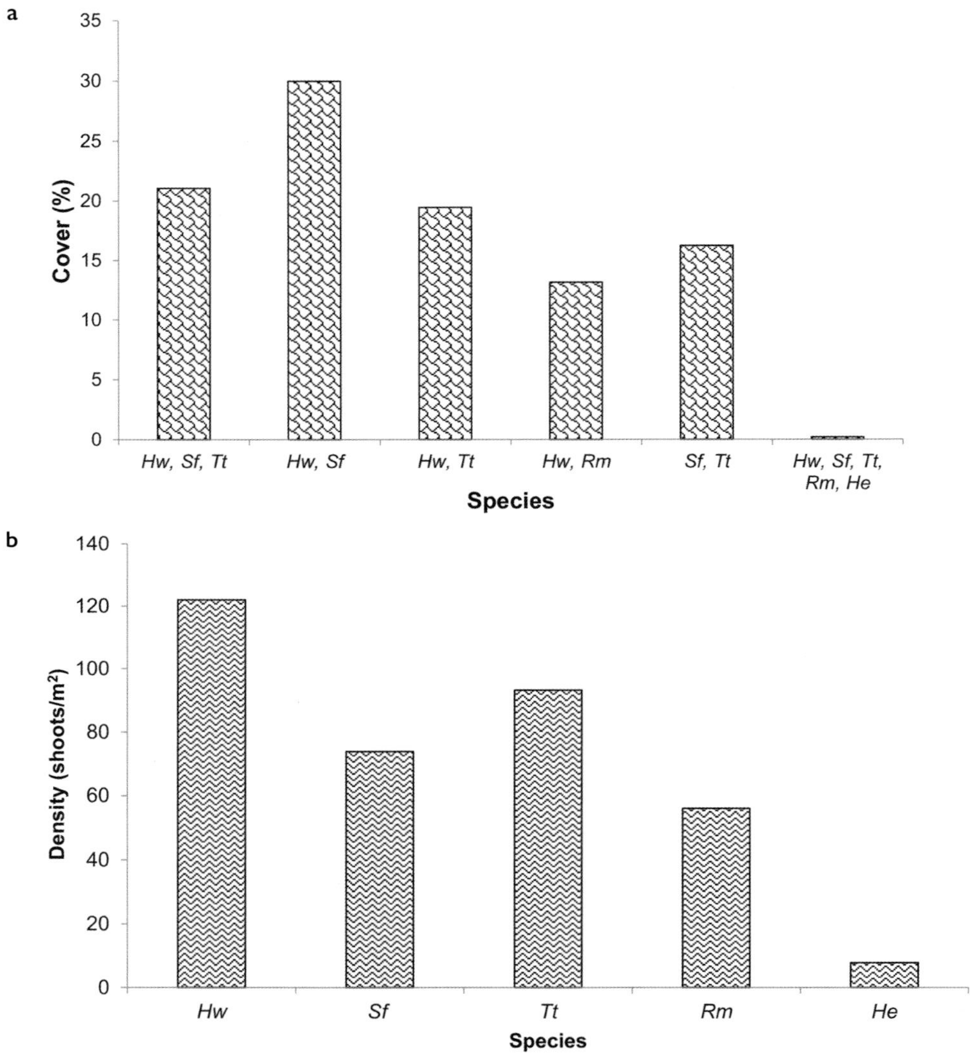

Figure 12.3. *Seagrass cover and species composition of meadows in the Laguna Madre de Tamaulipas: (a) five species are present in various associations; (b) shoot density by species. Values are averages from 625 sampling stations surveyed from January 2017 to October 2018:* Hw = *Shoal Grass;* Sf = *Manatee Grass;* Tt = *Turtlegrass;* Rm = *Wigeon Grass;* He = *Star Grass. ([b] from Arellano-Mendez, unpub. data)*

Shoal Grass, the most abundant seagrass in the LMT (fig. 12.3a), is present in about one-third of the lagoon covered by seagrass beds. It is the main food source for the Redhead, a migratory species that generates economic benefits to local communities through hunting and birding. The seagrass occurs in shallow areas with generally less than 0.5 m (1.7 ft) depth but is sometimes found at depths between 0.5 and 1.0 m (1.75 and 3.3 feet). It frequently coexists with Manatee Grass and Turtlegrass, but near Punta de Alambre it exists within small patches of Widgeon Grass along the barrier island. Historically, Widgeon Grass was found only along the western margin of the lagoon in association with freshwater discharges (Zamora-Tovar and Mora-Olivo 2014), but lower salinity in many locations within the lagoon modified its distribution. The density of Shoal Grass averages 122 shoots/m^2 (fig. 12.3b) with a biomass of 134 g/m^2 (0.028 lb/ft^2; fig. 12.4), slightly lower than previously reported by Zamora-Tovar and Mora-Olivo (2014).

Turtlegrass has increased within the LMT, particularly around the passes. This supports the hypothesis that temperate ecosystems in the Gulf of Mexico are undergoing tropicalization, but it could also be related to ecological succession within seagrass meadows. Turtlegrass occurs primarily in the central zone, always in multispecific beds with high percentage of cover (see fig. 12.3a), densities of 93 shoots/m^2 (see fig. 12.3b), and a total biomass of 126 g/m^2 (0.026 lb/ft^2; see fig. 12.4).

Manatee Grass is currently distributed in all zones of the lagoon, and its areal coverage has increased based on a comparison to an older report (Mora-Olivo 1996). Changes are likely related to dredging activity near the settlements of Carboneras, Punta de Alambre, and Punta Piedras in the central zone. Dredging produces greater water circulation and exchange and lower salinity, particularly near Carboneras, where Manatee Grass dominates. Manatee Grass beds always include Shoal Grass or Caribbean Turtlegrass with a diverse macroalgae assemblage. The average density of Manatee Grass was 74 shoots/m^2 (see fig. 12.3b) with a total biomass of 92 g/m^2 (0.019 lb/ft^2; see fig. 12.4).

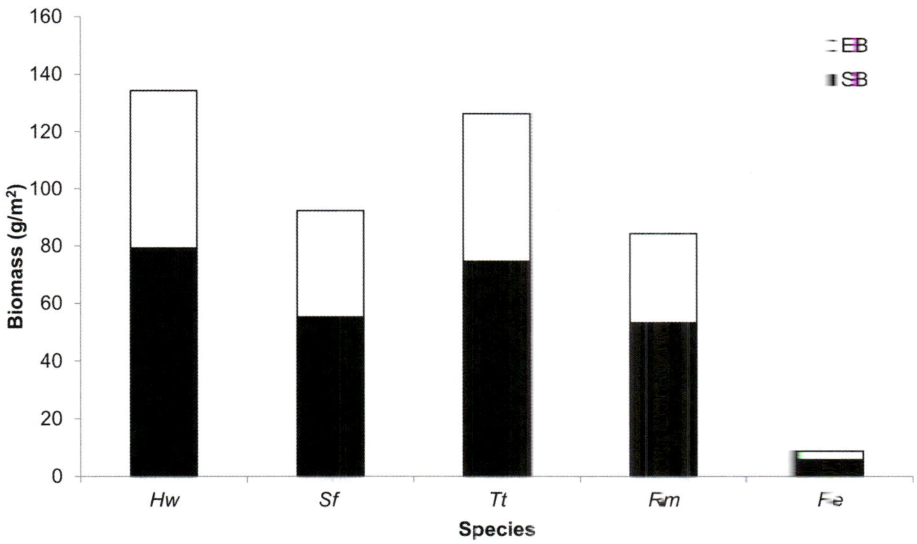

Figure 12.4. *Dry biomass per area unit of five seagrass species in Laguna Madre de Tamaulipas: EB = aboveground biomass (shoots); SB = belowground biomass (roots and rhizomes). See figure 12.3 for other details and abbreviations.*

Marine phanerogam distribution in the LMT has components of tropical regions, including the Caribbean (Short et al. 2001), but there are species-specific differences in areal cover, plant structure, or community composition (Arellano-Méndez 2011). Since 1977, seagrass cover has declined within the LMT, which is in alignment with ongoing global trends. Cornelius (1975) observed that seagrasses covered 42,000 ha (103,784 ac), but the extent of coverage decreased to 34,000 ha (84,016 ac) by 1996 (DUMAC 1996) and to 31,000 ha (76,603 ac) in 2001 (Withers 2002a). Our recent estimates of 14,459 ha (35,729 ac) of seagrasses in the LMT indicate a net loss of 19,541 ha (48,287 ac) over 22 years.

Carbon storage in seagrass meadows is a significant ecosystem service, but the decline in seagrasses is obviously affecting the capacity of the lagoon to provide this service. No information on blue carbon stocks exist for the LMT, but a back-of-the-envelope estimate based only on Turtlegrass suggests that this species stores an aboveground stock of 0.05 Mg C/ha based on a shoot biomass of 126 g/m^2. This is more than 10 times lower than estimates for this species in Bahía de la Ascensión, Quintana Roo (14.5 Mg C/ha) (Arellano-Méndez et al. 2015). However, these data represent an incomplete approximation. It will be necessary to conduct studies that include all species and the organic carbon stored in sediments. Recent increases in human activities within the LMT, such as periodic dredging to maintain passes and channels, the opening of new passes, and the use of trawl nets and other damaging fishing gear, are likely contributing to the decline of seagrass meadows.

Mangroves and Salt Marshes The management plan of the protected area in the LMT identified both mangrove and salt marsh ecosystems as main conservation targets. Salt marshes extend over 50,800 ha (~125,530 ac) within the LMT (SEMARNAT 2015) and cover ample portions of floodplains on the mainland margins of the lagoon, the inner shores of barrier islands, and the system of lagoonal islands. The north zone hosts the greatest extent of salt marshes, where the most common herbaceous species are Saltwort, Perennial Glasswort, Bush Seepweed, Gulf Cordgrass, Sea Oxeye, Seashore Dropseed, Shoregrass, Saltgrass, and Sea Purslane. Although the floristic compositions are similar, intertidal salt marshes in the central and south zones cover less area than their northern counterpart. In the interior of barrier islands, these species associate with halophytes (herbaceous and woody) typical of coastal dunes (Zamora-Tovar 2007; SEMARNAT 2015). Salinity determines the distribution and assemblage of LMT salt marsh species. Contreras and Castañeda (2004) recognized that under high-salinity conditions Bush Seepweed and Perennial Glasswort dominate the association but are replaced by Saltwort in soils with lower salinities and by the halophyte grasses, Saltgrass and Shoregrass, in less flooded soils. In slightly more well-drained plains, Gulf Cordgrass and Denseflower Cordgrass are also present (Zamora-Tovar, unpub. data). At higher elevations, Gulf Cordgrass becomes dominant under more intermittent flooding, whereas Perennial Glasswort and Sea Oxeye are most common where floods are periodic.

The mangrove ecosystem of the LMT is represented by natural and reforested mangrove stands. At the southernmost point of the LMT (at or near the Tropic of Cancer), the Río Soto la Marina estuary is the transitional zone between the typical tropical mangal, which is structurally more diverse and developed, and the northern counterparts are characterized by monospecific scattered scrubs of Black Mangrove (Tunnell 2002). However, our more recent observations confirm the presence of the four species reported for the northern Gulf of Mexico: Black Mangrove, Buttonwood, White Mangrove, and Red Mangrove (Zamora-Tovar, unpub. data). All four species

Figure 12.5. *Planted Black Mangrove and Red Mangrove in the north zone of Laguna Madre de Tamaulipas (Isla Leonardo, Matamoros municipality) four years after establishment in 2011 with the participation of the local community*

are used successfully in reforestation programs. The region has increasingly favorable conditions for tropical species, a consequence of climate change and what is now referred as the "tropicalization" of the Gulf of Mexico (Comeaux et al. 2012; Bianchi et al. 2013; Cavanaugh et al. 2014; Yáñez-Arancibia et al. 2014). Mangrove reforestation establishes new stands on previously mangrove-occupied sites and new areas (Zamora-Tovar et al. 2011; fig. 12.5). Mangrove expansion in the LMT is also partially attributable to the ecological resilience of these plants (Lugo et al. 2014) and the role that salt marsh species play in the facilitation of mangrove seedling establishment and growth (Milbrandt and Tinsley 2006; Peterson and Bell 2015). Consequently, mangroves have also expanded into salt marsh communities (Perry and Mendelssohn 2009; Guo et al. 2013; Lunt et al. 2013; Scheffel et al. 2013; Saintilan et al. 2014; Armitage et al. 2015; Rodríguez et al. 2016).

The natural (rather than planted or reforested) mangrove community in the north zone is an almost pure scrub of Black Mangroves associated with salt marsh species. In the central zone, the same species is most common along the mainland shoreline, where it is associated with salt marsh and mesquite vegetation. Some islands, such as El Garzón, El Viborero, and La Yegua, are covered by Black Mangrove, representing relict communities and habitats. The south zone is more diverse and has the tallest mangrove trees (fig. 12.6). All four species are present in the margins of Río Soto la Marina and other nearby coastal lagoons (Laguna de Morales and Laguna de Almagre) and intermix with a diverse assemblage of plant species from the dry deciduous forest, salt marshes, sand dunes, and freshwater aquatic plants. On all of the barrier islands, Buttonwood exists as a low, creeping shrub in isolated patches on the upper part of dunes, while scattered shrubby individuals of Black Mangrove occupy the lagoon shoreline (landward) of the barrier islands. All mangrove species and their habitats are protected under Mexican federal legislation, regardless of their natural or reforested origin. Furthermore, the Comisión Nacional para el Conocimiento y Uso de la Biodiversidad (CONABIO) recognizes 81 mangrove priority sites in the country. Three of these sites in the LMT (GM37, Delta del Río Bravo; GM41, La Pesca; and GM44, Laguna de Morales) encompass about 210 ha (~519 ac) of mangrove habitat (CONABIO 2009; Valderrama-Landeros et al. 2017).

Figure 12.6. *More diverse and developed mangrove association, characteristic of the south zone of Laguna Madre de Tamaulipas. The pneumatophores of Black Mangroves are obvious.*

Natural and reforested Black Mangrove stands from the protected areas (including Laguna de Morales) have taller trees in the south zone but with narrower crowns as a result of intraspecific competition. Buttonwood trees are taller and wider at lower latitudes, as they occupy dunes and areas of higher elevation than do other mangrove species. Naturally established individuals of Red Mangrove were recently found on the Rio Grande Delta and may represent the northernmost distribution for the species in Mexico.

The latitudinal location of the LMT makes possible the development of both tropical vegetation (mangroves) and temperate affinity communities (salt marshes). Coexistence of these plant associations constitutes a dynamic ecotone, which fluctuates in response to both biotic and abiotic factors. It is also influenced by anthropogenic activities causing climate change and its consequences but also by conservation and restoration efforts (Zamora-Tovar et al. 2011; Armitage et al. 2015; Peterson and Bell 2015; Rodríguez et al. 2016). Climate change can cause poleward migration of tropical plant species, such as mangroves (Cavanaugh et al. 2014; Saintilan et al. 2014; Yáñez-Arancibia et al. 2014; Armitage et al. 2015). Recent Black Mangrove expansion into cordgrass-dominated salt marshes on the western Gulf of Mexico coastline is evident. Distribution and abundance changes of both mangroves and salt marshes can fundamentally modify the structure and services provided by these ecosystems and transform the regional landscape of the LMT (Perry and Mendelssohn 2009; Comeaux et al. 2012; Bianchi et al. 2013; Guo et al. 2013; Lunt et al. 2013; Scheffel et al. 2013; Armitage et al. 2015).

Aquatic Invertebrates

Crustaceans are abundant and diverse in the LMT. Intensive sampling in 1989–90 in the LMT was conducted over three seasons: dry March–May, rainy June–September, and the cooler "norther" season (because of prevalent north winds). The sampling yielded 17 species of amphipods (table 12.2), which dominated the community in the 11 localities sampled. The next most abundant taxa were mysids (5 species) and isopods (10 species) (Barba and Sánchez 2005). More recently, two cosmopolitan caprellids, which may be invasive species, were recorded for the first time in the LMT by Rodríguez-Almaraz and Ortega-Vidales (2013). The species richness of benthic amphipods was recently updated by Raz-Guzmán and Soto (2017) to include a total of 24 species. The isopod community was dominated by two species, *Cymadusa compta* (61.7%) and *Elasmopus levis* (20.6%), but seven species represented new records for the lagoon. The two dominant species are the same ones reported decades ago, along with *Harrieta faxoni* (Barba and Sánchez 2005).

Table 12.2. *Non-decapod (e.g., pericarid) crustaceans documented in Laguna Madre de Tamaulipas*

Classification	**Scientific Name**
Class Branchiopoda	
Order Anostraca	
Family Artemiidae	*Artemia* sp.
Class Malacostraca	
Subclass Eumalacostraca	
Order Amphipoda	
Suborder Gammaridea	
Family Ampeliscidae	*Ampelisca* sp. *Ampelisca vadorum*
Family Ampithoidae	*Ampithoe longimana* *Ampithoe valida* *Cymadusa compta*
Family Hyalidae	*Apohyale prevostii* *Hyale nilssoni* *Parhyale hawaiensis*
Family Bateidae	*Batea catharinensis*
Family Ischyroceridae	*Cerapus tubularis* *Erichthonius brasiliensis* *Jassa falcata*
Family Melitidae	*Elasmopus levis* *Elasmopus pectenicrus* *Elasmopus rapax* *Melita nitida*
Family Phoxocephalidae	*Eobrolgus spinosus*
Family Gammaridae	*Gammarus mucronatus*
Family Aoridae	*Grandidierella bonnieroides* *Lembos websteri*
Family Amphilochidae	*Hourstonius laguna*
Family Isaeidae	*Microprotopus raneyi*
Family Corophiidae	*Monocorophium tuberculatum*

Table 12.2. *Continued*

Classification	Scientific Name
Family Dexaminidae	*Nototropis minikoi*
Family Taltridae	*Orchestia gammarella* *Orchestia grillus*
Suborder Caprellidea	
Family Caprellidae	*Caprella* sp.
Family Pariambidae	*Deutella californica* *Paracaprella* sp. *Paracaprella tenuis*
Order Isopoda	
Suborder Flabellifera	
Family Cymothoidae	*Aegathoa medialis*
Family Ancinidae	*Ancinus depressus*
Family Sphaeromatidae	*Exosphaeroma productatelson* *Harrieta faxoni* *Sphaeroma quadridentatum*
Suborder Cymothoida	
Family Cymathoidae	*Cymathoa oestrum*
Family Aegidae	*Rocinela americana*
Suborder Valvifera	
Family Idoteidae	*Erichsonella attenuata* *Erichsonella filiformis*
Suborder Oniscidea	
Family Ligiidae	*Ligia exotica*
Order Mysida	
Family Mysidae	*Americamysis almyra* *Americamysis bahia* *Bowmaniella brasiliensis* *Bowmaniella flordana* *Mysidopsis sp.*
Subclass Hoplocarida	
Order Stomapoda	
Family Squillidae	*Squilla chydaea* *Squilla empusa*

Source: Barba and Sánchez 2005; Raz-Guzmán and Soto 2017; Raz-Guzmán and Villegas 2018.
Note: These organisms do not have common names for the most part.

The first published study surveying thoracican shallow-water barnacles from the Tamaulipas coast was performed by Celis et al. (2007), who found 13 species. All but one occurred within the LMT (table 12.3). An acorn barnacle (*Megabalanus stultus*), which was never before recorded in Mexican waters, apparently is an obligate symbiont of *Millepora* spp. corals. Raz-Guzmán et al. (2004) collected hermit crabs in the LMT over a 13-year period in 76 localities and reported the presence of six species (table 12.4). The Thinstripe Hermit Crab and the Fat-claw Hermit Crab were the most abundant. With the exception of the Thinstripe Hermit Crab, all were new records for Mexico. Sixty-two species of Crustacea are listed for the LMT excluding shrimp.

Table 12.3. *Shallow-water barnacles (Cirripedia: Thoracica) documented in Laguna Madre de Tamaulipas by zone*

Species	Zone
Order Sessilia	
Family Balanidae	
Amphibalanus eburneus	C, S
Amphibalanus subalbidus	C, S
Amphibalanus venustus	C, S
Balanus trigonus	S
Megabalanus coccopoma	S
Megabalanus stultus	N, C, S
Megabalanus tintinnabulum	S
Family Chthamalidae	
Chthamalus fragilis	C
Chthamalus proteus	C
Family Tetraclitidae	
Tetraclita stalactifera	C, S
Order Pendunculata	
Family Lepadidae	
Lepas anatifera	N, C, S
Lepas anserifera	N, C, S

Source: Celis et al. 2007.

Note: These organisms do not have common names for the most part. Zones (see text for geographic extent): N = north, C = central, and S = south.

Table 12.4. *Decapod (Order: Decapoda) crustaceans (excluding non-caridean shrimp) documented in Laguna Madre de Tamaulipas*

Classification	**Common Name**
Hermit crabs and allies (Infraorder Anomura)	
Land Hermit Crabs (Family Coenobitidae)	Land Hermit
Left-handed hermit crabs (Family Diogenidae)	Orangeclaw Hermit Mexican Hermit Thinstripe Hermit Stareye Hermit Surf Hermit Bandeye Hermit NCN *Paguristes hummi* Giant Hermit
Right-handed hermit crabs (Family Paguridae)	NCN *Pagurus annulipes* NCN *Pagurus maclaughlinae* Flatclaw Hermit
Porcelain crabs (Porcellanidae)	Green Porcelain Crab Spotted Porcelain Crab Striped Porcelain Crab
Mole crabs (Family Albuneidae)	NCN *Lepidopa benedicti*
Sand crabs (Family Hippidae)	Atlantic Sand Crab
True crabs (Infraorder Brachyura)	
Swimming crabs (Family Portunidae)	Speckled Swimming Crab Dana Swimming Crab Sharptoothed Swimming Crab Blue Crab Lesser Blue Crab Iridescent Swimming Crab Blotched Swimming Crab NCN *Portunus floridanus*
Box crabs (Families Calappidae, Hepatidae)	Flame Box Crab Yellow Box Crab Calico Box Crab
Spider crabs (Families Inachidae, Pisidae, Mithracidae)	Portly Spider Crab Longnose Spider Crab Channelnose Spider Crab Furcate Spider Crab Tenspine Spider Crab Yellowline Arrow Crab Spongy Decorator Crab

Table 12.4. *Continued*

Classification	Common Name
Fiddler and ghost crabs (Family Ocypodidae)	Olmec Fiddler Mudflat Fiddler Laguna Madre Fiddler Hairback Fiddler
Purse crabs (Family Leucosiidae)	Mottled Purse Crab
NCN (Family Plagusiidae)	Tidal Spray Crab
NCN (Family Sesarmidae)	Humic Marsh Crab
Frog crabs (Family Raninidae)	Gulf Frog Crab
Sponge crabs (Family Dromiidae)	Hairy Sponge Crab
NCN (Family Menippidae)	Gulf Stone Crab Redfinger Rubble Crab
Marsh and shore crabs (Family Grapsidae)	Sally Lightfoot Crab Dark Shore Crab Mottled Shore Crab
Land Crabs (Family Gecarcinidae)	Blackback Land Crab Blue Land Crab
Mud Crabs (Family Panopeidae)	Gulf Grassflat Crab
Caridean shrimp (Infraorder Caridea)	
Grass shrimp (Family Palaemonidae)	Brackish Grass Shrimp Daggerblade Grass Shrimp
Spiny and slipper lobsters (Infraorder Palinura)	
Slipper lobsters (Family Scyllaridae)	Scaled Slipper Lobster Ridged Slipper Lobster

Source: Raz-Guzmán et al. 2004; RAMSAR 2018.

Note: NCN = no common name.

Not surprisingly, shrimp of commercial importance have been studied in more detail than other invertebrate groups. The shrimp fisheries of the Tamaulipas coast target the Brown Shrimp, which constitutes more than 80% of the catch (Anonymous 2005). However, Pink Shrimp are also abundant (Wakida-Kusunoki et al. 2008; Pérez-Castañeda et al. 2012). In 2005, Pérez-Castañeda et al. (2010) found Brown and Pink Shrimp in similar abundance. The authors suggested that seagrass biomass facilitates the abundance of both species and recommended further study to evaluate differential use of seagrass beds by the two species throughout the day.

During the year, Brown Shrimp are more abundant during spring, Pink Shrimp are more abundant in winter, and White Shrimp are more common in summer (Wakida-Kusunoki et al. 2008). In two heavily fished areas within the lagoon (Carrizal and Carboneras), Brown and Pink Shrimp were found year-round, but the presence of White Shrimp was not recorded during May–July (Pérez-Castañeda et al. 2012). During that study, 80.4% of the shrimp captured were subadults and 15.2% were juvenile organisms, but only 4.4% were adults.

Presence of the infectious hypodermal and hematopoietic necrosis virus (IHHNV) and Taura syndrome virus (TSV) in wild populations of Brown and Pink Shrimp within the LMT at La Pesca estuary is a recent concern. A sanitary monitoring program should be established to prevent the dispersion of diseases to shrimp farms in the region (Guzmán-Sáenz et al. 2009). Similarly, the shrimp sampled at Carboneras and Carrizal tested positive for necrotizing hepatopancreatitis (NHP). This was the first report of NHP presence in wild shrimp in the Gulf of Mexico (Aguirre-Guzmán et al. 2010).

Nine species of caridean shrimp are present in the LMT. Zostera Shrimp is the most abundant species (~54%), followed by Arrow Shrimp (~28%) and Daggerblade Grass Shrimp (~11.2%). The other six taxa are Brackish Grass Shrimp, a grass shrimp (*Palaemonetes* spp.); Common Shore Shrimp; Marsh Grass Shrimp; Slender Sargassum Shrimp; Bigclaw Snapping Shrimp; and Sand Snapping Shrimp. A processid shrimp (*Ambidexter symmetricus*) is also present. Salinity and submerged aquatic vegetation appear to be key environmental parameters determining the distribution of caridean shrimp in the LMT (Barba et al. 2005).

Blue Crab fisheries are fifth in volume for the country and occur along the Tamaulipas coast, including the LMT. Blue Crab captures range from 1,284 to 3,879 tonnes (1,415–4,276 tons) (Velazquez de la Cruz et al. 2012). Possible causes for periodic fluctuations in captures of Blue Crabs include reproduction dynamics of the species and the climatic changes affecting those dynamics, changes in salinity and temperature affecting migration patterns, imprecise capture statistics, excessive numbers of individuals smaller that the permitted size, overexploitation, juvenile mortality associated with illegal shrimp fishing, and lack of incentives from buyers to favor capture of larger sizes. Although a general consensus exists among local fishermen to not keep gravid females, about 23% of the individuals landed between February and June 2009 were below the allowable size (Velazquez de la Cruz et al. 2012).

Correa-Sandoval and Rodríguez-Castro (2013) recorded 131 species of molluscs from the LMT and compiled a complete list of taxa by combining their data with previously published species lists. The Sunray Venus and Smooth Duck Clam were new records for Tamaulipas. Some of the most abundant molluscs were the Grass Cerith, Gulf Dovesnail, Scorched Mussel, Eastern Oyster, Cross-barred Venus, and Dwarf Surf Clam (Reguero and Raz-Guzmán 2018).

Of the 131 molluscs, 24 species are considered of commercial importance (table 12.5). Twenty-one species of bivalves and 36 gastropods were reported in RAMSAR (2018). The list also included two squids: Longfin Inshore Squid and Atlantic Brief Squid; and four octopuses: Caribbean Armstripe Octopus, Pigmy Octopus, White-spotted Octopus, and Scuttle.

Table 12.5. *Bivalves and gastropods (Mollusca: Bivalvia and Gastropoda) documented in Laguna Madre de Tamaulipas by zone (when available)*

Classification	**Zone**
Class Gastropoda	
Family Fissurellidae	
Cayenne Keyhole Limpet	—
Family Neritidae	
Olive Nerite	S
Virgin Nerite	N, S
Emerald Nerite	N
Family Batillariidae	
West Indian False Cerith	N
Family Cerithiidae	
Grass Cerith	N
Dark Cerith	N
Ivory Cerith	S
Variable Cerith	S
Family Potamidae	
Plicate Hornsnail	S
Family Modulidae	
Buttonsnail	—
Family Elachisinidae	
NCN *Elachisina floridana*	N
Family Littorinidae	
Cloudy Periwinkle	—
Interrupted Periwinkle	—
Family Rissoidae	
Catesby's Risso	—
Family Strombidae	
Florida Fighting Conch	—
Hawkwing Conch	—
Family Xenophoridae	
American Carrier Shell	—
Family Calyptraeidae	
Convex Slippersnail	N
Eastern White Slippersnail	N, C, S
Common Atlantic Slippersnail	N (CE)
Spotted Slippersnail	N
Family Cypraeidae	
Atlantic Deer Cowrie	—
Family Ovulidae	
Flamingo Tongue	—
Family Naticidae	
False Shark Eye	N, C, S
Miniature Moonsnail	N, S
Shark Eye	—
Family Ranellidae	
Atlantic Trumpet Triton	—
Angular Triton	—
Giant Triton	—
Family Cassidae	
Scotch Bonnet	—
Family Tonnidae	
Giant Tun	—
Family Muricidae	
Giant Eastern Murex	—
Florida Rocksnail	—
Family Columbellidae	
Lunar Dovesnail	N, C, S
Greedy Dovesnail	N
West Indian Dovesnail	N
Fat Dovesnail	N, S
Semiplicate Dovesnail	N, C

Table 12.5. Continued

Classification	Zone
Family Fasciolariidae	
Banded Tulip	—
True Tulip	—
Branham's Tulip	C, S (CE)
Horse Conch	— (CE?)
Family Melongenidae	
Lightning Whelk	C (CE)
Turnip Whelk	S (CE)
Shouldered Pearwhelk	S (CE?)
Family Turbinellidae	
West Indian Chank	
Family Nassariidae	
Sharp Nassa	N, S
Bruised Nassa	N
Family Olivellidae	
Caribbean Olive	—
Family Terebridae	
Salle's Auger	—
Common American Auger	—
Family Turridae	
Plicate Mangelia	N
Family Pyramidellidae	
Interrupted Turbonille	
Family Acteonidae	
Pitted Baby-bubble	N
Family Bullidae	
Striate Bubble	N
Family Haminoeidae	
Conrad's Paperbubble	N
Family Cylichnidae	
Channeled Barrel-bubble	N
Family Elobiidae	
Eastern Melampus	—
Coffee Melampus	—

Classification	Zone
Class Bivalvia	
Family Nuculidae	
Atlantic Nut Clam	N, C, S
Family Nuculanidae	
Pointed Nut Clam	C, S
Concentric Nut Clam	N, C, S
Family Arcidae	
Baughman's Ark	N, C, S
Incongruous Ark	C, S
Cut-ribbed Ark	N, C, S
Chemnitz Ark	C, S
Eared Ark	—
Blood Ark	— (CE)
Transverse Ark	S (CE)
Mossy Ark	S (CE)
Turkey Wing	S
White Miniature Ark	N, C, S
Family Noetiidae	
Adams Miniature Ark	N, S
Ponderous Ark	S (CE)
Family Glycymeridae	
Giant American Bittersweet	S
Family Mytilidae	
Atlantic Paper Mussel	—
Scorched Mussel	C, S (CE)
Hooked Mussel	S (CE)
Southern Ribbed Mussel	C, S
Scissor Date Mussel	S
Black Date Mussel	—
American Horse Mussel	S
Brown Mussel	C, S
Family Pteriidae	
Atlantic Wing Oyster	—
Family Isognomonidae	
Flat Tree Oyster	—
Two-toned Tree Oyster	—

Table 12.5. *Continued*

Classification	Zone
Family Ostreidae	
Mangrove Oyster	C, S
Eastern Oyster	C, S (CE)
Dwarf Oyster	—
Family Pinnidae	
Stiff Pen Shell	S
Half-naked Pen Shell	S
Saw-toothed Pen Shell	S
Family Pectinidae	
Atlantic Calico Scallop	C, S (CE)
Western Bay Scallop	S (CE)
Northern Lion's Paw	S
Rough Scallop	S
Paper Scallop	—
Family Spondylidae	
Atlantic Thorny Oyster	S
Family Plictulidae	
Atlantic Kitten's Paw	N, S
Family Anomiidae	
Common Jingle	N, S
Family Crassatellidae	
Lunate Crassinella	—
Family Carditidae	
Broad-ribbed Carditid	S
West Indian Cardita	S
Family Pandoridae	
Bush's Keeled Pandora	N, S
Family Periplomatidae	
Unequal Spoonclam	N
Family Verticordiidae	
Ornate Verticord	N, C, S
Family Cuspidariidae	
Ornate Cardiomya	N, C, S
Rostrate Cardiomya	N, C, S

Classification	Zone
Family Lucinidae	
Buttercup Lucine	C, S
Tiger Lucine	C, S
Thick Lucine	S (CE)
Chalky Lucine	C, S
Family Chamidae	
Florida Spiny Jewelbox	N, C, S
Spiny Jewelbox	S
Atlantic Jewelbox	—
Family Lasaeidae	
Texas Aligena	—
Family Corbiculidae	
Carolina Marsh Clam	—
Family Cardiidae	
Yellow Prickly Cockle	S
Giant Atlantic Cockle	C, S (CE)
Yellow Egg Cockle	C
Egg Cockle	S
Prickly Cockle	C, S (CE)
Florida Prickly Cockle	— (CE)
Dyed Micro-cockle	N, C, S
Family Veneridae	
Pointed Venus	C
Texas Venus	S
Florida Cross-barred Venus	N, C, S (CE)
Atlantic Cyclinella	—
Disk Dosinia	C, S
Elegant Dosinia	S
Rigid Venus	C, S
Atlantic Gould Clam	N, C, S
Imperial Venus	S
Paphia Imperial Clam	N, C, S
Calico Clam	S
Sunray Venus	C (CE)
Southern Quahog	C, S (CE)
False Angelwing	S
Lady-in-Waiting Venus	S
Gray Pygmy-venus	N, C, S

Table 12.5. Continued

Classification	Zone
Family Tellinidae	
Alternate Tellin	C, S
Taylor's Alternate Tellin	S
Angulate Tellin	—
Tampa Tellin	—
Sunrise Tellin	—
Striate Tellin	N, C, S
Constricted Macoma	—
Tagelus-like Macoma	S
Elongated Macoma	C, S
Matagorda Macoma	—
Family Donacidae	
Texas Coquina	—
Variable Coquina	C, S
Giant False Coquina	C, S
Family Semelidae	
Common Atlantic Abra	N, C, S
Cancellate Semele	S
Purplish Semele	—
Family Solecurtidae	
Corrugated Razor Clam	S
Purplish Tagelus	—
Stout Tagelus	S

Classification	Zone
Family Mactridae	
Smooth Duck Clam	C
Fragile Surf Clam	S
Dwarf Surf Clam	—
Channeled Duck Clam	S
Atlantic Rangia	—
Brown Rangia	— (CE)
Family Corbulidae	
Caribbean Corbula	N, C, S
Contracted Corbula	—
Barratt Corbula	N, S
NCN *Varicorbula krebsiana*	N, C, S
Family Pholadidae	
Angelwing	S
Wedge Piddock	S
Striate Piddock	—
Campeche Angelwing	S

Note: Zones (see text for geographic extent): N = north, C = central, and S = south. NCN = no common name; CE = commercially exploited.

Eastern Oyster fisheries from the LMT constitute 79% of the total capture in Tamaulipas, yielding 2,896 tonnes (3,192 tons) in 1993 and 3,196 tonnes (3,523 tons) in 1994 (Téllez et al. 1999). Mexican federal regulation for the State of Tamaulipas is lacking, but the Norma Oficial Mexicana (NOM) NOM-015-PESC-1994 establishes a minimum shell length of 7.0 cm (2.75 in) for the State of Tabasco (DOF 1994a). Using that standard, J. Cruz (2003) determined that more than 50% of the oysters collected in the LMT were hatchlings, 26% were juveniles, and less than 25% were adults. *Salmonella* spp. and *Vibrio cholera* were not detected in samples tested by Téllez et al. (1999), and all of the measured bacteriological contents were below the sanitary levels established by Mexican regulations.

Bay Scallop harvests in the LMT are generally for personal consumption and tend to be opportunistic. Fishermen in El Mezquital and Carboneras keep them while targeting oysters (Wakida-Kusunoki 2009). Twenty-eight polychaetes (annelids), 17 cnidarians, and 16 echinoderm species have also been recorded in the LMT (Leija-Tristán et al. 2000; RAMSAR 2018).

Aquatic Vertebrates

Fishes Of the 83 species of fishes recorded in the LMT (table 12.6), 93% are temporal marine visitors, and of these, 69.7% are euryhaline and 23.2% are stenohaline (Raz-Guzmán and Huidobro 2002). Pinfish, Gulf Pipefish, Code Goby, Silver Jenny, and Mexican Mojarra were the most abundant species.

Larvae of 39 fishes were recorded within the LMT. Seven of these species, the Atlantic Thread Herring, Green Goby, Skilletfish, Gulf Stream Flounder, Highfin Blenny, Freshwater Goby, and Cowfish, were new records for the area. The last four species were also newly reported for the Laguna Madre of Texas (Sánchez-Ramírez and Ocaña-Luna 2015). The Gobiidae and Sciaenidae families were represented by the highest species richness with eight and six species, respectively. The greatest number of larvae were collected in May, and the lowest number occurred in October. The most abundant species were Bay Anchovy, Striped Anchovy, Finescale Menhaden, Rough Silverside, Naked Goby, and Fat Sleeper.

The Striped Mullet fishery in the LMT is the most important by volume in Mexico (Gómez Ortiz 2012). The fishing season for mullet is the second most important in the lagoon and opens immediately after the most important fishery (shrimp) closes. Striped Mullet migrate seaward to reproduce during the last quarter of the year (Ibáñez-Aguirre and Gallardo-Cabello 2004). The species is extensively targeted at that time, because their gonads (mainly the eggs) are prized by Mexican consumers (Gómez Ortiz 2012). Thus, in 1994 the federal government imposed a regulation to limit harvest of Striped Mullet and White Mullet in an attempt to balance the demand for their eggs with the potential for population sustainability (DOF 1994b; Gómez Ortiz 2012).

The average capture of Striped Mullet in the LMT from 1993 to 2010 was 3,204 tonnes (3,532 tons), which represented about 93% of the Tamaulipas production. About 75% of the Striped Mullet fishery comes from Tamaulipas, which exceeds that of all other Mexican states bordering the Gulf of Mexico (Gómez Ortiz 2012). The 1994 regulations established a minimum size of 30 cm (11.8 in), and Gómez Ortiz proposed a series of management policies designed to further protect the species.

Table 12.6. *Fish species recorded in Laguna Madre de Tamaulipas with ecology and life history information (when available)*

Classification	Life History and Ecology
Superclass Chondrichthyes	
Family Carcharhinidae	
NCN *Carcharhinus* sp.	—
Family Dasyatidae	
Atlantic Stingray	R
Family Mobulidae	
Atlantic Devil Ray	—
Family Myliobatidae	
Spotted Eagle Ray	—
Family Rajidae	
Roundel Skate	—
Family Rhinobatidae	
Atlantic Guitarfish	—
Cownose Ray	—
Superclass Osteichthyes	
Family Achiridae	
Lined Sole	L, N, E
Fringed Sole	—
Family Anguillidae	
American Eel	—
Family Antennariidae	
Sargassum Frogfish	—
Family Ariidae	
Hardhead Catfish	E
Gafftopsail Catfish	E
Estuarine Seacatfish	R
Dark Seacatfish	E
Family Atherinopsidae	
Rough Silversides	L, R, E
Inland Silversides	L, R, E
Tidewater Silversides	—
Family Batrachoididae	
Gulf Toadfish	E
Family Belonidae	
Atlantic Needlefish	L, T
Family Blennidae	
Highfin Blenny	L, R

Classification	Life History and Ecology
Family Bothidae	
Sash Flounder	S
Family Carangidae	
Crevalle Jack	E
Horse-eye Jack	R
Atlantic Bumper	
Leatherjack	L, T, E
Atlantic Moonfish	
Lookdown	E
Florida Pompano	
Permit	S
Family Centropmidae	
Smallscale Fat Snook	E
Common Snook	E
Family Chaetodontidae	
Spotfin Butterflyfish	S
Family Characidae	
Banded Astyanax	—
Family Cichlidae	
American cichlids	—
Speckled Tilapia	—
Family Clupeidae	
Finescale Menhaden	L, N
Gulf Menhaden	
Gizzard Shad	
Scaled Sardine	L, T
Atlantic Thread Herring	L, T
Family Cynoglossidae	
Offshore Tonguefish	S
Double-spot Tonguefish	
Blackcheek Tonguefish	E
Family Cyprinidae	
Common Carp	—
Rio Grande Shiner	—
Family Cyprinidontidae	
Sheepshead Minnow	—

Table 12.6. Continued

Classification	Life History and Ecology
Family Diodontidae	
Striped Burrfish	E
Family Echeneidae	
Sharksucker	S
Family Eleotridae	
Fat Sleeper	L, N, R
Family Elopidae	
Ladyfish	L, T
Family Engraulidae	
Striped Anchovy	L, N
Dusky Anchovy	
Bay Anchovy	L, R, E
Family Ephippidae	
Atlantic Spadefish	E
Family Fundulidae	
Gulf Killifish	—
Longnose Killifish	—
Rainwater Killifish	R
Family Gadidae	
Southern Hake	—
Family Gerreidae	
Irish Pompano	E
Rhomboid Mojarra	L, T, E
Spotfin Mojarra	E
Silver Jenny	E
Mottled Mojarra	L, T, E
Flagfin Mojarra	E
Yellowfin Mojarra	R, E
Family Gobiesocidae	
Skilletfish	L, R
Family Gobiidae	
River Goby	—
Frilled Goby	L, N, E
Darter Goby	L, R, E
Freshwater Goby	L, N, T
Lyre Goby	L, T
Blackfin Goby	R

Classification	Life History and Ecology
Highfin Goby	L, T
Naked Goby	L, R, E
Code Goby	L, R, E
Clown Goby	R
Green Goby	L, T
Family Haemulidae	
Black Margate	S
Barred Grunt	S
Pigfish	E
Family Hemramphidae	
Halfbeak	L, T, E
Family Ictaluridae	
Blue Catfish	—
Channel Catfish	—
Family Labrisomidae	
Hairy Blenny	S
Family Lepisosteidae	
Longnose Gar	—
Family Lutjanidae	
Gray Snapper	E
Dog Snapper	—
Lane Snapper	S
Family Monacanthidae	
Dotterel Filefish	—
Planehead Filefish	S
Pygmy Filefish	—
Family Mugilidae	
Striped Mullet	E
White Mullet	E
Liza	E
Family Nomeidae	
Man-of-War Fish	—
Family Ophichthidae	
Speckled Worm Eel	L, T, E
Family Ophidiidae	
Bank Cusk-eel	—

Table 12.6. *Continued*

Classification	Life History and Ecology
Family Ostraciidae	
Scrawled Cowfish	L, T
Family Paralichthyidae	
Gulf Stream Flounder	L, T
Spotted Whiff	S
Bay Whiff	E
Fringed Flounder	S
Southern Flounder	—
Family Poeciliidae	
Western Mosquitofish	—
Sailfin Molly	—
Family Polynemidae	
Atlantic Threadfin	S
Family Pomacentridae	
Sergeant Major	—
Family Sciaenidae	
Silver Perch	L, N, E
Ground Drummer	E
Sand Seatrout	E
Spotted Seatrout	L, N, E
Silver Seatrout	R, E
Spot	L, T, E
Black Drum	E
Red Drum	—
Star Drum	L, T, E
Southern Kingfish	E
Gulf Kingfish	L, T
Northern Kingfish	R, E
Atlantic Croaker	E
Family Scorpeanidae	
Spotted Scorpionfish	E

Classification	Life History and Ecology
Family Serranidae	
Goliath Grouper	S
Black Grouper	—
Family Sparidae	
Sheepshead	E
Pinfish	L, T, E
Family Sphyraenidae	
Great Barracuda	—
Family Stromateidae	
Northern Harvestfish	S
Family Syngnathidae	
Dwarf Seahorse	L, T, S
Dusky Pipefish	L, T, E
Sargassum Pipefish	E
Gulf Pipefish	L, T, E
Bull Pipefish	E
Family Synodontidae	
Inshore Lizardfish	L, T, E
Family Tetraodontidae	
Smooth Puffer	S
Northern Puffer	—
Least Puffer	E
Family Triglidae	
Bigeye Searobin	—
Bighead Searobin	E
Family Trichiuridae	
Atlantic Cutlassfish	S
Family Uranoscopidae	
Southern Stargazer	S

Source: Raz-Guzman and Huidobro 2002; Sánchez-Ramírez and Ocaña-Luna 2015; RAMSAR 2018.

Note: L = present as ichthyoplankton and may also be present as juveniles or adults; N = nursery (juveniles); R = permanent resident (adults); T = transient (juveniles to adults); E = euryhaline member of marine assemblage; S = stenohaline member of the marine assemblage; NCN = no common name.

Amphibians and Reptiles The most important reptilian conservation efforts in the LMT focus on the Kemp's Ridley Sea Turtle (see chapter 16). Other species of conservation interest in LMT shoreline habitats are the Texas Tortoise, Spiny Softshell, Red-eared Slider, and Texas Horned Lizard (SEMARNAT 2015).

Birds The 275 species of birds recorded for the LMT are dominated by Passeriformes (35%) and Charadriiformes (20.9%) (Garza Torres and Navarro Sigüenza 2001). Of these species 26 are protected, 12 are threatened, 9 are under special protection, 3 are regarded as rare, and 2 are considered in danger of extinction under Mexican regulations. The largest number of aquatic and semiaquatic bird species (134) occupy Laguna la Nacha. Reproductive colonies of 26 species occur throughout the LMT, but Los Potros was the most used, followed by El Anillo and La Tabalina.

At least 24 species of shorebirds frequent the coastal wetlands and barrier islands, and 16 species regularly visit the LMT (Withers 2002b). Tidal flats and barrier islands in the lagoon are used in a similar way to their counterparts in Texas, where the majority of the species on tidal flats are members of the shallow-probing or surface-feeding guilds. On barrier islands, shorebirds tend to concentrate along the "swash" and damp zones to forage on invertebrates, but a few forage along the berm and wrack line. Plovers use the Laguna Madre in Texas and Tamaulipas as wintering grounds, but further studies are needed to determine the spatial and temporal variations among species (Mabee et al. 2001).

During the winter of 1997–98, the following abundances (number of individuals) of shorebirds were recorded in the LMT: 2,589 Black-bellied Plovers, 1,191 Snowy Plovers, 542 Semipalmated Plovers, 365 Piping Plovers, and 11 Wilson's Plovers (Mabee et al. 2001).

Other birds of high conservation interest in the protected area include the Yellow-headed Amazon, Reddish Egret, Common Black Hawk, Grey-lined Hawk, White-bellied Wren, Redhead, Northern Pintail, Whimbrel, Short-billed Dowitcher, Sanderling, and American White Pelican (SEMARNAT 2015).

Mammals Species of conservation importance within the protected area include Jaguar, Margay, Jaguarundi, Mountain Lion, Ocelot, and Bobcat. White-tailed Deer is a species of recreational hunting importance. Endemic species include the Spotted Ground Squirrel, Gulf Coast Kangaroo Rat, and Ord's Kangaroo Rat. The Common Bottlenose Dolphin is frequently observed around the jettied passes of El Mezquital, Enramadas, and La Pesca (SEMARNAT 2015).

Range Expansions and Biological Invasions

In addition to human-mediated invasions, many species new to the LMT appear to be expanding their range amid a tropicalizing trend caused by climate change. Relatively fewer organisms have been reported as invasive species in the LMT than in the majority of the priority conservation regions considered by the North American Commission for Environmental Cooperation (Mendoza et al. 2011).

Seagrasses and some common macrophytes have new competitors in *Udotea flabellum* and *Hydroclathrus clathratus* (Zamora-Tovar, pers. observ.). Both of these species have tropical Caribbean affinity, and it is possible that their distribution range is expanding northward because of the tropicalization of the Gulf of Mexico (Yáñez-Arancibia et al. 2014). They may also be reestablishing in their former range in response to salinity changes, as reported for *Penicillus capitatus* in the Lower Laguna Madre of Texas (Kowalski et al. 2007).

Recent aquaculture development worldwide has facilitated the movement of many aquatic species that can adapt to the climatic conditions at the introduction site (Semmens et al. 2004). Many of these species escape into the natural environment, but few will survive, reproduce, and form ex-situ populations. A recent example of a successful escapee from aquaculture is the Asian Tiger Shrimp. The shrimp was imported for aquaculture in South Carolina in 1988 and escaped (Rodríguez-Almaraz and García Madrigal 2014; Avalos 2015). It was first reported off the northern Tamaulipas coast in 2012 when it was captured in a shrimp trawl not far from the open passes of the LMT (Wakida-Kusunoki et al. 2013). Soon thereafter, the shrimp was recorded from the central-south Veracruz coast (Morán-Silva et al. 2014), and multiple reports of it from various locations in LMT (Soto la Marina estuary, Carboneras, and Punta de Alambre) have been generated by local fishermen.

Other potentially invasive invertebrates include two cosmopolitan caprellid amphipods that were recorded for the first time in the LMT by Rodríguez-Almaraz and Ortega-Vidales (2013). Thirteen specimens of *Caprella scaura*, including three ovigerous females, were found in *Ulva lactuca* beds, and 31 individuals of *Caprella penantis* were collected in Shoal Grass beds.

Three exotic barnacles are also present in the LMT. *Balanus trigonus*, a barnacle of Pacific origins, might have been introduced to the Gulf of Mexico sometime before 1961 when a marine paint test station in Florida first reported its presence on test panels (Moore and McPherson 1963). Its congener from the eastern tropical Pacific, the Titan Acorn Barnacle, probably colonized the LMT area sometime in the last 50 years (Celis et al. 2007). *Megabalanus tintinnabulum*, originally from the Indo-Pacific, should be considered a cryptogenic species (Rodríguez-Almaraz and García-Madrigal 2014).

The Red Lionfish offers the most notable example of a fish invasion. Originally a tropical and subtropical species from the Indian Ocean, its abundance is increasing in the Caribbean, western Atlantic, and Gulf of Mexico (Morris and Whitfield 2009; Côté et al. 2013; Raymond et al. 2015). The Red Lionfish, a predator with high reproductive and growth rates, has the potential to eliminate native fish populations and generate structural and functional changes in fish communities (Green et al. 2012; Albins 2013). The first reports of the species in Mexican waters came from the Yucatán Peninsula in 2009 and then from the coasts of Veracruz and Tabasco (Aguilar-Perera and Tuz-Sulub 2010; Santander-Monsalvo et al. 2012; Wakida-Kusunoki and Amador del Ángel 2015). The first reported occurrence for the Tamaulipas coast and the LMT was in October 2016 from La Pesca, where a local fisherman captured a few specimens (Arellano-Méndez et al. 2017). The latest report from the LMT (March 2018) was from Punta de Alambre in the central zone of the lagoon. Red Lionfish are mostly associated with coral reefs or other hard substrates, but they also are habitat generalists that invade mangroves and seagrass meadows within estuarine systems (Claydon et al. 2012; Jud and Layman 2012).

Another infamous invasive fish is the armored catfish (Loricariidae). About a dozen species, including *Hypostomus guacari*, *H. plecostomus*, *Loricaria flava*, and *P. brasiliensis*, are all of South American origin and listed with the same common name. The first report of the armored catfish in Mexico was from the Río Mezcala in the State of Guerrero in 1995. Since then, these species have rapidly invaded most freshwater and mixohaline systems in the country. These species are highly adaptable, and some exhibit high tolerance to brackish conditions. A specimen with a total length of 70 cm (~27.5 in) was captured in the Rio Grande Delta (fig. 12.7). These fishes have been found in freshwater and brackish sites in all three zones of the LMT. Although these catfish are considered potentially harmful to native species and habitats, their

Figure 12.7. *Record armored catfish (*Plecostomus *spp.) captured in the north zone of the Laguna Madre de Tamaulipas (Rio Grande Delta area) in 2014.*

eradication has been impossible in other parts of the country (Martínez-Palacios et al. 2010). These invaders likely will cause structural and functional changes in the LMT.

Conservation

On 14 April 2005, the Mexican federal government declared the totality of the LMT, all associated wetlands and islands, and significant portions of the adjacent terrestrial habitats as a natural protected area, encompassing more than 572,800 ha (1,415,420 ac) (DOF 2005). The protected area is limited by the Rio Grande to the north, the Barra de Ostiones estuary to the south, and the Gulf of Mexico to the east. The official name of the protected area is Area de Protección de Flora y Fauna, Laguna Madre y Delta del Río Bravo. It has a level of protection equivalent to a Category VI (managed resource protected area), as defined by the International Union for Conservation of Nature. The LMT has been catalogued as a wetland of international importance since 2004 (RAMSAR Site no. 1362) (RAMSAR 2018). In addition, UNESCO (2006), through its "Man and the Biosphere" program, incorporated the protected area into its global network of biosphere reserves in 2006. Other designations are Region for Priority Terrestrial Conservation (RTP-83), Region for Priority Marine Conservation (RMP-44), and Region for Priority Hydrological Conservation (RHP-41 and RHP-72). The LMT is also considered within the priorities of the North American Wetland Conservation Act Plan, by including three Areas of Importance for Bird Conservation (AICAS Nos. 67, 230, and 235). It is also considered an internationally important site for the Western Hemisphere Shorebird Reserve Network and is recognized by Birdlife International as one of the Endemic Bird Areas of the World (Zamora-Tovar et al. 2011; SEMARNAT 2015).

The Secretaría de Medio Ambiente y Recursos Naturales recently published a management program for the protected area. It is a detailed planning and regulation-guiding document, considered the most important conservation instrument based on knowledge and use of natural resources and inherent problems. It calls for the organization, prioritization, and coordination of actions to attain the creation objectives of this natural protected area (SEMARNAT 2015). These considerations have allowed the development of conservation and restoration programs by multiple stakeholders, including educational and research institutions, governmental and nongovernmental entities, environmental organizations, and society in general (mostly from Mexico and the United States). Some actively engaged institutions are the Tamaulipas state government, Pronatura Noreste AC, The Nature Conservancy of Texas, Texas Center for Policy Studies, Universidad Autónoma de Nuevo León, and Universidad Autónoma de Tamaulipas (UAT).

Significant community engagement in restoration efforts has occurred in the Burgos Basin region in the north and north-central zones of the LMT (Zamora-Tovar et al. 2011). About 315 ha (~778 ac) were planted with White, Black, and Red Mangroves. Local cooperation has resulted in an effective methodological strategy, incorporating community opinions and decisions, which along with ecological concepts are promoting an encouraging perception of environmental conservation and sustainable management.

The promotion of this work has allowed the continuation of community mangrove reforestation activities in the LMT, and there are many cases of success in the three zones. Particularly, governmental entities and nongovernmental organizations have worked together to promote the production of mangrove seedlings and their subsequent use in reforestation. This multifaceted collaboration is based on the understanding that conservation and restoration are much more than the simple maintenance of natural spaces. They also transform the communities' perception of their natural heritage and sustainable use of resources, ensuring that present and future generations have available the necessary services and ecosystem functions.

13 Redheads and Other Wintering Waterfowl

BART M. BALLARD AND
ELIZABETH H. SMITH

The lagoon systems of the lower Texas and northeastern Mexico coasts are a special place for waterbirds, as during each season they support a different suite of species. For example, approximately 26 species of colonial waterbirds (wading birds, pelicans, cormorants, ibises, gulls, and terns) breed within the lagoons during late spring and summer (see chapter 14). These species rely on the productive fishery to feed their young and recruit offspring into the population. During autumn and spring, countless shorebirds use the expansive algal flats and shallow waters to fuel their migrations, some of which traverse the length of the globe (see chapter 15). However, during winter, the lagoons are dominated by waterfowl. Even the casual observer cannot help being amazed by the immense flocks of ducks that often stretch for miles.

The Laguna Madre of Texas and Tamaulipas is a seagrass-dominated system with unique hydrological characteristics as a result of low average precipitation, high evaporation rates, and limited freshwater inflow. A combination of the semiarid climate and relatively frequent tropical storm activity creates a dynamic environment of alternating wet and dry periods that influences salinities within the lagoons and availability of coastal ponds on mainland areas adjacent to the lagoons (McAdams 1987). Abundant food resources in the lagoons in combination with available fresh water for drinking in inland coastal ponds collectively provide a favorable landscape for migrating and wintering waterfowl.

Laguna Madre of Texas

The Laguna Madre is the focus of waterfowl distribution along the lower Texas coast for several reasons. First, the density of freshwater wetlands (typically referred to as coastal ponds) is greater close to the coast (McAdams 1987; Muehl 1994), thus providing more habitat for waterfowl near the lagoon. Second, the immense size of the Laguna Madre and its relatively shallow water equate to large areas of foraging habitat (Kiel 1957). The lagoon has abundant seagrass meadows where waterfowl forage on parts of seagrasses, invertebrates, and small fish; thus, food is generally plentiful. Third, the stability (within and among years) of these food resources is much greater than resources in more ephemeral inland wetlands. This stability can be important for those species that exhibit high winter-site fidelity. Fourth, the large expanse of the Laguna Madre and its relatively shallow water provide many areas where waterfowl can rest and escape disturbance. Large flocks of ducks often rest on the shoreline or mudflats where the water is too shallow for boat traffic and predators can be detected for long distances in any direction. This is also apparent in regard to nocturnal roost sites. For example, waterfowl that do not forage in the Laguna Madre (geese) frequently use large, shallow areas as nocturnal roost sites. Thus, the Laguna Madre offers several attributes that make it attractive to wintering waterfowl.

The Redhead is by far the most abundant species of waterfowl in the Laguna Madre of Texas during winter (fig. 13.1). This species is endemic to North America and has a relatively broad breeding distribution with greatest concentrations throughout the Northern Great Plains and Intermountain West regions (Baldassarre 2014). Redheads are largely associated with estuaries of the Gulf of Mexico during winter, and the percentage of the continental population wintering in these areas has increased in recent years (Michot 2000). Within the Gulf of Mexico region, it appears that Redheads capitalize on wintering sites closest to the breeding areas, as wintering densities (number of Redheads per unit area of foraging habitat) decrease farther from breeding areas (Michot 2000). However, by far the largest numbers of Redheads winter in the Laguna Madre of Texas, followed by Tamaulipas, Florida, and Louisiana (fig. 13.2).

Figure 13.1. *Male Redheads have distinctive plumage: males have obvious red heads; females have brown heads. (Photograph by Kevin Bercaw, Wikimedia Commons)*

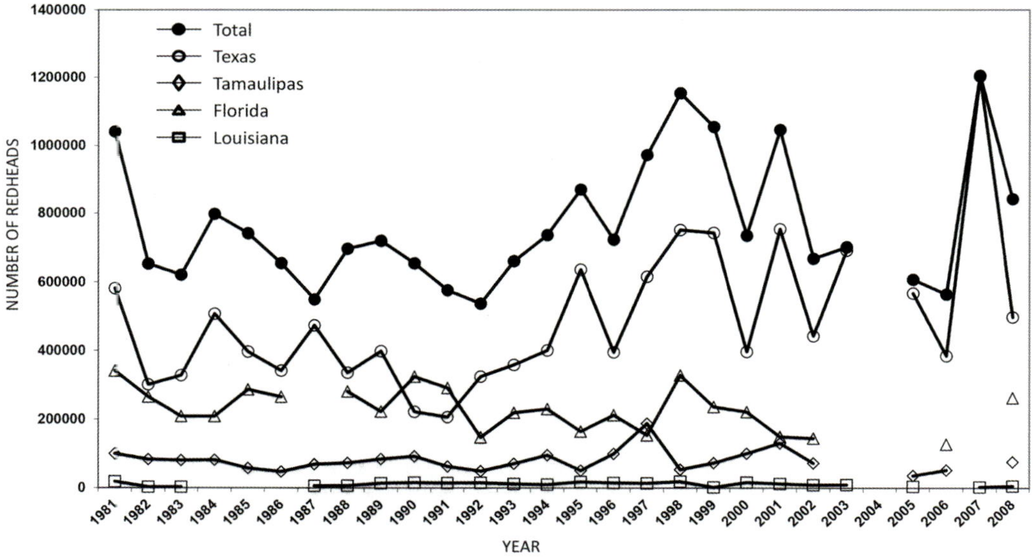

Figure 13.2. *Estimates of Redhead abundance in Gulf of Mexico states from 1981 to 2008, based on a special winter Redhead survey conducted by the US Geological Survey (unpub. data)*

This distribution holds for most winters, with most recent surveys suggesting that the proportion of Redheads wintering in Texas has increased. Although the percentage of the continental population that spends its winter in the Laguna Madre of Texas and Tamaulipas varies from year to year, the average proportion is 76% but is as high as 86% in some years (Bellrose 1980). The great reliance by Redheads on the Laguna Madre of Texas and Tamaulipas provides a situation where impacts to habitats in this region could have measurable effects on the continental population. Thus there is no doubt that the Laguna Madre of Texas and Tamaulipas is critical to the sustainability of the Redhead in North America (Clark 1961).

Redheads and many other wintering waterfowl start arriving at the Laguna Madre of Texas during early October in most years, and influxes continue until late November (Custer et al. 1997; James 2006). The arrival of Redheads is often related to the passage of weather fronts—flocks arrive the day of frontal passage and for the following two days (Custer et al. 1997). Their abundance increases exponentially to late November and remains high until early to mid-February (fig. 13.3). In some years there appear to be large shifts of birds between the Laguna Madre of Texas and Tamaulipas during midwinter. Numbers decline quickly from mid-February to mid-March as birds depart to northern breeding areas.

Other species of waterfowl that are regular winter inhabitants of the Laguna Madre include the Northern Pintail, American Wigeon, Gadwall, Green-winged Teal, Blue-winged Teal, Northern Shoveler, Lesser Scaup, Bufflehead, Ruddy Duck, and Red-breasted Merganser (table 13.1). The Northern Pintail is the second most abundant species in most years in both the Texas and Tamaulipan Laguna Madre (table 13.2), but Lesser Scaup can be quite abundant in some years. Abundance of the American Wigeon also varies greatly from year to year. Buffleheads and Red-breasted Mergansers are locally abundant in scattered small flocks (Bolen and Chapman 1981). Blue-winged Teal migrate through the Laguna Madre region in relatively large numbers during early autumn before wintering in regions farther south; few overwinter in the Laguna Madre.

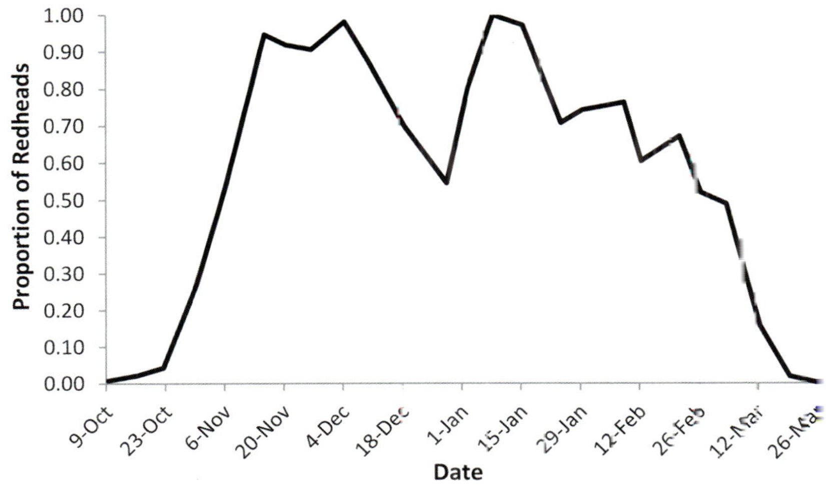

Figure 13.3. *Proportion of peak winter count of Redheads using the Laguna Madre of Texas averaged from aerial surveys during winters 2000–2001, 2001–2002, and 2002–2003 (Adapted from James 2006)*

Table 13.1. *Number of waterfowl using the Laguna Madre Initiative Area of the Gulf Coast Joint Venture, 1991–92 and 1992–93*

	1991–92				**1992–93**			
	Sept.	**Nov.**	**Jan.**	**Mar.**	**Sept.**	**Nov.**	**Jan.**	**Mar.**
Black-bellied Whistling-Duck	2,365	17,131	0	8,084	11,865	18,563	9,248	83,641
Fulvous Whistling-Duck	0	0	274	1,917	1,189	255	0	16,956
Green-winged Teal	45	2,520	6,200	274	7,109	41,675	6,706	185
American Wigeon	87	28,641	9,477	18,496	648	15,466	10,157	5,055
Blue-winged Teal	55,233	587	4,260	31,371	105,625	30,119	22,525	15,696
Gadwall	4	1,275	3,039	548	31	5,931	5,872	460
Mottled Duck	803	1,059	951	4,111	86,477	28,887	15,129	15,211
Northern Pintail	104	12,905	17,989	0	20,620	59,869	23,527	0
Northern Shoveler	72	4,526	10,617	14,661	2,409	15,711	32,489	10,473
Bufflehead	0	344	969	0	0	255	1,380	2,067
Canvasback	0	9,683	25,139	0	0	23,364	48,104	0
Redhead	0	585,382	445,775	0	52	301,690	312,724	672
Lesser Scaup	0	1,214	73	0	0	0	50	46
Ruddy Duck	822	21,797	11,561	822	326	10,633	5,537	5,763
Red-breasted Merganser	0	20	71	0	0	0	69	0
Snow and Ross' Goose	0	—	81,771	0	0	0	0	0
Greater White-fronted Goose	0	—	0	0	0	703	0	0
Canada Goose	0	—	685	0	0	0	0	0

Source: Adapted from Muehl 1994.

Note: Extrapolated from aerial surveys over the Laguna Madre, Texas, and inland wetland surveys during September, November, January, and March 1991–92 and 1992–93.

Table 13.2. *Average waterfowl counts by decade from the midwinter waterfowl survey conducted in the Laguna Madre de Tamaulipas, 1947–2000*

	1947–49 (3)[a]	1950–59 (9)	1960–67 (8)	1970–79 (5)	1980–8[illegible]	1991–2000 (4)
Whistling ducks	0	2	28	15	4[illegible]	17
Dabbling ducks						
Blue-winged Teal	174	363	1,429	649	1,9[illegible]	322
Green-winged Teal	3,095	424	954	2,895	6,33[illegible]	2,085
Northern Shoveler	3,270	1,430	5,115	3,503	5,03[illegible]	2,403
Gadwall	16,128	1,186	1,497	8,985	11,53[illegible]	2,961
American Wigeon	17,743	1,683	2,253	40,065	49,42[illegible]	5,817
Mallard	83	2	7	1	1[illegible]	1
Mottled duck	212	2	44	259	1,2[illegible]	317
Northern Pintail	166,814	33,079	5,125	107,268	98,4[illegible]	18,137
Total	**207,519**	**38,169**	**16,424**	**163,625**	**173,9[illegible]**	**32,043**
Diving ducks						
Redhead	263,513	88,489	18,496	160,476	268,6[illegible]	244,073
Canvasback	1,410	943	463	3,080	6,5[illegible]	758
Ring-necked Duck	0	176	1,412	1,312	13[illegible]	6
Lesser Scaup	44,115	3,068	5,769	70,233	19,7[illegible]	9,862
Bufflehead	47	15	15	91	13	250
Common Goldeneye	17	2	0	0	1[illegible]	44
Ruddy Duck	1,353	1,224	2,754	3,991	4,97[illegible]	5,456
Mergansers	0	0	155	278	53[illegible]	516
Total	**310,455**	**93,917**	**29,064**	**239,461**	**300,7[illegible]**	**260,965**
Geese						
Light geese	1,952	752	474	12	8,6[illegible]	5,860
Dark geese	700	2,418	2,029	978	11,2[illegible]	1,144
Total	**2,652**	**3,170**	**2,503**	**990**	**19,8[illegible]**	**7,004**

Source: US Fish and Wildlife Service, unpub. data.

[a] Numbers in parentheses indicate the number of years within that decade that the midwinter waterfowl survey was flown for the Rio Grande Delta region.

Several species that commonly occur in the region are more associated with inland freshwater wetlands than the lagoon itself, including the Black-bellied Whistling-Duck, Fulvous Whistling-Duck, Blue-winged Teal, Canvasback, and Ruddy Duck. Flocks of the Snow Goose, Ross' Goose, and Greater White-fronted Goose often use shallow arms of Baffin Bay and the Laguna Madre for nocturnal roost sites, particularly in dry years when larger inland freshwater wetlands are not available. For example, about half of 472 nocturnal roost locations of Greater White-fronted Geese marked with GPS tracking devices in southern Texas were in estuarine habitats associated with the Laguna Madre and Baffin Bay (B. Ballard, unpub. data).

The Mottled Duck is a resident of the Laguna Madre region and uses the lagoon primarily during winter. It can be found nesting in low numbers in the lagoon on dredge-spoil islands vegetated with dense grass cover; however, most nests are associated with the mainland and barrier islands where ducklings have access to freshwater habitats (Moorman et al. 1991). The Black-bellied Whistling-Duck and Fulvous Whistling-Duck breed in the region but spend the winter in Mexico and Central America (Baldassarre 2014). Fulvous Whistling-Ducks are more ephemeral in their occurrence during the breeding period and are more abundant in wet years. Black-bellied Whistling-Ducks occasionally nest on dredged-material islands in the Laguna Madre, sometimes in association with wading-bird colonies (Ballard 2002).

Although Redheads are diving ducks, they forage primarily by dipping or tipping in shallow water in the Laguna Madre; thus, their distribution is regulated by tide fluctuations and the presence of shallow Shoal Grass meadows (Custer et al. 1997). Consequently, Redheads frequently move throughout the Laguna Madre during winter in response to changing water levels to remain in optimal water depths for efficient foraging (McMahan 1967a; Mitchell 1991, 1992; Custer et al. 1997). Mitchell (1992) found that water depths of 12–30 cm (5–12 in) were a reliable predictor of Redhead flock locations in the Laguna Madre during winter. In fact, 90% of the time when flocks were present at a study site, water depth was 12–30 cm; Redheads avoided water depths outside this range. In the Laguna Madre of Texas and Tamaulipas, distribution may also be governed by location of available freshwater drinking sites on the adjacent mainland. In winters when more coastal ponds are available, Redheads and other waterfowl are likely distributed over more foraging habitat in the Laguna Madre than in dry winters when the availability and distribution of coastal ponds are reduced (Ballard et al. 2010).

During aerial surveys in the Laguna Madre, Custer et al. (1997) found that 71% of area used by flocks was over monotypic stands of Shoal Grass, and only 4% was in areas with no observable seagrasses. Over the entire winter, Redheads used 62% of the Shoal Grass habitat, but 90% of that was used less than 7% of the winter. Use of foraging sites may also be related to biomass of Shoal Grass rhizomes, as Redheads appear to reduce use of sites when rhizome density falls below a threshold of biomass (Mitchell et al. 1992). Areas of potential foraging habitat that Redheads tend to avoid are those in deeper water and those near the Gulf Intracoastal Waterway or ports with frequent boat traffic (Custer et al. 1997).

The distribution of Redheads in the Laguna Madre is highly variable during winter because they capitalize on the shallowest foraging habitats as they become available due to changing water levels (James 2006). In the Upper Laguna Madre, except for the very northern portion, Redheads rarely use the area west of the Gulf Intracoastal Waterway, likely because of the deeper water and limited amount of Shoal Grass. Most use occurs on the east side from the JFK Causeway to Bird Island Basin, across from the mouth of Baffin Bay to Yarborough Pass, and in Nine Mile Hole. In the Lower Laguna Madre Redheads mostly use the area on the east side from the Mansfield Cut south to Green Island, west of the Gulf Intracoastal Waterway in Peyton Bay, and Rattlesnake Bay south to Laguna Vista Bay (James 2006). The Lower Laguna Madre tends to receive more use by Redheads (57%–69%) than the upper basin.

Rio Grande Delta

Historically, the Rio Grande Delta likely provided substantial wintering waterfowl habitat since the Rio Grande supplied water to the vast landscape of marshes and lagoons at its mouth (Saunders and Saunders 1981; Baldassarre et al. [illegible]). Because of considerable alterations to the flow of the Rio Grande from damming and water-diversion efforts, the quantity and quality of wetland habitat in the delta have greatly diminished. As a result, the numbers of waterfowl wintering in the region have declined over the years. Despite the reduced river flows, the region can support large numbers of waterfowl following significant precipitation events (Saunders and Saunders 1981). The Rio Grande Delta has likely always had highly fluctuating wintering waterfowl abundances due to the dynamic nature of flooding events. During the [illegible], when conditions were wet, the Rio Grande Delta held as many as a half million ducks and 50,000 geese. However, when conditions were dry during those same decades, waterfowl populations could drop by 90% from the drastic change in habitat conditions.

The Rio Grande Delta supports greater goose abundance than any other area along the Mexican Gulf Coast, and the proportion of all wintering goose populations increased in Tamaulipas over other Gulf areas during the 1970s and 1980s (table 13.3). Over half of the Lesser Canada Goose populations along the Mexican Gulf Coast were documented in the Rio Grande Delta during 1951–64 (Saunders and Saunders 1981) and increased even more in the later surveys (Baldassare et al. [illegible]) as a result of increased agriculture, in particular, grain sorghum. Greater White-fronted Goose populations were highest in the delta as well, with over 75% of the species wintering in Mexico being documented along the Gulf Coast during 1961–64. Prior to the 1950s, only a few thousand Greater White-fronted Geese were present, but numbers increased as land use was converted to sorghum fields (Saunders and Saunders 1981). As more geese were observed in the Rio Grande Delta, aerial survey coverage also intensified, and greater numbers were documented in the region. One-third of the Lesser Snow Geese documented along the Mexican Gulf Coast are typically counted within the Rio Grande Delta.

According to midwinter survey tallies, waterfowl numbers in the Rio Grande Delta were highest in the 1970s and 1980s. Abundance of most species was similar across these two decades, except for Green-winged Teal and mergansers, which were observed in greater abundance during the 1970s, and the Canvasbacks and Ruddy Ducks, which were observed in greater abundance during the 1980s. There was a general trend of increasing abundance from the initiation of the survey in the 1940s to the 1970s across taxa, but the numbers of waterfowl observed between the 1980s and 1990s in the Rio Grande Delta declined sharply. In general, the Northern Pintail was the most abundant species across years, followed by Northern Shoveler. In certain decades, American Wigeon, Green-winged Teal, and Gadwall also were observed in relatively large numbers. In relation to the entire east coast of Mexico, the Rio Grande Delta historically supported a significant proportion of Northern Shovelers and Green-winged Teal (Saunders and Saunders 1981).

Table 13.3. Average waterfowl counts by decade from the midwinter waterfowl survey conducted in the Rio Grande Delta, 1947–2000

	1947–49 (3)[a]	1950–59 (9)	1960–67 (8)	1970–79 (5)	1980–88 (5)	1991–2000 (4)
Whistling ducks	0	0	877	4,507	3,262	16
Dabbling ducks						
Blue-winged Teal	73	1,940	727	2,239	1,524	210
Green-winged Teal	171	2,411	7,179	27,261	11,864	1,679
Northern Shoveler	3,966	3,059	3,834	23,664	21,123	10,870
Gadwall	3,695	1,502	6,199	10,927	13,366	2,108
American Wigeon	1,585	799	4,097	22,339	19,048	629
Mallard	99	13	116	95	165	39
Mottled Duck	98	11	229	1,800	1,866	472
Northern Pintail	2,401	17,207	15,668	33,273	27,918	1,198
Total	**12,088**	**26,942**	**38,049**	**121,598**	**96,874**	**17,205**
Diving ducks						
Redhead	1,617	11	91	162	208	59
Canvasback	0	7	139	383	1,594	65
Ring-necked Duck	0	22	1,058	419	138	71
Lesser Scaup	510	358	950	1,169	1,529	55
Bufflehead	0	0	47	127	237	52
Common Goldeneye	0	0	0	0	129	38
Ruddy Duck	10	22	29	2,480	5,938	1,120
Mergansers	0	0	51	1,249	300	93
Total	**2,137**	**420**	**2,365**	**5,989**	**10,073**	**1,553**
Geese						
Light geese	1,757	625	1,730	14,164	7,303	2,535
Dark geese	318	295	11,339	20,507	6,315	270
Total	**2,075**	**920**	**13,069**	**34,671**	**13,618**	**2,805**

Source: US Fish and Wildlife Service, unpub. data.

[a] Numbers in parentheses indicate the number of years within that decade that the midwinter waterfowl survey was flown for the Rio Grande Delta region.

Laguna Madre de Tamaulipas

The east coast of Mexico contains relatively larger areas of wetland habitat than those in other regions of Mexico. In fact, this region contains more waterfowl habitat than the Interior Highlands, west coast of Mexico, and Baja California regions combined (Saunders and Saunders 1981). Within the east coast region, the Laguna Madre de Tamaulipas by far holds the most waterfowl during winter at ~54% of total waterfowl (DUMAC 1990a, 1990b). Similar to the lower Texas coast, the inland area adjacent to the Tamaulipan Laguna Madre is sparsely populated by people with a dominant land use of cattle ranching. Within this area, there are 36,330 ha (89,773 ac) of freshwater wetlands used by waterfowl (Carrera and de la Fuente 1996), and many ducks forage in the Laguna and use these areas as drinking sites.

Severe droughts in the early 1930s fueled considerable habitat loss in North America, much of which occurred in areas important for waterfowl. Consequently, declines in North American waterfowl populations became a serious concern among biologists at that time. Increased monitoring efforts in breeding and wintering areas were initiated in an attempt to evaluate waterfowl trends. The importance of waterfowl habitats in Mexico was highlighted at this time when the US Fish and Wildlife Service in cooperation with Mexico extended the midwinter survey into portions of Mexico (Saunders and Saunders 1981). The first of these surveys that included Mexico occurred in 1937–38 as part of George B. Saunders's survey to estimate waterfowl species utilizing the Central Flyway in North America. Based on his recommendation to conduct yearly, low-level aerial surveys, these surveys were formally initiated in 1947 in Mexico. Midwinter waterfowl surveys were conducted in Mexico until 2000, with later years of the survey being somewhat sporadic.

Redhead and Northern Pintail are two of the most abundant duck species wintering along the east coast of Mexico, and by far the largest proportions winter in Laguna Madre de Tamaulipas. Any major changes in duck abundance are primarily driven by changes in these two species. Redheads and Northern Pintails, on average, composed 74% of the duck community wintering in the Laguna Madre de Tamaulipas and ranged above 95% during some years of the survey. Duck abundance in the Laguna Madre de Tamaulipas was greatest in the 1940s, when midwinter waterfowl surveys tallied on average over a half million ducks (see table 13.2). Gadwall, American Wigeon, and Lesser Scaup were observed in relatively high abundances as well. This was apparently a time when food resources within the lagoon were high. As it is in the Texas Laguna Madre, Shoal Grass is the principal waterfowl food and covers approximately 30% of the basin (Saunders and Saunders 1981; Baldassarre et al. 1989).

Wintering waterfowl populations in the Laguna Madre de Tamaulipas declined drastically after the highs observed in the 1940s. Overall duck abundance declined by 74% during the 1950s, and by the 1960s, the decade average was only 9% of those tallied in the 1940s (see table 13.2). Redhead abundance was extremely low during the 1960s, with tallies below 5,000 birds during three years of the survey during that decade. Wintering populations of the Redhead in the Laguna Madre de Tamaulipas appear to fluctuate more than those in the Texas Laguna Madre. Historically, larger wintering populations of Redheads were documented south of the Laguna Madre de Tamaulipas stretching to Laguna Morales and Laguna San Andres, but in more recent years the great majority of Redheads winter in the Laguna Madre de Tamaulipas. This has likely occurred because the abundance and quality of waterfowl habitat have varied over the years. The natural opening and closing of passes to the Gulf of Mexico, often as a result of hurricanes, have had large impacts on water levels and salinities within the lagoon (Saunders and Saunders 1981). Also, the Río San Fernando (also known as Río Conchos), the largest potential source of freshwater inflow into the Laguna

Madre de Tamaulipas, historically provided considerable amounts of fresh water to the lagoon. More recently, however, freshwater inflow has been drastically reduced as withdrawals for agricultural irrigation increased. In the 1947 midwinter survey, over 300,000 waterfowl were observed in the delta as water from this river emptied into the Laguna Madre de Tamaulipas. However, waterfowl abundance rebounded in the 1980s, ranking second over the survey period. The largest average Redhead tally occurred during this period, and there were relatively large numbers of Northern Pintail, American Wigeon, Gadwall, and Lesser Scaup.

Wintering goose abundance has declined drastically along the Gulf Coast of Texas and Tamaulipas over the last 20 years. Overall goose abundance tallied on winter surveys in the Laguna Madre de Tamaulipas region declined by 65% from the 1980s to the 1990s (see table 13.2). Winter distributions of geese in North America have shifted northward as milder winter temperatures have occurred in areas farther north. Lakes in Arkansas, Kansas, and Missouri often remain unfrozen and can be used as roost sites, and less snow covering agricultural fields allows access to food (primarily waste corn). Through the 1980s, Lesser Canada Geese used the Laguna Madre de Tamaulipas and adjacent coastal ponds and wetlands, primarily in areas where cultivated fields were in close proximity to the lagoon and available for foraging. Lesser Snow Geese and Greater White-fronted Geese represented 21% and 16% of total Gulf population, respectively, in the Laguna Madre area during the 1940s–60s. The Greater White-fronted Goose is the most widely distributed goose in Mexico, but it was found in fewer numbers on the Laguna Madre and adjacent freshwater ponds prior to the 1980s (Saunders and Saunders 1981; Baldassarre et al. 1989).

Redhead Foraging Behavior and Influence on Resources

Redheads spend about 45% of the daytime feeding, which is most intense immediately after sunrise and is unrelated to tide cycles (Mitchell et al. 1992). Feeding occurs predominantly in shallow water by dipping (58% of time) and less frequently by tipping (25%), diving (16%), or gleaning (0.1%). Over and above the considerable amount of shallow foraging habitat available to waterfowl in the Laguna Madre, foraging by dipping has benefits because a dipping feeding sequence takes considerably less time than tipping or diving. Similarly, the energetic costs of dipping are likely considerably less than the other foraging behaviors, making it more efficient in both time and energy.

James (2006) estimated that approximately 17,000 ha (~42,000 ac) of Shoal Grass dominated substrate in the Laguna Madre of Texas that was ≤30 cm (≤12 in) at some point during winter and considered this to be the extent of optimal foraging habitat for Redheads. Based on 11,620 benthic core samples throughout the Laguna Madre, he estimated Shoal Grass rhizome biomass to be 7,543 tonnes (8,314 tons) at arrival of Redheads in October 2001 and 6,114 tonnes (6,739 tons) at arrival in October 2002. Using exclosures to estimate declines in biomass unrelated to Redhead foraging, James estimated that Redheads reduced Shoal Grass rhizome biomass by 26% to 33% during winter across the entire Laguna Madre. This was less than half the 75% decline in rhizome biomass estimated by Mitchell et al. (1992) across winter at foraging sites used by Redheads in the Lower Laguna Madre. This difference is likely related to the spatial extent of sampling and the variation in rhizome depletion by Redheads at different foraging sites. In fact, James (2006) derived a similar estimate of 68% reduction when using his data to examine roughly the same area sampled by Mitchell and colleagues. To provide further insight into the relationship between Shoal Grass resources and Redhead populations, James developed an energetics model based on the estimated amount of energy provided by Shoal Grass resources in the Laguna Madre (supply) and the energy requirements of the wintering Redhead population (demand). Based on model simulations, Redheads should have consumed 77.5%, 49.6%, and 100% of Shoal Grass rhizomes in shallow foraging habitat throughout the Laguna Madre during the three years of study. The energetics model appeared to underestimate energy available or overestimate energy demand by Redheads, or both. This was especially apparent during 2002–3, when the model estimated that by late January the energy supplied by Shoal Grass decreased to levels lower than the demand by Redheads. Conversely, analysis of body condition of Redheads indicated that both lipid and protein stores maintained an increasing trend through March that same year, suggesting that resources were not limiting to Redheads. If the nearly 17,000 ha (42,000 ac) of foraging habitat was considered to be available throughout the entirety of winter, Shoal Grass resources in the Laguna Madre would support enough energy for an estimated 54.6 million duck-use-days. This equates to supporting approximately 364,000 ducks across the entire winter period (15 October to 15 March).

There is some indication that the Redheads' alternate use of foraging areas between consecutive winters: sites that are heavily grazed in one year are used much less the following year (Mitchell et al. 1992; James 2006). Shoal Grass rhizome biomass of 10g dry mass/m^2 appears to be a threshold at which Redheads terminate foraging, presumably because of the decrease in foraging efficiency when biomass reaches this density. This giving-up density (GUD) by Redheads is similar to that for swans foraging on submersed aquatic vegetation in shallow water (12 g/m^2; Nolet et al. 2006) and for Mallards foraging on tubers in moist-soil impoundments (7 g/m^2; McKenzie 1987). The 10 g/m^2 threshold also appears to be the level at which Shoal Grass meadows are unable to recover to initial biomass the following year (Mitchell et al. 1992).

Diets of waterfowl wintering in the Laguna Madre generally include parts of seagrasses (primarily Shoal Grass and Wigeon Grass) and animal foods associated with seagrass meadows. In particular, Shoal Grass has been a dominant food found in diets of most waterfowl investigated in the Laguna Madre (Koenig 1969; McMahan 1970; Woodin 1996; Ballard et al. 2004). Redheads rely heavily on Shoal Grass resources in the Laguna Madre during winter. Based on esophageal contents, their diet comprises up to 94% Shoal Grass rhizomes (Marsh 1979; Woodin 1996). Northern Pintails wintering in the Laguna Madre consume a variety of foods, but their diets comprise predominantly Shoal Grass rhizomes, Wigeon Grass drupelets, amphipods, Dwarf Surf Clams, and Variable Bittiums (Ballard et al. 2004). Unlike Redheads, Northern Pintails also feed in freshwater wetlands adjacent to the coast (White and James 1978). Seeds of freshwater plants such as sedges, spike rushes, bulrushes, smartweeds, and pondweeds are also consumed, and during wet winters, large numbers of pintails can be found inland. American Wigeons are highly herbivorous (Mini et al. 2014), and their diet in the Laguna Madre is consistent with this, comprising 97% Shoal Grass rhizomes (Koenig 1969). Lesser Scaup feed on Shoal Grass to a lesser degree (~20% of diet) than American Wigeons, Northern Pintails, and Redheads and consume primarily animal material such as small clams, snails, and fish (Harmon 1962; Rogers and Korschgen 1966; McMahan 1970).

Use of Coastal Ponds

The Laguna Madre is well-known for supporting Redheads throughout winter, and its expansive, shallow Shoal Grass meadows provide abundant foraging opportunities. However, freshwater coastal ponds on the mainland and Padre Island that are adjacent to the lagoon are another habitat critical to Redheads (Adair 1990; Mitchell et al. 1992; Adair et al. 1996; Woodin 1994). These ponds provide fresh water that allows

Figure 13.4. *Redheads on a freshwater pond adjacent to the Upper Laguna Madre (Photograph by Bart Ballard)*

Redheads and other waterfowl that forage in the salty waters of the Laguna Madre to purge high salt concentrations ingested while foraging and to maintain their osmotic equilibrium (fig. 13.4). During winters when conditions are dry and fewer coastal ponds are inundated with water, there tends to be a strong correlation between the number of coastal ponds used by Redheads and the number of Redheads in the region (Ballard et al. 2010). Additionally, Redheads forage at coastal ponds farther inland during these dry winters. In average to wet winters, Redheads tend to be dispersed across the greater availability of coastal ponds unrelated to the number of Redheads in the region and tend to concentrate their activity closer to lagoon foraging areas.

James (2006) found that Redheads concentrate their foraging activities within the Laguna Madre to areas that are closer to the limited number of coastal ponds during a dry year. In a wet year, Redhead foraging activity was more dispersed. Thus, the availability and distribution of coastal ponds adjacent to the Laguna Madre may have implications concerning the severity of foraging pressure on Shoal Grass meadows within the Laguna Madre. Foraging pressure by Redheads on Shoal Grass meadows can be severe at times. Mitchell et al. (1992) found that about two-thirds of the Shoal Grass meadows that incurred high foraging pressure by Redheads did not recover to the same rhizome biomass recorded prior to Redhead arrival. Thus, the distribution and abundance of freshwater coastal ponds adjacent to the Laguna Madre can have impacts on Shoal Grass resources in the Laguna Madre.

Some species of ducks use their supraorbital nasal gland to excrete excess salts and maintain osmotic equilibrium when foraging in saline habitats (Schmidt-Nielsen and Kim 1964; Bennett and Hughes 2003). The gland acclimates as a bird transitions from freshwater to saltwater habitats and becomes more efficient at extracting salts. Mitchell et al. (1992) suggested that the greater efficiency of salt extraction as winter progresses allowed Redheads to become less reliant on coastal ponds to assist with osmotic regulation. However, the reduction in use of coastal ponds by Redheads could have resulted from birds moving to coastal ponds outside the study area. Other research investigating larger portions of the Laguna Madre (Adair et al. 1996; Ballard et al. 2010) found that Redheads maintain their reliance on coastal ponds throughout winter or even increased the amount of time they spent drinking in late winter/early spring (J. Moore 1991). Thus, the acclimation of the nasal gland to the saline environment in the Laguna Madre does not appear to influence the reliance of wintering Redheads on adjacent freshwater sources.

Redheads select coastal ponds for drinking sites based on characteristics related to three primary drivers: water salinity, safety, and energetics (Adair 1990; Adair et al. 1996; Ballard et al. 2010). Adair et al. (1996) investigated factors that separated coastal ponds that experienced high use by Redheads from those ponds that had low use. They found that highly used ponds had greater water depths and surface areas and lower salinities and distances to nearest estuary feeding sites than ponds experiencing low use. Based on observations from 63 aerial surveys across three years, Ballard et al. (2010) compared characteristics of all coastal ponds used by Redheads adjacent to the entire Laguna Madre to the nearest pond that was not observed to be used. They also found that ponds used by Redheads had greater water depths and surface areas, as well as a greater percentage of open water. Redheads avoid coastal ponds dominated by emergent vegetation during winter, possibly because these vegetated areas provide cover for predators. Water salinity was lower in coastal ponds used by Redheads than in those that exhibited no observed use. A common characteristic of ponds used by Redheads is a relatively continuous source of fresh water, mostly in the form of windmills or artesian wells, which appear to lower the water salinity in coastal ponds. Redheads also concentrate their freshwater use to coastal ponds that are positioned <9.0 km (<5.5 mi) from the nearest foraging habitat. However, this distance can be longer in dry winters when the availability and distribution of coastal ponds adjacent to the Laguna Madre are reduced.

Redheads arrive on ponds and typically drink before initiating any other behavior (Mitchell et al. 1992; Woodin 1994; Adair et al. 1996). For example, J. Moore (1991) found that in 489 out of 490 observations of individually marked Redheads, drinking was the initial behavior upon arrival on ponds (fig. 13.5). Following drinking, Redheads typically transition into some sort of maintenance behavior, such as preening or bathing. Both Redheads and Lesser Scaup spend a significant proportion of their time

Figure 13.5. *Redheads visit coastal ponds to obtain fresh water for drinking. (Photograph by Bart Ballard)*

resting on coastal ponds before returning to foraging sites in the lagoon (J. Moore 1991; Adair et al. 1996). Although foods consumed by Redheads during the post-breeding period are available in some of the ponds visited during winter, they conduct nearly all their feeding activities in Shoal Grass meadows in the Laguna Madre and rarely feed in coastal ponds (Mitchell et al. 1992; Woodin 1994; Adair et al. 1996).

Large concentrations of Redheads during winter influence the water quality of coastal ponds adjacent to the Laguna Madre (Adair 1990; Skoruppa and Woodin 2000; Ballard et al. 2010). Temporary increases in nitrate ($[NO_2 + NO_3]$–N), ammonium (NH_4–N), and chlorophyll-*a* in coastal ponds following use by Redheads has been tied to fecal deposition by the birds. This is likely most pronounced in dry years when coastal pond availability is reduced and Redheads and other waterfowl are concentrated on the limited number of available ponds. Increases in bacteria and nutrients have been tied to waterfowl use in other systems as well (Brandvold et al. 1976; Hussong et al. 1979). The presence of Redheads on coastal ponds also increases other water-quality metrics, such as turbidity and salinity (Ballard et al. 2010). Although some of these parameters (chlorophyll-α and nitrates) appear to decrease relatively quickly after birds discontinue use (Skoruppa and Woodin 2000), others (salinity) are likely more persistent and are dependent on inputs of fresh water for the water quality to improve for waterfowl use.

Conservation Issues

The reliance of much of the continental Redhead population on a single food source within a small geographic range creates a tenuous situation should Shoal Grass resources severely decline within the Laguna Madre. Because shallow Shoal Grass habitat is fundamental to the sustainability of wintering waterfowl populations along the lower Texas coast, particularly to Redheads, monitoring and protection of Shoal Grass meadows should be a high priority for waterfowl management in the region. Also, the relationship between shallow Shoal Grass meadows in the Laguna Madre and coastal ponds on the mainland is strong for waterfowl in this system. Thus, any proposed development or alteration of habitats that directly or indirectly impacts the lagoon or mainland within 10 km (~6 mi) of the lagoon should be thoroughly investigated to understand how these alterations will impact waterbirds, including wintering waterfowl populations.

The Laguna Madre is protected from development to a large degree because it is surrounded by federally protected lands (Padre Island National Seashore, Laguna Atascosa National Wildlife Refuge) and large private ranches (e.g., King Ranch, Kenedy Ranch, Yturria Ranch). Because the dominant land uses on these properties are compatible with relatively natural vegetation communities, the lower Texas coast has been touted as the longest undeveloped shoreline in the contiguous United States (Fulbright and Bryant 2002). Nonetheless, with a growing human population, it is inevitable that development interests will eventually threaten this status and impact the natural resources of the region. C. Lange (2014) developed a spatial model that prioritizes coastal ponds and adjacent areas based on several metrics of the ponds (use by Redheads, hydroperiod, amount of foraging habitat in close proximity, and isolation). This model provides a starting point for conservation efforts that protect the integrity of existing coastal ponds and for actions to develop freshwater sources to increase the availability of certain ponds during years when they would typically be dry. These types of conservation efforts have many positive effects on Redheads and the resources on which they rely. For example, in dry years Redheads may overgraze Shoal Grass meadows in close proximity to the limited number of coastal ponds available (Mitchell 1991; Ballard et al. 2010). Targeting specific wetlands that allow Redheads to distribute foraging effort across a larger area in the Laguna Madre should allow for lower impacts to Shoal Grass resources. Additionally, by dispersing Redheads across more coastal ponds and reducing densities on individual ponds, water quality at drinking sites should improve (Skoruppa and Woodin 2000; Ballard et al. 2010).

Wind farms are a relatively recent but significant threat to Redheads in the Laguna Madre. Although most research investigating the impacts of wind farms on waterfowl has found little effect, these studies may not be completely comparable to wind farms near the Laguna Madre. For instance, most studies have investigated offshore wind farms or have looked at impacts to long-lived species of waterfowl (sea ducks, geese, and swans). How representative these studies are to the shorter-lived Redhead and on discrete wetlands near the coast is debatable. Large, open tracts of land with minimal local human populations are attractive for certain types of development, such as wind farms. With relatively consistent winds, the wind industry has targeted the lower Texas coast as a priority site for wind farm development. Construction of the first wind farm was initiated in 2006 on the Kenedy Ranch adjacent to the Laguna Madre. The infrastructure of this wind farm (roads, etc.) appears to have altered the hydrology of nearby coastal ponds and impacted their availability to wintering waterfowl (Lange et al. 2018). Comparing water availability within the wind farm between preconstruction and postconstruction, Lange et al. (2018) estimated that 88% of original ponds were available during wet winters, 57% were available during winters of average wetness, and 7% were available during winters of drought. The reduced availability of coastal ponds has obvious effects on the physiology of wintering Redheads as well as grazing intensity on Shoal Grass meadows in proximity to the remaining ponds.

Another threat to the waterfowl habitat of the region is sea-level rise related to global climate change (C. Lange 2014). The Texas coast is already experiencing rates of sea-level rise higher than global rates (Church and White 2006). Although there is considerable uncertainty related to sea-level rise models, many predict similar amplitudes of rising sea levels within the next 100 years (Horton et al. 2008; Pfeffer et al. 2008; Jevrejeva et al. 2010). C. Lange (2014) modeled a 1.0 m (3.3 ft) rise in sea level and investigated its impacts on waterfowl habitats in the Laguna Madre region. He concluded that number of coastal ponds available as drinking sites for waterfowl using the Laguna Madre would decline by 60% by 2100 as a result of inundation with salt water. However, developing quality drinking sites for Redheads and other waterfowl is not challenging because the need to manage for specific vegetation communities or complex hydrological regimes is not necessary (Adair et al. 1996; Ballard et al. 2010). Many sites that Redheads extensively use are simple dugouts developed to water livestock. Thus, the ability to mitigate loss of coastal ponds to sea-level rise appears possible.

The conservation implications for ensuring the continued stability of Shoal Grass resources in the Laguna Madre of Texas and Tamaulipas are huge. The great reliance of Redheads, Northern Pintails, and other waterfowl on relatively shallow Shoal Grass meadows in these lagoons during winter provides a need to ensure protection of these sites. It also emphasizes the need for frequent monitoring of Shoal Grass distribution and health. The effects of other anthropogenic activities, such as dredging channels, placement of dredged material, increased inflows of nutrients from agricultural runoff, and excessive scarring of seagrass beds by propellers of shallow-draft boats (Onuf 1994, 1996b), remain serious threats to seagrasses in the Laguna Madre. Beyond waterfowl, most species using these systems rely to some degree on seagrass meadows at some point in their life cycle.

14 Colonial Waterbirds and Rookery Islands

ELIZABETH H. SMITH, BRIAN R. CHAPMAN, AND BEAU HARDEGREE

Early on a bright June morning in 1926, J. J. Carroll and two companions piloted a small skiff from the shore of Flour Bluff to Bird Island (South Bird Island) in the Laguna Madre with hopes of observing nesting waterbirds. After landing at his destination, Carroll discovered an island replete with numerous colonial waterbirds, including Black Skimmers, Laughing Gulls, Caspian Terns, Royal Terns, Sandwich Terns, Brown Pelicans, Great Blue Herons, Tricolored Herons, and Reddish Egrets. On Carroll's visit the following May and in three successive springs (1928–30), these species were accompanied by nesting American White Pelicans, Black-crowned Night-Herons, and Gull-billed Terns (Carroll 1927, 1930a, 1930b; fig. 14.1). Carroll, however, was not the first to document waterbird colonies in the Laguna Madre. Several ornithologists had previously visited the lagoon and documented the abundance of colony-nesting birds (Sennett 1878, 1879; T. Pearson 1921; Cahn 1923; Pemberton 1922a, 1922b; Bent 1926). Made during an era when many waterbird species were wantonly slaughtered to supply feathers for the millinery trade, these reports were the first of many that described the rich colonial waterbird fauna of the Laguna Madre.

When Carroll's ventures were undertaken, five natural islands—Dimmit, Crane, Pita, North Bird, and South Bird—provided the only isolated nesting habitats for colonial waterbirds in the Upper Laguna Madre. Green Island and Three Islands were the only islands permanently surrounded by water in the Lower Laguna Madre

Figure 14.1. *The diversity of nesting waterbirds seen in this 1974 photograph of South Bird Island approximates what J. J. Carroll may have seen when he visited the island in 1928. (Photograph by Brian R. Chapman)*

of Texas, but some *lomas* (usually clay dunes in South Texas, but can also be a more generic "hill") encircled by intermittently flooded wind-tidal flats bore island-like names. Farther south, some of the several hundred natural islands in the Laguna Madre de Tamaulipas also supported colonial waterbird rookeries.

During the late 1940s, the US Army Corps of Engineers created several hundred islands in the Laguna Madre of Texas when they dredged the Gulf Intracoastal Waterway (GIWW). This navigational channel bisected the lagoon's entire length. Numerous shallow tributary channels were subsequently excavated to provide access to marinas and oil and gas well drilling sites. During the dredging process, slushy materials removed from channel bottoms were deposited in emergent domes alongside the channels (see fig. 2.10). These substrates eventually solidified to form islands where a succession of salt-tolerant grass, forb, and shrub assemblages gradually colonized (Coover and Rechenthin 1965; Chaney et al. 1978).

Barren dredged-material islands become important loafing or roosting sites for colonial waterbirds. Terns and skimmers construct shallow nest depressions (scrapes) in open sandy or shelly substrates and usually are the first species to use the islands as rookeries (Smith and Cox 1998). Low vegetation affords nesting habitat for Laughing Gulls, Caspian Terns, Royal Terns, and Gull-billed Terns (fig. 14.2a). Invasions of woody shrubs and cacti eventually provide elevated platforms upon which herons and egrets build flimsy stick nests (figs. 14.2b, c).

a

b

c

Figure 14.2. *Nesting habitats on dredged-material islands in the Laguna Madre vary with age, elevation, and degree of disturbance. Islands with fresh deposits are largely barren (a) whereas slightly older islands develop a blanket of low vegetation (b). Some undisturbed islands eventually develop thickets of shrubs and low trees that offer elevated nesting locations for herons and egrets (c). (Photographs by Allan H. Chaney [a, b] and Brian R. Chapman [c])*

Natural and created islands isolated from most disturbances are critical for the continued survival of colonial waterbirds along the lower Texas coast, where trees are rare. Since most waterbird colonies are concentrated on islands in this region, the birds serve as excellent indicators of ecosystem health (Soots and Landin 1978). Islands are also important to resident and migratory waterbirds, shorebirds, songbirds, and raptors for resting, roosting, and feeding.

Advantages and Disadvantages of Colonial Nesting

A safe nesting site is essential for reproductive success in most avian species (Collias and Collias 1984). The nests of birds in a colony (rookery) are spaced closely and do not represent a pattern of random dispersal throughout suitable habitat (Gaston 2004). Colonial nesting occurs in only 13% of all species, but it is typical of most seabirds and waterbirds (Lack 1968). The behavior likely evolved in response to a shortage of nesting sites located away from predators and where the proximity and abundance of food resources made the maintenance of separate feeding territories unprofitable (Siegel-Causey and Kharitonov 1990; Gawlik 2002). Coastal rookeries are most often located on islands with appropriate nesting substrate or vegetation (Soots and Parnell 1975).

Colonial nesting offers both advantages and disadvantages (Gaston 2004). Some colony members are always alert, thereby decreasing the probability that an approaching predator (or human visitor) can arrive undetected. Once an unwanted caller is observed, colony members join forces to mob the trespasser: gulls typically excrete a foul deluge of "white rain" from above an interloper; terns and skimmers often dive from the air to deliver blows with their beaks or pluck tufts of hair (Pemberton 1922a); and herons, egrets, and cormorants usually remain in their tree nests to regurgitate partially digested crop contents. Group defense of the nesting area is a prime example of the "safety in numbers" concept.

Colonies may also function as information centers where adults watch the arrivals of successful foragers to obtain information about the location of food resources (Ward and Zahavi 1973). Such information may not always be required in the Laguna Madre because the shallow open waters adjacent to rookeries usually provide an abundance of accessible fish and crustaceans. Although Great Blue Herons and Reddish Egrets often forage in groups near the colony where both species defend small, individual feeding territories, a change in wind speed or direction, periodic droughts, and increases in human activity can alter the location of foraging habitats (Simersky 1971; Bates and Ballard 2014; Bates et al. 2016). Blooms of Texas Brown Tide caused by a microscopic alga may temporarily reduce or eliminate seagrasses, rendering nearby habitats unsuitable for juvenile or adult fishes and some shellfish (see chapter 18). When long-distance travel is required, reliance on cues about productive foraging sites might be indispensable.

Heron and egret chicks respond to reductions in food supply in a unique way—the largest nestlings dominate their nest mates to obtain most of the prey brought by adults, while the others are left to starve. In extreme cases, the dominant nestling engages in siblicide by pecking its siblings to death (Mock et al. 1990). This seemingly cruel behavior allows some young to survive during a period of environmental stress.

Parasites and diseases are easily transmitted among birds in dense associations (Tella 2002), and large gatherings of raucous birds often attract predators (Rogers 1987; Tella 2002). On small islands, a predator can overwhelm the numerical advantages associated with colonial nesting. A single Raccoon, Coyote, or Feral Pig can create enough havoc to cause colony abandonment. Human disturbances, even those by well-meaning nature observers, fun-loving boaters, or nearby anglers, often are just as detrimental (Carney and Sydeman 1999).

Nesting activities in densely populated waterbird colonies can alter vegetation composition, density, and succession, rendering a site unsuitable for years (Soots and Parnell 1975). Early-arriving large terns, such as Royal or Caspian Terns, can limit the establishment and growth of annual vegetation by trampling and overfertilization and prevent the terns and other ground-nesting species from using the site in subsequent years. Similarly, concentrations of nitrogen-rich droppings beneath heron and cormorant nests can kill trees and eliminate nesting habitats.

Habitat Characteristics of Laguna Madre Islands

Stabilized by well-established vegetation, the natural islands in the Laguna Madre have likely existed for centuries. Unless disturbed, these islands are more likely than the dredged-material islands to support shrubs and low trees. The patterns of vegetation on dredged-material islands are similar except that shrub zones are less developed on dredged-material islands in the Lower Laguna Madre than on islands farther north (Chaney et al. 1978). Many dredged-material and natural islands contain waterbird rookeries ranging in size from <0.5 to 5.6 ha (1.2 to 13.8 ac).

The soils deposited by dredging activities range from fine sand, clay, and silt throughout most of the lagoon area to reef material and clay in the Baffin Bay region (DePue 1974). Dragline dredges used in the 1940s and more recent suction dredges deposited excavated sediments in a dome adjacent to the channel under construction. Over many years, wind, tidal action, and rainfall gradually erodes an island, reducing the elevation of the dome as the island expands in circumference. Vegetation colonizes the fine soils slowly, and it sometimes takes 5 to 10 years for plants to appear. The first plants to become established typically are halophytes, including Sea Purslane, Glasswort, Saltwort, Seashore Dropseed, and Seashore Paspalum. These initially colonize the ridge of sand and decaying seagrass detritus just above the high-tide zone before gradually spreading inward to form a wide belt of vegetation around the island's periphery. On low islands, such as North and South Bird, halophytes extend to the island centers.

Vegetation on the slopes of domes and at the centers of higher islands must tolerate limited freshwater availability (Barnes 1971). Development of plant associations on the slopes may require a decade or more depending on how rapidly salt is leached from the soil, but the increased water-holding capacities of fine soils may allow succession to proceed faster once a pioneer community is established (Chaney et al. 1978). Dense to sparse associations of low shrubs and grasses such as Sea Oxeye, Gulfdune Paspalum, and Bushy Bluestem often occupy mid-elevations. On some islands, these plants initially become established in erosion channels and in the troughs cut by dredged-material runoff when the island was formed. Occupancy of barren slopes by nesting terns in a previous year appears to increase the rate of vegetative development (Ortiz 1976; Soots and Parnell 1975). Introduction of nutrients into the soil from bird guano likely spurs plant survival and growth.

The crowns of dredged-material islands often remain barren for 15 to 30 years. Where elevations exceed about 1.5 m (5.0 ft), larger species such as Honey Mesquite, Texas Prickly Pear, and White Leadtree develop as isolates or sometimes form dense thickets. Green Island, a natural island in the Lower Laguna Madre, is blanketed by Honey Mesquite, Snakewood, and Spanish Dagger, a thorny vegetative association that Cahn (1923) described as "a huge pincushion armed with a million needle points."

Erosion often causes adjacent islands to coalesce, forming peninsula-like fingers linked by low, barren strands. Island erosion and windblown sands from Padre Island's roving dunes also decrease the water depth between the barrier island and some dredged-material islands, allowing access for mammalian predators and rodents from Padre Island to waterbird colonies (Jones and Frey 2013).

Beginning in the mid-1980s, the US Army Corps of Engineers began depositing dredged material in diked containment areas to minimize siltation and elimination of seagrass beds. Containment areas often integrate a series of dredged-material islands, but some are large mainland shoreline sites. Barren substrates inside the dikes occasionally serve as nesting sites for terns, but other species generally avoid them.

Significant Waterbird Colonies

Upper Laguna Madre of Texas Observers for the Texas Colonial Waterbird Survey (TCWS) found 34 active colonies on islands or groups of islands in the Upper Laguna Madre from 1973 to 2015 (TCWS 1982, 2017). Twenty-two species were documented during the surveys (table 14.1), and some species nested most years. Over time, the number of nesting pairs has decreased in the Upper Laguna Madre. During the same period, the number of colonies decreased slightly.

Historically, the majority of nesting waterbirds in the Upper Laguna Madre used seven sites comprising individual islands or island groups. By 2009 most waterbirds were restricted to four sites, with 75% of breeding taking place at just three locations (table 14.2). For example, South Bird Island remained a significant colony site until 2005 and continued to host a few birds through 2009. Mammalian predators reached the island and caused disruption and abandonment. Four rookery sites in the Upper Laguna Madre, including the three with the largest populations of nesting pairs, are briefly described here.

Naval Air Station Islands Approximately 30 islands located between the Kennedy Causeway and Corpus Christi Bay represent a single colony location in the TWCS survey data. Dimmit Island, a natural island near the bay to which dredged material has been added, is more than 4.6 ha (9.0 ac) in size. Other islands in this area formed from dredged material are much smaller, and some are reduced to mere spits. During the 2011–15 census years, an average of 1,238 pairs representing 15 species nested on the island group. One dredged-material island near the center of the area once hosted a large rookery of Royal Terns, but their nesting zones were eliminated by vegetative succession. Laughing Gulls are now the most abundant nesting species in the island group.

Kennedy Causeway Islands Channels dredged for oil and gas exploration resulted in a number of small islands south of the Kennedy Causeway. Many of these eroded away, but approximately 25 remain and provide rookery sites for an average of 5,502 pairs representing 17 species per year. Zig Zag Island, an elongated strip that once contained a road connecting five widened areas for oil wells, is slowly eroding. Large patches of Texas Prickly Pear provide nesting substrate for herons and egrets, but the most abundant nesting species are Laughing Gulls, Royal Terns, and Sandwich Terns.

Pita Island/Humble Channel Islands Pita Island is a 5.0 ha (13 ac) natural island that was blanketed by shrubs and low trees. Beginning in the 1950s, the habitat on Pita Island was altered by oil and gas activities that removed vegetation and filled wetlands. The island has been reworked several times since the 1970s by petroleum companies. The shrubby vegetation on the interior of the island was removed, and only the outer leveed portions of the island now contain cacti and shrubs that support

Table 14.1. *Temporal comparison of colony sites occupied by colonial waterbirds in the Upper and Lower Laguna Madre of Texas and the Laguna Madre de Tamaulipas*

Species	Number of Colonies				
	ULM		LLM		LMdT
	1973–96	2011–15	1973–96	2011–15	1999–2001
American White Pelican	4	1	0	0	1
Brown Pelican	0	1	0	3	1
Neotropic Cormorant	1	0	0	2	0
Great Egret	14	10	11	12	1
Snowy Egret	22	11	12	9	5
Cattle Egret	11	2	12	3	1
Great Blue Heron	26	18	14	12	3
Reddish Egret	23	16	14	10	3
Tricolored Heron	26	15	13	9	3
Little Blue Heron	15	5	9	6	0
Black-crowned Night-Heron	10	5	5	5	2
Yellow-crowned Night-Heron	0	0	2	0	1
White Ibis	4	2	9	4	1
White-faced Ibis	14	1	8	0	0
Roseate Spoonbill	8	5	6	4	1
Laughing Gull	25	21	17	9	7
Least Tern	22	3	11	2	3
Forster's Tern	20	14	9	1	0
Sandwich Tern	14	5	14	6	2
Royal Tern	19	13	15	9	4
Caspian Tern	23	13	0	6	4
Gull-billed Tern	31	21	18	6	1
Sooty Tern	8	2	9	1	1
Black Skimmer	32	24	19	7	3
Number of Active Colonies	34	27	22	13	12

Note: ULM = Upper Laguna Madre; LLM = Lower Laguna Madre; LMdT = Laguna Madre de Tamaulipas.

Table 14.2. *Summary information for all Upper Laguna Madre, Texas, waterbird colonies, 1973–96 and 2011–15*

Colony Number	Colony Name	Number of Islands	Species Richness		Cumulative Survey Total		Bird Use (# of years)		Average Pairs per Year	
			1973–96	2011–15	1973–96	2011–15	1973–96	2011–15	1973–96	2011–15
221	Naval Air Station Islands	30 DMI and NI	19	15	42,070	6,640	19	5	2,214	1,328
222	Kennedy Causeway Islands	27 DMI and NI	16	17	48,541	27,512	19	5	2,555	5,502
240	Marker 13–35 Spoil Islands	14 DMI	14	8	7,934	127	16	5	567	25
241	Crane Islands	4 DMI	0	9	0	1,734	0	5	0	346
300	Pita Island/Humble Channel	1 NI, 10 DMI	18	15	27,302	17,378	24	5	1,138	3,476
301	Marker 65–74 Spoil Islands	w5 DMI	8	6	652	290	18	4	36	73
302	Marker 75 Spoil Islands	1 DMI	7	8	1,565	169	21	5	75	34
303	Marker 79–85 Spoil Islands	2 DMI	4	0	198	0	20	0	10	0
304	N of Bird Island (Mkr 87–91)	5 DMI	14	5	2,550	184	22	4	116	46
305	W of N Bird Island (Mkr 93–103)	7 DMI	8	2	630	30	19	1	32	30
306	North Bird Island	1 NI	9	0	4,023	0	21	0	192	0
307	West Side Spoil Islands	3 DMI	3	0	53	0	16	0	3	0
320	Padre Island Development Fill	Open fill	1	0	24	0	3	0	8	0
340	South Bird Island	1 NI	19	1	84,019	333	23	1	3,653	333
341	S of South Bird Island	4 DMI	21	1	136,006	1	24	1	5,667	1
342	Marker 127–131 Spoil Islands	2 DMI	14	10	2,056	434	21	5	98	87
343	Marker 141 Spoil Island	1 DMI	13	8	2,892	828	21	5	138	165

Table 14.2. Continued

Colony Number	Colony Name	Number of Islands	Species Richness		Cumulative Survey Total		Bird Use (# of years)		Average Pairs per Year	
			1973–96	2011–15	1973–96	2011–15	1973–96	2011–15	1973–96	2011–15
344	Marker 152 Spoil Island	1 DMI	9	7	1,711	359	20	4	86	90
345	Marker 163 Spoil Island	1 DMI	20	10	24,455	1,836	24	5	1,019	367
346	Marker 155 Spoil Island	1 DMI	9	2	418	22	16	1	26	22
348	Marker 165 Spoil Island	1 DMI	11	3	584	100	15	2	39	50
348	Yellow House Spoil (Mkr 162)	1 DMI	2	14	55	882	1	5	55	176
360	Marker 185 Spoil Island	1 DMI	14	0	593	0	17	0	35	0
361	Marker 207–221	7 DMI	19	13	41,153	1,968	24	5	1,715	394
362	South Baffin Bay Islands	4 DMI	17	12	15,377	4,907	24	5	614	981
363	Diked Island	1 DMI	5	3	849	24	16	3	54	8
364	Side Channel Islands	3 DMI	3	5	71	25	7	2	10	12
365	Chaney Island	1 DMI	0	6	0	270	0	3	0	90
380	Marker 21–35 Spoil	8 DMI	12	11	15,194	3,084	21	5	760	617
381	Yarborough Pass	1 DMI	10	0	1,358	0	12	0	113	0
382	N Yarborough Pass (Mkr 37–39)	1 DMI	9	5	2,353	970	17	5	138	194
383	S Yarborough Pass [illegible]	1 DMI	10	6	1,688	202	18	4	94	51
384	The Hole	4 DMI	11	6	961	564	46	5	60	564
385	Padre Island Spoil	9 DMI	12	0	745	0	15	0	50	0
400	East of Potrero Grande	2 DMI	5	0	599	0	8	0	75	0
401	East of Potrero Cortado	3 DMI	5	0	435	0	8	0	54	0

Note: All colony numbers begin with "614-." DMI = dredged material island, NI = natural island.

Table 14.3. *Summary information for all Lower Laguna Madre, Texas, waterbird colonies for 1973–96 and 2011–15*

Colony Number	Colony Name	Number of Islands	Species Richness		Cumulative Survey Total		Bird Use (# of years)		Average Pairs per Year	
			1973–96	2011–15	1973–96	2011–15	1973–96	2011–15	1973–96	2011–15
100	South Land Cut	8 DMI	18	1	34,136	4	21	1	1626	4
120	East Flats Spoil	1 DMI	11	11	5572	6338	13	4	429	1585
121	Mansfield Odd Spoil	Several DMI	10	0	7992	0	16	0	499	0
122	East Marker 265 Spoil	1 DMI	4	5	191	525	9	1	21	525
140	NE Mansfield Inter-section	1 DMI	15	11	39,285	1540	24	5	1,637	308
141	SW Mansfield Inter-section	1 DMI	11	0	5149	0	24	0	210	0
142	SE Mansfield Intersection	1 DMI	2	0	344	0	12	0	28	0
143	Green Hill Spoil Island	1 DMI chain	20	16	31,035	11,013	20	5	1,293	2,202
144	Mansfield Channel Spoil		2	0	80	0	6	0	13	0
160	Green Island Cut Spoil	7 DMI	16	12	32,715	10,561	21	5	2,112	1,558
161	Green Island	1 NI	16	11	54,652	12,146	21	5	2,602	2,429
162	Spoil SW of Green Island		1	0	60	1	8	0	7	0
180	Arroyo Colorado Int. Spoil	1 DMI chain	17	14	40,168	5,926	17	5	2,114	1,185
181	East Arroyo Spoil	8 DMI	9	0	4,135	0	13	0	318	0
182	Three Island Spoil	1 DMI chain	18	7	193,187	31	21	3	9,199	10

Table 14.3. *Continued*

Colony Number	Colony Name	Number of Islands	Species Richness		Cumulative Survey Total		Bird Use (# of years)		Average Pairs per Year	
			1973–96	2011–15	1973–96	2011–15	1973–96	2011–15	1973–96	2011–15
183	South Three Islands	5 DMI	15	2	5,311	9	14	2	379	4
184	Four Islands		8	2	8,334	105	10	1	833	105
200	Port Isabel Spoil	1 DMI chain	13	0	13,996	0	18	0	777	0
201	Padre Beach Estates	Sand beach	1	0	71	0	3	0	24	0
220	Laguna Vista Spoil	9 DMI	16	16	42,139	14,033	17	5	2,479	2,806
240	Dead Pecker Hill	Mangrove Island	11	6	4,522	70	14	1	323	70
250	Bahia Grande Bird Island		0	18	0	10,294	0	5	0	2,058

Note: All colony numbers begin with "618-." DMI = dredged-material island; NI = natural island.

herons and egrets. The interior portions of the island support robust populations of Laughing Gulls.

The Humble Channel Spoil Islands comprise 12 small dredged-material islands. All islands in this group possess a combination of shrubs and barren zones and collectively provide nesting habitat to an average of 3,476 pairs representing 15 species. Herons, egrets, and Gull-billed Terns commonly nest on the islands, but Laughing Gulls are by far the most abundant.

Marker 163 Spoil Island The erosion history of Marker 163 Spoil Island and its effect on the nesting of colonial waterbirds is emblematic of many islands in the Laguna Madre. Located adjacent to the GIWW within the boundary of Padre Island National Seashore, the dredged-material island encompassed about 1.7 ha (4.2 ac) in the 1970s (TCWS 1982), but erosion decreased its size to about 1.1 ha (2.8 ac) by 1986 (Coste and Skoruppa 1989). Nesting habitats still include shrubs, halophytes, and bare sand, but exposed areas suitable for ground nesters are greatly diminished. Black Skimmers, Gull-billed Terns, Laughing Gulls, Great Blue Herons, and 400–800 White Pelicans still nest on the island.

Lower Laguna Madre of Texas Twenty-four colonies were present in the Lower Laguna Madre during 1973–96, but only 14 were in use during the 2011–15 surveys (table 14.3). Species composition changed slightly in the intervening years. Nesting Brown Pelicans were completely absent from TCWS data in the Lower Laguna Madre until 2005, when six pairs were observed. Since then, nesting Brown Pelicans have become more common. Caspian Terns and Neotropic Cormorants continued to nest in several of the colonies during the 2011–15 seasons, whereas Yellow-crowned Night-Herons and White-faced Ibises were notably absent (see table 14.1). Nest counts of all colonial waterbirds in the Lower Laguna Madre averaged 14,506 pairs per year during 2011–15, a substantial decline from 25,992 pairs per year observed during 1973–96 (summarized from TCWS data).

Seven waterbird colonies in the Lower Laguna Madre provided habitat for more than 82% of all nesting pairs during 2011–15; more than 2,000 pairs nested annually in five of the colonies. Four of these also supported large nesting associations in 1973–96.

Green Hill Spoil Island Two small but heavily vegetated islands provide nest substrates for 16 species, including large numbers of Laughing Gulls. Herons, egrets, Roseate Spoonbills, Neotropic Cormorants, and White Ibises nest in the cactus and shrubs. More than 100 pairs of Reddish Egrets occupy nests on the islands during most years (McMurry 1971). Black Skimmers, Royal Terns, and Caspian Terns occasionally find enough open ground on the beaches of a finger-like extension on the southernmost of the two islands, although Coyotes and Raccoons sometimes cause disruptions (Coste and Skoruppa 1989), and both islands are threatened by erosion.

Green Island Cut Spoil Prior to 1973, a channel was dredged eastward from the GIWW toward Padre Island to service an oil and gas operation. Of the 14 dredged-material islands deposited alongside the channel, only 11 remain. Seven of the islands consist of barren exposures of clay, but four have patches of cactus and shrubby vegetation. Despite poor substrate for ground-nesting birds, hundreds of pairs of Royal, Caspian, Sandwich, and Gull-billed Terns and Black Skimmers nest on the islands each year. Reddish Egrets, Great Blue Herons, and Tricolored Herons raise young in the shrub patches.

Green Island Long known for its dense, thorny mezquital habitat and high populations of shrub-nesting wading birds, Green Island is a natural island adjacent to Laguna Atascosa National Wildlife Refuge. The area of the colony was estimated at 10.0 ha (24.7 ac) or less in 1980, with a maximum elevation of 5 m (~17 ft). The rookery has supported a diversity of shrub-nesting wading birds since ornithologists first visited the region (Sennett 1878, 1879; Pemberton 1922b). More than 2,400 wading birds nest annually on the island, and this total includes hundreds of Reddish Egret, Great Egret, Snowy Egret, Tricolored Heron, Roseate Spoonbill, and White Ibis pairs (figs. 14.3a, b). Laughing Gulls, Great Blue Herons, Little Blue Herons, and

a b

Figure 14.3. *The density of vegetation on Green Island is evident in the photographs of a Black-crowned Night-Heron (a) and White Ibis (b) perched near their nests concealed in the foliage. (Photographs by Beau Hardegree)*

Black-crowned Night-Herons also nest on Green Island, albeit in somewhat lesser numbers. Although the island is more than 0.8 km (0.5 mi) from shore, mammalian predators occasionally reach this site and cause colony disruptions.

Laguna Vista Spoil Two islands located north of the Laguna Vista community compose the colony site known as Laguna Vista Spoil. The northern half of the largest island is less than 3 m (~8 ft) above sea level and supports a mesquite and thorn brush community. Shrubs and barren ground are also abundant on the islands. Royal Terns, Sandwich Terns, and Laughing Gulls are the most numerous of the 16 nesting species. During 2011–15, the number of nests recorded annually on the islands averaged more than 2,800. Erosion is reducing the size of the large island, and an expanding carpet of invasive Guinea Grass threatens to eliminate the open habitats used by gulls and terns.

Bahia Grande Bird Island Bahia Grande is a large interior lagoon located between Laguna Vista and Brownsville. Its connection to the Laguna Madre was severed in the 1930s by various construction projects, and the degraded wetland remained largely dry for over 70 years. The Laguna Atascosa National Wildlife Refuge acquired the Bahia Grande basin and the surrounding area in 2000 and began efforts to restore the wetland. Its connection to the Laguna Madre was restored in 2005, allowing water to reflood the area and isolate four islands. Birds began to nest on the islands in 2008. By the 2011–15 censuses, an average of 2,068 pairs representing 18 species of waterbirds nested on the islands. Laughing Gulls are the most abundant nesting species, but in most years the islands also are home to one of the largest Gull-billed Tern colonies on the Texas coast (Molina et al. 2010).

Laguna Madre de Tamaulipas Population surveys of wading birds and other colonially nesting species were conducted from the Rio Grande south to Laguna de San Andreas in 1973–76 (Sprunt and Knoder 1980). Waterbird colonies in the Laguna Madre de Tamaulipas were not extensive, although significant populations of American White Pelicans and Reddish Egrets were documented (fig. 14.4). The only coastal breeding colony of American White Pelicans in Mexico is located in the northern

Figure 14.4. *The southernmost colony of American White Pelicans nests on islands in the Laguna Madre de Tamaulipas as seen in 1977. (Photograph by Brian R. Chapman)*

portion of the Laguna Madre de Tamaulipas (Chapman 1938). Recent nest counts on islands of the Mexican lagoon are unavailable, but Garza-Torres and Navarro-Sigüenza (2001) describe nesting by 19 colonial waterbird species on 12 islands (see table 14.1). A photograph in the report depicts a large colony of Royal and Sandwich Terns.

Selected Colonial Waterbirds of the Laguna Madre

Reproductive behavior occurs on rookery islands throughout most of the year (January through October), although the timing varies depending on weather conditions. Most nesting activity begins in March and is concluded by mid-August although Great Blue Herons usually initiate courtship and nest building in early January. Because the nesting season peak varies among species, colonial waterbird censuses may not accurately represent all species.

Laguna Madre colonies are surveyed annually from the last week of May to the first week of June. During each census, observers record the number of adults, nests, and breeding pairs. Where possible, birds and nests in small colonies are counted from a boat. Larger colonies and those with complex vegetation that restricts vision are surveyed on foot. Census data from aerial surveys conducted annually from 1973 to 1992 augmented the colony counts (E. Smith 2002; Hackney 2016). Although the methodology for counting nests or nesting birds was not standardized during most early years of the TCWS, the data were deemed sufficient for analysis of population trends (Chaney et al. 1996).

Habitats on an island determine which species might colonize. In general terms, colonial waterbirds may be categorized as either ground nesters, species that select nesting substrates ranging from bare ground to dense herbs and shrubs, or arboreal nesters, which build an elevated nest in supporting vegetation. Some species prefer one type of nesting substrate but use an alternative if necessary. Generally, pelicans, skimmers, gulls, and terns are ground nesters, but some Brown Pelican nests are slightly elevated. Spoonbills, herons, egrets, cormorants, and ibises generally are arboreal nesters, but herons, egrets, and spoonbills sometimes construct ground nests. Some species also prefer dead vegetation, litter, or drift line flotsam as low nesting locations. Ground nests are subject to flooding by storm tides and boat wakes.

Ground-Nesting Species Some ground-nesting species establish territories in colonies where nests are densely clustered on small patches of suitable habitat (figs. 14.5a–d). Royal and Sandwich Terns often share densely packed colonies, where they defend hexagonal "pecking distance" territories surrounding scrape nests (fig. 14.6a, b). The territories allow adjacent birds to peck at, but barely miss, the terns on surrounding nests (Buckley and Buckley 1977). Other ground nesters, including Black Skimmers, Forster's Terns, Caspian Terns, Sooty Terns, and Least Terns, nest in loosely spaced colonies often interspersed among Laughing Gulls and other tern species.

Black Skimmers typically nest in low-lying, unvegetated areas of loose sand and shell that often are subject to tidal inundation (Burger and Gochfeld 1990). If nests are swamped early in the breeding season, skimmers usually renest and raise young in the same year (Bent 1921; Tompkins 1933). Nests with eggs are rarely left unattended, but if neither parent is incubating the eggs, at least one stands nearby. Black Skimmers sometimes nest in association with terns, especially Least Terns and Gull-billed Terns (see fig. 14.6b). Declines in Black Skimmer populations have been steady in recent years (Hackney 2016). Possible causes of these declines are attributed to organochlorine contaminants, erosion, and human disturbance (White et al. 1985; Custer and Mitchell 1987b; Chaney et al. 1996). Human disturbance frequently results in lethal

Figure 14.5. *Ground-nesting waterbird nests of (a) Royal and Sandwich Terns and (b) Least Terns; (c) Gull-billed Terns dig shallow depressions (scrapes) for their nests; and (d) Forster's Terns construct nests on vegetative debris. (Photographs by Allan H. Chaney)*

a

b

c

d

a

b

Figure 14.6. *Views of nesting colonial waterbirds on Laguna Madre dredged-material islands: (a) a mixed-species tern and gull colony; and (b) a Black Skimmer and Gull-billed Tern colony (Photographs by Brian R. Chapman [a] and Beau Hardegree [b])*

exposure of eggs or nestlings to the sun, disruption of feeding, egg theft by Laughing Gulls, or nest abandonment (DePue 1974; Chaney et al. 1996).

About 60% of the eastern subspecies of Gull-billed Terns occur in Texas (Molina et al. 2010). Pemberton (1922a) discovered a colony of several thousand Gull-billed Terns in the Bahia Grande, a basin of the Lower Laguna Madre. Since then, the species has experienced a long-term decline in Texas. Between 1997 and 2007 the total nesting population in the Lower Laguna Madre was about 100 pairs each year (TCWS 2017), but soon thereafter the population rebounded to an average of around 500 nesting pairs annually. This increase corresponded to the 2005 reflooding of the Bahia Grande basin, which isolated islands from predators.

Laughing Gull populations have been relatively stable since 1973 and constitute a large majority of the total population in many rookeries (Hackney 2016). The species adapts readily to new food resources (thus earning the nickname "Sky Roach"), nesting habitats, and the presence of humans. Laughing Gulls engage in kleptoparasitism (food piracy) by chasing conspecifics, Brown Pelicans, or terns to make them drop a fish (Carroll and Cramer 1985). The species is also an opportunistic predator of eggs and chicks in mixed-species rookeries, and its presence can negatively impact reproductive success, particularly of Black Skimmers and terns. Although nesting pairs have declined in some rookeries, they remain abundant in the Laguna Madre. Statewide, the range of population variability indicates that there may be large fluctuations both interannually and within years among rookeries (Hackney 2016).

American White Pelican colonies in the Laguna Madre and Laguna Madre de Tamaulipas represent the southernmost breeding records for the species (Chapman 1988). After abandoning South Bird Island when predators arrived, the pelicans shifted their colonies to four islands in the Upper Laguna Madre for several years. Since 1983, all of the White Pelicans in the Upper Laguna Madre have nested on Marker 163 Spoil Island. Each year several hundred pairs are counted on the island.

Eggshell thinning caused by DDT and other organochlorine pesticides reduced reproductive success of the Brown Pelican (King et al. 1985). The population declined rapidly during the 1950s to a statewide total of about 50 birds in the early 1970s (King et al. 1977). Brown Pelicans no longer nest on North or South Bird Island, but a few pairs nest periodically in Lower Laguna Madre locations.

Arboreal-Nesting Species The semiarid subtropical climate of southern Texas, the low elevation of most islands, and the hypersalinity of Laguna Madre waters combine to restrict the growth of trees and woody shrubs on most islands in the lagoon system. Although the few islands with woody plant communities are heavily used by wading birds, the Laguna Madre is less important to arboreal-nesting wading birds, with the exception of the Reddish Egret, than other regions of the Texas coast (figs. 14.7a, b).

Great Blue Herons begin selecting nest sites earlier than other species, possibly to avoid competition for nest sites in tall, shrubby vegetation. Nests are constructed in a variety of species, including Honey Mesquite, Texas Prickly Pear, Sea Oxeye, Spanish Dagger, and Common Cattail (Bent 1926; Simersky 1971). Artificial nest platforms are used when available, and lacking elevated sites, Great Blue Herons nest on the ground. The same nest may be used several years in succession (Palmer 1962; Simersky 1971), but new mates are selected each year. Although the number of breeding pairs has fluctuated annually, the population of Great Blue Herons in the Laguna Madre has experienced steady declines.

Figure 14.7. *(a) Tricolored Herons and most other herons and egrets construct stick nests in trees or shrubs but also nest on the ground. (b) Prickly Pear offers well-protected, elevated platforms for Great Blue Heron nests. (Photographs by Brian R. Chapman [a] and Beau Hardegree [b])*

a

b

Among long-legged wading birds in Texas, Reddish Egrets have one of the most restricted distributions (Oberholser 1974): the coastal bays and marshes along the Gulf of Mexico and Caribbean coastlines (Bent 1926; Palmer 1962; Bates et al. 2009; Bates and Ballard 2014). Populations of Reddish Egrets on the Texas coast recovered after suffering a precipitous decline at the turn of the century. Plume hunters decimated the heronries (a more restrictive term than "rookery"), and the birds produced no young from around 1900 to 1918. Reddish Egrets construct nests in trees, Spanish Dagger, Texas Prickly Pear, Sea Oxeye, Sea Purslane, Common Ragweed, or bare sand. Initiation of nesting is determined by availability of preferred vegetation, and when drought conditions limit shrub growth, few individuals nest (Simersky 1971). Often described as the rarest heron in North America, Reddish Egrets are more abundant in the Laguna Madre than anywhere else on the Texas coast (Bates et al. 2009). Their populations fluctuate but do not appear to be declining. The Laguna Madre de Tamaulipas is an important wintering area for Reddish Egrets. The flats are often flooded during successive "northers" (cold fronts), providing excellent winter habitat for this species. Individuals begin southward migration from rookeries in the Laguna Madre of Texas to Laguna Madre de Tamaulipas in early fall (Allen 1954, 1955; Bates et al. 2015; Geary et al. 2015).

Plume hunters also decimated breeding populations of Great Egrets, Snowy Egrets, Tricolored Herons, and Roseate Spoonbills to satisfy the demands of the fashion industry. They completely eliminated the Roseate Spoonbill from Texas by 1895, but the species began recolonizing the coast by the 1920s. All of these wading birds nested in mixed-species colonies, often in association with Reddish Egrets, Little Blue Herons, and Black-crowned Night-Herons. The species not hunted for their plumaceous feathers likely suffered nest destruction or abandonment caused by disturbance. Fewer pairs of all these species inhabit the Laguna Madre now than when the annual censuses began in 1973, with the possible exception of the Little Blue Heron. The possible increase of Little Blue Heron breeding pairs might be attributed to loss of habitat in inland areas, which precipitated movement to nesting islands in the Upper Laguna Madre. Great Egret, Snowy Egret, and Tricolored Heron populations may have declined as a result of habitat degradation (including predators). These species may also be experiencing increased competition for remaining nest space on islands slowly eroding into oblivion. Black-crowned Night-Heron and Little Blue Heron numbers also may be inaccurate because these species nest low in the vegetation and are difficult to detect during surveys.

Cattle Egret populations on the Texas coast have declined in recent years, possibly due to migration inland during wet years to nest around freshwater ponds with emergent shrubs. In addition, this exotic (not native) species nests later than other herons and egrets; therefore, the availability of nest sites in existing island colonies may be scarce when Cattle Egrets initiate breeding activity (Telfair et al. 2000).

Prospects for the Future

Construction of the GIWW, Humble Channel, and numerous other small channels in the Laguna Madre produced hundreds of dredged-material islands. Over time, vegetation flourished on the emergent substrates, providing an abundance of new nesting locations for colonial waterbirds. Reproductive success was high on the islands surrounded by rich food sources and largely isolated from predators.

Almost as soon as the islands were deposited, winds and waves in the lagoon began to erode them. About a third of the islands that once were present in the

Figure 14.8. *Red Imported Fire Ants have attacked a Snowy Egret hatchling emerging from the egg. (Photograph by Brian R. Chapman)*

Laguna Madre have simply disappeared. Others eroded outward, joining adjacent islands and filling the water barrier between Padre Island and the dredged-material islands that prevented predator access to waterbird colonies. Mammalian predators subsequently caused abandonment of many rookeries. The combination of island loss and increases in predator and human disturbance are significant contributing factors to the continuing decline of colonial waterbird populations in the Laguna Madre. Recently, a new predator has invaded many islands. The Red Imported Fire Ant, an invasive South American species, can also affect waterbird reproductive success (Lockley 1995). The ants sometimes attack and devour hatchlings as they emerge from the egg (fig. 14.8).

Invasive plants are colonizing some islands. Brazilian Peppertree, a species native to South America, was introduced to Texas as an ornamental plant, and birds dispersed the seeds to the islands. It grows as a sprawling shrub or small tree, and its dense foliage shades out native plants. The impact of Brazilian Peppertree on rookeries in the Laguna Madre is yet to be determined, but a few Great Blue Herons and Great Egrets nest in this invasive plant on one island north of the Kennedy

Figure 14.9. *A pair of Great Blue Herons near their nest in a Brazilian Peppertree, an invasive species now abundant on an island north of the Kennedy Causeway. (Photograph by Beau Hardegree)*

Causeway (fig. 14.9). Australian Pines completely cover several islands just south of the Kennedy Causeway, but the weak limbs of this tree and its ability to harbor predators causes these islands to rarely be used by nesting waterbirds. Another invasive plant, Guinea Grass, which is native to Africa and the Middle East, grows in dense clumps and often suppresses or displaces midlevel plant communities on islands. It can rapidly colonize barren and sparsely vegetated areas, rendering the habitat unsuitable for gulls and terns.

Recent changes in the distribution and abundance of Black Mangrove, a native plant species, in the Laguna Madre also may affect colonial waterbird populations. Its high tolerance for salinity allows Black Mangrove to thrive in the hypersaline environment (Walsh 1974), but it is susceptible to freezing temperatures (Sherrod and McMillan 1985; Pickens and Hester 2011). In recent decades, the increased frequency of freeze-free winters has allowed Black Mangroves to multiply, especially in the Lower Laguna Madre where dense mangrove thickets (mangals) are expanding (see chapter 19). Most mangrove encroachment is occurring on the western shorelines of dredged-material islands where mangroves are protected from wave action caused by prevailing southeasterly winds (Yeager et al. 2010). Although once they become established, mangroves reduce the threat of shoreline erosion caused by wind and wave action (Yeager et al. 2010; Guo et al. 2013), the impact of the increasing abundance of mangroves on shrimp, fish, and waterbirds is uncertain at this time. However, arboreal-nesting waterbirds will likely exploit the shrubby vegetation. Recent increases in the White Ibis population might be attributed in part to the expansion of Black Mangrove and its use as nesting habitat.

Human disturbance in the Laguna Madre has increased over time. With the advent of faster boats and motors that can traverse nearly every place covered with water, areas that were once remote are now easily accessible. This easy, quick access means that colonies are disturbed much more frequently throughout the nesting season by fisherman boating around and fishing too close to nesting birds. Resource agencies and nonprofit organizations continually engage in education outreach to user groups to help reduce the level of human disturbance, but the effectiveness of these activities is uncertain and colonies continue to be impacted. Repeated human disturbance to colonies is likely one of the factors driving declines observed in species such as the Black Skimmer (Safina and Burger 1983; Burger and Gochfield 1990; Carney and Sydeman 1999).

Aside from human disturbance, perhaps the most significant threat facing colonial waterbirds is the potential consequences of sea-level rise (R. Davis 2011). Projections of future impacts are difficult, but many low islands that now support colonial waterbird rookeries will undoubtedly disappear or be substantially reduced in size during the current century (Montagna et al. 2007; Starek 2016). The future does not bode well for the islands that support colonial waterbirds in the Laguna Madre unless the islands are replenished with fresh dredged material. Fortunately, the current Dredged Material Management Plan calls for deposition of dredged material on some islands where bird use has declined due to plant overgrowth or erosion (USACE 2002). Addition of dredged material to overgrown or heavily eroded islands restores nesting habitat for terns and skimmers, especially if the sediments contain shells and shell fragments (Guilfoyle et al. 2006; Golder et al. 2008). Care should be taken to ensure that tall shrubs and trees are not eliminated or damaged. Herbicide treatment and prescribed burning are also management considerations where needed to control invasive Guinea Grass.

15 Shorebirds and Wading Birds

KIM WITHERS AND BRIAN R. CHAPMAN

The term "shorebird" brings to mind images of tiny, white birds skittering along the Padre Island shoreline, abruptly racing seaward to follow a retreating wave (fig. 15.1). True to their name, most shorebirds (Charadrii) assemble along the water's edge on sandy beaches, exposed mudflats, sandbars, or island margins. Shorebirds also frequent inland habitats, including coastal ponds, washover passes, rainwater-filled depressions, dredged-material placement areas (Briggs and Everett 1983; EHA 1997; Zonick et al. 1998), and other wetlands (Chaney 1981; Bacak-Clements 1988; Bauer 1993). Many species are small and difficult to identify, but shorebirds in Texas vary in size from the 15 cm (6 in) Least Sandpiper to the 61 cm (24 in) Long-billed Curlew. Identification is also seasonally difficult because many species change from colorful breeding plumage to dull browns and grays during the fall and winter months, and these alterations among winter migrants often occur when the birds are on the Texas coast. Consequently, many observers simply refer to all small shorebirds as "sandpipers," "peeps," or "wind birds" (Matthiessen 1967). Older references sometimes describe shorebirds collectively as "waders" or "wading birds," but now those names more often refer to herons and egrets.

Most shorebirds breed in northern latitudes where environmental conditions and seasonal food abundance favor reproductive success (Colwell 2010). But the breeding season is short and conditions quickly become inhospitable, so shorebirds migrate to spend the winter in temperate latitudes (Burger and Olla 1984). For migrating and wintering shorebirds using the Central Flyway, barrier beaches and washover passes along the northern Gulf of Mexico and the wind-tidal flats bordering the Laguna Madre of Texas and Mexico represent the largest continuous expanse of suitable habitats between northern breeding grounds and more distant wintering grounds in South America (Withers 2002a). This region contains some of the largest undisturbed wetland complexes in the Western Hemisphere (Baldassarre et al. 1989) and is one of the most significant areas for aquatic bird life on the Gulf of Mexico coast (Senner and

Figure 15.1. *A flock of sanderlings retreat in front of an oncoming wave, a sight familiar to most beachgoers. (Photograph by Waterguy, Wikimedia Commons)*

Howe 1984; Morrison et al. 1993; Pérez-Arteaga et al. 2002; Burger 2018). For example, more than 70,000 shorebirds were counted on a single day at Laguna Atascosa in the Lower Laguna Madre (Brush 1995). The Laguna Madre de Tamaulipas had 65% of all shorebirds counted on the Mexican Gulf/Caribbean coast during 1993 (Morrison et al. 1993) and remains just as significant for migrating and wintering shorebirds today (Pérez-Arteaga et al. 2002). The Laguna Madre provides especially important wintering habitats for species such as the Snowy Plover and Piping Plover, which are suffering population declines (Gorman and Haig 2002). Snowy Plovers, Wilson's Plovers, Black-necked Stilts, and Willets also nest within the region.

The Laguna Madre of Texas provides plentiful shallow-water habitat for wading birds (Pelecaniformes), where they feed on fish and invertebrates. Many species of herons and egrets breed in colonies located on natural and dredged-material islands in the region (see chapter 14). Nearby flooded tidal flats, seagrass meadows, shallow-bay margins, and coastal ponds are used as foraging habitats, and the waders do not have to travel far to find an abundance of fish, crabs, and other invertebrates. Throughout the year, wading birds also share loafing sites on shorelines, wind-tidal flats, and sand flats with shorebirds, gulls, terns, cormorants, and pelicans.

Shorebird Abundance and Habitat Use: Texas

Barrier Island Beaches Shorebirds, wading birds, gulls, and terns use barrier island beaches for foraging, resting, and roosting. More than 30 shorebird species (table 15.1) and 20 gull and tern species regularly use these habitats on Padre Island National Seashore (B. Chapman 1984; Ecoservices 1993a). Few herons and egrets regularly visit barrier island beaches, but flocks of shorebirds, gulls, and terns are common. The most abundant shorebird, gull, and tern were Sanderlings, Laughing Gulls, and Royal Terns. Annual abundance patterns are bimodal (fig. 15.2) with a large peak during fall migration (August–September) and a smaller peak during spring (March–April). Numbers were lowest in early winter (December–January) and June. Gulls and terns constitute at least 50% of the total during all seasons except spring, when shorebirds dominated (B. Chapman 1984). Birds concentrate in the damp foreshore, where shorebirds forage for benthic invertebrates and gulls and terns loiter. Piping Plovers, Snowy Plovers, Ruddy Turnstones, and several other species forage by picking at debris along the wrack lines and berm (fig. 15.3). The backshore is rarely used as either foraging or resting habitat, but when onshore winds are strong, shorebirds often cluster behind coppice dunes or large pieces of driftwood on the backshore to avoid windblown sand. Barrier island beaches in Texas do not appear to support nesting shorebirds, but some Snowy Plovers and Willets nest along the vegetated edges of washover passes.

Wind-Tidal Flats Shorebirds forage in abundance on exposed wind-tidal flats in the Laguna Madre during winter and migratory periods (Senner and Howe 1984; Bauer 1993; Ecoservices 1993b; Haig and Plissner 1993; Withers 1994, 1996a, 1998; Brush 1995; Lee 1995; Withers and Tunnell 1998). A few shorebirds also nest on the flats, notably Snowy and Wilson's Plovers (Hood and Dinsmore 2007a, 2007b). Most wind-tidal flats in the Laguna Madre are at least partially covered with mats of cyanobacteria (blue-green algae), but vegetated and unvegetated sand flats are also present. Shorebird abundance and habitat use, as well as prey abundance and biomass, have been studied extensively on many blue-green algal flats in the Upper Laguna Madre, but nonalgal flats have received little attention. There have been few studies of shorebird ecology in the Lower Laguna Madre (Drake et al. 2001; Hood and Dinsmore 2007a).

Table 15.1. *Relative abundances of shorebirds in Texas on tidal flats bordering the Laguna Madre, beaches of Padre and Mustang Islands, and Padre Island washover passes*

Species	Tidal Flats	Beaches	Passes
American Avocet (*Recurvirostra americana*)	C, U	*	C
American Oystercatcher (*Haematopus palliatus*)	R	R	
Baird's Sandpiper (*Calidris bairdii*)	R	*	
Black-bellied Plover (*Pluvialis squatorola*)	C	A	C
Black-necked Stilt (*Himantopus mexicanus*)	U	R	R
Buff-breasted Sandpiper (*Tryngites subruficollis*)		*	
Dowitchers (*Limnodromous* spp.)	C	R	A
Dunlin (*Calidris alpina*)	A	*	A
Greater Yellowlegs (*Tringa melanoleucus*)	U	R	R
Killdeer (*Charadrius vociferus*)	R	*	
Least Sandpiper (*Calidris minutilla*)	A	R	A
Lesser Golden Plover (*Pluvialis dominica*)		*	
Lesser Yellowlegs (*Tringa flavipes*)	U	R	U
Long-billed Curlew (*Numenius americanus*)	C, U	C	R
Marbled Godwit (*Limosa fedoa*)	U	R	R
Pectoral Sandpiper (*Calidris melanotus*)	R	*	
Piping Plover (*Charadrius melodus*)	C, U	A	U
Red Knot (*Calidris canutus*)	C, U	A	A
Ruddy Turnstone (*Arenaria interpres*)	C, U	A	U
Sanderling (*Calidris alba*)	C	A	A
Semipalmated Plover (*Charadrius semipalmatus*)	C, U	R	R
Semipalmated Sandpiper (*Calidris semipalmatus*)	A	R	A
Snowy Plover (*Charadrius alexandrinus*)	C, U	C	C
Solitary Sandpiper (*Tringa solitaria*)	R	*	
Spotted Sandpiper (*Actitis macularia*)	R	R	R
Stilt Sandpiper (*Calidris himantopus*)	U, R	R	U
Upland Sandpiper (*Bartramia longicauda*)		*	
Western Sandpiper (*Calidris mauri*)	A	R	A
Whimbrel (*Numenius phaeopus*)	R	*	R
White-rumped Sandpiper (*Calidris fuscicollis*)	R	*	U
Willet (*Cataptrophorus semipalmatus*)	C	A	C
Wilson's Phalarope (*Phalaropus tricolor*)	R	*	R
Wilson's Plover (*Charadrius wilsonia*)	U	A	U

Source: Tidal flats: Ecoservices 1993b; Withers 1994, 1996a, 1996c, 1998; Withers and Tunnell 1998. Beaches: B. Chapman 1984; Ecoservices 1993a. Washover passes: Ecoservices 1993b.

Note: A = abundant; C = common; U = uncommon; R = rare; * = present but no abundance data available.

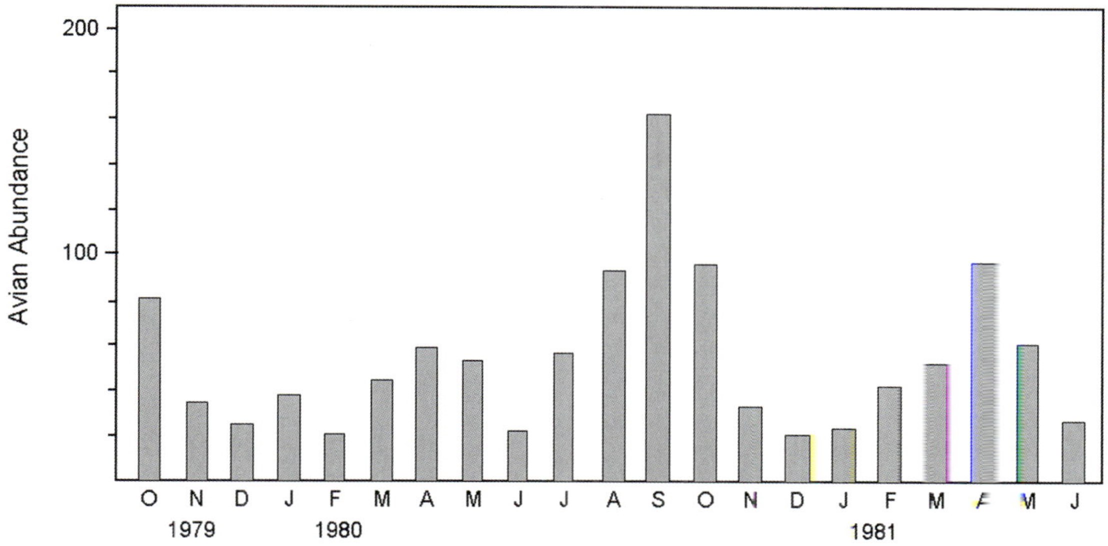

Figure 15.2. *Average monthly total avian abundance along an 8.1 km (5.0 mi) stretch of beach north of Malaquite Beach (Padre Island National Seashore), October 1979–June 1981 (Chapman 1984)*

Figure 15.3. *Decaying marine organisms accumulate at wrack lines on Gulf beaches and Laguna Madre shorelines, providing food for Sanderlings, Ruddy Turnstones, and other scavenging shorebirds. (Photograph by Brian R. Chapman)*

Figure 15.4. *Black-necked Stilts frequent the shallow waters of flooded wind-tidal flats to feed on abundant invertebrates. (Photograph by Brian R. Chapman)*

Most shorebirds using wind-tidal flats are shallow-probing, surface-feeding species, a guild that relies primarily on visual cues to locate prey. Members of this guild avoided direct competition by selecting distinct microhabitats on which to search for prey (Withers 1994). Each microhabitat offers slightly different substrate characteristics than those on adjacent areas on a wind-tidal flat (see also chapter 9). Sandpipers and Black-bellied Plovers prefer to feed in wet microhabitats—areas with wet or saturated sediments. Piping, Snowy, and Semipalmated Plovers frequent damp microhabitats where surfaces appear damp or superficially dry but are moist just beneath the dehydrated crust. Longer-legged, longer-billed shorebirds, such as Dunlins, Marbled Godwits, dowitchers, and Willets, select open-water microhabitats where the substrate was covered with at least 5.0 cm (2.0 in) of water. The Western Sandpiper and the Black-necked Stilt are the only species that regularly occupied inundated intertidal microhabitats where less than 4.0 cm (1.6 in) of water cover the substrate, but they also feed in the wet microhabitat (fig. 15.4). American Avocets feed by sweeping their upturned bill from side to side through shallow water to catch swimming invertebrates, thus avoiding competition with shorebirds that probe for prey.

Upper Laguna Madre As many as 37 species of shorebirds frequent wind-tidal flats on the central Texas Gulf Coast (Rappole and Blacklock 1985). Large congregations of shorebirds occupy mainland mudflat areas from August to May (Bauer 1993). On barrier island blue-green algal flats, shorebirds are most abundant from late fall (November) through early spring (March–April; fig. 15.5) (Withers 1994, 1996a, 1996b, 1998; Withers and Tunnell 1998). Sandpipers (*Calidris* spp.), notably Least Sandpipers and Dunlins, usually are the most abundant species (see table 15.1), followed by Piping and Snowy Plovers. Highest populations of shorebirds are commonly observed after the February low tide through March, when exposed flats are used as a staging area for northbound migrants. Throughout winter and spring, peak densities of the most abundant species are often staggered (Withers 1994), suggesting that northbound migratory movements are timed to reduce competition for space and food in staging areas (Recher 1966).

Numbers of shorebirds are typically low but variable between May and October (Withers 1994). Rising water levels through fall limit space available to birds for foraging. Sandpipers, particularly Western Sandpipers, were the only short-legged shorebirds that were relatively abundant on algal flats during fall. Willets and Black-bellied Plovers usually are the most abundant species of shorebirds on vegetated sandflats throughout the year (Lee 1995). Willets feed in submerged habitats on the flat, but when falling tides increase the extent of exposed habitat, significantly more shorebirds occupy the sand flat than nearby beach zones.

Nearly all Texas wading-bird species occasionally frequent wind-tidal flats in the Upper Laguna Madre, but Reddish Egrets, Great Blue Herons, and Great Egrets are the most recurrently observed (Pulich et al. 1982; Farmer 1991, 1996; Ecoservices 1993b; Brush 1995; Withers 1996c). The Reddish Egret is the most abundant species. Great Blue Herons and Reddish Egrets are residents in the system, and most other species are migratory and move south during winter (Farmer 1991).

Wading birds are most abundant on wind-tidal flats in the Upper Laguna Madre from April to November (Farmer 1996; Withers 1996c; fig. 15.6). On the lagoon side of Padre Island between Yarborough Pass and Mansfield Pass, Reddish Egrets

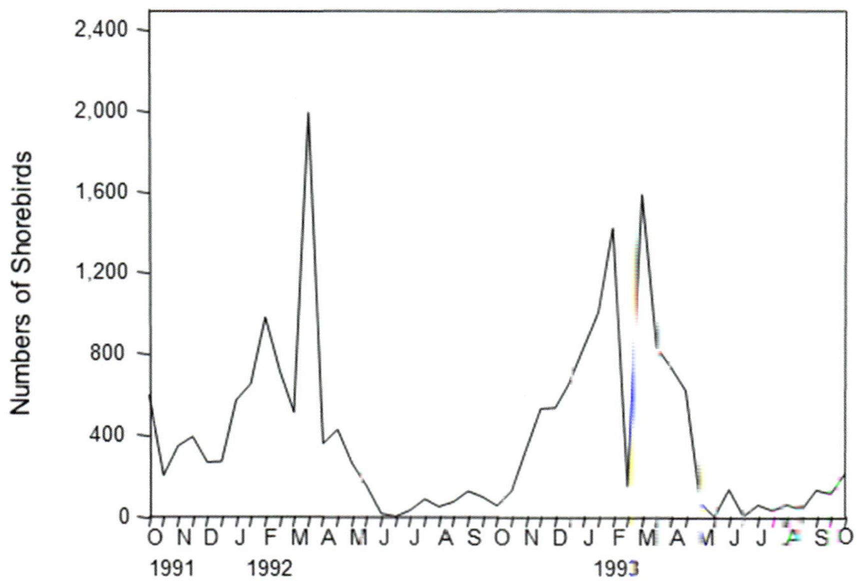

Figure 15.5. *Total number of shorebirds censused on a blue-green algal flat on North Padre Island, October 1991–93 (Withers 1994)*

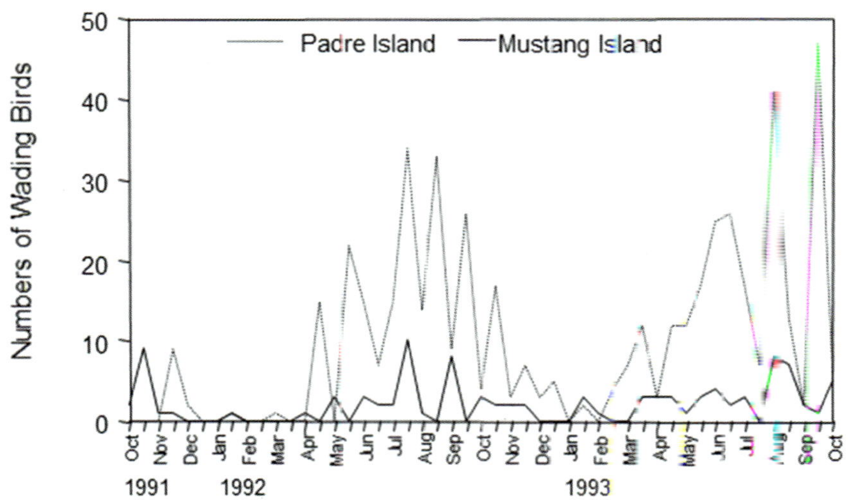

Figure 15.6. *Total numbers of wading birds censused in shallow waters adjacent to blue-green algal flats in the Upper Laguna Madre (Withers 1996c)*

are the most abundant species, composing 55%–100% of all observations of wading birds (Ecoservices 1993b). Overall, wading birds are most abundant in July and least abundant during October. Populations remain fairly steady from November to February and then begin to increase slightly in April. Wading birds use flooded wind-tidal flat habitats along with lagoon margins, grassflats, and washover channels containing water. Habitat availability for Reddish Egrets in the Upper Laguna Madre is reduced during late fall and winter because persistent north winds raise water levels above their preferred foraging depth (Farmer 1996).

Land-Cut and Lower Laguna Madre The areal extent of flat habitats suitable for shorebirds within the boundaries of Padre Island National Seashore is smaller than in other parts of the Laguna Madre. The majority of shorebirds concentrate in wind-tidal flats between Yarborough Pass and Mile 32. These areas offer excellent staging habitats, as indicated by the 86,776 shorebirds of 32 species observed during February 1993 (fig. 15.7). Many birds also use washover channels containing water in the area south of Mile 32 (Ecoservices 1993b).

The wind-tidal flat habitats in and near Laguna Atascosa National Wildlife Refuge in the Lower Laguna Madre form a complex of enclosed shallow-water and exposed, firm mudflats. Several distinct microhabitats are present: high algal flat—wind-tidal flats with a well-developed algal mat; low mudflat—shoreline and frequently flooded mudflats and a seagrass area; and low seagrass flat—a regularly inundated low mudflat area alternately exposed by daily tides, half of which was covered by Turtlegrass. Of the 23 species observed on these wind-tidal flats during November–February 1992–93 and 1993–94, 11 species were recorded in all four microhabitats (Brush 1995). Shorebird

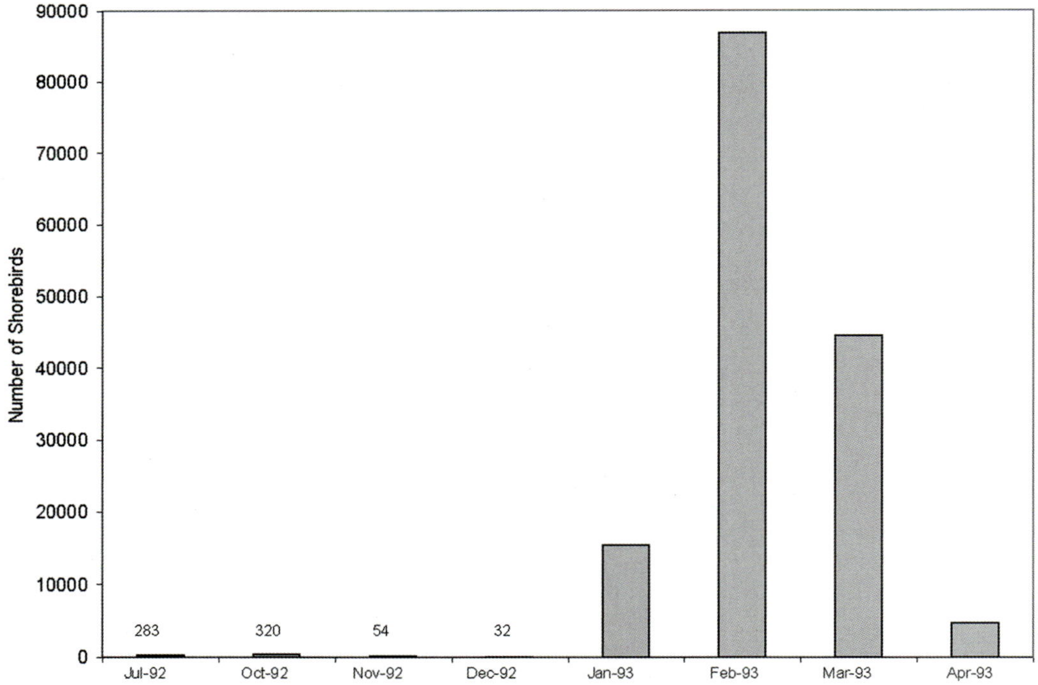

Figure 15.7. *Total number of birds counted on tidal flats between Yarborough Pass and Mansfield Pass during eight surveys in 1992–93 (Data from Ecoservices 1993b)*

numbers varied between years and months but were highest during December and January. The high mudflat/pool habitat attracted the most species and individuals, but few shorebirds used the low seagrass flat. Populations of some species, notably Western Sandpipers, fluctuated in response to changing water levels, and as substrates dried, shorebird use declined.

Brush (1995) grouped shorebirds into three guilds based on habitat preferences: shallow water, exposed mudflat, and water/mudflat. The species in the shallow-water and exposed mudflat guilds were recorded foraging in those habitats on ≥ 95% of observations. Long-legged, long-billed shorebirds, such as American Avocets and Stilt Sandpipers, occupied the shallow water exclusively. Visual foragers, such as Wilson's Plovers, Ruddy Turnstones, Black-bellied Plovers, Least Sandpipers, and Sanderlings, constituted the exposed mudflat guild. Tactile foragers such as Willets, Dunlins, and dowitchers and some, such as Western Sandpipers, that include a tactile component in their foraging strategy confined most of their feeding activities to the water/mudflat habitat in the study area.

Shorebird density was also determined on a tidal flat just south of the Land-Cut (Mile Marker 45) during December 1997–April 1998 (Withers 1998). This wind-tidal flat differed from similar habitats in the Upper Laguna Madre because the cyanobacterial mat was poorly developed. Only seven shorebird species were observed. Densities ranged from under 5/ha (5/2.5 ac) during December, March, and April to nearly 10/ha (10/2.5 ac) during January and February.

In the Lower Laguna Madre, many Reddish Egrets and Snowy Egrets occupy high mudflat/pool habitats, and some Reddish Egrets and Tricolored Herons are usually found in low seagrass flats (Brush 1995). Great Blue Heron and Reddish Egret abundance in the Lower Laguna Madre increases from March to early December and then declines sharply in late December (Farmer 1991). Few Little Blue Herons and Great Egrets are present, and they appear to be resident species. The other three common species, Snowy Egrets, Tricolored Herons, and White Ibises, are most abundant during summer (May–August).

Other Habitats

Washover Passes Washover passes are used regularly by numerous shorebird species, including Piping and Snowy Plovers (Ecoservices 1993a; Zonick 1997a, 1997b; see table 15.1). Passes are most commonly inhabited by shorebirds when high bayside tides inundate mudflats during spring migration (Zonick 1997a, 1997b). Washover passes also support large breeding colonies of Snowy Plovers and smaller breeding populations of other shorebirds, notably Wilson's Plovers (Rupert 1997; Zonick 1997a, 1997b). Both species construct scrape nests in or adjacent to sparsely vegetated areas along the margins of a pass.

Dredged-Material Placement Areas Habitat in dredged-material placement areas range from well-developed uplands to mudflats and open water. Seventeen shorebird species occupied dredged-material placement areas in the Lower Laguna Madre during February 1997 (EHA 1997). Willets and Sanderlings were the most frequently observed species. Snowy Plovers and several other shorebird species sometimes nest within these areas (Zonick et al. 1998).

Coastal Ponds and Wetlands Over 100 species of birds frequent coastal ponds and wetlands near the Laguna Madre (Chaney 1988). At least 15 species of shorebirds use these habitats, and Black-necked Stilts often nest along pond shorelines. Peak shorebird abundances coincide with fall and spring migratory periods (Bacak-Clements 1988; Bauer 1993), and the majority of shorebirds using the ponds are winter residents (Bauer 1993).

Great Blue Herons, Great Egrets, Snowy Egrets, Tricolored Herons, and White Ibises often skulk along the shorelines and shallow waters of ponds and other small wetlands. These birds may display intraspecific territoriality in small ponds, but in larger wetlands, several birds of the same species search for prey without aggression. Although the herons and egrets usually seek small fishes, they also consume frogs, snakes, and other animals that they happen to encounter. Small groups of Roseate Spoonbills sometimes frequent ponds to filter tiny invertebrates from the water column.

Shorebird and Wading Bird Abundance and Habitat Use: Tamaulipas

The Laguna Madre de Tamaulipas supports greater numbers of wintering shorebirds than any other coastal wetland or lagoon system on the Mexican Gulf and Caribbean coasts (Morrison et al. 1993; Mabee et al. 2001). The 24 species that regularly winter in coastal habitats south of the Rio Grande primarily occupy the wind-tidal flats on the lagoon side of the barrier island and the mainland. Sandpipers, plovers, and dowitchers constitute the most numerous species. Sanderlings are the most abundant shorebirds wintering on barrier island beaches adjacent to the Laguna Madre de Tamaulipas (Morrison et al. 1993). Several shorebirds, including Snowy and Wilson's Plovers, nest in the broad coppice dune zone on beaches between the Rio Grande and Mezquital, Tamaulipas (Zonick 1997a).

Seasonality and residency status of some of the more common species have been described. Perales-Flores and Contreras-Balderas (1986) divided the lagoon into three zones. The northern zone, consisting of isolated mudflats and ephemeral wetlands between the Rio Grande and the Northern Laguna Madre, was used by many species of shorebirds. Farmer and Carrera-Gonzalez (1993) describe this area as broad tidal flat but without regular tidal inundation, with most flooding occurring after heavy rainfall and persisting only if rainfall was regular. The northern lagoon was designated as the central zone, and the southern zone consisted of the southern lagoon. Black-bellied Plovers, Willets, Sanderlings, Western Sandpipers, and Least Sandpipers occurred on mudflats throughout the lagoon. A few species, including Snowy and Wilson's Plovers, were found only in the southern zone. Eleven species were listed as year-round residents. Wilson's Plovers breed in the area and were found only during summer. Sixteen shorebird species were observed between Carboneras and La Pesca during fall, but only nine of these species were also seen during the winter (Contreras-Balderas et al. 1990).

Fourteen species of wading birds regularly frequent the Laguna Madre de Tamaulipas (Perales-Flores and Contreras-Balderas 1986; Contreras-Balderas et al. 1990; Garza-Torres and Navarro-Sigüenza 2001), and of these, 10 were considered residents (Perales-Flores and Contreras-Balderas 1986; Garza-Torres and Navarro-Sigüenza 2001). Seasonal occurrence of the other species varies with their migratory ecology, but none reside there during summer. The resident species—Great Blue Herons, Great Egrets, Snowy Egrets, Tricolored Herons, Reddish Egrets, Little Blue Herons, Cattle Egrets, White Ibises, and White-faced Ibises—occur throughout the lagoon (fig. 15.8).

Figure 15.8. *Reddish Egrets are resident in the Laguna Madre de Tamaulipas and occur throughout the lagoon. (Photograph by Brian R. Chapman)*

Green Herons occupy only the northern zone of wetlands between the Rio Grande and the Northern Laguna Madre. Black-crowned Night-Herons, Yellow-crowned Night-Herons, and Roseate Spoonbills inhabit only the northern lagoon. No wading birds occur exclusively in the southern lagoon, where Little Blue Herons are present during fall and Reddish Egrets and Black-crowned Night-Herons are absent during winter (Contreras-Balderas et al. 1990).

Winter and Migratory Ecology of Shorebirds

Shorebirds spend the majority of the year on their wintering grounds or in migratory staging areas and temporary stopover sites. Most regularly used stopover sites are separated by fairly large distances (Goss-Custard and Durell 1990). Individuals of many species exhibit remarkable site fidelity by returning to the same staging areas and wintering grounds year after year (Pitelka et al. 1980; Johnson and Baldassare 1988; Haig et al. 2005). At these locations, migrating and wintering shorebirds spend the majority of their time foraging (Senner and Howe 1984).

The primary food source for wintering and migrating shorebirds is macrobenthic invertebrates inhabiting intertidal flats. However, the distribution and abundance of these organisms vary temporally and spatially (Goss-Custard and Charman 1976), and shorebirds respond by concentrating in areas where food resources are plentiful (Goss-Custard 1983).

Densities and distributions of shorebirds have been linked to distribution and density of their prey in both Europe (e.g., Wolff 1969; Goss-Custard 1970, 1983; Goss-Custard et al. 1977; Bryant 1979; Meire and Kuyken 1984) and North America (e.g., Hicklin and Smith 1984; Kelsey and Hassall 1989; Wilson 1990; Colwell and Landrum 1993; Withers 1994). Wunderle et al. (1989) noted that northbound birds appeared to bypass the Caribbean, possibly because the January–May dry season limits invertebrate populations on mudflats dependent on rainfall runoff. In fact, the circular or elliptical migration patterns exhibited by many shorebird species likely evolved in response to spatial and temporal changes in food abundance and availability (Cooke 1910; McNeil and Burton 1977; Williams et al. 1977; Harrington and Morrison 1979; Morrison 1984; Hicklin 1987; Myers et al. 1987; Myers et al. 1990; Withers and Chapman 1993).

As shorebirds concentrate in areas of resource abundance, habitat quality may decrease because of declines in prey availability. Concentrations of birds may affect prey availability by driving prey organisms deeper into the substrate and out of reach of probing bills (Goss-Custard 1970) or through depletion (e.g., Baker and Baker 1973; Bengston et al. 1976; O'Conner and Brown 1977; Schneider 1973; Evans et al. 1979; Schneider and Harrington 1981; Quammen 1984; Withers 1994). Shorebirds typically remove 20%–40% of the invertebrate standing crop in temperate areas of Europe and North America (Goss-Custard 1980). Agonistic or competitive interactions between birds often results in younger, inexperienced birds being forced into marginal feeding habitats (Goss-Custard and Durrell 1990). Most mortality occurs during winter and is largely related to food shortages (Goss-Custard 1979, 1985). Shorebirds may starve or suffer increased predation or disease as a result of malnutrition. Morphologically similar species (those with similar bill sizes and/or similar foraging behavior) sometimes circumvent direct interference by timing their migratory movements to avoid being present in the same area at the same time (Recher 1966). Even small increases in winter mortality due to competition, prey depletion, or habitat degradation can potentially result in large shorebird population declines (Goss-Custard 1979; Goss-Custard and Durell 1990).

Avian Ecology on Tidal Flats in the Texas Laguna Madre

The cycle of flooding and exposure on wind-tidal flats determines the extent of biological development, especially abundance and persistence of benthic invertebrate communities (see chapter 9). Frequent flooding is essential to maintain invertebrate communities, but the flats must be exposed for most shorebirds to use them (Withers 1994). On flats that are frequently flooded, invertebrates are usually most abundant during winter and spring, although a brief summer flush may occur during June. Hot, dry, and hypersaline conditions during periods of summer low tides sometimes extirpate communities on most flats.

Shorebird abundance usually reflects the richness of benthic organisms on wind-tidal flats, but the distribution of shorebirds may be modified locally as patches of prey organisms become depleted. Fluctuations in shorebird numbers also occur in response to local and seasonal changes in water levels. Few shorebirds are present on tidal flats during fall, not only because fall high tides reduce the foraging area but also because summer low tides result in reductions in prey. Barrier island beaches are typically used by shorebirds when preferred lagoon-associated habitats are unavailable (Zonick and Ryan 1994). Withers and Chapman (1993) suggested that high resource availability during winter and early spring on tidal flats of the central Texas coast were logistically important in the evolution of migration routes of some shorebird species.

Relative abundance and species composition of prey communities on a wind-tidal flat vary by microhabitat. Most shorebirds foraging on wind-tidal flats in the Laguna Madre are shallow-probing, surface-feeding birds. These species reduce interspecific competition by selecting slightly different microhabitats on which to feed (Withers 1994). Even though the greatest prey density was found in the intertidal microhabitat, it was of little use when inundated, suggesting that shallow-probing, surface-feeding shorebirds are inefficient foragers in areas covered by water. Only the Western Sandpiper, a species that likely includes a tactile element in its foraging strategy, exhibited a preference for this microhabitat. This may explain its presence during fall high tides when wind-tidal flats are mostly covered by shallow water. Saturated and damp microhabitats were used more than expected based on prey availability. Thus competition for space and limited food resources may be intense in preferred microhabitats since interference tends to be more pronounced among visual foragers (Goss-Custard 1980). These results were contrary to the lack of competition inferred from large niche breadths and overlaps among the most abundant species in nearby Oso Bay (Withers and Chapman 1993).

Significant associations between the abundance of wet-guild shorebirds and the density of small, shrimp-like crustaceans (tanaids) and between the abundance of damp-wet-guild shorebirds and adult insect density were exposed when Withers (1994) examined the relationships using association analysis (X^2). Regression analysis of benthic variables, tide height, and shorebird numbers revealed some strong relationships at several levels of spatial and temporal scales. Typically, shorebird abundance on tidal flats appeared more strongly dependent on benthic density than benthic biomass. At the landscape level, sufficient area as well as adequate numbers of prey were important determinants of shorebird densities within the estuary. Tide height, an indirect measure of the area available for foraging, was negatively related to shorebird abundance, reflecting an estuary-wide decrease in foraging area as tide height increased. On a specific tidal flat, however, tide height was often positively related to shorebird density, a seemingly contradictory finding. On individual sites, shorebirds appeared to be responding to the frequency and extent of flooding, the primary means by which benthic organisms are recruited to flats and the primary factor governing invertebrate community persistence. Seasonally, the best predictors of shorebird numbers were densities of aquatic-adapted organisms, such as polychaetes. Taken together, these results indicated that overall shorebird numbers on tidal flats were determined by adequate total estuary-wide area containing well-developed invertebrate communities and seasonal flooding of sites adequate enough to maintain local populations of invertebrates.

Species of Concern: Abundance, Habitat Use, and Ecology

Piping Plover Most of the threatened Great Plains population of the Piping Plover likely winters along the South Texas coast (Haig and Plissner 1993; Haig et al. 2005), where both barrier islands and Laguna Madre tidal flats are essential habitats (Withers 1994; Zonick and Ryan 1994; LeDee 2005; LeDee et al. 2008; Arvin 2010). The latest International Piping Plover Census estimated that the wintering Piping Plover population consisted of only 3,973 birds, of which 54% were found on Texas shorelines (Elliott-Smith et al. 2015). In 1991, more than 43% of the wintering Piping Plovers frequented tidal flats (Haig and Plissner 1993). Many more Piping Plovers likely winter south of the Rio Grande in the Laguna Madre de Tamaulipas. During winter censuses in Tamaulipas, several hundred Piping Plovers were counted on barrier islands and tidal flats adjacent to the lagoon (Mabee et al. 2001).

Piping Plovers are prominent members of the shorebird communities wintering on wind-tidal flats in the Upper Laguna Madre where cyanobacteria flourish (Withers 1994). The plovers typically appear on the flats in late September, and their numbers increase to a peak in December or January before declining steadily until April, when they depart for the breeding grounds. Polychaete density (or biomass) and density of insect larvae and tanaids are the best predictors of Piping Plover abundance on blue-green algal flats.

On South Padre Island, Piping Plovers are present during all months but are most abundant between September and January (Garza 1997). They usually return to the same sites in successive years, where they disperse into many habitats (Johnson and Baldassarre 1988; Noel and Chandler 2008), but densities are highest on mudflats and algal flats (Drake et al. 2001). Piping Plovers usually forage within 20 m (66 ft) of the water's edge and maintain well-defined territories (Drake et al. 2001; Newstead and Vale 2014). Sandflats are used mostly for roosting and are occupied by the plovers more often in the winter than in the fall and spring (Nichols and Baldassare 1990; Drake et al. 2001).

Piping Plovers occur at higher densities on the upper Texas coast than in the Laguna Madre (Zonick and Ryan 1994). Prey types differed markedly between the two systems (upper coast dominated by polychaetes and lower coast dominated by crustaceans and insect larvae), but foraging efficiency was similar. Piping Plovers are more gregarious in the Laguna Madre and exhibit lower levels of intraspecific aggression and territoriality than their upper coast counterparts. The greatest densities of Piping Plovers occur where three habitat types (tidal flats, washover channels or passes, and beaches) are in close proximity. They are more common on barrier island tidal flats than on mainland tidal flats.

Tidal amplitude is a strong determinant of Piping Plover use of tidal flats; numbers decrease as tides inundate flat areas (Zonick and Ryan 1994). More Piping Plovers use lagoonal and bayshore flats on the upper Texas coast when all habitat types are available. Prey availability, bayshore habitat availability, beach habitat availability, and human disturbance combine to affect Piping Plover abundance (LeDee et al. 2008). However, availability of protected beach habitat has a greater effect on its abundance than the availability of tidal flats or other habitats.

Wintering Piping Plovers in the Lower Laguna Madre occupy relatively large home ranges, with most moving among a suite of widely spaced locations (Zonick et al. 1998; Drake et al. 2001). Dredged-material disposal areas are rarely used. Piping Plovers regularly move between locations in response to flooding and exposure of preferred bayshore habitats. Although beaches are preferred secondary habitat on the upper and middle coast, mainland flats and washover passes are preferred over beaches in the Lower Laguna Madre. Barrier island beaches, mainland flats, and washover passes in the Lower Laguna Madre are critical high-tide refuges and roosting areas. A complex of bayshore habitats (sandflats, mudflats) with small inlets (e.g., washover passes) and beaches appears essential to Piping Plovers in their wintering range (Johnson and Baldassarre 1988; Nichols and Baldassarre 1990; Zonick and Ryan 1994; Zonick et al. 1998; LeDee 2005; LeDee et al. 2008). These findings underscore the need for a comprehensive, landscape approach to preserving habitat for Piping Plovers in the Laguna Madre.

Snowy Plover Habitat use of wind-tidal flats and adjacent habitats in the Laguna Madre by Snowy Plovers and Piping Plovers abundance are similar (Ecoservices

1993b; Withers 1994, 1996c, 1998; Brush 1995; Withers and Tunnell 19[illegible]). Snowy Plovers in the Lower Laguna Madre roosted almost exclusively on high-flat habitats and appeared to require multiple roosts because they moved among multiple feeding areas (Zonick et al. 1998).

Snowy Plovers begin breeding and nesting activities in late March (Rupert 1997). Nests are situated in a variety of habitats, including high sand flats, washover passes, and dredged-material placement areas (Rupert 1997; Zonick 1997a; Zonick et al. 1998; Hood and Dinsmore 2007a, 2007b). Snowy Plovers typically construct nests in shell-covered areas and line their nests with shell fragments (Zonick 1997a). Nests situated next to a conspicuous piece of debris increase the probability of nest survival (Hood and Dinsmore 2007b), but in areas where debris is uncommon, predators are more likely to locate the eggs or young (Page et al. 1985). Snowy Plovers often nest in association with Least Terns and may experience reduced predation losses due to the proximity (Rupert 1997; Zonick 1997a). Many plover nests in dredged-material placement areas were constructed on the sides of berm-like perimeter levees, often in areas with accumulations of stones and shell fragments (Zonick et al. 1998). The Lower Laguna Madre is an important breeding area for Snowy Plovers and Wilson's Plovers (Hood and Dinsmore 2007a, 2007b). In Tamaulipas, Snowy Plovers nest primarily in broad coppice dune fields bordering the beach shoreline (Zonick 19[illegible]a). Hatch rates and nest success were generally low (Rupert 1997; Zonick 1997a). High rates of mammalian predation were the primary cause of nest failure

Reddish Egret Reddish Egrets are more dependent than other wading birds on the shallow waters of coastal areas (Bent 1926; Palmer 1962; Oberholser 1974). The largest concentration of Reddish Egrets in the world is in the western Gulf of Mexico (Farmer and Carrera-Gonzalez 1993), where they are the most common wading bird throughout the Laguna Madre in both Texas and Tamaulipas (Farmer 1991, 1996; Ecoservices 1993b; Farmer and Carrera-Gonzalez 1993).

Most Reddish Egrets winter in the Laguna Madre de Tamaulipas but some remained in the Upper and Lower Laguna Madre of Texas. Reddish Egrets occupy broad, wind-tidal flats of the Laguna Madre at all water levels. These habitats are not used extensively by other wading-bird species, which largely use wind-tidal flats as "traps" when water levels recede and leave fish stranded in isolated pools (Farmer 1991). Fewer than one Reddish Egret per kilometer (0.6 mi) inhabits narrow bay-margin habitats in the same areas (Farmer 1991, 1996; Farmer and Carrera-Gonzalez 1993; Garza-Torres and Navarro-Sigüenza 2001). Post-breeding adults sometimes migrate between Texas and Tamaulipas to take advantage of habitat availability resulting from changing water levels.

Reddish Egrets in the Lower Laguna Madre usually fed in flocks, a characteristic not previously noted in the literature (Farmer 1991). This tendency to aggregate, as well as the number and size of flocks, varied by site and by season. In spring, few groupings of Reddish Egrets occupied any site, but during fall, 59%–8[illegible]% of the birds occur in flocks consisting of 93–207 birds. In some locations, 17–64 flocks occupy a site. Results are variable during summer, but in areas where birds aggregate, the majority of birds are members of relatively large flocks consisting of 98–203 birds. The combination of wind-tidal flat topography and hydrology within the bird's narrow water-depth requirement restricts available habitats. Consequently, small changes in water depth can greatly increase or decrease amount of foraging habitat available for Reddish Egrets, forcing the birds to congregate in flocks.

Conservation Issues

The primary conservation issue for shorebirds and wading birds is habitat preservation. Because many shorebirds return to the same highly productive sites along migratory routes and in wintering areas every year, they are especially vulnerable to habitat degradation and loss (Smith and Houghton 1984; Johnson and Baldassare 1988). Tidal flats are the preferred habitat for wintering and migrating shorebirds as well as resident wading birds, and issues pertaining to their loss and degradation are discussed fully in chapter 9. Habitat preservation cannot be limited only to sites used by large numbers of birds. Both shorebirds and wading birds move throughout the system as changing water levels cause changes in both foraging space and prey availability. For shorebirds in general, habitats exhibiting a wide range of "quality," defined as amount of foraging space and quantity of prey, are needed for populations to remain stable. In high-quality areas where both space and prey are abundant, crowding and/or prey depletion cause habitat quality to decline, so there is a need for lower-quality and even marginal areas for birds to move into. Thus, preserving the gradient of habitat quality is just as important as preserving habitats that are consistently used by birds year after year. For Piping Plovers, this appears to require preserving habitat complexes containing both tidal-flat or other bayshore and beach habitats for foraging, as well as washover passes for resting and roosting. In some cases, wetland habitats that mimic natural habitats can be created to replace lost shorebird foraging areas (Brusati et al. 2001).

The most important habitat conservation issue for wading birds, including the Reddish Egret, is hydrological maintenance of broad tidal flats. Tidal flats in the Lower Laguna Madre and Laguna Madre de Tamaulipas support dense aggregations of Reddish Egrets (Farmer 1991, 1996; Farmer and Carrera-Gonzalez 1993). Channelization and dredged-material disposal have substantially altered the hydrology of broad flats in some portions of the Lower Laguna Madre (Farmer 1991), and now only about 40% of the 1920s area of tidal flats remain in the South Bay region. The construction of ship channels in the Lower Laguna Madre between 1934 and 1951 coincided with a decline in the breeding Reddish Egret population. Much of the loss of habitat seen in the Lower Laguna Madre is the direct result of dredged-material placement and its effect on tidal-flat hydrology.

16 Sea Turtles

DONNA J. SHAVER, ELIZABETH H. SMITH,
AND SUSAN A. CHILDS

Five species of marine turtles are documented in the western Gulf of Mexico: Green Sea Turtle, Kemp's Ridley Sea Turtle, Loggerhead Sea Turtle, Hawksbill Sea Turtle, and Leatherback Sea Turtle (fig. 16.1). All are currently listed as threatened or endangered under the US Endangered Species Act of 1973. Of these five, Green Sea Turtles are by far most numerous in the Laguna Madre, although the other four species also occasionally occur there.

Figure 16.1. *Five species of threatened and endangered sea turtles that occur in marine waters of Texas and Mexico (Drawings and photographs courtesy of NOAA)*

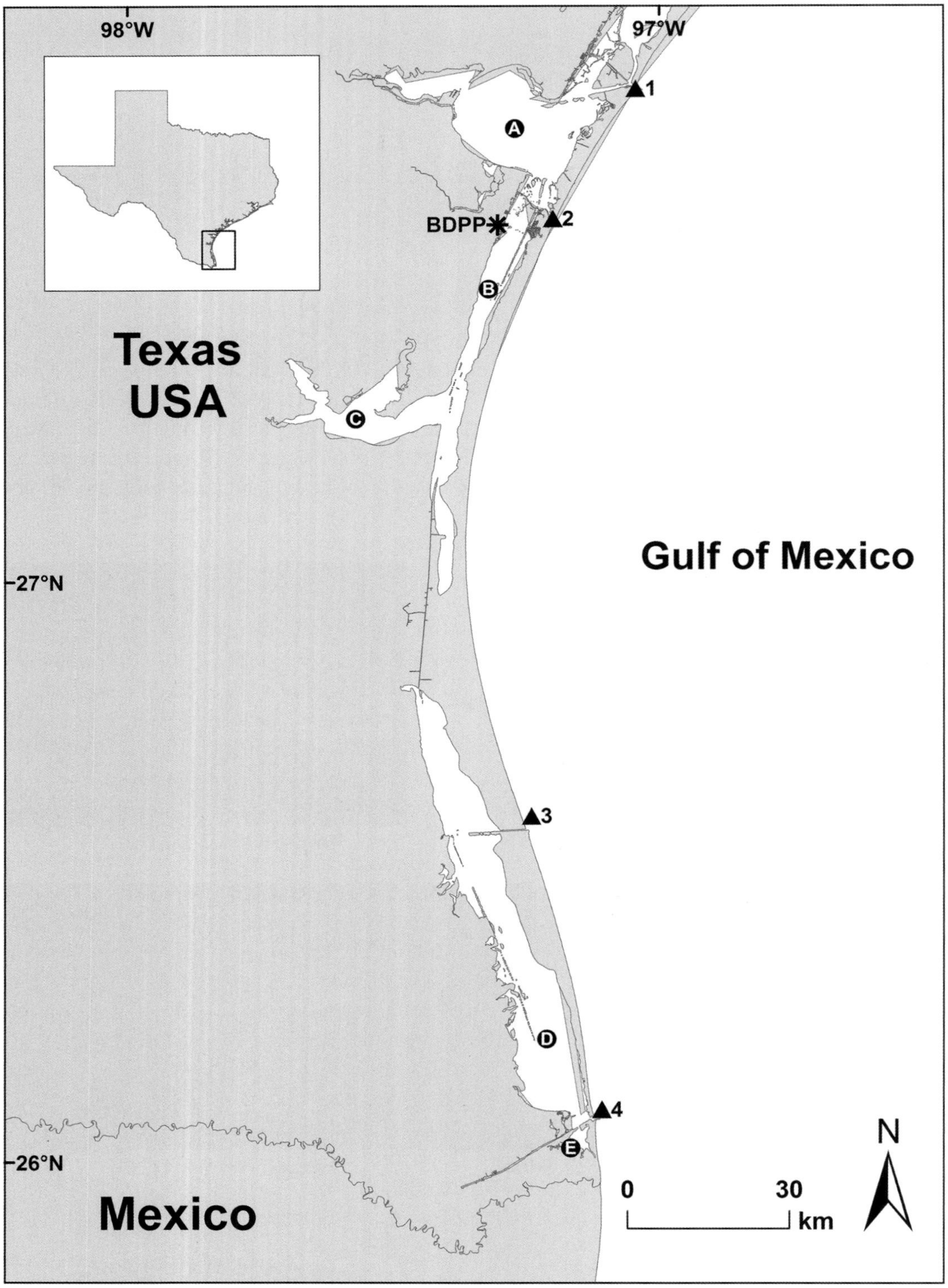

Figure 16.2. *Map of the South Texas coast. The bays are labeled from north to south: (A) Corpus Christi Bay; (B) Upper Laguna Madre; (C) Baffin Bay; (D) Lower Laguna Madre; and (E) South Bay. The channels and passes are designated from north to south: (1) Aransas Pass Ship Channel; (2) Packery Channel; (3) Mansfield Channel; and (4) Brazos Santiago Pass. The Barney Davis Power Plant (BDPP) is also indicated on the map.*

Sea turtle distribution varies by species and is influenced by habitat preferences of the species (Landry et al. 1992). The Laguna Madre of Texas (fig. 16.2) and Tamaulipas contain lush seagrass beds that provide important developmental habitat to juvenile Green Sea Turtles, which feed mostly on seagrasses and algae (Bjorndal 1980, 1982; NMFS and USFWS 1991; Avens et al. 2012). Passes connecting the Gulf of Mexico and Laguna Madre are armored with jetties, and the algae growing on jetty boulders also provide food for Green Sea Turtles (fig. 16.3). Gulf of Mexico waters off the southern Texas and northern Mexico coasts provide important foraging and migratory habitat for sea turtles, but the other four species rarely enter the Laguna Madre through the adjoining passes and bays. Information about the occurrence of Green Sea Turtles in the Laguna Madre is restricted primarily to the Laguna Madre of Texas. Comparatively little is known about sea turtle activities in the Laguna Madre de Tamaulipas, but biologists report similarity in sea turtle habitat use there (E. Zapada, pers. comm.).

Information regarding the recent and current status and habitat use of the Laguna Madre by sea turtles has been obtained through research involving directed-capture netting and satellite tracking (Renaud et al. 1993, 1996; Shaver 1994a, 2000; Metz and Landry 2013; Shaver, Hart, Fujisaki, et al. 2013). Based on these studies, evidence suggests that Green Sea Turtles often overwinter in the Laguna Madre during warmer winters and migrate from the lagoon when waters cool.

The most robust assessment of sea turtles in the Laguna Madre has been through the Sea Turtle Stranding and Salvage Network (STSSN) in the United States (Teas 1993; Shaver 1998a, 1998b; Shaver and Teas 1999). The STSSN was established in 1980 and since that time has documented sea turtles found stranded (washed ashore, alive or dead) on the US Atlantic and Gulf Coasts. Sea turtles captured as a result of active

Figure 16.3. *The green algae that serves as food for juvenile Green Sea Turtles is evident on jetty boulders during periods of low tide. (Photograph by Brian R. Chapman)*

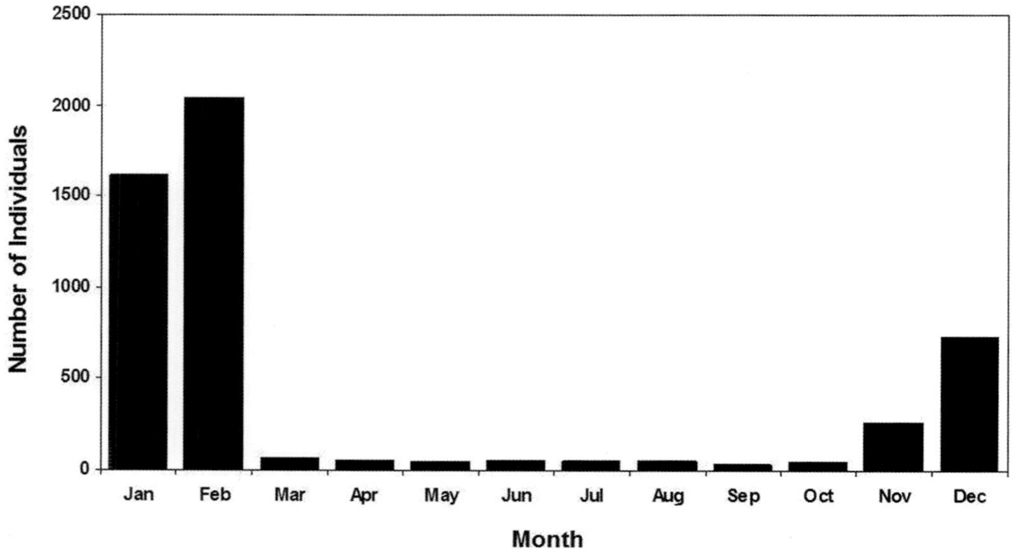

Figure 16.4. *Total numbers of Green Sea Turtles recorded stranded or incidentally captured in the Laguna Madre and adjoining passes and bays during each month of the year, 1980–2017*

fishing and dredging operations or in power-plant intake canals are categorized as incidental captures (Shaver and Teas 1999). Each stranded and incidentally captured sea turtle is documented on a standardized form in records maintained by state and national coordinators. Live turtles are taken to rehabilitation facilities to receive medical care, and those that survive rehabilitation are repatriated to the wild after they regain health. Those found dead or that succumb during rehabilitation are often salvaged for necropsy and study.

From 1980 through 2017, 5,200 sea turtles were found stranded or incidentally captured in the Laguna Madre, adjoining passes to the Gulf of Mexico (Packery Channel, Mansfield Channel, and Brazos Santiago Pass), Baffin Bay, and South Bay. The incidental captures included in this analysis were from the intake canal at the Barney Davis Power Plant near Corpus Christi, Texas (Murray and Jinette 1978). The total includes 5,066 Green Sea Turtles, 57 Loggerhead Sea Turtles, 27 Kemp's Ridley Sea Turtles, 24 Hawksbill Sea Turtles, 1 Leatherback Sea Turtle, and 25 unidentified sea turtles. In accordance with STSSN protocols, hatchlings and post-hatchlings (less than 10 cm [~4 in] straight-line carapace length) were excluded from these totals. Strandings and incidental captures of Green Sea Turtles within the Laguna Madre and adjoining passes and bays were most numerous in recent years, especially during the months of November through February, when hypothermic stunning typically occurs (Shaver, Tissot, et al. 2017; fig. 16.4). Strandings and incidental captures of other sea turtle species there were rare and relatively constant through months and years.

Kemp's Ridley Sea Turtles, which consist of one genetic stock (Bowen et al. 1991; Wallace et al. 2010), nest almost exclusively on the western Gulf of Mexico beachfront. By far, the Kemp's Ridley Sea Turtle is the most frequent nester on the Texas coast (fig. 16.5). Most individuals of the other four sea turtle species inhabiting Texas waters were probably not hatched in Texas because the other sea turtle species rarely nest on Texas beaches. Sea turtle nests from the other four species confirmed in Texas from 1980 to 2017 include 95 of the Green Sea Turtle (on North and South Padre Islands), 95 of the Loggerhead Sea Turtle (statewide but mostly on North Padre Island), 1 of

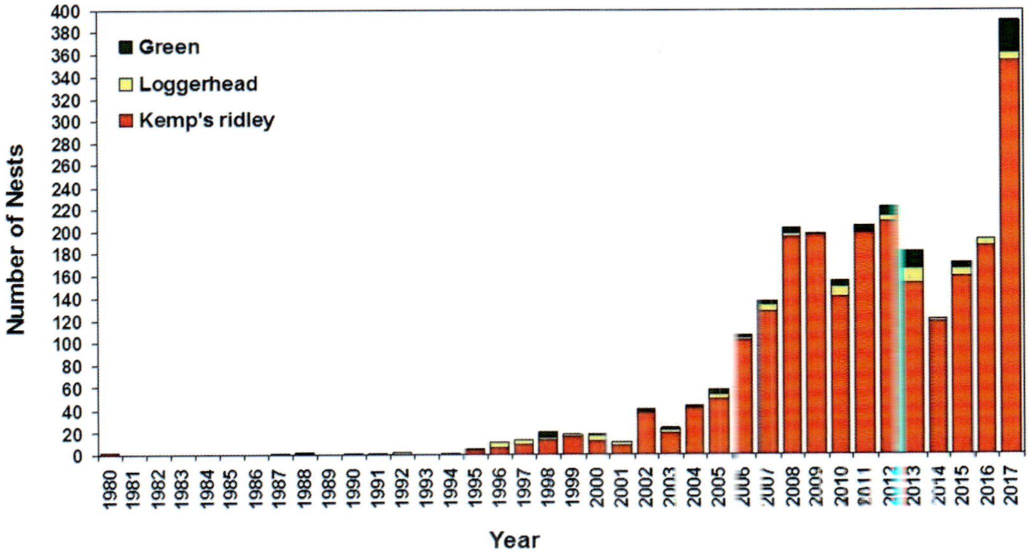

Figure 16.5. *Annual numbers of sea turtle nests recorded in Texas 1980–2017. In addition to the numbers of nests portrayed, one Leatherback Sea Turtle nest was recorded in Texas during 2008 and one Hawksbill Sea Turtle nest during 1998.*

the Hawksbill Sea Turtle (at Padre Island National Seashore [PAIS] in 1998), and 1 of the Leatherback Sea Turtle (at PAIS in 2008).

Based on genetic studies, most of the Green Sea Turtles inhabiting Texas waters were likely hatched in Veracruz and Tamaulipas, Mexico, but a few hatched in Florida and elsewhere in the Caribbean (Anderson et al. 2013; Shamblin et al. 2016). The Loggerheads likely hatched in Florida and the Caribbean (Dodd 1988; Bowen et al. 2004), and the Hawksbills likely came from nests in Mexico and the Caribbean (Bowen et al. 2007). The origins of the Leatherback Sea Turtles inhabiting Texas waters are unknown. Sea turtles nest more frequently along the Gulf of Mexico coast in Mexico than in Texas, but the origin of sea turtles inhabiting the Laguna Madre de Tamaulipas and Gulf of Mexico waters off Mexico is limited.

Species Descriptions

Marine turtles vary greatly in size, weight, and physical description. The Green Sea Turtle is a medium to large, brownish sea turtle with radiating or mottled patterns on the shell (NMFS and USFWS 1991). In comparison to those of other sea turtles, the head is small and the biting edge of the lower jaw is serrated. Adult weights average ~150 kg (330 lb).

Kemp's Ridley Sea Turtle is the smallest sea turtle, ranging in weight from 36 to 45 kg (79 to 99 lb) (NMFS, USFWS, and SEMARNAT 2011). Its color changes significantly during development from the gray-black dorsum (back) and venter (belly) of hatchlings to the lighter gray-olive carapace (back shell) and cream-white or yellowish plastron (belly shell) of adults.

Adults and subadults of Loggerhead Sea Turtles possess a large reddish-brown carapace, and dorsal (back) scales along the shell edge are reddish-brown with light yellow margins (NMFS and USFWS 2008). Adults in the southeastern United States weigh up to 113 kg (248 lb). Unlike adults, hatchlings possess three dorsal and two plastral keels (ridges), lack the reddish tinge, and vary from light to dark brown on the back.

The Hawksbill Sea Turtle is a small to medium sea turtle with a colored shell of thick overlapping scales from which the term "tortoise shell" is derived. Adults weigh approximately 45–90 kg (100–200 lb) (NMFS and USFWS 1993).

Leatherback Sea Turtles are the largest living marine turtles worldwide and can attain a weight up to 696 kg (~1,500 lb). The black shell is covered with white blotches. Lacking scales, the carapace is covered by firm, rubbery skin with seven longitudinal ridges or keels (NMFS and USFWS 1992). The front flippers are proportionally longer than in other sea turtles, and adults may measure 270 cm (106 in). Hatchlings are covered with tiny polygonal, bead-like scales and are black dorsally. Flippers are margined in white, and rows of white scales resembling stripes occur along the length of the carapace.

Ecological Requirements

After emerging from their nests, the hatchlings of most sea turtle species enter the sea, drift with the currents in pelagic (open-ocean) waters for their first few years of life, and become associated with drift lines of *Sargassum*, a sinewy brown macroalga, and debris (Witherington et al. 2012). Juvenile sea turtles usually move into developmental habitats (Musick and Limpus 1997; Meylan et al. 2011). They may move seasonally to avoid cold temperatures, which can cause hypothermic stunning (Spotila and Standora 1985; Witherington and Ehrhart 1989; Shaver 1990a, 2000; Burke and Standora 1991; Morreale et al. 1992; Davenport 1997; Spotila et al. 1997; Still et al. 2005; Foley et al. 2007; Godfrey 2013; Roberts et al. 2014; Shaver, Tissot, et al. 2017), and to seek areas containing preferred food items (Bjorndal 1997, 1999; Williams et al. 2013). After reaching maturity, sea turtles inhabit adult foraging areas, which may or may not be at the same locations as juvenile foraging areas. Adults migrate between adult foraging areas and nesting beaches (Baudouin et al. 2015).

Green Sea Turtles are widely distributed globally in tropical and temperate waters. After their pelagic hatchling phase, Green Sea Turtles settle into developmental

Figure 16.6. *An adult Green Sea Turtle feeding in a seagrass meadow (Photograph by P. Lindgren, Wikimedia Commons)*

habitats containing their preferred food items of seagrasses and algae (Bjorndal 1980, 1982; NMFS and USFWS 1991; Avens et al. 2012). Juvenile Green Sea Turtles occupy nearshore Gulf of Mexico waters, passes, and bays, particularly in South Texas (Shaver 1990b, 1990c, 1994a, 2000; Coyne 1994; Renaud, Carpenter, and Williams 1995; Arms 1996; Gorga 2010; Metz and Landry 2013; Howell et al. 2016; Shaver, Tissot, et al. 2017). The Laguna Madre of Texas and its adjoining passes provide important developmental habitat for juvenile Green Sea Turtles and are the epicenters of abundance of this species in Texas inshore waters (Shaver, Tissot, et al. 2017). The young turtles feed on algae growing on the jetties armoring Texas passes, and adults favor the lush seagrasses in the shallow-water meadows in South Texas bays (Kieslich 1977; Armstrong 1987; Quammen and Onuf 1993; Kaldy et al. 1995; Lehman 1999, 2013; Pulich and Onuf 2007; Wynne 2009; Fikes and Lehman 2010; fig. 16.6).

Movement patterns and behavior of juvenile Green Sea Turtles were investigated in Brazos Santiago Pass (Renaud, Carpenter, and Williams 1995) and the Mansfield Channel (Shaver 2000; Shaver, Hart, Fujisaki, et al. 2013). Individuals grazed on algal growth and wedged themselves between rocks for resting. Renaud, Carpenter, and Williams (1995) found that tagged individuals used jetties more than other available habitats, suggesting that juveniles congregate in these areas for algal food resources. Some individuals exhibited site fidelity to the jetty area, rarely moving far from this habitat during the study. In contrast, Shaver, Hart, Fujisaki, et al. (2013) found that some individuals exhibited site fidelity to the Mansfield Channel, but others traveled southward in Gulf of Mexico waters, and some used the Mansfield Channel as a passageway between the Gulf of Mexico and the Laguna Madre. Individuals that were larger than 40 cm (~16 in) CL (carapace length) were captured in seagrass meadows in the Lower Laguna Madre (Metz and Landry 2013).

Hundreds to more than 1,000 Green Sea Turtles were stranded during hypothermic stunning events in Texas from 1980 through 2017 (fig. 16.7); the highest numbers were

Figure 16.7. *Cold-stunned Green Sea Turtles in seagrass wrack on the Upper Laguna Madre shoreline (Photo courtesy of the National Park Service)*

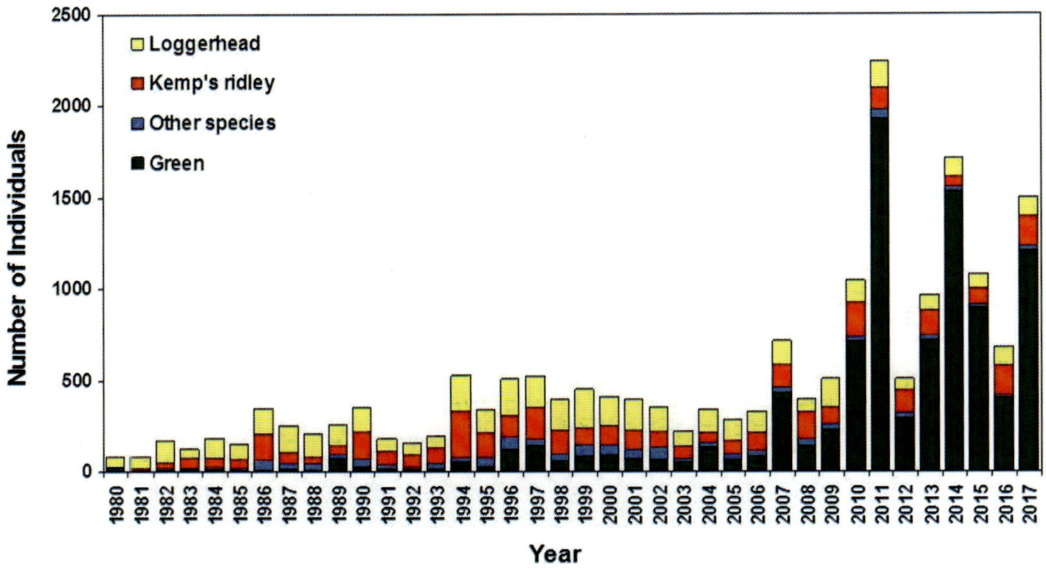

Figure 16.8. *Annual number of sea turtles recorded stranded in Texas, 1980–2017. Other species include the Leatherback Sea Turtle, Hawksbill Sea Turtle, and several unidentified sea turtles.*

recorded in the Laguna Madre (Shaver 1990b; Shaver, Tissot, et al. 2017). The largest stranding events that the STSSN recorded were between 2007 and 2017 (fig. 16.8). The hypothermic stunning season in Texas typically extends from November through early March, and these events shape the monthly pattern of strandings and incidental captures recorded in the Laguna Madre.

Kemp's Ridley Sea Turtle has the most restricted distribution of the five species, occurring mainly in coastal waters of the Gulf of Mexico and northwestern Atlantic Ocean. Kemp's Ridley Sea Turtle hatchlings, post-hatchlings, and juveniles tend to forage on the surface in pelagic environments, whereas subadults and adults are more likely to frequent benthic (bottom) habitats (Ogren 1989). Young individuals are usually found over crab-rich sand or mud bottoms in bays, river mouths, coastal lagoons, and Gulf coastal habitats. Juvenile Kemp's Ridley Sea Turtles tend to move along the Gulf shoreline, usually within 15 km (9 mi) of shore in waters less than 18 m (60 ft) deep (Renaud 1995). Kemp's Ridley Sea Turtles only occasionally occur in the Laguna Madre and its adjoining passes since they prefer more open, muddy bay bottoms for foraging. Juveniles consume a varied diet of crabs and other invertebrates, whereas adults forage almost exclusively on crabs (Ogren 1989; Shaver 1991) and also consume fisheries bycatch when available.

Most adult Kemp's Ridley Sea Turtles remain offshore until they are ready to mate, which occurs approximately 30 days before the first clutch of the year is laid. A pair of mating Kemp's Ridley Sea Turtle turtles were sighted within the Mansfield Channel, Texas, on 3 June 1991 (Shaver 1992), but nests were not found that year. Generally, nesting occurs from March through August, and females may nest two to three times during a nesting season, 21 days apart, every 2 years (NMFS, USFWS, and SEMARNAT 2011). The epicenter of nesting for the species is at Rancho Nuevo, Tamaulipas, Mexico (Marquez et al. 1989; Rostal et al. 1997). Kemp's Ridley Sea Turtle nesting also occurs elsewhere in Tamaulipas and Veracruz, Mexico, and South Texas (fig. 16.9), which is the northernmost extent of the documented historic nesting range (Shaver

Figure 16.9. *Kemp's Ridley Sea Turtle nesting at Padre Island National Seashore, Texas (Photo courtesy of the National Park Service)*

and Caillouet 2015). Between nesting attempts during a season, the turtles remain 6–11 km (4–7 mi) offshore in waters with a mean depth of 14–19 m (46–63 ft) (Shaver, Hart, et al. 2017). After depositing eggs for the last time in a nesting season, females travel in nearshore Gulf of Mexico waters to reach foraging areas (Shaver and Rubio 2008; Shaver, Hart, Rubio, et al. 2013; Shaver, Hart, et al. 2016a, 2016b). The migratory corridor for adult females spans nearshore waters of the entire Gulf of Mexico (Dobie et al. 1961; Shaver, Hart, et al. 2016a, 2016b).

Foraging areas for adult females are distributed in nearshore Gulf waters, especially in the northern Gulf of Mexico. Lesser used adult female foraging areas are distributed off the western Gulf of Mexico nesting beaches, the west coast of Florida, and the Yucatán Peninsula (Shaver, Hart, Rubio, et al. 2013). Both foraging areas and migratory habitat are in a mean water depth of about 20 m (66 ft) located within 25–33 km (16–21 mi) from shore (Shaver, Hart, Rubio, et al. 2013; Shaver, Hart, et al. 2016a, 2016b).

Migration from northern Gulf foraging areas to western Gulf nesting beaches begins in November or December. Migration from southern Gulf of Mexico foraging areas to the nesting beaches begins during February (Renaud et al. 1996; Shaver, Hart, et al. 2016b). Adult males appear to remain in nearshore waters of the western Gulf of Mexico.

Loggerhead Sea Turtles are distributed in temperate and tropical waters of the world, often inhabiting continental shelves and estuarine environments along the margins of the Atlantic, Pacific, and Indian Oceans (Carr 1986; NMFS and USFWS 2008). During the summer months, Loggerhead Sea Turtles often inhabit waters adjacent to oil platforms, rock reefs, and other obstructions (Hildebrand 1982, 1983). Hatchlings feed on *Sargassum* and the macroplankton, gastropods, and small fish associated with the *Sargassum* mats, whereas juveniles and adult loggerheads feed on crabs, jellyfish, molluscs, and other invertebrates (Plotkin et al. 1993; NMFS and USFWS 2008). Loggerhead Sea Turtles only occasionally enter the Laguna Madre, preferring Gulf waters for foraging and migration.

The Hawksbill Sea Turtle is a tropical, reef-dwelling turtle that forages principally on sponges (NMFS and USFWS 1993). Hawksbills are rare in Texas but are sometimes found at South Texas jettied passes where sponges occur. They rarely enter the Laguna Madre; most records in Texas are of post-hatchlings and juveniles found stranded on the Gulf of Mexico beachfront. Four small (<20 cm [8 in] CL) Hawksbill Sea Turtles were observed off the Aransas Pass Inlet jetties in late summer from 1970 to 1975 (Rabalais and Rabalais 1980). During 25 years studying sea turtle nesting habits, Hildebrand (1982) observed only two Hawksbills, which were captured in healthy condition. Other observations made of this species were on small individuals that washed up dead on the beach. Shaver (1994a) captured one juvenile Hawksbill Sea Turtle at the Mansfield Channel jetties during sea turtle directed-capture netting research conducted there from 1989 through 1992.

Leatherback Sea Turtles are a pelagic species that undertake long migrations and deep dives and feed primarily on jellyfish (NMFS and USFWS 1992). They are not a Laguna Madre inhabitant. Information on the Leatherback Sea Turtle is limited in the western Gulf of Mexico (Hildebrand 1982). Leatherback Sea Turtles are widely distributed in Mexican waters, with aggregations concentrated off Barra de San Pedro from August to November. In Texas, an estimated 100 leatherbacks were recorded in December 1956 along a 50 km (31 mi) line extending north from Port Aransas (Leary 1957). Collectively, from 1980 to 2017, stranded Leatherback Sea Turtles were reported along all Texas coastal counties except Chambers (D. Shaver, pers. data).

Status and Trends

Populations of sea turtles are monitored through nesting beach surveys, in-water studies, aerial surveys, and STSSN activities.

Green Sea Turtle Historic accounts suggest that Green Sea Turtles were once abundant in Texas bays (Hildebrand 1982; Doughty 1984). The Karankawa Indians hunted turtles, which they boiled in earthen pots or roasted in the ashes of fire (Doughty 1984). Beginning early in the nineteenth century, Green Sea Turtles were slaughtered for meat, oil, skins, and shells all along the Texas coast (Witzell 1994a, 1994b). Turtle canneries were developed at four sites during the mid-1800s (Doughty 1984), and by 1890, sea turtles ranked fifth among 46 fishing products caught in Texas waters and tenth for the entire Gulf. One of the canneries, established in 1881 at Fulton near Aransas Bay, processed turtles from local seagrass feeding grounds in Aransas Bay, Matagorda Bay, and the Lower Laguna Madre. From 1880 to 1900, about 10,909 kg (24,000 lb) of Green Sea Turtles were landed in Texas, and about 13,636 kg (30,000 lb) were taken in Louisiana. Turtle catches in Texas reached a record high in 1890, with the take of Green Sea Turtles that year increasing more than 20-fold over that of 1880.

To capture turtles, nets and seines were set in shallow lagoons and channels, and fishing vessels were equipped with trawls. The Green Sea Turtle became the leading marine product by weight in the Laguna Madre and lower Rio Grande tidal zone. During that time, turtle meat was inexpensive and readily available in Texas. However, overexploitation and periodic freezes that resulted in the hypothermic stunning of many individuals soon led to a sharp decline in the Green Sea Turtle population. By 1900, the catch declined to an extent that turtle fishing ceased and the processing industry collapsed (Hildebrand 1963). After passage of the US Endangered Species Act, commercial exploitation of sea turtles was outlawed in the United States. The

Green Sea Turtle is currently listed as threatened in Texas (NMFS and USFWS 1991, 2007; Seminoff et al. 2015).

Green Sea Turtle numbers remained low in Texas waters until recent years. Only 10 Green Sea Turtles were among the 259 sea turtles found stranded in South Texas between September 1976 and September 1979 (Rabalais and Rabalais 1980). After 2006, the juvenile Green Sea Turtle population in Texas began to increase (Shaver 2000; Metz and Landry 2013; Shaver, Tissot, et al. 2017). Loggerhead and Kemp's Ridley Sea Turtles were stranded more frequently than Green Sea Turtles in most early years of the STSSN. However, from 1980 to 2017 the Green Sea Turtle was the most frequent species recorded among sea turtle strandings in Texas. The Green Sea Turtle was also the most frequent species found stranded and incidentally captured in the Laguna Madre and its adjoining passes and bays; 97.4% of the sea turtles recorded there from 1980 through 2017 were Green Sea Turtles.

Hypothermic stunning was the most significant cause of Green Sea Turtle stranding in Texas. However, illegal gill netting, washing ashore in association with *Sargassum*, boat strikes, ingestion of marine debris (Shaver and Plotkin 1998), and entanglement were also stranding causes. Fibropapillomatosis (FP) was first documented in Green Sea Turtles in Texas in 2010 (Tristan et al. 2010). The incidence of FP appears to be increasing in Texas, and efforts to monitor prevalence and treat affected stranded turtles should continue.

The nesting grounds of Green Sea Turtles that historically occupied Texas waters are not known. Based on the low numbers of Green Sea Turtle nests located in Texas during the past 30 years, it is unlikely that the Green Sea Turtles currently inhabiting Texas waters emerged from eggs laid on the Texas coast. The first confirmed Green Sea Turtle nest located in Texas was at PAIS in 1987 (Shaver 1989). From 1987 through 2017, 95 Green Sea Turtle nests were confirmed in Texas, all on Padre Island (both North and South). Some nests likely went undetected since Green Sea Turtle nesting can continue into September and nest-detection patrols conducted in Texas typically ended in late July or August. Green Sea Turtle nesting has increased in Tamaulipas and Veracruz, Mexico, in recent years (Xavier et al. 2006; Zavaleta-Lizárraga and Morales-Mávil 2013). Annual numbers of Green Sea Turtle nests found in Texas have increased slightly in recent years but based on trends on other Gulf of Mexico nesting beaches, nesting may continue to increase in Texas during upcoming years.

Kemp's Ridley Sea Turtle The Kemp's Ridley Sea Turtle was listed as endangered throughout its range on 2 December 1970, and the population underwent a precipitous decline from an estimated 121,517 nests in 1947 to a low of only 702 nests worldwide in 1985 (NMFS, USFWS, and SEMARNAT 2011; Bevan et al. 2016). After decades of conservation efforts, the population increased substantially, but from 2010 through 2016, the numbers of nests fluctuated, prompting renewed concern about species recovery (Bevan et al. 2016; Caillouet et al. 2016).

Hildebrand (1963) obtained an amateur film that showed an estimated 40,000 females nesting in a 1947 daytime *arribada* (diurnal, synchronized landings) at Rancho Nuevo. Based on a recent analysis of this film, the *arribada* produced 26,906 nests and there was a total of 121,517 nests during the 1947 season (Bevan et al. 2016). By 1961, the population was severely depleted due to exploitation of both eggs and adults. However, nest-poaching activities may not have been the only cause, for the decline also coincided with the buildup of the shrimp fishery industry during the late 1940s to 1950s. Shrimping activities in the United States and Mexico and the entrapment of turtles in shrimp trawls were identified as major causes for the decline of Kemp's Ridley Sea Turtle (NMFS, USFWS, and SEMARNAT 2011).

In 1978, the Kemp's Ridley Sea Turtle Restoration and Enhancement Project was undertaken by the Instituto Nacional de Pesca (INP) of Mexico, USFWS, National Marine Fisheries Service (NMFS), NPS, and Texas Parks and Wildlife Department (TPWD) in accordance with Section II of the Endangered Species Act. Binational participation in this project has continued since that time. The project has facilitated protection of the nesting turtles, nests, and habitat at Rancho Nuevo. An aim of the project is to establish a secondary breeding population of Kemp's Ridley Sea Turtles at PAIS as a safeguard against extinction (Shaver and Caillouet 2015). PAIS was chosen because the species nested there historically and the NPS could protect the nesting turtles, eggs, hatchlings, and habitat in the event that a breeding population became reestablished.

This 10-year experimental attempt to establish a secondary breeding colony at PAIS through "imprinting" involved transfer of 22,507 eggs to PAIS for incubation. More than 12,000 turtles were ultimately released after 9–11 months of rearing in captivity at the NMFS Laboratory in Galveston, Texas. Most of these "head-started" Kemp's Ridley Sea Turtles were released offshore from Mustang and North Padre Islands (Shaver and Wibbels 2007).

Nest and nesting turtle protection efforts have been ongoing at Rancho Nuevo since the mid-1960s. Protection of the beach used for nesting in the vicinity of Rancho Nuevo has increased significantly over the past four decades. Other beaches in Tamaulipas and Veracruz are now being used as nesting areas, and nest-detection programs are under way there (NMFS and USFWS 2015). Poaching of nesting turtles and eggs has been significantly reduced as a result of intensive vigilance of the binational protection team, adequate motorized beach patrols, and presence of armed Mexican marine guards. For decades, eggs from nearly all nests in Mexico were transferred to protective corrals (large screened enclosures), which protected those eggs from predation and poaching. This resulted in large increases in survival of nesting turtles and nests and exponential increases in nesting from the 1990s through 2009. Because of the dramatic increase in nests, some nests are now left in situ, which subjects them to predation and inundation and reduces their success (NMFS, USFWS, and SEMARNAT 2011).

Efforts to detect, document, and protect nesting by head-started and wild stock Kemp's Ridley Sea Turtles through an extensive program of beach patrols and public education began at PAIS in 1986 and expanded to other beaches in Texas during the late 1990s (NMFS, USFWS, and SEMARNAT 2011; Shaver and Caillouet 2015). Some turtles from the imprinting and head-starting project have returned to nest, but most turtles nesting in South Texas have been from the wild stock (Shaver 2005; Shaver and Caillouet 2015). Operation of the PAIS incubation facility continues, and PAIS incubates most sea turtle eggs found on the Texas coast from North Padre Island northward in that facility and a corral located at PAIS; eggs found on South Padre Island and Boca Chica Beach have been incubated in a corral on South Padre Island (Shaver and Caillouet 2015). Nesting also increased exponentially in Texas through the 2009 nesting season. Numbers of nests found from 2010 through 2016 fluctuated but then surged during 2017. It is vital that monitoring continue in future years so that long-term trends can be documented.

The most significant factor affecting the reproductively valuable large juveniles and adults is shrimp trawling in the Gulf (Magnuson et al. 1990). However, since the mandatory use of turtle excluder devices (TEDs) starting in 1987, establishment of an area off the South Texas coast that is seasonally closed to shrimp trawling in 2000, and decline of shrimping effort in South Texas in recent years, the threat of incidental capture from shrimping in Gulf of Mexico waters off the South Texas coast has been reduced (Gallaway, Gazey, Caillouet, et al. 2016; Gallaway, Gazey, Wibbels, et al. 2016). Consequently, Kemp's Ridley Sea Turtle stranding stabilized in Texas in the 1990s, although the population increased through time.

Nesting trends in Texas and Mexico are highly correlated (Shaver, Rubio, et al. 2016; fig. 16.10). In contrast to the nesting increases predicted in the Kemp's Ridley Sea Turtle Recovery Plan (NMFS, USFWS, and SEMARNAT 2011), the numbers of nests found in both areas have fluctuated. Various hypotheses proposed to explain the decline from 2010 through 2016 included mortality from the Deepwater Horizon Oil Spill, detrimental fisheries interactions, lower fecundity due to cold winters, reduced freshwater inflow, reductions in fisheries bycatch resources (Shaver 1991), density-dependent factors, and various combinations of the above (Caillouet 2014; Plotkin and Bernardo 2014; Gallaway, Gazey, Caillouet, et al. 2016; Gallaway, Gazey, Wibbels, et al. 2016). The availability of bycatch has likely been reduced since 1988 following the decline of shrimping efforts and the mandatory use of bycatch reduction devices

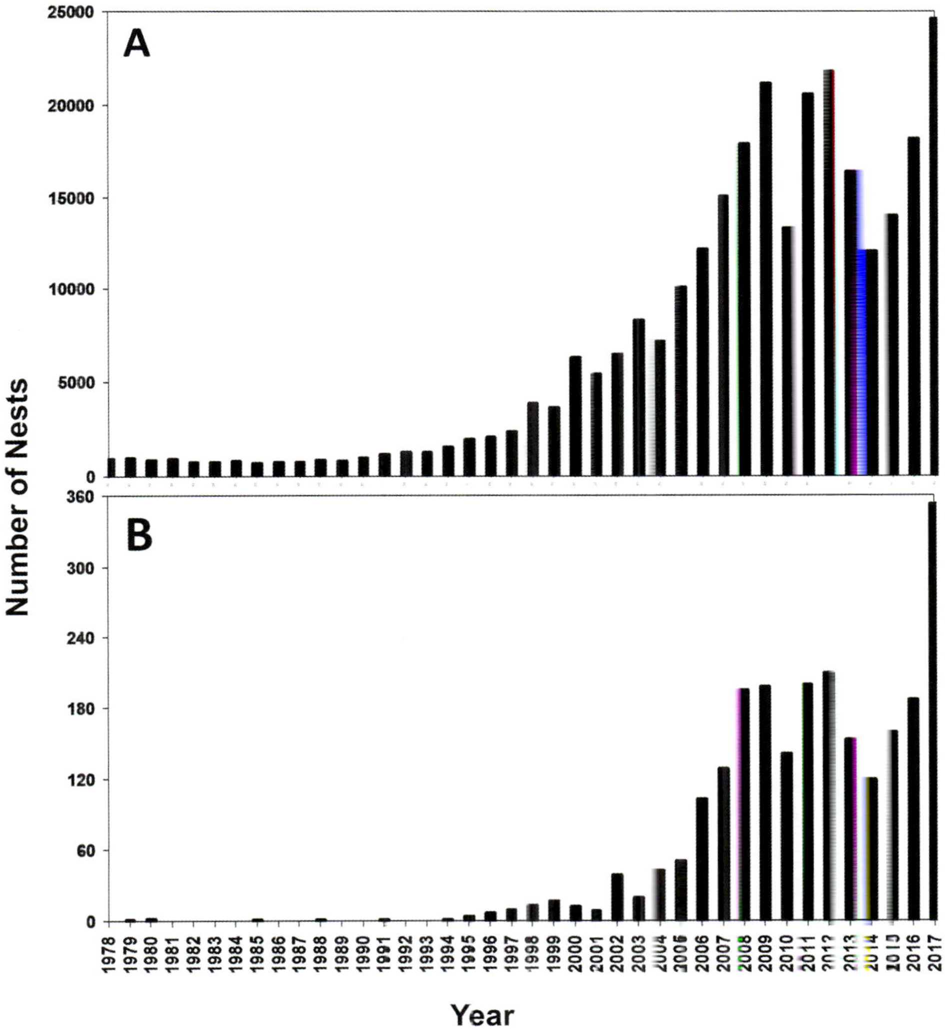

Figure 16.10. *Annual number of Kemp's Ridley Sea Turtle nests recorded in (A) Mexico and (B) Texas, 1978–2017 (Data from Mexico courtesy of Gladys Porter Zoo and CONANP)*

(BRDs) on shrimping vessels operating in US Gulf of Mexico waters (Gallaway, Gazey, Caillouet, et al. 2016; Gallaway, Gazey, Wibbels, et al. 2016; Shaver, Rubio, et al. 2016). Also, Blue Crab populations may have declined in recent years (Gallaway, Gazey, Caillouet, et al. 2016; Gallaway, Gazey, Wibbels, et al. 2016). Based on tag returns from the long-term mark-recapture tagging program conducted in Texas since the 1990s, Shaver, Rubio, et al. (2016) found a reduction in the percentage of turtles nesting that were neophytes and an increase in the remigration interval for nesting turtles. Additionally, Avens et al. (2017) found a reduction of growth rate of individual Kemp's Ridley Sea Turtle turtles in the Gulf of Mexico. These findings highlight key demographic processes that may be involved in the nesting reduction and call for continued research (Shaver, Rubio, et al. 2016).

Other Species On 28 July 1978, the Loggerhead Sea Turtle, which nests in temperate zones and subtropical areas, was listed as threatened under the Endangered Species Act of 1973 (NMFS and USFWS 2008). Geographic subpopulations of nesting Loggerheads were designated using genetic analyses (Shamblin et al. 2010; Shamblin et al. 2012; Shamblin et al. 2014), but the turtles nesting in Texas were not included in these analyses. Nesting by turtles from the Peninsular Florida Atlantic Recovery Unit appears to be increasing (Ehrhart et al. 2014), Dry Tortugas and Northern Gulf of Mexico Recovery Units remain low and declining, and the status of the Greater Caribbean Recovery Unit is unknown (NMFS and USFWS 2008). Only 95 confirmed nests were documented on the Texas coast from 1980 through 2017. Although these nests were distributed statewide, most were found on North Padre Island. Annual numbers of nests found in Texas remained relatively stable through these years.

Some fishery activities threaten Loggerheads in Texas, but Loggerhead Sea Turtles also suffer mortality from various other causes attributable to human activities. Dead turtles were observed on the beaches of South Texas, but no counts were made prior to or during the early 1950s (Hildebrand 1982). For decades, strandings were greatest during fall and spring, periods corresponding with peak inshore and nearshore shrimping (Rabalais and Rabalais 1980; Caillouet et al. 1991; Caillouet et al. 1996; Shaver 1994b, 1995). However, strandings due to incidental capture in shrimping nets declined after the implementation and enforcement of TED regulations, establishment of an area closed to shrimp trawling seasonally off the South Texas coast (Lewison et al. 2003), and reduction of shrimping in Texas waters (Gallaway, Gazey, Caillouet, et al. 2016; Gallaway, Gazey, Wibbels, et al. 2016).

The Hawksbill Sea Turtle is listed as endangered throughout the world. Hawksbills typically nest in tropical areas, and South Texas beaches probably always were insignificant as nesting sites (Lund 1974; Hildebrand 1982). Through 2017, only one Hawksbill Sea Turtle nest was recorded in Texas, at PAIS in 1998. A small population of Hawksbills was reported near Veracruz, Mexico (Xavier et al. 2006). Fishermen reported nests from Isla de Lobos in northern Veracruz southward to Anton Lizardo. The northern part of the state around Cabo Rojo and the area from Isla de Lobos to Alvarado provide suitable nesting and feeding ground for sea turtles, including Hawksbills (Hildebrand 1982).

Hawksbill populations face significant threats in the marine environment from human pressure and exploitation, including gathering of eggs for food and capture of turtles of all sizes to make curios, jewelry, and other products. Shaver (1994a) captured one juvenile Hawksbill Sea Turtle at the Mansfield Channel, Texas, during a sea turtle research project conducted there from 1989 through 1992.

Few Hawksbill Sea Turtle strandings have been reported from inshore waters along the Texas coast. Most individuals found stranded along the Texas coast are post-hatchlings and small juveniles. Small individuals are found entangled in marine debris or injured but are usually alive; entanglement and young washing ashore with strong currents are cited as the most common reasons for Hawksbill strandings (D. Shaver, pers. data). From 1 January to 31 May 1994, one Hawksbill Sea Turtle was included among 227 sea turtles reportedly stranded along the Texas coast (Shaver 1994b). Of the 120 sea turtles (five species) stranded from 1983 to 1985, 12 were juvenile Hawksbills.

The Leatherback Sea Turtle is considered endangered throughout its global range and was listed as endangered under the authority of the Endangered Species Act by the US Department of the Interior on 2 June 1970. Nesting grounds of Leatherback Sea Turtles are distributed around the world (NMFS and USFWS 1992). Declines in the number of nesting females were documented and reported worldwide, and it is currently not known whether Leatherback populations within the United States are stable, increasing, or declining. In the past, this species has nested at Rancho Nuevo, Tamaulipas (Hildebrand 1982). Nesting occurs in Florida and occasionally in other states on the Atlantic Coast. Nesting was documented on the Texas coast in the vicinity of Little Shell beach on PAIS in the 1920s and 1930s, but no records of nesting in Texas exist since the 1940s (Hildebrand 1982) except one record of a Leatherback Sea Turtle nest found at PAIS in 2008.

Leatherback strandings on US shores are generally of adult or subadult size, which may indicate the importance of pelagic habitat under US jurisdiction to turtles breeding in tropical and subtropical latitudes (NMFS and USFWS 1992). Evidence from tag returns and strandings in the western Atlantic suggests that adults engage in routine migrations between boreal, temperate, and tropical waters. Thirty-six Leatherback Sea Turtle strandings were recorded in Texas coastal counties from January 1981 through December 1989 (Judd et al. 1991).

Conservation and Management

All five sea turtle species are susceptible to perturbations in their marine and nesting beach habitats. However, actions are under way in the United States and Mexico to minimize and mitigate for many threats in the marine environment and on nesting beaches through implementing and enforcing regulations, establishing observer programs, and conducting protection and public education efforts. Continuation of these efforts is vital to population recovery.

Threats in the marine and estuarine environment include dredging, pollution, oil and gas exploration, oil spills, Red Tide, disease (including FP), increases in marine debris resulting in ingestion or entanglement, contamination (Faust et al. 2014), and loss or degradation of habitat and food resources. Localized impacts from gill netting (illegal in Texas), shrimping, hook and line fisheries, longline fisheries, and improperly installed or absent TEDs result in deaths and/or injuries to sea turtles (Hildebrand 1982; NMFS and USFWS 1991, 1992, 2008, 2015; NMFS, USFWS, and SEMARNAT 2011). Other threats identified as harming sea turtles include the explosive removal of obsolete oil platforms (Klima et al. 1988), collisions of hulls or boat propellers, power plant entrapment, and disturbance in foraging grounds (Magnuson et al. 1990).

Investigations of suggested and known stranding factors are ongoing. Government agencies and their personnel, in addition to private organizations, have cooperated to document and investigate strandings in the United States through the STSSN since 1980. Efforts to document stranded sea turtles are also ongoing at the Kemp's Ridley Sea Turtle conservation camps in Tamaulipas, Mexico. Sea turtle strandings along coastal shorelines of the southeastern United States serve as an indicator of mortality due to shrimping and other fisheries (Magnuson et al. 1990; Epperly et al. 1996). Increases in sea turtle strandings during commercial shrimping seasons and decreases with the closing of these seasons are documented on the coasts of both the Atlantic and Gulf of Mexico (Magnuson et al. 1990; Caillouet et al. 1991; Caillouet et al. 1996; Shaver 1994b, 1995, 1998a, 1998b; Renaud et al. 1997). There is unequivocal documentation that sea turtles are caught in shrimp trawls (Hillestad et al. 1978; Magnuson et al. 1990). An important management effort is the implementation of TEDs by the US shrimp fleet. The publication of regulations requiring TEDs in 1987 was a major accomplishment in protecting sea turtles in their foraging and migratory habitats. Their use reduces mortality of most of the sea turtles captured (Renaud et al. 1997). TEDs are also now mandatory aboard the Mexican shrimp fleet. However, since mandatory use and enforcement of TEDs, establishment of an area seasonally closed to shrimp trawling off the South Texas coast (Lewison et al. 2003), and decline of shrimping effort in Texas (Gallaway, Gazey, Caillouet, et al. 2016; Gallaway, Gazey, Wibbels, et al. 2016), strandings of Loggerhead and Kemp's Ridley Sea Turtles in Texas have stabilized.

Hypothermic stunning occurs in Texas nearly every winter and sometimes occurs simultaneously in Tamaulipas. Rapid rescue of hypothermic stunned turtles and transport for care are critical to saving these turtles so that they can ultimately be released back into the wild and contribute to population recovery. In recent years, the recovery strategy has been to hold affected turtles only long enough to stabilize their health and then release them in the Gulf of Mexico surf as soon as it reaches 13 °C (55 °F). Most surf releases in Texas have occurred off North and South Padre Islands and have been successful, with few washbacks (Shaver, Tissot, et al. 2017).

Nesting turtles, nests, and nesting habitat are in danger from human disturbance, development pressures, and natural threats. Sea turtle nest-detection and protection efforts, vital to species recovery, are under way in Texas and Mexico and must be continued. Poaching was historically a significant threat to Kemp's Ridley Sea Turtles but has been controlled on many nesting beaches. Research is ongoing at Kemp's Ridley Sea Turtle nesting beaches and nesting beaches for some other species. Continued human population growth and expanding development pressure will increase development threats such as beach armoring, beach nourishment, artificial lighting, mechanical beach maintenance, recreational beach equipment, and beach vehicular driving. Several state and federal laws are designed to protect beach and dune habitat, but some have had varying degrees of success at maintaining stable nesting sites for Loggerhead Sea Turtles in the southeastern United States. Continued protection of nesting beach habitat, monitoring of nest numbers, research, and protection of nests are necessary for conservation.

Many education activities and efforts are under way to enhance public awareness and understanding of sea turtle conservation. Social media, signs, posters, websites, public-service announcements, and other educational methods have helped reduce human-related threats. Personnel conducting nesting turtle patrol projects are in touch daily with tourists on how to minimize disturbance to nesting turtles. In addition, state and federal parks that conduct public awareness of sea turtle status provide information to visitors, and signs have been posted on many beaches informing people of the laws protecting sea turtles and providing a local or hotline number to report nesting and stranding observations (NMFS and USFWS 2008). Public support for sea turtle conservation efforts is essential for the long-term success of conservation programs.

Conservation efforts for the protection and restoration of all sea turtle populations have been implemented both nationally and internationally. One of the most significant measures achieved is the ratification of Protocol to the Cartagena Convention concerning specially protected areas and wildlife. Parties to the Convention for the Protection and Development of the Marine Environment of the Wider Caribbean Region (Cartagena Convention) adopted the Protocol for Specially Protected Areas and Wildlife in January 1990. Annex II of the Protocol includes all six sea turtle species found in the wider Caribbean. It prohibits taking, possession, killing, or commercial trade in sea turtles, their eggs, parts, or products, as well as disturbance particularly during periods of breeding, incubation, migration, and other periods of biological stress (NMFS and USFWS 2008).

17 *Fishes and Fisheries Organisms*

KIM WITHERS, SUZANNE J. (DILWORTH) BATES, AND JAMES TOLAN

Early descriptions of the fishes and fisheries of the Laguna Madre of Texas depict a remarkably productive system despite its relative isolation from the Gulf of Mexico. Its location in a semiarid region, restricted seawater exchange with the Gulf, limited freshwater inflow, shallow waters, and high evaporation rates contributed to frequent and often prolonged periods of hypersalinity. Although high salinities in bays and lagoons are usually associated with low primary productivity and reduced populations of marine organisms (Copeland and Jones 1965; Copeland 1967), seagrasses, fishes, waterfowl, and invertebrate marine organisms generally flourish in the Laguna Madre. During prolonged droughts, populations of marine organisms decline, but after tropical storms open passes and deluge the area with heavy rainfall, production of fish and shrimp surges for several years (Hildebrand 1958, 1969, 1980).

This boom and bust character of the fisheries has been altered by efforts to ameliorate hypersalinity. Although salinities in the Laguna Madre are still not characterized as truly "estuarine," they are markedly lower now than they were prior to the dredging of the Gulf Intracoastal Waterway (GIWW) in Texas in 1949 and pass stabilization with rock jetties in both Texas and Tamaulipas. Jetty stabilization at Brazos Santiago Pass was completed in 1935, and the shorter Mansfield Pass jetties were completed in 1957. Packery Channel, the third jettied pass connecting the Upper Laguna Madre to the Gulf, was completed in the northernmost region in 2005 (Palmer et al. 2013). These passes, combined with the GIWW, various channels dredged to access oil and gas wells, and a prevailing southeast wind helped ameliorate the hypersalinity for which the region was historically known. Today the Upper and Lower Laguna Madre remain the most productive bays for finfish in Texas. Both embayments support important recreational fisheries for Spotted Seatrout and Red Drum and a commercial Black Drum fishery. Baffin Bay, extending inland from the Upper Laguna Madre, has a well-deserved reputation among recreational anglers for producing trophy Spotted Seatrout.

Few data are available about the fishes and fisheries in the Laguna Madre de Tamaulipas. Poor record keeping and a dearth of scientific studies resulted in uncertainty about the health and resources of the Mexican lagoon, but evidence of overfishing and habitat alteration led to development of fishing regulations and restrictions. Unfortunately, Mexico does not have the resources to enforce these laws, so the fate of the ecosystem currently lies primarily in the hands of commercial and recreational fisherman.

Effects of Hypersalinity on Fish and Fish Communities

Effects of Hypersalinity on Fish Hypersalinity affects fishes directly by disturbing their osmotic balance and indirectly by decreasing the available dissolved oxygen content of the water, especially during high summer temperatures. Water with salinities exceeding 70 PSU (practical salinity units) are not strictly comparable to seawater because several compounds precipitate out, significantly altering both the ionic environment and osmotic pressure (Copeland 1967). Most fishes found in the Laguna Madre are euryhaline and are probably better able to cope with the harsh conditions because they can osmoregulate in salinities varying from nearly fresh water to two times the marine content (Gunter 1967). For example, Gunter (1945a) seined Sheepshead Minnows and Tidewater Silversides in Baffin Bay when salinities registered 71.5 PSU (Stewart 2008) and found that the same species were still present more than two months later after heavy rains reduced salinities to 2.2 PSU. Strictly marine fish are rarely encountered in the lagoon even though they are able to withstand fairly high (35 PSU) but relatively stable salinities (Gunter 1967).

Salinities in the range of 40–50 PSU appear optimum for many species in the Upper Laguna Madre (Simmons 1957). However, at higher salinities larger individuals of some species, such as the Spotted Seatrout, are more abundant than smaller individuals, and at nonlethal salinities the average size of fishes is larger (Simmons 1957). Although well documented in invertebrates (Simmons 1957; Hedgpeth 1965), stunting due to hypersalinity does not appear to occur in fishes (Gunter 1967), but growth rates may be affected. During periods of hypersalinity, Black Drum in Baffin Bay grow slower and reach maturity younger—age two in Baffin Bay and age five in other Texas Bays (Bumguardner et al. 1996; Olsen et al. 2014).

Higgins and Lord (1926) were the first to document summer mortality in fishes in the Upper Laguna Madre as a result of extreme hypersalinity (80–110 PSU; Gunter 1967). Prior to the 1940s, catastrophic summer fish kills that killed nearly all fish in the Upper Laguna Madre occurred about every 10 years (Gunter 1967). Although during normal years fish often went blind and their eyes took on a "cooked" appearance due to elevated water temperatures, extreme hypersalinity compounded the effects of raised water temperatures on dissolved-oxygen content and resulted in massive kills. The last catastrophic kill due to hypersalinity occurred in the late 1930s, but small kills were documented into the 1960s and may still occur locally, particularly in Baffin Bay. The highest documented salinity value since the late 1970s was 92 PSU in Alazan Bay, a tributary of Baffin Bay. While this extreme hypersalinity event took place during a coastwide drought in 2013, no accompanying fish kill in Baffin Bay was noted (F. Grubbs, pers. comm.).

Effects of Hypersalinity on Fish Communities The fauna of hypersaline lagoons is usually derived from local brackish waters rather than the sea (Hedgpeth 1967). In both high- and low-salinity environments, depauperate faunas are the rule and in stable hypersaline areas species richness is generally low while the populations of some species may be extremely large (Parker 1959). However, as salinities decrease (to a point), species richness increases as abundances of individual species decrease. In the Upper Laguna Madre, total species richness was greatest between 40 and 50 PSU and decreased below 40 PSU (Simmons 1957). When low water temperatures coincide with high salinity, species not normally tolerant of hypersalinity may be present.

Table 17.1. *Numbers of families and species of fish found in Corpus Christi Bay and coastwide compared with the hypersaline Laguna Madre*

	Chondrichthyes		Osteichthyes	
	Families	**Species**	**Families**	**Species**
Texas Gulf Coast	21	50	116	498
Corpus Christi Bay	9	13	47	102
Upper Laguna Madre/Baffin Bay	5	7	37	79
Lower Laguna Madre and South Bay	4	5	54	89
Laguna Madre de Tamaulipas	5	18	44	122

Source: Coastwide (including bays and shallow nearshore environments): Murdy 1983. Corpus Christi Bay, Upper Laguna Madre/Baffin Bay: Tunnell and Alvarado 1996. Lower Laguna Madre and South Bay: Breuer 1962; Hook 1991. Laguna Madre de Tamaulipas: Leija-Tristán et al. 2000.

The effect of hypersalinity on diversity can be seen by comparing the numbers of families and species present in the Texas Laguna Madre and Laguna Madre de Tamaulipas to that of Corpus Christi Bay, which has a more moderate salinity, and to coastwide (both estuarine and shallow nearshore environments) totals from Texas (table 17.1). Fewer families are found in the Laguna Madre than in Corpus Christi Bay or coastwide. The Laguna Madre de Tamaulipas harbors more species than the other lagoons, probably because 24 strictly tropical species are present (Leija-Tristán et al. 2000). Most of the species (54%) in the Laguna Madre de Tamaulipas are widely distributed in temperate and tropical waters of the Atlantic Ocean and Gulf of Mexico.

Although few studies describe the Lower Laguna Madre fish fauna, recent reports indicate that numbers are not substantially different from those in the Upper Laguna Madre. About half of all families (2 Elasmobranchiomorphi [sharks and rays] and 19 Osteichthyes [bony fishes]) were found in all bay and lagoon categories listed in table 17.1, but 9 families were found only in Corpus Christi Bay (3 Elasmobranchiomorphi and 6 Osteichthyes). Two families restricted to both the Lower Laguna Madre and Laguna Madre de Tamaulipas are generally tropical in their overall distributions. Family Sciaenidae (drums) was represented by the greatest number of species, with 10 species in the Laguna Madre of Texas and 13 in the Laguna Madre de Tamaulipas.

Ecology of Important Fish Species in Texas

Twenty-six of the 31 species of fish identified from Gulf of Mexico estuaries by NOAA's Estuarine Living Marine Resources program (Nelson 1992; Patillo et al. 1997) as having ecological, commercial, or recreational value occur within the Laguna Madre of Texas (table 17.2). Ecological value is defined by an organism's habitat use and trophic level. For example, Spotted Seatrout are top-level carnivores ecologically categorized as an estuarine-dependent species that spawns within bays and migrates through passes to use the seagrass beds of the Laguna Madre as nursery areas (Tolan 1994). Both habitat use and trophic level provide information describing the importance of the Laguna Madre in a species' life cycle, as well as the species' position within the ecosystem.

Equally important are the commercial and recreational values of many species found within the Laguna Madre. These values are categorized as commercially important

Table 17.2. Economic role, use as an indicator species, habitat use, trophic level, and qualitative abundances of major life stages for 10 invertebrate and 27 fish species in the Laguna Madre of Texas

Common Name	Economic	Indicator	Ecological Habitat/Trophic Level	Abundance				
				A	S	J	L	E
Invertebrates								
Eastern Oyster	CR	FPM	R/FW-f	+	+	+	+	+
Bay Scallop	C*R	H	G/FW-f	+	+	+	+	+
Atlantic Brief Squid	I		W/C	++	+−	++	−+	++
Blue Crab	CR	PMO	W/FW-o	−++	+−	+++	+++	+++
Brown Shrimp	CR	HPMO	GP/FW-o	+		+++	+	
Grass Shrimp	F	HPMS	G/FW-d	++++	++−+	++++	−−++	++++
Gulf Stone Crab	CR	TB	ST/FW-p	+	+	+	+	+
Pink Shrimp	CR	PMO	GP/FW-o	++		+++	+	
Caribbean Spiny Lobster	I		N/C	+		+		
White Shrimp	CR	PM	GP/FW-o	++		+++	+	
Fish								
Atlantic Croaker	CR	MO	NF/C	−++		+++	+++	
Bay Anchovy	I	W	W/FW-c	++++	++−+	++++	−−++	++++
Black Drum	CR		NF/C	++	+−	++	−+	++
Bull Shark	R	M	O/P			+		
Code Goby	F		S/P	−++	+++	+++	+++	+++
Crevalle Jack	R		NF/P	++		++		
Florida Pompano	R		NF/C	+		+		
Gizzard Shad	I		O/FW-o	+				
Gray Snapper	CR		NF/C	+		++		
Gulf Flounder	CR		O/C	+		+		
Gulf Killifish	F	HOA	S/FW-po	+++	+++	−++	+++	+++
Gulf Menhaden	CF	MO	NOP/FW-f	++		−++	−+	
Hardhead Catfish		M	W/C	+++	+++	++++	+++	+++
Pinfish	IF	H	SC/FW-po	+++		++++		
Red Drum	C*R	M	NP/P	+		++		
Sand Seatrout	C*R		W/P	+		+		
Sheepshead	IR		NP/P	++		−++		
Sheepshead Minnow	F	O	S/FW-c	++++	++++	++++	+−++	++++

Table 17.2. *Continued*

Common Name	**Economic**	**Indicator**	**Ecological Habitat/Trophic Level**	**Abundance**				
				A	**S**	**J**	**L**	**E**
Silversides	F	HI	S/FW-c	++++	++++	++++	++++	++++
Silver Perch	I	P	W/C	++	++	+++	++	++
Snook	C*R		W/P	+	+	++	+	+
Southern Flounder	CR		O/P	++		+++		
Spot	I	PMOI	W/C	+++		++++		
Spotted Seatrout	C*R	POI	NGP/P	++	++	++	++	++
Striped Mullet	CRF	HP	W/FW-h	++	++	+++	++	++
Tarpon	R	P	NOP/C	+		+		

Source: Nelson 1992; Patillo et al. 1997.

Note: Category codes:

Economic

C = commercial importance

C* = historical fishery species

R = recreational

F = forage or bait

I = incidental commercial or recreational use

Indicator

H = hydrocarbons

F = fecal coliform

P = pesticides

M = metals

O = organics

I = inorganics

W = water quality

T = toxicity

A = acidified water

Ecological Habitat

N = nursery

G = grassbeds

R = reef

W = widespread depending on life stage

O = open waters, including offshore

P = estuarine-dependent organisms that migrate through passes

S = shallow waters, widespread

Trophic Level

FW = food web linkage

f = filter feeder

d = detritivore

h = herbivore

o = omnivore

c = generalized carnivore

p = predator

C = generalized top-level carnivore, often demersal feeder

P = top trophic-level predator

Abundance

+ = rare

++ = common, frequently encountered

+++ = abundant, substantial numbers

++++ = very abundant, numerically dominant

A = adults

S = subadults

J = juveniles

L = larvae

E = eggs

species, recreationally important species, forage or bait species, and incidental commercial or recreational-use species. Of the 26 species, 11 are or have been commercially important, 15 are recreationally important, 5 are forage or bait species, and 6 are incidental species. A more detailed overview of life history and abundance is provided for Red Drum, Spotted Seatrout, and Black Drum, the most important commercial and recreational species. Data on species abundances and sizes since 1974 used in this section were compiled from data collected by the Texas Parks and Wildlife Department (TPWD) Coastal Fisheries Management Program as summarized in Martinez-Andrade et al. (2005) and Green and Campbell (2008).

Red Drum The Red Drum (or Redfish) typifies an estuarine-dependent taxon. Early observers reported seasonal movements of fishes ready to spawn heading toward the Gulf and great schools of tiny Red Drum moving into inside waters through these passes (Higgins and Lord 1926). Spawning occurs in nearshore and inshore waters near barrier island passes from late summer to early winter (Patillo et al. 1997). Larvae and postlarvae are carried by tides into shallow, inside waters where they settle in quiet areas with muddy bottoms and patchy submerged vegetation that protects them from predators until they are able to swim (Rooker et al. 1998; Scharf 2000; Sheridan and Minello 2003). Juveniles disperse and become widely distributed in shallow, protected, open waters, such as secondary bay margins and quiet backwaters. Subadults move into deeper, more-open water in primary bays before migrating offshore after first spawning.

All free-swimming life stages are carnivorous. Food consists of copepods, mysid shrimp, amphipods, decapods, and fish, depending on life stage, size of prey, and availability (Patillo et al. 1997). Red Drum are long-lived top predators (up to 50 years; Matlock 1987; SEDAR 2009) that feed opportunistically on the bottom or in the water column at dawn and dusk. While larvae and juveniles are potential prey for other fish and wading birds, little is known about population-level effects of predation on Red Drum. However, Rooker et al. (1998) and Stunz and Minello (2001) demonstrated that hatchery-reared Red Drum larvae had higher prey vulnerability than wild-caught conspecifics. Additional natural mortality occurs as a result of Red and Texas Brown Tides, parasites, disease, and old age.

Anglers highly prize the Red Drum for its fight, and it is regarded as excellent table fare. Consequently, the Red Drum was regarded historically as the most important sciaenid species harvested for commercial and recreational purposes (Patillo et al. 1997). It still remains one of the most important sciaenid sport fish and one of the seven most sought-after gamefishes in the Gulf of Mexico. Red Drum was second only to oysters in amount and value of commercial harvest during 1890 in Refugio, Aransas, and Nueces Counties and generally ranked fourth behind shrimp, Spotted Seatrout, and Black Drum in pounds harvested in Aransas and Laguna Madre districts (Rockport to Brownsville) between 1942 and 1970 (Ponwith and Dokken 1996).

Destruction of nursery habitat, overfishing with gill nets within bays, and the offshore purse-seine fishery in the 1980s seriously impacted Red Drum populations to the point that Texas waters and all federal waters in the Gulf were closed to commercial fishing and sale of the species in 1987 (Gulf of Mexico Fishery Management Council 1987). Federal waters in the Gulf remain closed to Red Drum harvest. After these closures, the population of Red Drum in the Laguna Madre rebounded and reached maximum levels between 2004 and 2007 (fig. 17.1), before declining to population sizes typical of the mid- to late 1990s.

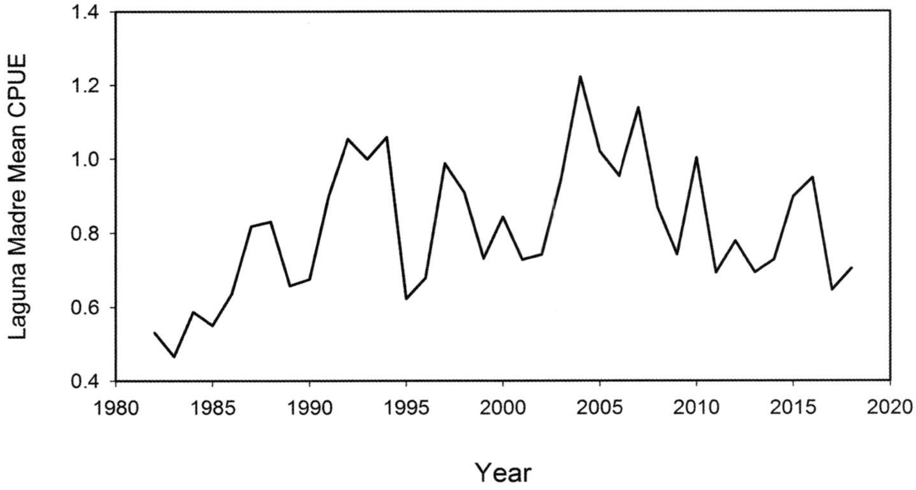

Figure 17.1. *Annual gill-net catch-per-unit effort (CPUE) of Red Drum in the Laguna Madre. Effort is defined as an overnight gill-net set, and values represent catch standardized to number per hour. (Data from the TPWD, Resource Monitoring Program)*

When commercial harvest was outlawed in the early 1980s, numerous commercial aquaculture operations were initiated to fill the market demand for Red Drum. However, a perceived difference in taste between farm-raised Red Drum and wild-caught stocks depressed the market. Only a few businesses are currently farming Red Drum.

TPWD immediately began working to restore depleted Red Drum populations in Texas after the commercial fishery was closed. Conventional management practices, such as net bans and bag and size limits, along with a stocking program that began in earnest in 1983, initiated a recovery in the fishery after the mid-1980s (McEachron et al. 1998). After recovery was under way, management strategies emphasized maintaining populations to ensure sustainable recruitment, to satisfy anglers with quality fish in sufficient numbers, and to add fish to the Gulf of Mexico spawning group to maintain stocks and genetic diversity. In all Gulf states, the Red Drum fishery is managed to allow for at least 30% of Red Drum to escape offshore even though each state is free to set its own bag and size limits in state waters. The 30% escapement goal is endorsed by the Gulf of Mexico Fishery Management Council and has been successful since its implementation in 1989. Since the stocking program began, millions of fry and fingerlings have been released into the Upper Laguna Madre (as well as other Texas bays and some inland lakes). Based on a genetic analysis by Karlsson et al. (2008), Red Drum released as early juveniles (25–40 mm [1.0–1.5 in] long) survive at least a year in Texas bays and estuaries and may compose a significant fraction of Red Drum in an individual bay or estuary. Overall, long-term increases in subadult and adult fish can be attributed to effects of the netting ban, more restrictive slot limits, and stocking by TPWD.

Recreational anglers can legally harvest Red Drum if they abide by strict regulations regarding size and bag limits. Size restrictions on Red Drum have existed since at least the early 1920s (J. Pearson 1929). Fish between 36 and 81cm (14 and 32 in) were the only salable fish from the 1920s to 1981, when the commercial fishery closed and sales were banned (Ponwith and Dokken 1996). Slot limits on sport fish have existed since that time and currently stand at 51–71 cm (20–28 in), although up to two oversized fish may be kept each year if properly tagged. A uniform three fish-per-day bag limit is applied to all Texas bays.

Spotted Seatrout The Spotted Seatrout (or Trout) is estuarine dependent, but unlike Red Drum, it completes the majority of its life cycle within inshore waters and exhibits limited migratory movements (Patillo et al. 1997). Robillard and Stunz (2012) conducted an acoustic-based study in the Laguna Madre and the accompanying outside Gulf beach to confirm or deny the existence of "tide runner trout" (ones thought to utilize tidal flow to move great distances along the Gulf beaches to visit neighboring bay systems). While year-round movements in and out of numerous passes could clearly be documented, most tagged individuals did not remain long in the bay and departed after only a short stay.

Spotted Seatrout spawn at dusk within bays and lagoons. Eggs are often released in shallow, grassy areas near passes or in deeper water where they can drift into grassy areas. Moderate to high salinities appear necessary for spawning, but excessive hypersalinity may lower recruitment of young fish into the population (Gray et al. 1991; Kucera et al. 2002). Other factors influencing reproductive timing may be related to increasing water temperatures or changing photoperiods. Spotted Seatrout occupy many underwater habitats ranging from unvegetated sandy, muddy, or shelly substrates to oil platforms and shell reefs, but juveniles and adults are often associated with seagrasses, particularly Shoal Grass and Turtlegrass. Adults, spawning adults, juveniles, larvae, and eggs of Spotted Seatrout are widely distributed in the Laguna Madre and Baffin Bay throughout the year (Nelson 1992; Lacson and Lee 1997).

Spotted Seatrout are top-level carnivores that probably play a significant role in structuring prey communities, especially in seagrass beds (Patillo et al. 1997). These visual predators feed almost entirely on free-swimming organisms at the surface and mid-depths. Prey choices change as fish grow and seasonally with abundance of prey organisms. Larvae favor copepods, and small juveniles favor shrimp; larger juveniles and adults consume mostly shrimp in warmer months and fish in cooler months when shrimp are not as readily available. Juveniles are preyed on by several species including larger trout, Red Drum, Hardhead Catfish, and Atlantic Croaker.

Destruction of seagrass habitats has been implicated as a cause of declining seatrout populations in some areas. A number of environmental conditions can result in catastrophic mortalities, including severe cold, hurricanes, extremely low salinities, and high turbidities. Populations of Spotted Seatrout in the Laguna Madre experienced sudden dips in annual abundance in 1984, 1990, and 2017 following an anomalous freeze event the prior year (fig. 17.2), but populations of Spotted Seatrout in the Laguna Madre continued to increase in abundance.

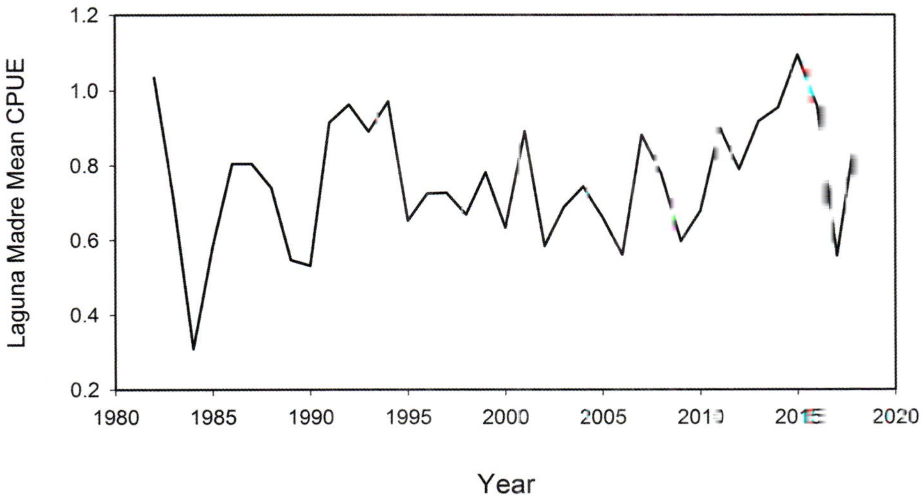

Figure 17.2. *Annual gill-net catch-per-unit effort (CPUE) of Spotted Seatrout in the Laguna Madre. Effort definition follows that in figure 17.1. (Data from the TPWD, Resource Monitoring Program)*

Although the commercial fishery for Spotted Seatrout was closed in Texas in 1981, commercial landings still occur throughout the year in a few Gulf states (Patillo et al. 1997). Spotted Seatrout is one of the top seven species sought by anglers, and the recreational catch far exceeds commercial catch. The commercial harvest ranked third behind Eastern Oysters and Red Drum in Refugio, Aransas, and Nueces Counties in 1890 and ranked second with Black Drum behind shrimp between 1942 and 1970 in Aransas and Laguna Madre districts (Ponwith and Dokken 1996). In the 1920s, the species was regarded as "the most valuable marine food fish in Texas" (J. Pearson 1929). Size restrictions on the Spotted Seatrout have been in place since the early 1920s; minimum size limits for salable fish were originally set at 30.5 cm (12.0 in). Today, the minimum size for retained seatrout is 38.1 cm (15.0 in).

Within the Lower Laguna Madre, TPWD recorded several poor gill-net years in a row during the early 2000s. Fearing that Spotted Seatrout populations were declining, the agency initiated a regional management approach and reduced the daily bag limit in the Lower Laguna from 10 to 5. This change was based on the biology of the species and the fact that Spotted Seatrout did not appear to migrate between bays. This approach was deemed successful, and in 2011 the five-fish daily bag limit was instituted for the remainder of the coast except Galveston Bay and Sabine Bay. An upper slot limit, similar to that permitted for Red Drum, allowed anglers to keep one fish over 65 cm (25 in) per day as part of their daily bag. This upper limit was implemented again because most Spotted Seatrout over 51 cm (20 in) are female. By limiting the number of females removed, the overall reproductive potential of the population is increased.

For the 2019 fishing season, a bag limit of five fish per day/one fish over 65 cm may become standard for the entire coast. Recovery in this species can be attributed to removal of Spotted Seatrout from the commercial market, improved management measures (increased minimum size limits and decreased bag and possession limits), and removal of nets from coastal waters.

Black Drum Like the Red Drum and the Spotted Seatrout, the Black Drum (or Drum) is estuarine dependent. Biologists once thought that mature individuals typically remained within bays, moved into passes to spawn, then returned quickly to preferred bay habitats. However, we now know that spawning transpires in deeper nearshore waters of the Gulf, passes, and bays. Peak spawning activity occurs from January to mid-April, but a secondary peak sometimes occurs in fall (Patillo et al. 1997). During seasons other than spring, spawning occurs in almost all areas of the Upper Laguna Madre, including creeks (Olsen 2014). Juveniles and young adults inhabit unvegetated estuarine habitats with muddy substrates but often move into the more open waters of passes and the nearshore Gulf as they mature. Adults frequent unvegetated sandy or muddy areas and oyster or worm reefs. Migratory movements are limited in fish less than four years old except during adverse environmental conditions.

Black Drum are carnivores throughout their lives. Prey preferences range from zooplankton to molluscs, depending on life stage (Patillo et al. 1997), but the proportions of soft-bodied benthic organisms in the diet decreases as pharyngeal teeth develop. Adults are demersal feeders, preying primarily on bivalves and crabs (Rubio et al. 2018). Sensitive chin barbels help them locate organisms burrowed into the bottom, and hardened pharyngeal teeth help crush molluscs and crabs. Foraging Black Drum often dig pits into the surface of flooded tidal flats in the Laguna Madre (Fisk 1959), and these "blow-outs" are easily recognized when the tides recede. Filter-feeding fish are potential predators on Black Drum eggs and larvae, but larger larvae and juveniles are probably preyed on by various large fishes (Patillo et al. 1997).

Rapid and extreme temperature fluctuations often cause mass mortalities in Black Drum (Olsen et al. 2014). They constituted the majority of fish killed by freezing weather in the Laguna Madre in 1983 and 1989 (Martin and McEachron 1996) and again in 2004 and 2017. Population growth may be limited by the extent of preferred habitat containing adequate food resources and competition with other fishes for benthic resources (Patillo et al. 1997). The Black Drum population may be approaching carrying capacity in the Upper Laguna Madre, as evidenced by an approximate fourfold increase in abundance since the late-1980s (Martinez-Andrade et al. 2005), coupled with an unusual starvation event linked to a die-off of Dwarf Surf Clams, one of their primary food sources (Olsen et al. 2014).

Adult and juvenile Black Drum are frequently encountered in the Laguna Madre throughout the year, whereas spawning adults, larvae, and eggs are found during spring (Nelson 1992). The temporal pattern is similar in Baffin Bay, but abundances are higher regardless of season or life stage. Adult and subadult fish larger than 300 mm (12 in) total length (J. Pearson 1929; Simmons and Breuer 1962) are caught throughout the Upper Laguna Madre with concentrations noted within Baffin Bay and north along the lagoon side of Padre Island.

Black Drum are harvested commercially from the Upper and Lower Laguna Madre, primarily with trotlines. Currently, Texas supplies up to 71% of the US commercial Black Drum landings, which is about the same percentage as in the 1920s (J. Pearson 1929). Although Black Drum was not tallied in the commercial harvest of 1890, it represented the majority of finfish harvested in the Laguna Madre in 18 of 23 years between 1942 and 1970 (Ponwith and Dokken 1996). During the 1920s, market value of Black Drum exceeded value of Red Drum and Spotted Seatrout combined. Between January 1926 and May 1927 more than 60% (432,000 lb [~196,000 kg]) of the total landings were Black Drum (J. Pearson 1929). Although Pearson advised placing size limits of 20.3–50.8 cm (8.0–20 in) on Black Drum to increase proportion of spawning adults, no size restrictions existed until 1988, when a slot limit of 35.6–76.2 cm (14–30 in) was adopted (Ponwith and Dokken 1996). However, Baffin and Alazan Bays and the area of the Upper Laguna Madre known as The Hole, described by Pearson as "some of the best drum-fishing waters along the coast of Texas," were closed to seine fishing in the 1920s.

Populations of Black Drum in the Laguna Madre rebounded after declining for a decade during a prolonged Texas Brown Tide event that reduced their population and those of their primary prey (Montagna et al. 1993; Whitledge 1993; fig. 17.3). Gill-net catch rates in the Upper and Lower Laguna were low until the beginning of the 1990s, when numbers began to increase (fig. 17.4). Black Drum became increasingly abundant in both sub-bays until 1996, when numbers of Black Drum fell off rapidly in the Lower Laguna. In the Upper Laguna Madre, Black Drum continued to slowly increase in abundance through the 2000s and are now at record levels for any bay on the Texas coast. The increase in abundance during the last decade may be pushing this species to the carrying capacity of this bay system. The upsurge in population was regarded as evidence that Texas Brown Tide was not adversely affecting populations of reproductive Black Drum. However, the size restrictions enacted during that year also played a role. For recreational anglers, the daily bag limit for Black Drum is now set at five fish with a slot limit of 14–30 in; only one fish over 52 in (132 cm) may be included as part of the daily bag.

Figure 17.3. *Annual gill-net catch-per-unit effort (CPUE) of Black Drum in the Laguna Madre. Effort definition follows that in figure 17.1. (Data from the TPWD, Resource Monitoring Program)*

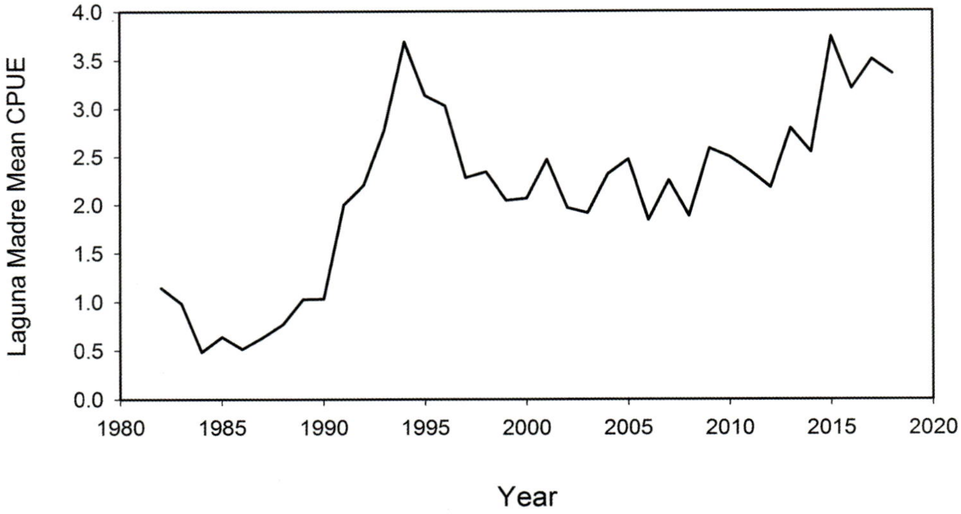

Figure 17.4. *Annual gill-net catch-per-unit effort (CPUE) of Black Drum, by season, in the Upper and Lower Laguna Madre. Effort definition follows that in figure 17.1. (Data from the TPWD, Resource Monitoring Program)*

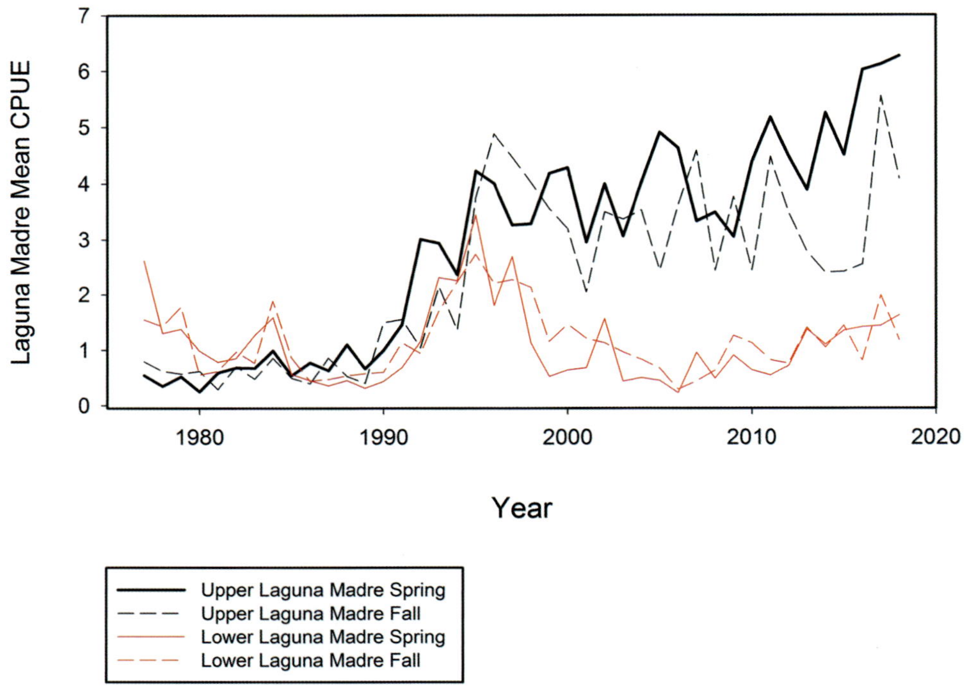

Ecology of Shrimp in Texas

Table 17.2 summarizes the economic and ecological attributes as well as general abundance patterns of 10 invertebrate fisheries organisms found in the Laguna Madre of Texas. Because penaeid shrimp are the most important commercial species, a more detailed overview of their ecology and abundance follows.

Substantial numbers of adult and juvenile Brown, White, and Pink Shrimp inhabit the Laguna Madre (Nelson 1992). Adults of all three species are present between March and May. Juvenile Pink Shrimp are present September–April but are most abundant during the late winter–spring period, whereas juvenile Brown and White Shrimp are present May–November and are most abundant during late summer–fall. In the Upper Laguna Madre, larger juvenile and subadult shrimp of all three species appear to be evenly distributed throughout the lagoon, but sometimes they seem to concentrate in Baffin Bay (Lacson and Lee 1997). Overall, bag-seine catch rates in the Lower Laguna Madre are higher than in the Upper Laguna, and both are dominated by Brown Shrimp (fig. 17.5). Pink and Brown Shrimp are the most widely distributed species in the Lower Laguna Madre. White Shrimp are generally limited to areas west of the GIWW from Port Mansfield to South Bay (Stokes 1974).

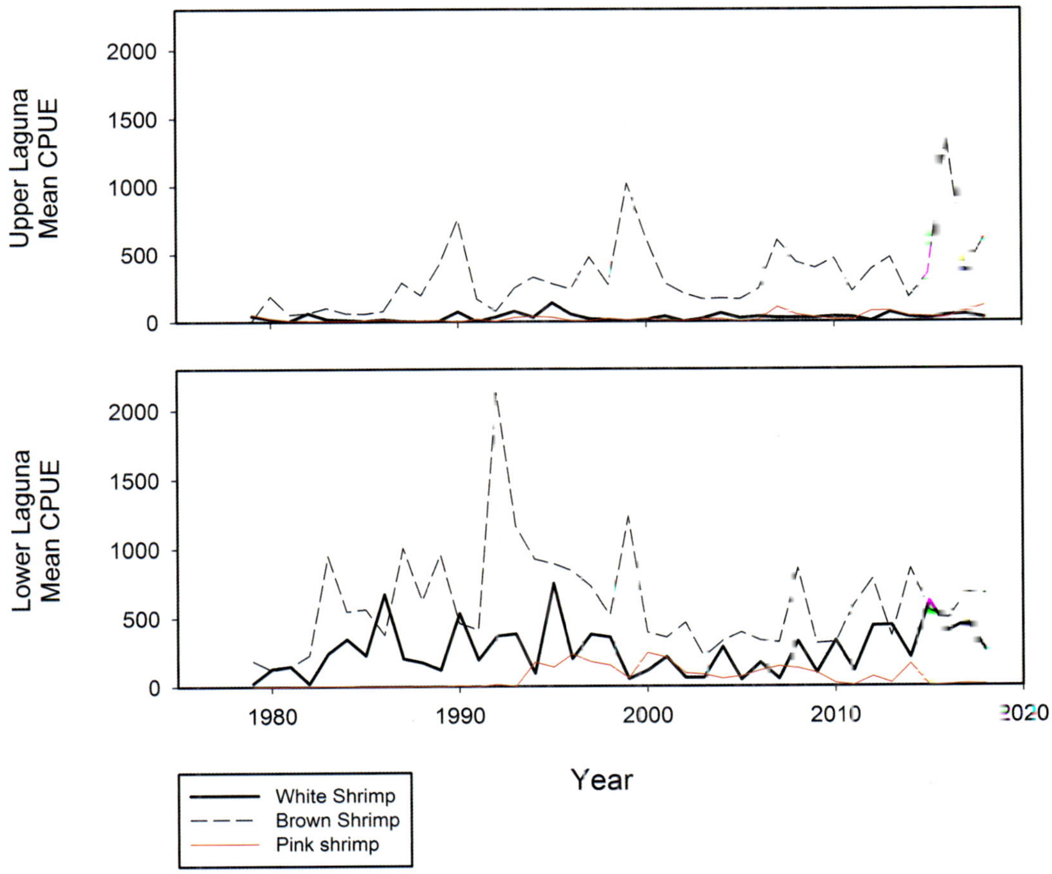

Figure 17.5. *Annual bag-seine catch-per-unit effort (CPUE) of shrimp in the Laguna Madre. Effort is defined as catch per area swept by the seine, and values represent catch standardized to number per hectare. (Data from the TPWD, Resource Monitoring Program)*

After mating within bays or lagoons, adult shrimp migrate in the spring and early summer to deep waters of the Gulf, where they spawn (Patillo et al. 1997). Their eggs and early larval stages are planktonic or demersal. Postlarval shrimp and early juvenile stages migrate through passes on incoming tides in February or March before dispersing into shallow vegetated areas, where they transform into juveniles. Shrimp exhibit a preference for seagrass habitats, especially dense Shoal Grass meadows that are flushed daily by tides. Postlarvae and juveniles of all three species burrow into substrates during the day and emerge from their burrows at night to feed and mate. Juveniles and adults are omnivorous and feed on algae, worms, and microscopic animals. Although shrimp generally prefer vegetated substrates, salinity, turbidity, and light conditions may cause them to move into unvegetated areas where they may be more vulnerable to predation. Shrimp play an important role in estuarine food webs because they transfer energy to top-level consumers such as finfish and large crustaceans (Patillo et al. 1997). Juvenile shrimp are particularly important as prey because they support juveniles of many commercially important fish species.

Predation by fish is the cause of most direct natural mortality in nursery areas. Episodic natural catastrophes (drought, freezes, hypoxia) and disease are the other major causes of natural mortality. Unlike estuarine-dependent finfish species, shrimp have only an 18- to 24-month life span and typically complete their estuarine recruitment, grow-out, and offshore movements within a single year. This makes them an extremely productive species, given the predation pressure by numerous finfish and the harvest by commercial operators. However, recruitment and population abundances estimated by annual surveys can fluctuate greatly under different environmental conditions. Shrimp are the most valuable shellfish industry in the Gulf of Mexico and make up the top three species by both value and poundage of landings (Tunnell 2017).

Shrimp are a favorite among seafood lovers, and the shrimp fishery is the most valuable commercial fishery on the Texas coast. Because shrimp are a favorite food of Red Drum and Spotted Seatrout, the most sought-after sport fish among anglers, a smaller, but significant bait fishery exists. Much of the bait shrimp harvested in the Upper Laguna Madre are caught using push nets and individual bait-shrimp trawls.

Shrimp stocks in the western Gulf of Mexico along the coasts of northern Mexico and South Texas probably represent a single entity and are managed accordingly (Gulf of Mexico Fishery Management Council 2015). Shrimp represented the largest proportion of the commercial catch of all species from Rockport to Brownsville (Aransas and Laguna Madre districts) from 1942 to 1970 (Ponwith and Dokken 1996). Brown Shrimp were the primary commercial species in Texas from 1956 to 1959, particularly from the Coastal Bend south, where they constituted 96% of average annual catch in waters between Rockport and the Land-Cut and 99% in the Lower Laguna Madre (Gunter 1961). Brown Shrimp dominated the live-bait fishery in the Lower Laguna Madre in 1970 and 1972, but nearly equal numbers of Brown and White Shrimp were taken in 1971 (Stokes 1974). Brown and White Shrimp remain the most significant commercial penaeid species on the central and southern Texas coast (Lacson and Lee 1997). The condition of each shrimp stock is monitored annually, and none has been classified as overfished or undergoing overfishing (Hart 2013).

Fish Community Structure in Texas

The relative abundances of important fish species within Texas bays can be determined and compared using two decades (1992–2012) of fisheries-independent data collected by TPWD (Tolan 2013). The TPWD uses several sampling methods to collect data, each targeting fishes of different sizes. Data from several sources can be used to describe the fish community structure in the Laguna Madre and compare it to that of Corpus Christi Bay and other major bays along the coast. The fish community structure can also be described for Baffin Bay, South Bay, and a seagrass bed in the Upper Laguna Madre by using other data sources.

Laguna Madre of Texas The TPWD uses gill nets to capture fish and collect population data on the large fish in Texas bays and lagoons during each fall and spring. Hardhead Catfish are the most frequently encountered fish in Laguna Madre gill nets, and the population trends exhibited by the species are similar to those of many other species. Although hardheads are more abundant in the spring, the trend lines for fall populations generally mirror those of the spring through time (fig. 17.6). The difference in fall to spring catch rates is often used as a measure of fishing pressure as a result of the catch by recreational anglers.

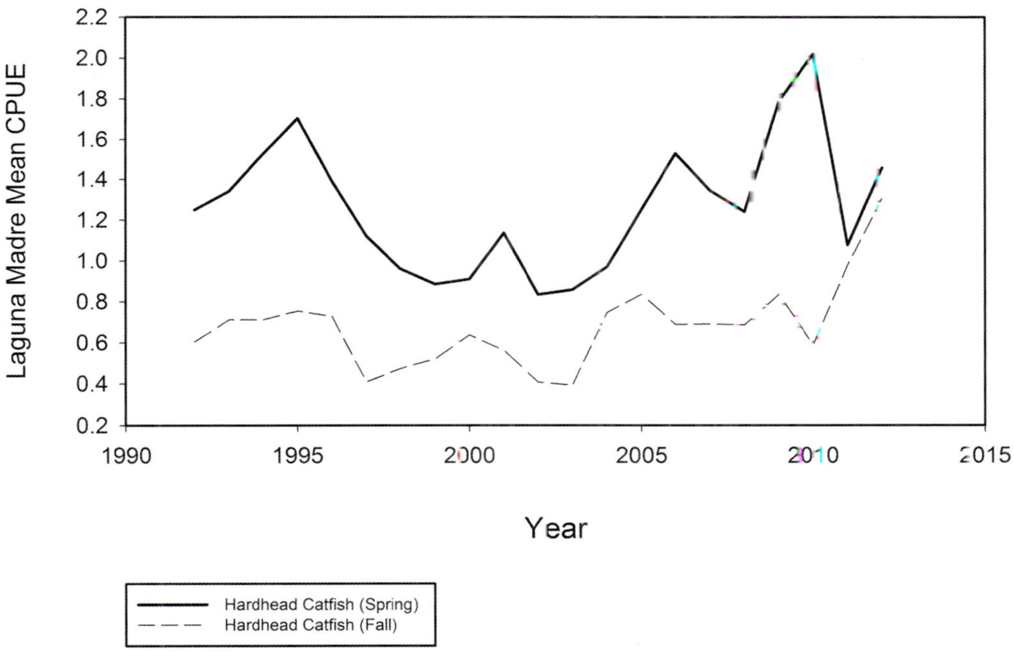

Figure 17.6. *Annual gill-net catch-per-unit effort (CPUE) of Hardhead Catfish in the Laguna Madre. Effort definition follows that in figure 17.1. (Data from the TPWD, Resource Monitoring Program)*

The commercially and recreationally important members of the drum family (Sciaenidae: Red Drum, Black Drum, Spotted Seatrout, Atlantic Croaker, and Spot) are important components of the fish community structure in all Texas bay systems. These species constitute a large proportion of fish captured coastwide (48.1%), within Corpus Christi Bay (51.2%), and in both the Upper (73.9%) and Lower (61.3%) Laguna Madre. The most striking difference is the dominance of Black Drum in the Upper Laguna Madre. As a result of the recent increases, Black Drum may be as much as three to four times more abundant in the Upper Laguna Madre than anywhere else along the coast. Regardless of season, the fish community structure in the Laguna Madre differs markedly from other coastwide systems. Lack of diversity in both lagoons is reflected by the smaller proportion of "other" species collected with the gill nets. A total of 249 unique taxa were found coastwide, while only 92 and 122 "other" taxa were found in the Upper Laguna Madre and the lower Laguna Madre, respectively.

Bag seines are used to collect data on the small fishes inhabiting the Laguna Madre and other Texas bays. Clear differences in community structure among the seasons are evident in the Lavaca-Colorado estuary (fig. 17.7a), with opposing seasons (winter–summer and spring–fall) being the most dissimilar. Even though the seasonal differences are dramatic, the species-composition catches from year to year within a season are stable through time, even under dramatically different environmental conditions. The species composition is remarkably similar whether the estuary is experiencing a flood or in the middle of a drought (fig. 17.7b). These patterns are evident in both

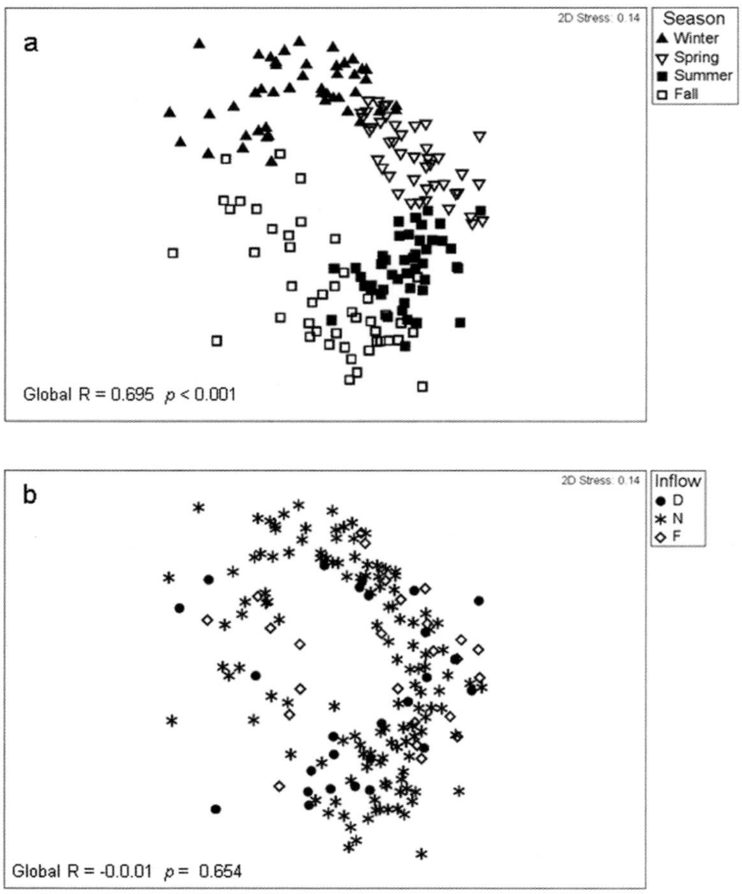

Figure 17.7. *Two-dimensional scaling configuration of community structure from a representative estuary (Lavaca-Colorado) overlaid with (a) Season and (b) Inflow Condition. Season of collection: Winter (Dec., Jan., Feb.); Spring (Mar., Apr., May); Summer (June, July, Aug.); and Fall (Sept., Oct., Nov.). Inflow Condition designations: D = Drought, N = Normal, F = Flood. Global R values for each Analysis of Similarity test included. (Data from Tolan 2013)*

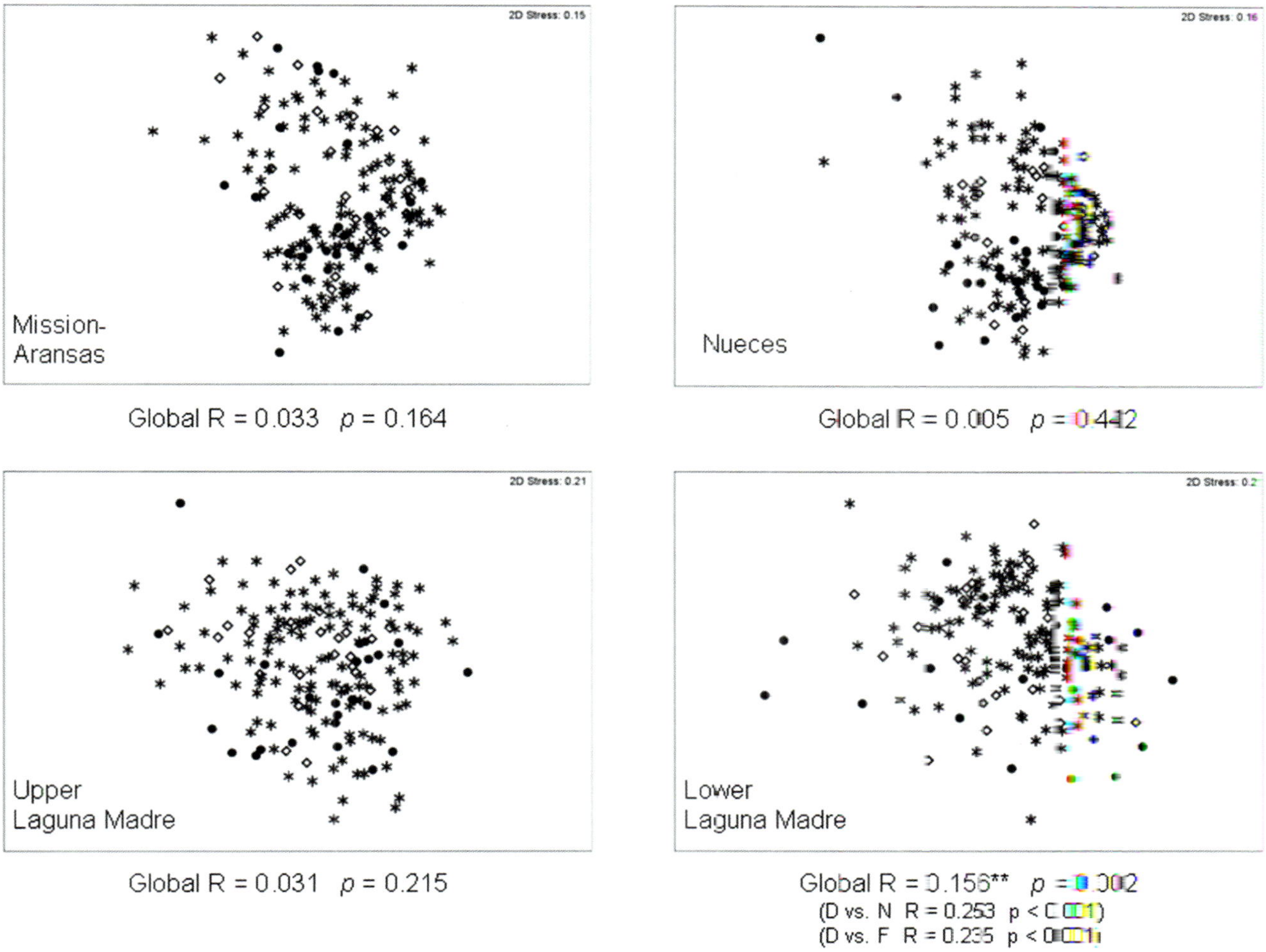

Figure 17.8. *Two-dimensional scaling configurations of community structure for South Texas estuaries identifying the categorical Inflow Condition, with the Global R values for the seasonal Analysis of Similarity test. Inflow Condition symbols follow those in figure 17.7. (Data from Tolan 2013)*

the Upper and Lower Laguna Madre, indicating high seasonal consistency. However, the seasonal differences in species composition are less than those observed in bays farther north (fig. 17.8). Reduced differences between winter minimum and maximum temperatures in the transitional temperate to tropic zone region of the Laguna Madre may account for the disparity in species-composition patterns (Tolan 2013).

Coastwide, Gulf Menhaden dominate the small-fish fauna, but they constitute only a small proportion of fish populations in southerly bays and lagoons. In both portions of the Laguna Madre, the Sheepshead Minnow is the most abundant small fish, although in the Lower Laguna Madre, Brown Shrimp are encountered more often than Sheepshead Minnows in bag seines. Juvenile Spot and Atlantic Croaker are the most abundant drum, especially in the Lower Laguna Madre, although in the Upper Laguna Madre Black Drum replaces Atlantic Croaker as the second-most abundant sciaenid. Overall, the small-fish community is far more diverse than the large-fish fauna. Coastwide, more than 400 unique taxa (fish and invertebrates) have been taken with bag seines but only 212 taxa in Corpus Christi Bay, 162 in the Upper Laguna

Madre, and 243 in the Lower Laguna Madre. Generally, composition of the small-fish fauna of the Lower Laguna Madre and Corpus Christi Bay is similar, differing mainly in the combined proportion of Brown Shrimp, Pinfish, Sheepshead Minnow, and Spot.

Upper Laguna Madre Shoal Grass Beds Chaney (1988) surveyed the fish associated with Shoal Grass beds near Bird Island Basin in the Upper Laguna Madre using a 6 mm seine (quarter-inch mesh). A total of 24 species in 15 families was collected. Killifish, primarily Rainwater Killifish (39.1%), and Tidewater Silversides were the most abundant species in all seasons. Four species of juvenile drum were collected, including Spotted Seatrout and Red Drum, but Spot (4%) were the most numerous. More fishes were collected during the day than at night, although the relative abundances of families were similar. More fish were caught during low tides in February, May, and August, and fewer fish were collected during fall and winter.

Baffin Bay The fish fauna of Baffin Bay has received more study than in any other part of the Laguna Madre system. The number of species in the resident fauna ranges from 6–10 (Breuer 1962; Fuls 1974; Krull 1976) and includes Striped Mullet, Gulf Menhaden, American Gizzard Shad, Spot, Atlantic Croaker, Black Drum, and Spotted Seatrout. Tinnin (1974) and Fuls (1974) studied the fish fauna in Laguna Salada (Alazan Bay) using trammel nets with a bag mesh size of 3.75 cm (1.5 in) (fig. 17.9). The relative abundance and number of species (eight to nine) of drum was greatest, but composition of the moderately sized fish fauna was split almost evenly between four families (Sciaenidae, Clupeidae [menhaden], Mugilidae [mullet], Ariidae [sea catfish]). Spot and Spotted Seatrout were the most abundant drum species (18%–19% of total). Using large trammel nets (bag mesh size 9.0 cm [3.5 in]), the relative abundance of drum captured in association with serpulid reefs throughout Baffin Bay was more than double that captured in smaller trammel nets (Hardegree 1997). Black Drum (20%), Spotted Seatrout (15%), and Spot (12%) were the dominant fauna. During the year after the heavy rainfall associated with Tropical Storm Fern (September 1971) reduced salinities to as low as 1 ppt, the small-fish fauna (5.0 mm [1.9 in] bag seines) in Laguna Salada was dominated by Striped Anchovies (Krull 1976). The fish community was much more diverse than the community described using similar gear throughout Baffin Bay during a period of normal hypersalinity (Hardegree 1997). The most striking difference between the faunas is the dominance of Gulf Menhaden in Hardegree's collections and their relative unimportance in Krull's. Although both species were present throughout Baffin Bay, Rough Silversides were more abundant in Laguna Salada at lower salinities, whereas Pinfish were more abundant during periods with "normal" salinities. Trawl and beach seine collections from Alazan Bay (Cornelius 1984) were much more diverse than collections with other gear. Although anchovies, silversides, and menhaden constituted 60% of the fish collected, killifish were important in these collections but not in the others. In addition, puffers (family Tetraodontidae) were more abundant than drum or mullet. Naked Gobies (64%) and Striped Anchovies (35%) dominated the larval assemblage in Alazan Bay (Dokken 1981).

South Bay The fish assemblage in South Bay was described using an otter trawl (25 cm [9.8 in] cod end mesh) in deep water at the bay entrance and a bag seine (8.0 mm [3.1 in] outside mesh, 4.0 mm [1.6 in] inside mesh) in shallow environments around the bay margins (Hook 1991). A total of 90 species in 55 families was collected, 10 species of which were collected only in trawl samples. Overall, the assemblage was

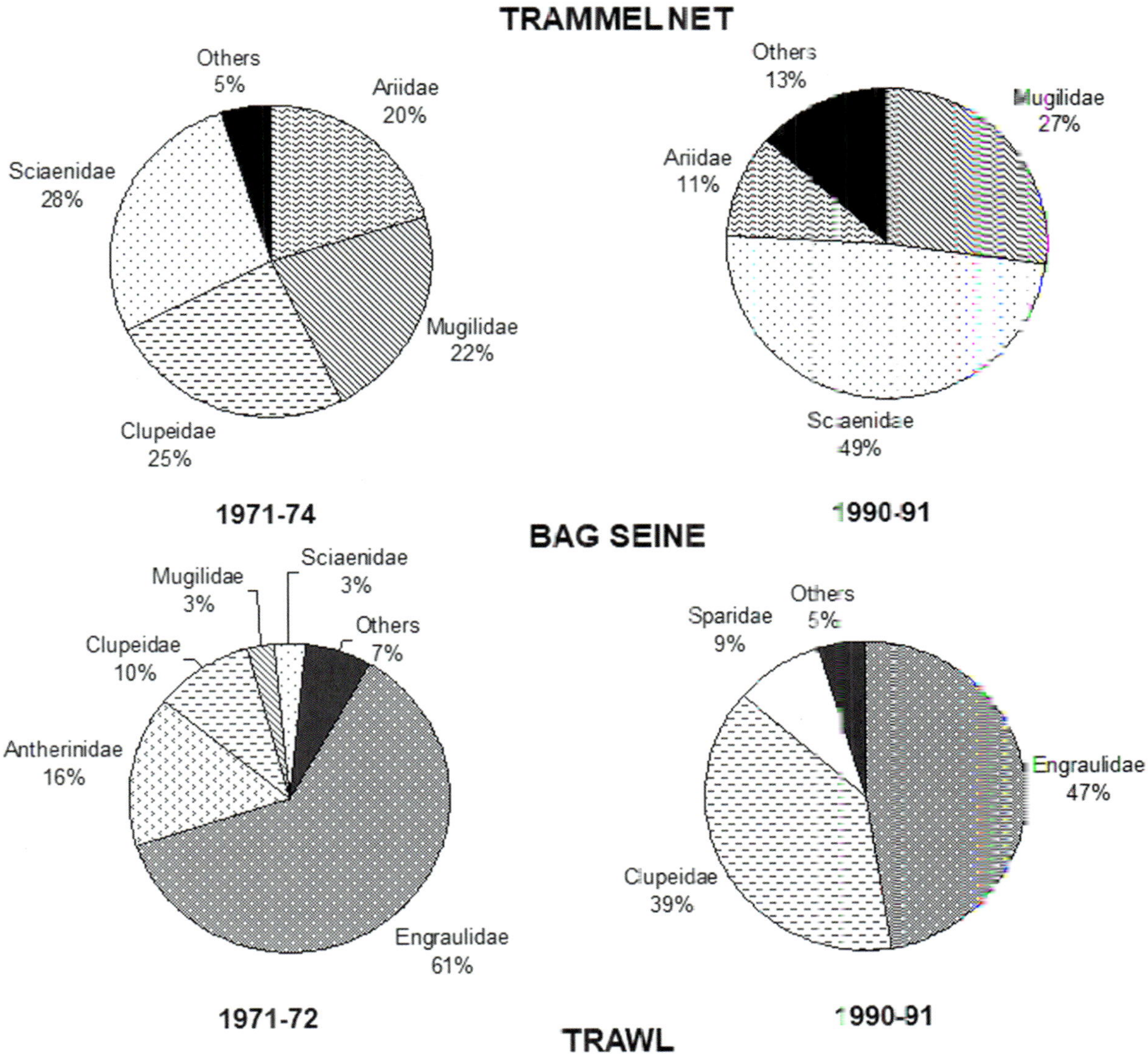

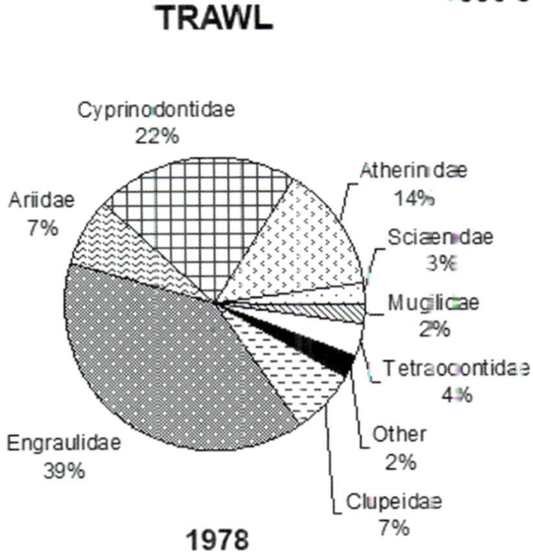

Figure 17.9. *Relative abundances of fishes in Baffin Bay. Trammel-net data for 1971–74 from Tinnin (1974) and Fuls (1974); for 1990–91 from Hardegree (1997). Bag-seine data for 1971–72 from Krull (1976); for 1990–91 from Hardegree (1997). Trawl data for 1978 from Cornelius (1984).*

diverse but dominated by drum, particularly Spot, and killifish, primarily Sheepshead Minnows. Seagrass habitats were dominated by Spot or Tidewater Silversides; a Black Mangrove channel by Sheepshead Minnows; a steep sloping, unvegetated shoreline by Tidewater Silversides; and a shallow, turbid area surrounding oyster beds by Striped Mullet. In addition to site-specific differences, seasonal variances were noted. The greatest abundances were recorded at all sites during January. Seasonally, dominance shifted between the four most important species: Gulf Menhaden in winter; Spot in spring; Sheepshead Minnow in summer; and Bay Anchovy in fall. Decreased abundances of some species at some sites were associated with major changes in temperature and salinity.

History of Commercial Harvest in Texas

In the 1890s and early 1900s, most fish were harvested using seines pulled by crews in thigh-deep water (Higgins and Lord 1926). The efficiency of this method resulted in the summer seining ban in 1914, and fishermen began using gill and trammel nets. Although the ban on seining was lifted in 1922, the change from seines to other nets was permanent. Availability of cracked ice to preserve fish that formerly would have spoiled before reaching market was another factor contributing to continued use of gill nets. Hook and line, cast nets, and gigs were also used to commercially harvest fish.

Although the Laguna Madre is a productive and flourishing ecosystem providing an excellent resource for commercial and recreational fishing, overfishing has long been a concern. Early in the twentieth century, Texas relied chiefly on size limits and closures to conserve fisheries resources (J. Pearson 1929). Higgins and Lord (1926) reported that as early as 1890 sport fishermen believed commercial fishermen were depleting fish supplies along the Texas coast. In 1925, scarcity of certain fish in markets aroused fear that fish populations were being exhausted. Conversely, fishermen complained that regulations prevented them from supplying the markets. These concerns were heard by state legislators, who prompted an investigation of Texas fisheries by the state fishery administration. After studying the limited records and talking with local fishermen and dealers, Higgins and Lord determined that coastwide yields were level or rose slightly between 1901 and 1924. However, yields from Corpus Christi (primarily Nueces Bay and the Upper Laguna Madre) and Port Isabel generally increased until 1916–18 during World War I, when no local statistics were kept (fig. 17.10). During that time, all fisheries regulations were suspended, after which the overall state yield increased and local yields likely increased as well. Corpus Christi ranked second in

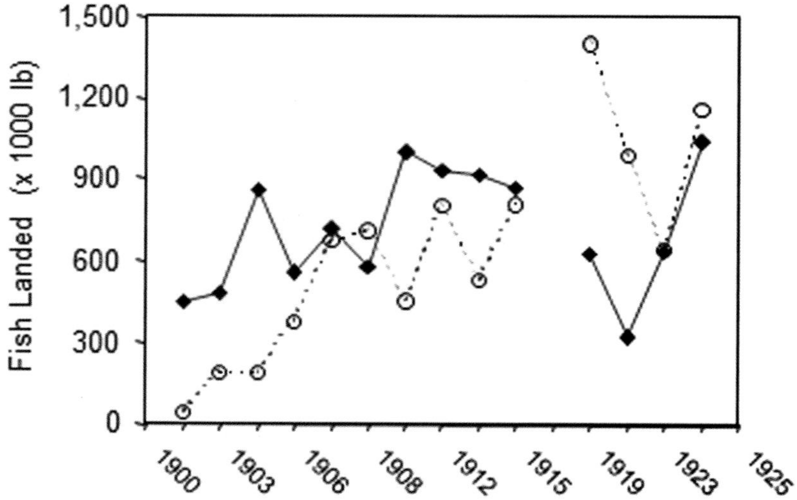

***Figure 17.10.** Fish landings from Corpus Christi and Port Isabel from 1900 to 1924 (Data from Higgins and Lord 1926)*

annual fishery production from 1901 to 1924. Port Isabel ranked fourth from 1901 to 1910 but became the most productive port on the coast between 1912 and 1924.

The commercial Texas finfish industry concentrated on Spotted Seatrout and Red Drum between 1890 and the end of World War I. Thereafter, Black Drum, previously considered an inferior species, equaled or slightly exceeded Red Drum landings (Higgins and Lord 1926). Spotted Seatrout and Red Drum harvests declined after the war, and the fact that Black Drum was marketed was evidence of the scarcity of both desired species in markets. Higgins and Lord emphasized the need for life history studies for several species because most regulations of the time were based on incomplete knowledge. They recommended a conservation program that included establishing a system whereby fishery statistics were uniformly kept and reported; initiating scientific investigations of species' life histories and ecological requirements; and reorganizing the Texas Game, Fish, and Oyster Commission (now the Texas Parks and Wildlife Department).

As a result of Higgins and Lord's report, a detailed study of the life history of Red Drum and other commercial sciaenids was initiated, as well as investigations of fish kills in the Upper Laguna Madre and Baffin Bay due to hypersalinity. Pearson's (1929) work in the waters between Copano Bay and the Land-Cut is the earliest work on many aspects of the life histories of Red Drum, Black Drum, Spotted Seatrout, Atlantic Croaker, and Spot. He recommended reopening closed waters to drum fishing but restricting legal size to 20.3–50.8 cm (8.0–20 in). Pearson especially recommended reopening the Upper Laguna Madre to both drum and seatrout fishing because cold and hypersalinity destroy "great quantities of fish with no benefit whatsoever to the State of Texas." He called for closure of waters in the vicinity of passes, especially the area around Corpus Christi Pass (now Packery Channel). He reiterated previous calls to collect fisheries statistics and continue biological research. However, not until 1934 was the first biologist hired and legislation enacted requiring record keeping by fishermen and dealers (Ponwith and Dokken 1996). The Texas Game, Fish, and Oyster Commission began keeping records in a fairly consistent format in 1942, but size limits were not enacted for Black Drum until 1988.

With the decline of Red Drum and Spotted Seatrout stocks in the late 1970s, many new restrictions were implemented to prevent further depletion and promote recovery of the populations. Currently, nine methods can be used to harvest nongame fish: hook and lines, cast nets, gigs, spears, minnow seines, perch traps, purse seines (for Gulf Menhaden only), trawls, and trotlines. Restrictions apply to all other fishing gear. Gamefish such as Red Drum can be harvested using only hook-and-line and sail-line methods. Size, bag, and possession limits for selected species are regulated.

Commercial shrimping in Texas bays is closely controlled. Commercial shrimping vessels may deploy either an otter trawl or beam trawl for harvesting shrimp in the large bays, but strict laws regulate net length, trawl door dimensions, and net mesh size. Shrimping hours and bag limits are enforced for each shrimping season for all commercial shrimp, including table and bait shrimp. Both the Upper and Lower Laguna Madre are designated as bait bays, so shrimp can legally be harvested only by persons holding a valid recreational fishing license and a saltwater fishing stamp endorsement. Shrimp may be taken using a cast net, a seine not longer than 6 m (20 ft), or an individual bait-shrimp trawl. Boat-pulled trawls are not permitted, and it is illegal to harvest shrimp in natural or human-made passes leading to the Gulf of Mexico. Shrimp captured for personal use (bait or food) cannot be sold commercially (TPWD 2019).

Fisheries in the Laguna Madre de Tamaulipas

Historically, fisheries of Laguna Madre de Tamaulipas alternated between relatively short, productive (boom) periods when salinities were marine or estuarine followed by increasing salinities that left the Laguna unproductive for years at a time (Hildebrand 1958, 1969, 1980). Major favorable changes resulted from singular, aperiodic tropical storms that lowered salinities dramatically either by opening passes or generating heavy rainfall. Changes caused by the storms were short-lived, but organisms living within the lagoon responded rapidly to favorable conditions following storms. Populations of both fish and shrimp increased dramatically in subsequent months. As the interval of dry spells between major storms increased, salt concentrations in the Mexican Laguna magnified. Consequently, the intensity of storms and amount of rainfall necessary to reduce the salinity also increased.

For almost half of the years (45) during a 100-year period (1880–1980), lagoon fishery was not productive (Hildebrand 1969, 1980). From 1880 to 1909, the passes were closed and the water levels so low that in 1902 and 1907 the area inhabitants petitioned the government to reopen the passes. Unfortunately, no records of the fisheries were kept during this period. The 1909 hurricane reopened all the passes, and the lagoon was productive from 1909 to 1914. The first record of commercial landings from Laguna Madre de Tamaulipas was in 1911, when the catch from a fishing camp at La Capilla was transported by mule team to Brownsville. Conflicting accounts for the period 1914–26 suggested that passes were closed during 1916–17. The Laguna was moderately productive during 1919–24. All passes were apparently completely closed in 1926, and the Laguna remained unproductive until 1933. During 1933, four hurricanes opened eight passes, and the Laguna remained moderately productive until 1960, although the northern area was occasionally extremely hypersaline.

Mexican Harvest Methods and Regulations Harvest regulations and methods in waters off the coast of Mexico are much different than those implemented and practiced in US waters. The most commonly used fishing method in the Laguna Madre de Tamaulipas is the shrimp trap, the *charanga* (Hildebrand 1980), and as many as 10,000 may be in use (Cruz-Nieto 2000). *Charangas* are V-shaped traps that terminate in a mesh-enclosed rectangular *yagual* or *matadero*. The traps may be used individually, but as many as eight often are connected together and set in channels,

Figure 17.11. Charangas *(fish traps) in shallow water near Mezquital in Northern Laguna Madre de Tamaulipas (Photograph by J. W. Tunnell Jr., February 1999)*

passes, or near shore (fig. 17.11) to capture shrimp on an ebbing current (Hildebrand 1980). Normally, *charangas* are fished only at night, but they are also used during the day when fishing for White Shrimp.

Chinchorros, which are now illegal (Cruz-Nieto 2000), were large drag seines that effectively captured White Shrimp (Hildebrand 1980). Originally, these devices were used to capture fish, but because of increases in shrimp populations in the Laguna Madre de Tamaulipas and their amplified marketability, the nets were modified to harvest shrimp. The nets had two wings, each up to 600 m (1,980 ft) in length, and a bag. They could be used as stop nets across an ebbing current or could be dragged between two sailboats for several hours, closed, and retrieved.

By 1980, several other fishing methods used by commercial fishermen in Laguna Madre de Tamaulipas were being restricted (Hildebrand 1980). Otter trawls were banned to prevent habitat destruction. The size of the mesh in *charangas* was also regulated because small mesh dimensions resulted in the deaths of unmarketable shrimp and fish, contributed to closing of passes and silting of channels, and hindered the movements of small juveniles to nursery grounds. Historically, however, fishing regulations have not been strictly enforced in Mexican waters.

Commercial Fisheries in the Laguna Madre de Tamaulipas Approximately 80% of fish species within the Laguna Madre de Tamaulipas are characterized as having some commercial use (Leija-Tristán et al. 2000). The shrimp fishery there expanded rapidly after 1972 (Hildebrand 1980), and currently about 90% of the fishing effort is dedicated to harvest of penaeid shrimp (Gomez Soto and Contreras Balderas 1987). Black Drum was the principal commercial finfish species, although Red Drum and Spotted Seatrout were the principal fishes exported to the United States between 1944 and 1962 (Hildebrand 1969). The following discussion of the finfish and shrimp fisheries in the Laguna Madre de Tamaulipas comes from the work of Hildebrand (1958, 1969, 1980).

Large-scale exploitation of the Laguna started around 1935, when it became feasible to haul fish to market. The commercial fishery was concentrated in the Northern Laguna until 1947. Between 1935 and 1941, 4–6 million kg (8.8–13.2 million lb) of fish, mostly Red Drum and Spotted Seatrout, were harvested each year from the Laguna Madre de Tamaulipas. Yields of both Red Drum and Spotted Seatrout were high during most years between 1944 and 1960 (figs. 17.12 and 17.13). In general, Red Drum exports were only about 30%–50% of the exports of Spotted Seatrout. Sport fishing was extremely good during those years. Hildebrand (1969) reports photographing 112 Red Drum weighing a total of 545 kg (1,200 lb) caught by five fishermen in only six hours in the northern lagoon. Shrimp were not important fisheries organisms, with catches remaining small during the 1950s, except during 1959.

From 1961 to 1967, Gulf passes were closed except for a brief period after Hurricane Carla in 1961. An 80% decline from the previous year in fish imports into the United States was evident in 1961 (see figs. 17.12 and 17.13). After good shrimp harvests in 1959, a large number of *charangas* were placed in the Laguna, especially around Boca Jesús María, and the resulting siltation was implicated as the reason the pass closed in 1960. Hurricane Beulah, accompanied by torrential rains, reopened passes in 1967. Productivity, especially of shrimp, was high between 1971 and 1978 (fig. 17.14). One report showed that as much as four tons (3.6 tonnes) of shrimp were caught in a *chinchorro* during a single trip in 1972. At the time, the goal of the Mexican fishery department appeared to be to maximize the shrimp fishery in the Laguna Madre de Tamaulipas without regard for the offshore fishery. By 1975, most *chinchorro* fishermen

Figure 17.12. *Total Red Drum imports at Port Isabel and Brownsville, Texas, from the Laguna Madre de Tamaulipas, 1944–62 (Data from Hildebrand 1969)*

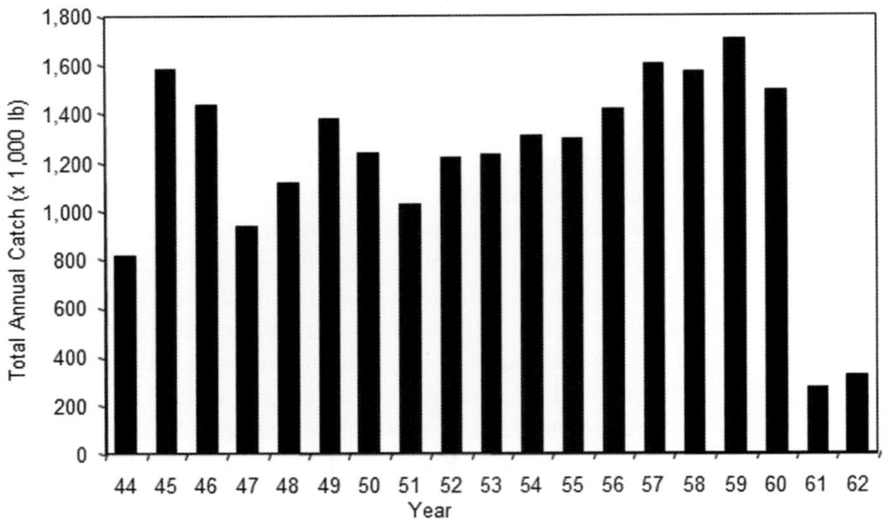

Figure 17.13. *Total Spotted Seatrout imports at Port Isabel and Brownsville, Texas, from the Laguna Madre de Tamaulipas, 1944–62 (Data from Hildebrand 1969)*

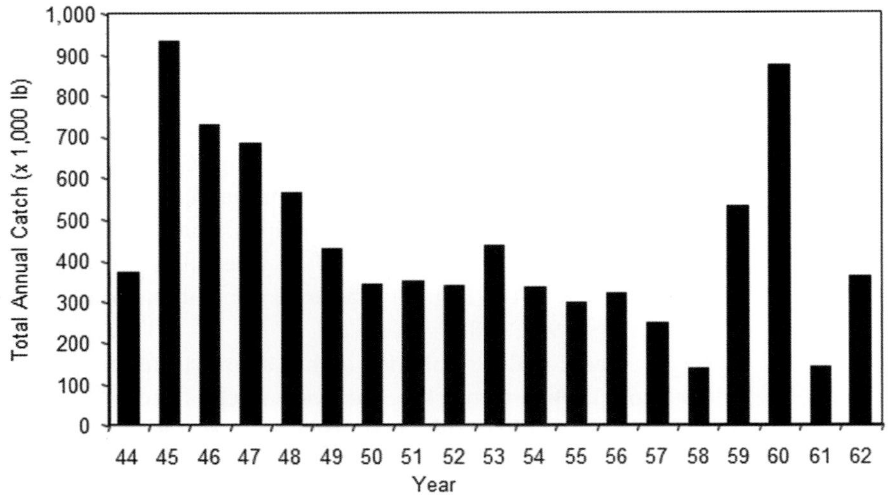

Figure 17.14. *Total shrimp catch in the Laguna Madre de Tamaulipas for 1971–78 (Data from Hildebrand 1980)*

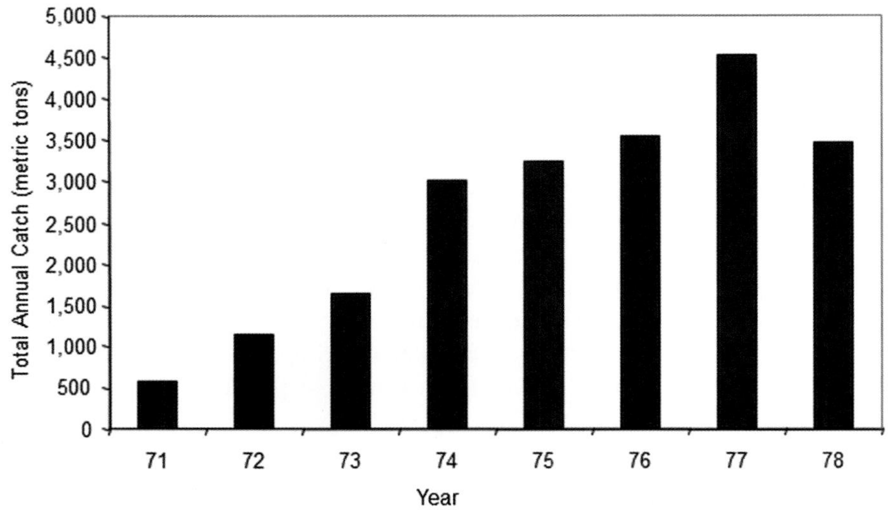

had reduced the mesh size of their nets and begun fishing primarily for shrimp. By 1976, conversion of fish nets to shrimp nets was essentially complete and resulted in the large increase in shrimp catches during 1977. Decline in shrimp landings in 1978 was attributed to actual shortages, possibly due to a combination of overfishing and environmental conditions rather than changes in price or effort.

Between 1990 and 1998, inshore shrimp landings were generally higher than during the 1970s (fig. 17.15) and consistently lower than the Texas offshore harvest (fig. 17.16). The general trend of higher and lower shrimp harvests was the same instance in Tamaulipas as it was offshore in Texas. Red Drum and Spotted Seatrout were also harvested (fig. 17.16), but at relatively low rates. Of interest is the harvest of Striped Mullet, which generally increased through the mid-1990s, and Eastern Oysters, a species reaped only from South Bay in the Lower Laguna Madre in Texas. Oyster harvest peaked in 1992, four years after Hurricane Gilbert made landfall near La Pesca in Laguna Madre de Tamaulipas, possibly freshening the system, but dropped dramatically in 1995. Oyster harvest appeared to be beginning another increase in 1997, possibly due to freshening by tropical systems that came near the northern coast (Tropical Storm Josephine) and made landfall near Tampico (Hurricane Dolly) in 1996.

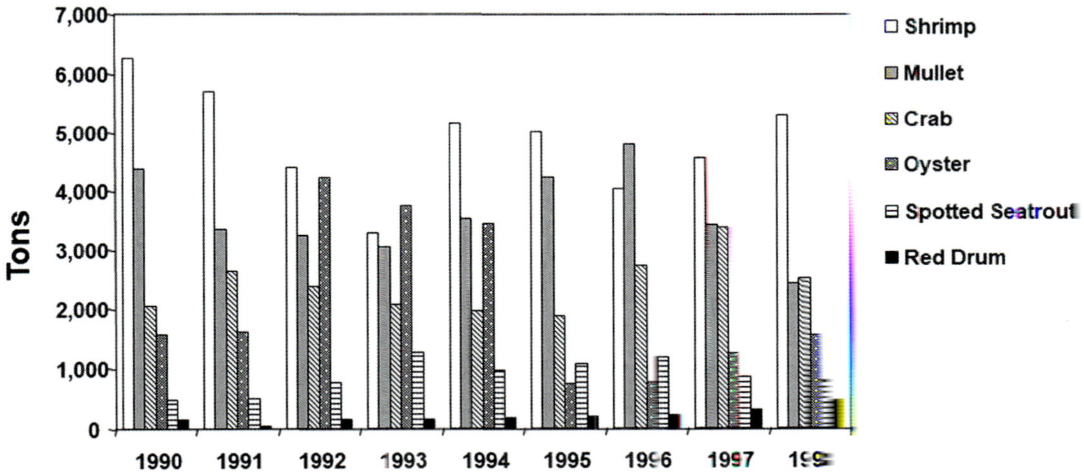

Figure 17.15. *Commercial landings from Laguna Madre de Tamaulipas, 1990–98 (J. Brenner, pers. comm., data from Secretaría de Medio Ambiente y Recursos Naturales [SEMARNAT])*

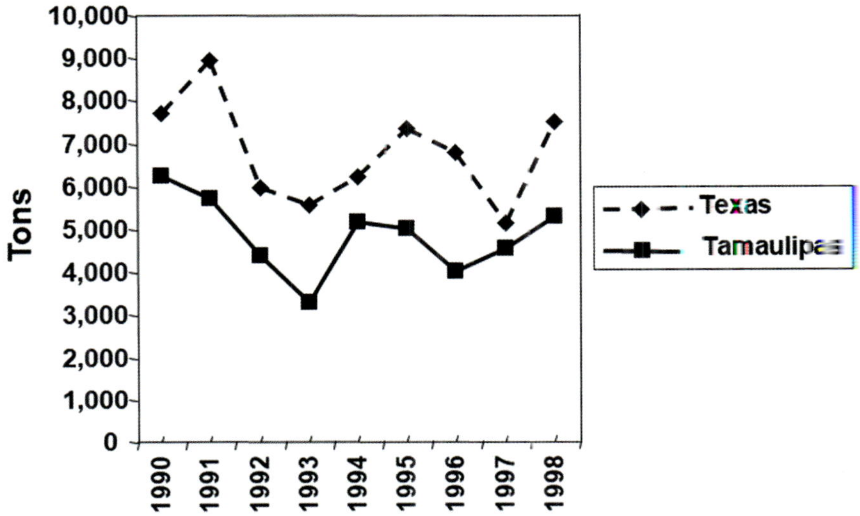

Figure 17.16. *Comparison of Texas Gulf and Laguna Madre de Tamaulipas inshore shrimp landings, 1990–98*

Part 3 Special Issues

18 Brown and Red Tides

KIM WITHERS, MICHAEL WETZ, AND PAUL V. ZIMBA

Brown and Red Tides are created by two types of harmful algal blooms that occur worldwide. The brown tide that occurs in the Laguna Madre is caused by blooms of a microscopic alga, *Aureoumbra lagunensis*. Generally called the Texas Brown Tide, this species can multiply rapidly under certain environmental conditions. Brown tides may alter benthic species abundance, distribution, and diversity as a result of light reduction. The alga may also inhibit feeding by benthic and planktonic organisms or by replacing more nutritious food sources and can cause changes throughout the estuarine food web (Street et al. 1997). A distinct but somewhat similar species, *Aureococcus anophagefferens*, is responsible for brown tide blooms in Rhode Island, Spain, and China (Sieburth et al. 1988; Zhang et al. 2012).

Several dinoflagellate taxa can form Red Tides, but *Karenia brevis*, sometimes known as Florida Red Tide, is the most prominent and harmful (Castle et al. 2013; Buttke et al. 2018). Red Tide blooms typically occur in late summer–autumn and while blooms are believed to initiate offshore, they frequently affect Texas estuaries and coastal waters (Magana et al. 2003). Red Tide produces a powerful toxin (brevetoxin) that can be deadly to fish and marine invertebrates and may cause health problems in humans either through ingestion of toxins contained in shellfish or via aerosols that may cause respiratory difficulties (Kirkpatrick et al. 2004). Along the Texas coast, animals, including Green Treefrogs, Coyotes, Spotted Ground Squirrels, Virginia Opossums, domestic dogs, and cormorants, have died in areas up to 2.0 km (1.2 mi) inland from the shoreline bloom site after breathing aerosolized brevetoxin ([illegible]afalski 2012; Castle et al. 2013; Buttke et al. 2018).

The Laguna Madre of Texas has been affected by brown tides regularly since 1989. In fact, the first bloom to be observed in the region, from late 1989 to late 1997, is one of the longest-lasting algal blooms ever documented. In contrast, it is rare to have Red Tide blooms in the Laguna Madre, in particular those caused by *K. brevis*. Historical occurrences of reddish- or brownish-colored "bad water" in the Laguna Madre have been noted (Gunter 1945a; Simmons 1957), but the causative agent was not identified. On sampling trips associated with an ongoing water-quality study, it is common to see brown, turbid water that is not associated with an algal bloom (M. Wetz, pers. comm.). Thus, the history of harmful algal blooms in the Laguna Madre remains ambiguous.

Brown Tide

The alga associated with the Texas Brown Tide is a tiny (3–5 μm diameter), morphologically indistinct phytoplankton species classified in division Pelagophyta (DeYoe et al. 1995; DeYoe et al. 1997; fig. 18.1). This organism is not endemic to the Laguna Madre and has been detected using immunofluorescence techniques in other bays along the Texas coast (Lopez-Barreiro 1998), three bays in South Florida (Lopez-Barreiro et al. 1998), and more recently in the Indian River Lagoon in Florida, as

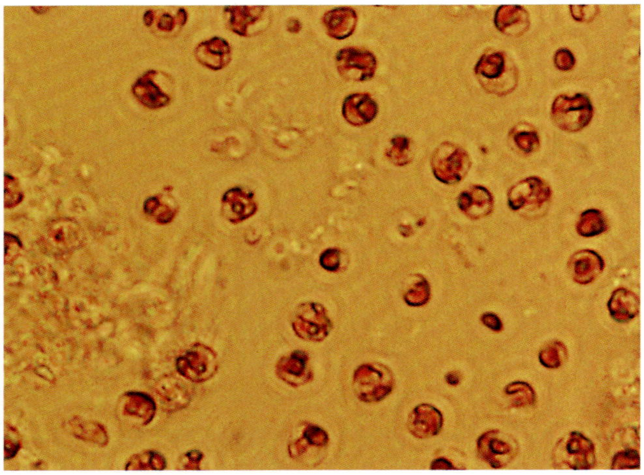

Figure 18.1. *Texas Brown Tide cells in a brown tide–laden water sample collected from Baffin Bay (Photograph courtesy of E. Cira)*

well as Guantanamo Bay in Cuba (Gobler et al. 2013; Koch et al. 2014). Apparently, the organism is present in low concentrations in many coastal bays and, because it is both minute and visibly unremarkable, is considered part of the "hidden flora" (Buskey et al. 1999).

Biology and Bloom Dynamics The first documented brown tide in Laguna Madre began in December 1989 and lasted until October 1997 (Buskey et al. 1997; Buskey et al. 1998). Significant abundances of *Aureoumbra lagunensis* were first noted in Baffin Bay in autumn 1989 (Buskey et al. 1997; Buskey et al. 1998), and by December, bloom conditions were present in the bay (Buskey et al. 1998). The bloom was restricted to Baffin Bay and its secondary bays until May 1990, when the bloom appeared in the adjacent Upper Laguna Madre (Buskey et al. 1997). The bloom spread into the Lower Laguna Madre in late 1990 when strong north winds associated with winter cold fronts pushed water south through the Land-Cut (Stockwell et al. 1993). Amazingly, this organism was present in bloom concentrations in at least some parts of the Upper and Lower Laguna Madre, especially Baffin Bay, for eight years, one of the longest continual algal blooms ever recorded (fig. 18.2).

Phytoplankton growth is typically dependent on light and nutrients and is also influenced by water temperature. The extreme range of salinities observed in Baffin Bay (0–70+ ppt) also influences phytoplankton growth. In the laboratory, maximum growth rates of *A. lagunensis* were observed at 25 °C (77 °F) and salinities of 30–50 ppt, but it was able to continue growing at salinities of 10–90 ppt (Buskey et al. 1998). It can grow in a range of light intensities and is not adapted to either low or high light (Buskey et al. 1998; Muhlstein and Villareal 2007). Growth rates of *A. lagunensis* are positively correlated with temperature, based on laboratory studies (Buskey et al. 1998) and field studies that documented lowest abundances in winter and highest abundances during warmer months of the year (Rhudy et al. 1999).

The nutrient conditions that support growth of *A. lagunensis* and persistence of blooms are complex. The alga is unable to use nitrate and instead relies on reduced nitrogen, such as ammonium or organic nitrogen (DeYoe and Suttle 1994; Muhlstein and Villareal 2007; Agostoni and Erdner 2011; Kang et al. 2015). In the field, Texas Brown Tide blooms have typically occurred when inorganic nitrogen concentrations are relatively low but concentrations of organic nitrogen are high (Gobler et al. 2013; Wetz et al. 2017). Given its nutrient acquisition capabilities and preferences, it appears that *A. lagunensis* is adapted to conditions, such as prolonged droughts, that

Figure 18.2. *Aerial image of the Laguna Madre near South Padre Island during an intense brown tide event (Photograph by Mike Weeks, TPWD)*

promote reductions in inorganic nitrogen—conditions that are common in Baffin Bay and the Laguna Madre (Rhudy et al. 1999; Wetz et al. 2017). When regenerated ammonium and organic nitrogen are the main forms of nitrogen available, *A. lagunensis* outcompetes other phytoplankton. In addition to being adapted to conditions of low inorganic nitrogen, Texas Brown Tide is also well adapted to the low inorganic phosphorus concentrations typically present in Upper Laguna Madre–Baffin Bay (Liu et al. 2001; Cotner et al. 2004). Furthermore, Sun et al. (2012) demonstrated that the organism employs alkaline phosphatase to obtain phosphorus from the larger organic phosphorus pool in the environment.

A bloom of phytoplankton can occur and persist only if phytoplankton growth exceeds losses to sinking, flushing, viral lysis, grazing, and other mortality (Buskey et al. 1997). The initiation of the 1989–90 brown tide co-occurred with two significant climatologic events: a drought that exacerbated hypersalinity in the system and a severe freeze that resulted in a major fish kill (DeYoe and Suttle 1994). Because of the extended drought that gripped South Texas in 1989, salinities exceeding 60 ppt were recorded within the upper reaches of Baffin Bay. In the months preceding the bloom, low microzooplankton abundances corresponded with the high salinities (Buskey et al. 1997; Buskey et al. 1998). Microzooplankton are the main grazer of phytoplankton similar in size to *A. lagunensis* (Hansen et al. 1994). Concomitantly in Baffin Bay, populations of Dwarf Surf Clam, a benthic filter feeder, were also greatly reduced (Montagna et al. 1993). Thus, the loss of top-down biological control was likely at least partially responsible for bloom initiation.

Then, in December 1989, two severe cold fronts within seven days resulted in freezing conditions along the coast. After the second front, water temperatures in Baffin Bay dropped to 0.2 °C (35.6 °F) and remained below 5 °C (41 °F) for seven days, whereas in the Upper Laguna Madre water temperatures dropped to -1.6 °C (29 °F) and remained below 0 °C (32 °F) for 36 hours. These two severe freezes resulted in a fish kill estimated at 965,000 fish in Baffin Bay and the Upper Laguna Madre (DeYoe and Suttle 1994). This, along with a recent freshwater pulse containing runoff from adjacent agricultural lands, resulted in a sharp increase in ammonium that would have been favorable to

A. lagunensis growth (Stockwell et al. 1993; Whitledge 1993). However, since there is evidence of bloom conditions prior to the freeze, nitrogen from the fish kill may not have initiated the bloom, but at the very least, these inputs of nitrogen probably intensified the growth phase of the bloom. Recently, Wetz et al. (2017) identified a large reservoir of dissolved organic nitrogen in Baffin Bay, suggesting that this pool provides a readily available nitrogen supply that can support bloom initiation, assuming other conditions (alleviation of top-down controls) are conducive.

Two major factors contributing to the persistence of the first bloom and subsequent blooms appear to be hypersalinity and favorable nutrient conditions (high organic nitrogen concentrations). Indeed, all subsequent blooms have occurred when salinities were high. Hypersalinity provides conditions under which *A. lagunensis* thrives but causes reduced growth in many other algae and potential grazer species. In addition, inorganic nutrient concentrations are low during drought conditions, hindering growth of non-brown-tide phytoplankton. Reductions in salinities (<30 ppt) with larger concentrations of inorganic nutrients following heavy rains have consistently led to a demise of *A. lagunensis* blooms, including the 1990–97 bloom (Buskey et al. 1998; Buskey et al. 2001) and the most recent bloom that started in 2012 and ended abruptly in 2015 with the onset of heavy spring rains (E. Cira and M. Wetz, unpub. data). Lower salinities may have favored both grazers and algal competitors while also being associated with increased flushing. Brown tides have not been recorded in the Laguna Madre de Tamaulipas. Because *A. lagunensis* is found in many coastal areas of the Gulf of Mexico, it is possible that blooms may have already occurred in Tamaulipas and/or may occur there in the future, especially considering that the salinity regime of the system is very similar to that of Baffin Bay (Yáñez-Arancibia and Day 2004) and sources of organic nitrogen are present in the system (Smith et al. 1997).

Effects on the Environment Brown tide blooms typically are noted for two significant impacts: long-term light reduction and alteration of food webs. Significant declines in Shoal Grass cover and biomass in the Upper Laguna Madre have been attributed to reduced penetration of sunlight (Dunton 1994; Onuf 1996a, 1996b). Additional information on the effect of Texas Brown Tide blooms on seagrass meadows can be found in chapter 7. Production by benthic microflora also decreased to nearly nothing because of shading during the 1989–97 brown tide bloom (Blanchard and Montagna 1992).

Declines in abundance and diversity of a variety of organisms, including micro- and mesozooplankton and benthic invertebrates, have also been noted (Buskey and Hyatt 1995; Street et al. 1997; Buskey et al. 1998; Gobler et al. 2013). The primary negative effect of *A. lagunensis* on grazers may have to do with the thick mucus layer that it produces, which may inhibit feeding (Liu and Buskey 2000) or prevent nutrient assimilation once ingested (Buskey and Hyatt 1995; Street et al. 1997; Ward et al. 2000). The Texas Brown Tide alga may be toxic to very young larval fish, but it does not appear to affect adult fish or macroinvertebrates (Buskey and Stockwell 1993; Buskey et al. 1996). To date, no chemical structure (toxin or other compound) has been identified

Red Tide

Two species of dinoflagellates that cause Red Tides, *K. brevis* (fig. 18.3) and *Alexandrium monilita*, are present in Laguna Madre phytoplankton communities (Gunter 1945b; Simmons 1957; Hildebrand and King 1978). Few occurrences of Red Tide caused by *K. brevis* have been documented in the Laguna Madre, but the blooms have occurred elsewhere on the Texas coast (fig. 18.4). Prevailing hypersalinity may restrict *K. brevis* from reaching bloom populations. Optimal salinity for *K. brevis* has been documented as 27–37 ppt, although the organism was able to survive salinities of 46 ppt for five weeks (Aldrich and Wilson 1960). A fish kill in late November 1990 alerted public health officials and marine scientists to a bloom in the Brownsville Ship Channel that lasted until April 1991 (Buskey et al. 1996). In addition, South Bay was closed to shellfish harvesting in March 1991 as a result of that bloom. Red Tide was also documented in the Upper Laguna Madre during September–October 1996, with some fish mortality, primarily Pinfish and Striped Mullet. In addition, a moderate fish kill of Striped Mullet, Atlantic Needlefish, and Hardhead Catfish during October 1996 near Green Island in the Lower Laguna Madre was attributed to Red Tide (Hensley and Fuls 1998). No Red Tides have been documented in the Laguna Madre de Tamaulipas; however, several blooms have occurred in the nearshore waters around La Pesca and may have spread into lagoonal waters (H. H. Hildebrand, pers. comm.). It is likely that blooms of *K. brevis* are transported through passes/ship channels into the Laguna Madre, as has occurred in Corpus Christi Bay when the blooms are seeded by offshore waters.

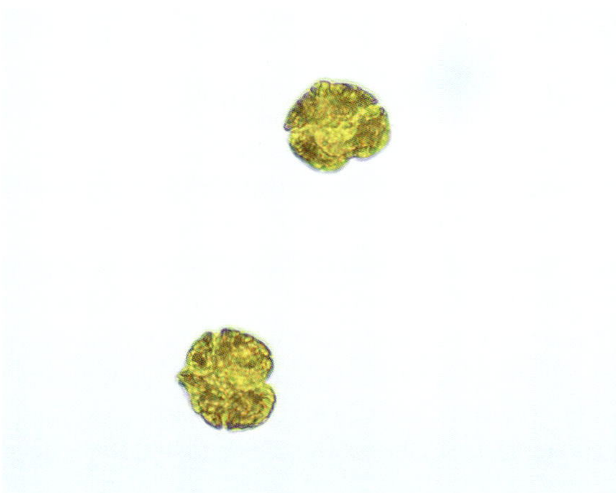

Figure 18.3. *Image of* Karenia brevis *taken at 400× magnification (Photograph courtesy of K. H[illegible])*

Figure 18.4. *Aerial image of a Red Tide bloom near Palacios, Texas, 28 October 201[illegible] (Photograph by Winston Denton, TPWD, https://tpwd.texas.gov/landwater/water/environconcerns/hab/redtide/photos.phtml)*

19 *Expansion of Mangrove Habitat*

C. EDWARD PROFFITT AND DONNA J. DEVLIN

Worldwide expansion of mangroves poleward into temperate regions is one of the ecological changes heralded by climate change (Cavanaugh et al. 2014; Saintilan et al. 2014). Historically, much of the intertidal Texas shoreline was dominated by salt marsh. For hundreds of years, mangroves occurred only as ephemeral, patchy components, sporadically suffering extensive dieback after hard freezes and recolonizing during periods of milder weather (Sherrod and McMillan 1981). Four mangrove species have colonized the shorelines of the northern Gulf of Mexico: Buttonwood, White Mangrove, Red Mangrove, and Black Mangrove (Sherrod and McMillan 1985). All except the White Mangrove have been reported from Texas.

For the last several decades global warming has allowed for considerable expansion of Texas mangrove populations—mainly Black Mangroves (Armitage et al. 2015). Mangroves were not recorded in surveys of the Laguna Madre shoreline along the Padre Island National Seashore (PAIS) between 1999 and 2004. Black Mangroves first appeared near the base of the John F. Kennedy (JFK) Causeway after 2008 (C. Sassine, pers. comm.). Black Mangroves have now colonized the Laguna Madre, and dense stands occur near passes connecting with the Gulf of Mexico. Global warming and reductions of salinity associated with the dredging of passes and the Gulf Intracoastal Waterway (GIWW) may be accelerating the encroachment of mangroves into the Laguna Madre (table 19.1). From the 1950s to 2003–4 mangrove cover expanded at a

Table 19.1. *Estuarine marsh and mangrove land cover in the Upper and Lower Laguna Madre of Texas*

Vegetation Type	Survey Period		
	1950s	1979	2003–4
Upper Laguna Madre			
Mangroves	0	0	85 (210)
Estuarine marsh	1,796 (4,551)	1,364 (3,246)	1,696 (3,874)
Lower Laguna Madre			
Mangroves	12 (30)	70 (173)	93 (230)
Estuarine marsh	584 (1,442)	612 (1,512)	604 (1,492)

Source: Padre Island National Seashore: White et al. 2007. South Padre Island: White et al. 2005.

Note: Land cover is expressed in hectares (acres).

Figure 19.1. *Black Mangroves are expanding their range in the Laguna Madre. This specimen is growing on the backshore of Padre Island at Yarborough Pass.*

rate of 1.7 ha/year (~4 ac/year) in the Lower Laguna Madre, reaching 93 ha (230 ac; White et al. 2006, 2007). In the Upper Laguna Madre, 85 ha (210 ac) of mangroves existed in 2002 (White et al. 2005). Mangrove population expansion continues in the Upper and Lower Laguna Madre and the Corpus Christi Bay complex of bays and estuaries (fig. 19.1).

In the Upper Laguna Madre, north of the JFK Causeway nearshore spoil islands and the south end of Mustang Island have dense stands of short, shrub-like mangroves (fig. 19.2). Farther south, many small tidal cuts on the Laguna side of PAIS now are already populated with dozens of small (but multi-year-old) mangroves, suggesting further expansion and spoil islands that have differing densities of mangroves (Yeager et al. 2010). We observed scattered Black Mangroves as far south as Yarborough Pass during a field trip with PAIS biologist Charles Sassine (table 19.2). In the Lower Laguna Madre, clusters of Red Mangroves have established and appear to be expanding. We recorded several Red Mangroves at Packery Channel adjacent to the northern edge of the Upper Laguna Madre. Buttonwood Mangroves are rare north of the Land-Cut but a fairly robust Buttonwood was found in Mustang Island State Park. In the Lower Laguna Madre, dense patches of mangroves occur along the shoreline from Boca Chica to Horse Island at the Arroyo Colorado. Large reproducing populations of Black Mangroves provide a continual source of propagules to bolster populations in the Upper and Lower Laguna.

All four mangrove species that occupy northern Gulf of Mexico estuarine habitats can be found along shorelines of the Laguna Madre de Tamaulipas (Sherrod and McMillan 1985). On shorelines of the Mexican Laguna these species form extensive

Figure 19.2. Black Mangrove located on an island north of the Kennedy Causeway in the Upper Laguna Madre (Photograph by Brian R. Chapman)

Table 19.2. *Densities and height ranges for Black Mangroves at different areas of the Upper Laguna Madre (ULM) and nearby Oso Bay*

Location	Survey Extent	Height Range (m)	Average Tree Density (no. per 0.1 ha)	Average Density Trees + Saplings (no. per 0.1 ha)	Seedling Density (no. per 0.1 ha)	Reference
ULM, dredged-material island ~11 km (6.8 mi) south of JFK Causeway (TAMUCC field station)	Entire island perimeter	0.5–1.2	0.3	—	3,250	Ziegler 2018
ULM, TAMUCC field station	Entire island perimeter	0.5–1.2	1.8	—	—	Unpub. data, 2013
ULM, 3 tidal inlets from south of Bird Island to Yarborough Pass	8 plots (1.5 m wide × varying distances), area estimated by distance walked × 1 m wide	0.25–1.0	310.3	967.8	—	Unpub. data from airboat survey with C. Sassine, HES 2018
Ward Island West, Oso Bay	*n* = 18, 3 × 3 m plots	1.5–2.7	789.7	—	6,000	Ziegler 2018
Ward Island East, Oso Bay	*n* = 5, 10 × 10 m plots	0.3–1.1	236	—	—	Unpub. data, 2013

mixed- or single-species stands south of the Río Soto la Marina at La Pesca. Between Río Soto la Marina and the northern limit of Laguna Madre de Tamaulipas, a few small patches of Black Mangroves are distributed along the shoreline (Sherrod and McMillan 1985; also see chapter 12). Few Red, White, or Buttonwood Mangroves occur farther north than the Río Soto la Marina in the Mexican Laguna.

The focus of this chapter is the expansion of mangroves in the Laguna Madre and Laguna Madre de Tamaulipas. However, to understand this expansion, it is imperative to consider mangrove characteristics, populations of mangroves in the Laguna Madre, and the possible consequences of regime shift from grassy or herbaceous estuarine habitats to woody mangrove ecosystems.

Mangrove Characteristics

Mangroves usually form dense intertidal forests (Tomlinson 2016). Most mangrove species occur between 30° N to 37° S latitude (Feller et al. 2010). "Mangrove" is a broad term encompassing at least 57 species in 21 families, or about 110 species if taxa from what some describe as "minor elements" and "mangrove associates" are included (Tomlinson 2016). Mangroves have the following characteristics: (1) evolutionary isolation from terrestrial relatives; (2) restriction to the intertidal environment; (3) possession of important morphological adaptations that promote existence in the marine environment (aerial roots, vivipary or cryptovivipary, mechanisms to deal with salt and flooded sediments); and (4) the capability of creating habitat for other plant and animal species (modified from Tomlinson 2016). Most mangroves are facultative halophytes and do not require saline water (Krauss and Ball 2013). They typically occupy shallow waters along shorelines of low wave energy in tropical and subtropical regions worldwide (J. H. Davis 1940; Lugo and Snedaker 1974). These forests develop on sand, mud, silt, clay, or peat substrates, where they form a natural barrier to storm surges and can reduce coastal subsidence and erosion.

Mangroves are highly adapted species that can withstand stressful conditions such as high salt concentrations and low dissolved oxygen. Black Mangroves are more salt tolerant than Red Mangroves (Tomlinson 2016) and are better adapted to the high salinities prevalent in the Laguna Madre. Black Mangroves take up saline water and extrude salt from specialized glands on leaves. Red Mangroves physiologically exclude most salt from uptake by the roots. Both species also sequester some salt in the epidermis and bark of stems and roots (Sanger 2002).

Mangrove roots are adapted to function in flooded, low-oxygen soils that promote the production of sulfides, which can adversely affect plants (Nickerson and Thibodeau 1985). Mangroves oxygenate the soil and reduce soil sulfide concentrations by absorbing air through lenticels (specialized pores) on the stems and aerial roots and circulating the air through aerenchyma (spongy tissue with large air spaces) into the belowground roots, which leak air into the soil and ameliorate stress on the plant (McKee et al. 1988). Prop roots extending to the soil from the trunk and branches of Red Mangroves and the long, shallow cable roots of Black Mangroves provide stability in soft, muddy soils (Sanger 2002; Tomlinson 2016).

Most mangroves are evergreen, even near their range limits (Sanger 2002). Their leaves possess several adaptations that allow survival in salty environments. Some mangroves adjust leaf angle to lower leaf temperatures and reduce transpiration in hot, dry (salty) conditions (Ball et al. 1988). The elastic properties of leaf cell walls in Black (but not Red) Mangroves (Suárez et al. 1998), small stomatal pore openings, and high leaf solute concentrations also reduce evaporative water loss in an environment where fresh water is limited (Sanger 2002).

High salinities curtail photosynthetic rates, leaf production rates, and leaf longevity (Suárez and Medina 2005). Maximum photosynthesis of Black Mangroves occurs in fresh water (0 psu [practical salinity units]); increasing salinities to 10 and 40 psu caused 21% and 37% reductions, respectively (Sobrado 1999). High porewater salinity increases leaf osmolality (an indication of reduced supply or use of water), and the size of leaves decreases, likely because of water-use stress (Medina and Francisco 1997). Soil composition can ameliorate the effects of hypersaline conditions. At high porewater salinities, increasing the clay content of sandy soil enhances growth and leaf color (McMillian 1975b).

Reproductive system and modes of dispersal are important defining characteristics of mangroves. The flowers of Black and White Mangroves are insect pollinated (fig. 19.3), whereas Red Mangroves are primarily self-pollinating, although some outcrossing occurs, likely by wind pollination (Tomlinson 2016; Proffitt and Travis 2014). In many mangrove species there is no true seed; the embryo breaks through the seed coat while on the tree (Tomlinson 2016). Moreover, propagules cannot withstand extended periods of dry or cold conditions; thus, there is no persistent seed bank in the soil. Red Mangrove embryos break through the seed coat and fruit wall (viviparity) and are nourished by the maternal tree for months before dropping. Mature Red Mangrove propagules are large, averaging 25 cm (9.8 in) long and contain abundant starch reserves that enable seedlings to remain viable while floating for about a year (Rabinowitz 1978). Black Mangroves exhibit "cryptovivipary," meaning that the embryo typically breaks through the seed coat but not the fruit wall while on the maternal tree (Tomlinson 2016). Black Mangrove propagules dispersing in Corpus Christi Bay and into the Laguna Madre average 2.4 cm (±0.4 cm SD, or just under 1.0 in) in length (authors' unpub. data). Although much smaller than Red Mangrove propagules, Black Mangrove propagules contain a substantial amount of

Figure 19.3. *Many species of insects, including this snout butterfly and ladybug, transfer mangrove pollen as they wander from bush to bush to feed.*

Figure 19.4. *Even small copses of Black Mangrove trees can produce hundreds of seeds.*

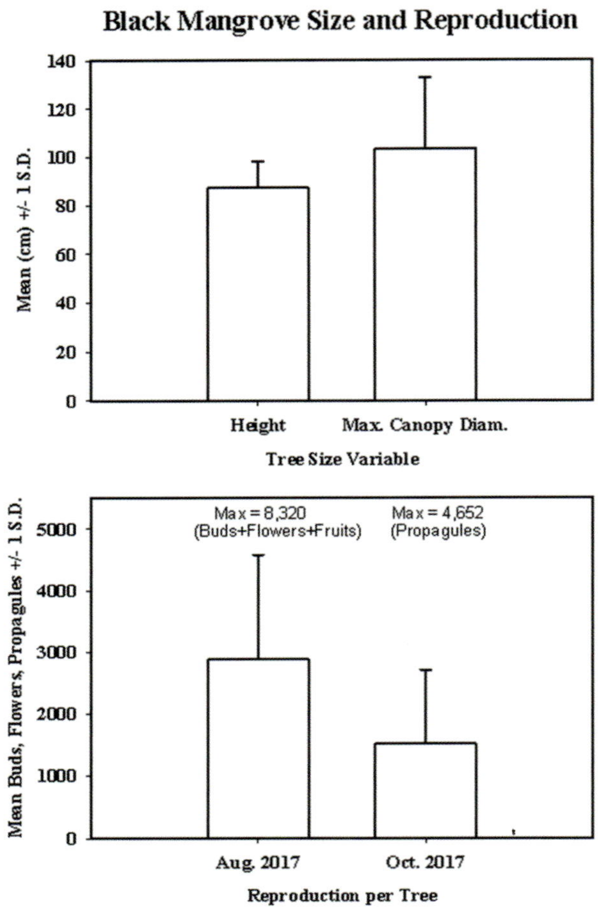

Figure 19.5. *Seeds of Black Mangrove float and are transported by wind and water currents to shorelines, where they sometimes accumulate in large numbers.*

stored reserves provided by the maternal tree and remain viable while floating for up to 110 days (Alleman and Hester 2011). The smaller White Mangrove propagules are viable for only 35 days (Rabinowitz 1978), which may be one reason they have not yet colonized Texas shores. The small fruits (about 1.0 to 1.5 cm [0.3–0.6 in]) of Buttonwood Mangrove float and disperse seeds via water currents, although seed longevity is not well-known (Tomlinson 2016). Mangrove propagules can be dispersed relatively long distances by ocean currents (Rabinowitz 1978; Clarke et al. 2001; Kennedy et al. 2016), which fosters colonization of distant shores. However, most recruitment is via propagules produced from nearby plants (Sengupta et al. 2005). In some areas of the Laguna Madre stranded propagules can be found in high densities along drift lines (figs. 19.4 and 19.5).

Few of the many propagules produced by mangroves survive to the rooted seedling stage. Physical stress, especially that caused by low precipitation levels coupled with very hot, dry conditions in the summer, results in high mortality (McKee 1995). Biological interactions also affect survival of propagules and rooted seedlings. Sources of mortality include herbivory, parasitism (Feller 1995; Devlin 2004), and burial by burrowing fiddler crabs (D. J. Devlin, unpub. data). Conversely, certain estuarine species facilitate mangroves by trapping propagules and thus enhance the survival of recruits (fig. 19.6; McKee, Rooth, et al. 2007).

Figure 19.6. *These Black Mangrove seeds became entangled in a mat of Shoregrass during a high tide. Some may eventually germinate.*

The few dispersing propagules that do establish can found rapidly expanding population nodes, because mangroves produce offspring within one to three years of age. For example, on the west side of Ward Island in the "Blind Oso," tidal flat was mostly devoid of estuarine or mangrove vegetation in 1998 (Withers and Tunnell 1998). After construction of a bridge, water exchange with Corpus Christi Bay increased and salt marsh vegetation began to colonize the tidal flats (K. Withers, pers. comm.). Therefore, essentially all mangroves recruited in the last 20 years. In 2018, we measured the heights and population density of Black Mangrove trees in plots in this area. Relatively tall mangroves (1.5 to >2.7 m [4.6 to >8.0 ft]) occupied the densest plot (authors' unpub. data). This mangrove population was highly reproductive—mangroves with heights between 0.4 and 2.0 m (1–6 ft) produced an average of 268 (324.8 SD) propagules per tree with a range of 0 to 1,520 propagules per tree (numbers varied in proportion to tree size). In another population on south Mustang Island, the average number of propagules per tree was 208 (294.5 SD), with a range of 14 to 602 (authors' unpub. data).

Tree size, architectural form, and distribution vary with local conditions in the Laguna Madre and other South Texas estuaries. In the Laguna Madre, most Black Mangroves would be described as "shrubs" rather than "trees." At Boca Chica near the south end of Laguna Madre, the mangroves closest to the shore are located on a small ridge and are of the shrub form, comparable in height with nearby cacti. However, a few meters back, several Black Mangroves reach 4–5 m (13–16 ft) in height (fig. 19.7). Black Mangroves also form dense stands at the Port Isabel bridge and in South Bay (authors' pers. observ.). Many islands at the Laguna Atascosa National Wildlife Refuge have expanding populations of Black Mangroves (Sherrod and McMillan 1981, 1985; Lonard and Judd 1985; Everitt et al. 2007; Everitt et al. 2009; Everitt et al. 2010) that have been increasing in height and canopy diameter in recent years.

Figure 19.7. *The height and density of the Black Mangroves in the Lower Laguna Madre are evident when compared to the woman standing behind the thicket.*

A small population of large (estimated at about 5 m [16 ft] tall) Red Mangroves exists near the mouth of the Rio Grande (D. J. Devlin, pers. observ.) and at Packery Channel bordering the Laguna Madre (see fig. 19.1). Propagules of Red Mangroves often wash ashore on North Padre Island shorelines in viable condition (authors' unpub. data), but only a few have established in the Laguna Madre and Corpus Christi Bay (Sherrod et al. 1986). Red Mangroves have been documented as far north in Texas as Port O'Connor (Richard 2016).

Ecosystem Functions and Services of Mangroves

Mangroves are ecological engineers because they build physical structure above water (stems and leaves), in the water (aerial roots), and belowground (subsurface roots). They are also foundation species because this structure provides refuge, foraging, reproductive, and nursery habitats for many marine (fish, crabs, shrimp, gastropods, etc.) and terrestrial (birds, insects, etc.) fauna (Bruno and Bertness 2001). Mangroves stabilize coastlines and reduce erosion, thereby enhancing economies in many regions of the world (Barbier 2013; Friess 2016).

The high levels of primary productivity in mangrove forests (Clark et al. 2001; Ewe et al. 2006) drive extensive food webs (Cannicci et al. 2008), including fisheries (Turner 1977), and further enhance biodiversity. Mangroves sequester more belowground carbon, mainly from root production, than most other terrestrial forests (Bouillon et al. 2008; Xiong et al. 2018; Kauffman et al. 2020). Wetland productivity can affect relative sea-level rise through biogenic accretion of sediment (Parkinson et al. 1994; McKee, Cahoon, et al. 2007).

The accumulation of peat (from cycles of root growth, death, and decomposition) and trapping of sediments carried by surface waters allow some mangrove forests to increase elevation at a rate that offsets the effects of sea-level rise (Ellison 1993; McKee, Cahoon, et al. 2007). In the rapidly subsiding Mississippi Delta, both salt marsh vegetation and mangroves build soil surfaces vertically through accumulation of inorganic sediment and organic matter at approximately the same rate, although neither keeps up with sea-level rise (McKee and Vervaeke 2018). In the Laguna Madre, the current rate of relative sea-level rise (3.4 mm/year [0.14 in/year]) exceeds historical sedimentation rates for most of the tidal flats (White et al. 2002). As sea-level rise continues, the wind-tidal flats and other shoreline habitats will become more frequently flooded and may eventually be permanently submerged (Morton and Holmes 2009). The effect of colonizing mangroves on wetland elevation in the Laguna Madre is poorly understood. Currently, Sediment Elevation Table methods are being used to assess subsidence, erosion, and accretion in the Upper and Lower Laguna Madre.

As mangroves expand into other shoreline habitats, the associated ecosystem functions and services of existing intertidal habitats will be altered (Coldren et al. 2016; Perry and Mendelssohn 2009). Consequently, ecologists are attempting to determine how mangrove expansion into salt marshes will affect ecosystem processes (Perry and Mendelssohn 2009; Comeaux et al. 2012).

Factors Affecting Mangrove Distribution in Texas

The distribution and abundance of mangrove forests are influenced by a combination of factors, including climatic conditions (Gabler et al. 2017), salinity, nutrients, interactions with marsh plants (McKee, Rooth, et al. 2007; Coldren and Proffitt 2017), and genetic selection (Sherrod and McMillan 1985). Warmer temperatures and decreases in the frequency and intensity of winter freeze events in the future may create conditions favorable to expansion of mangrove forests (Osland et al. 2013).

The extreme freeze of 1983 killed many Red and Black Mangroves throughout Texas; in Brownsville the freeze occurred for 54 hours, reaching temperatures of -6.1 °C (21 °F) (Sherrod and McMillan 1985; Sherrod et al. 1986). Black Mangrove mortality along the coast was 95% on Galveston Island, 85% at Harbor Island, and 60%–70% in Corpus Christi (Sherrod and McMillan 1985). Two groups of planted Red Mangroves in the Lower Laguna Madre also succumbed during the 1983 freeze (Sherrod et al. 1986). A hard freeze in 1989 may have had additional negative effects.

A freeze with a low temperature of -3.5 °C (25.7 °F) in January 2018 did not result in loss of many mangroves in the Laguna Madre or Corpus Christi Bay. Farther north, lower temperatures produced greater negative effects, although some trees on Galveston Island recovered via epicormic shoots (C. Weaver and C. Proffitt, unpub. data).

Recruiting Black Mangrove seedlings are more sensitive to temperature than established seedlings and adult trees. Rooting by seedlings was inhibited below 5 °C (59 °F) and above 37 °C (98.6 °F) (McMillan 1971, 1975a). All 45 propagules collected from the ground at Galveston Island after the 2018 freeze perished prior to rooting (authors' unpub. data).

Mangroves in the Laguna Madre also contend with stress from limited precipitation and elevated salinities. Red Mangroves may be less likely to colonize much of the Laguna Madre because hypersaline conditions and low rainfall may prevent seedling establishment (Lin and Sternberg 1992). Hypersaline conditions reduce survival and stunt growth of Black Mangroves, especially when coupled with high summer heat and drought (Lugo et al. 1981; Naidoo 1987). These conditions result in the shrub stature of mangroves in South Texas, while at this latitude in Florida, trees can attain heights greater than 9 m (~30 ft). Based on an inverse relationship between mangrove height and soil salinities in arid southern Puerto Rico, Cintron et al. (1978) suggested that mangroves may not establish a stand where salinities >90 psu persist. McMillan (1971) found that Black Mangroves root in salinities ranging from 0 to 65 psu, but rooting is reduced to 10% at 75 psu. It is possible that most recruitment in the Laguna Madre may occur during wetter-than-normal (higher rainfall and runoff, lower salinities) years. On the shoreline of a dredged-material island in the Laguna Madre, established adult Black Mangrove shrubs did not display signs of stress, such as leaf loss or chlorosis when porewater salinity was >100 psu in the mangrove root zone and 40 psu in nearby surface water, but planted seedlings from both Ward Island and the Lower Laguna Madre (Boca Chica) failed to recruit (Ziegler 2018). Alternatively, there may be selection for seedling genotypes that are more salt tolerant than the average Black Mangrove (McMillan 1975b).

Survival of the Black Mangrove also varies with soil type under hypersaline conditions. In the closely related Grey Mangrove, hypersalinity induced a loss of the potassium required for photosynthesis from soil via exchange of sodium and potassium ions—loss of potassium was inhibited with increasing clay content (Ball 1987). This mechanism may explain the results of McMillan's (1975b) experiment in which there was high mortality of Black Mangrove seedlings planted in sand, but 100% survival when 10% clay was added.

Burial by wrack deposited by high tides and storms can have a negative effect on salt marsh plants (Valiela and Rietsma 1995). In contrast, decomposing wrack can provide an energy supplement to intertidal vegetation (Orr et al. 2005). In some areas of the Upper Laguna Madre thick deposits of seagrass wrack cover mangrove propagules and protect them from desiccation (fig. 19.8). Those that emerge through the wrack have a chance for survival (Ziegler 2018). Seagrass wrack also provides refuge and forage

Figure 19.8. *Seagrass wrack lines along the Laguna Madre shorelines trap Black Mangrove seeds. Many trapped atop the wrack can survive and germinate, but seedlings developing from seeds buried beneath the wrack usually do not survive.*

for Coffee Bean Snails in dense mangrove stands along the barrier island shoreline of Laguna Atascosa National Wildlife Refuge in the Lower Laguna Madre (authors' field observ.).

Increasing nutrient concentrations cause shifts in competitive dominance among marsh plants (Levine et al. 1998), which may impact marsh-mangrove interactions. The expansion of Black Mangrove populations in the Laguna Madre and Corpus Christi Bay region will likely be enhanced in areas of high nutrients. At the mangrove–salt marsh ecotone in Texas and Florida, nutrient enrichment increases the rate of mangrove encroachment into salt marsh (Weaver and Armitage 2018; Dangremond et al. 2019). Moreover, initial results from a fertilization experiment on Ward Island indicate that after two years, Black Mangroves respond to nitrogen addition with increased growth in canopy diameter and reproductive output (authors' unpub. data). Elevated nutrients can also increase the damage from storms (Feller et al. 2015).

Stress and disturbance caused by storms are projected to increase with climate change. Hurricanes and tropical storms adversely affect salt marsh and mangrove communities by breaking or defoliating vegetation, causing erosion or deposition of sediment and extended flooding (Craighead and Gilbert 1962; Cahoon et al. 2003). Alternatively, moderate levels of sediment deposition can provide a positive effect (McKee and Cherry 2009). At Harbor Island winds and waves from Hurricane Harvey in 2017 caused minor damage to mature mangroves near the water's edge (C. Weaver, pers. comm.). However, there were few adverse effects to mature trees in the relatively well-protected habitat at Ward Island, and the five days of continuous flooding following Harvey did not cause mortality of six-month-old seedlings (D. J. Devlin, unpub. data).

Salt Marsh Communities

Salt marshes are coastal wetlands that are periodically flooded and drained by wind-driven or lunar tides. Salt marsh ecosystems support many commercial and recreational species, such as Red Drum, Blue Crab, and Brown Shrimp (Turner 1977), and provide feeding and roosting habitats for resident and migratory shorebirds and wading birds (Nagelkerken et al. 2008; also see chapters 14 and 15). Salt marshes also perform other ecological services: improvement of water quality, reduction of storm damage and erosion, and provision of nutrients to the marine environment. The assemblage of avian species attracted to salt marshes also encourages a vigorous tourism industry (Boesch and Turner 1984; Moulton et al. 1997; Barbier 2013).

In the high-precipitation environments north of Corpus Christi Bay, many salt marshes are dominated by graminoids with patchily distributed woody shrubs, such as Marsh-Elder and Groundsel Bush. The grass family (Poaceae) is well represented in salt and brackish marshes by Smooth Cordgrass, Saltmeadow Cordgrass, Gulf Cordgrass, Saltgrass, Shoregrass, Knotgrass, Switchgrass, Seashore Dropseed, and many other less abundant species. Most can reproduce clonally as well as by wind-pollinated seeds.

In the semiarid Laguna Madre south of Corpus Christi Bay, grass-dominated salt marshes are rare and barren wind-tidal flats are common (see chapter 9). Saltwort, a succulent forb, usually forms a dense vegetative band closest to the water's edge, and slightly farther inland, mixed groupings of succulent plants (Sea Purslane, Saltwort, and glasswort) dominate (Onuf 2006). In infrequently flooded areas, Seablite, Sea Lavender, and Carolina Wolfberry, often mixed with Saltgrass, are common inhabitants. Groundsel Bush, Sea Oxeye, Marsh-Elder, and Seacoast Goldenrod frequently occupy the highest elevations.

Replacement of Salt Marsh by Mangroves

The outcomes of interactions among salt marsh and mangrove species are complex and vary with life stage. Survival and growth of seedling mangroves are affected by marsh species. Saltwort facilitated survival of Black Mangrove seedlings in southern peninsular Florida (Milbrandt and Tinsley 2006). However, in a greenhouse study conducted in temperatures of -5 °C (23 °F), mixtures of Saltgrass and Sea Purslane competed with Black Mangrove seedlings, resulting in high mortality of the seedlings (Coldren and Proffitt 2017). Similarly, under stressful conditions near the edge of the mangrove latitudinal range in Louisiana where temperatures are more often at or below -5 °C, Smooth Cordgrass reduced the survival and growth of seedling Black Mangroves (Patterson et al. 1993). Thus, marsh species compete with, rather than facilitate, mangrove seedlings under intense cold stress, counter to the predictions of the stress-gradient hypothesis (SGH).

In the Laguna Madre and throughout Texas, as mangroves grow from seedlings to saplings, the effects of salt marsh vegetation on mangroves shifts from competitive to neutral (Guo et al. 2013). As mangrove stands mature further and become densely populated, increased shading can have severe consequences for marsh plants. As mangrove canopy closed at Ward Island, marsh plants were eliminated (authors' unpub. data). Shading is likely one major mechanism by which mangroves replace salt marsh, but belowground root competition and allelopathy cannot be ruled out.

Over decades, most mangrove stands show reduced densities due to inter- and intraspecific competition among trees (Proffitt and Devlin 2005b). Mangrove forests in the Laguna Madre do not show signs of thinning, and there are no standing dead trees. Likewise, at Ward Island, even high-density (3,000 trees/0.1 ha [0.25 ac]; up to 2.74 m [9 ft] tall) plots do not show evidence of thinning via competition (authors' unpub. data).

Broad spatial-scale studies do not always show patterns consistent with more direct studies of interactions. Mapping via Landsat indicated that little of the 24% marsh loss was associated with the 74% increase of Black Mangrove habitat in Texas between 1990 and 2010 (Armitage et al. 2015). Rather, most loss appeared to result from sea-level rise. Further studies of interactions and spatial mapping will be needed to adequately assess the effects of mangrove expansion and sea-level rise.

Effects on the Faunal Community and Ecosystem Functions

Although salt marshes and mangrove forests are sometimes considered ecological equivalents, they clearly are not. Kelleway et al. (2017) theorize that encroachment of mangroves into salt marsh will increase nutrient reservoirs and protection from storm surge but adversely impact some migratory species and fauna with special dietary requirements. In the Lower Laguna Madre, the Seaside Sparrow persists where Black Mangrove has replaced its usual salt marsh habitat; however, the population of a possible competitor, the Mangrove Warbler, is expanding in association with Black Mangrove (D. Newstead, pers. comm.). There is concern that the endangered Whooping Crane will decline if mangroves are allowed to encroach into their preferred salt marsh habitat at the Aransas National Wildlife Refuge. Colonization by mangrove endemics such as the Mangrove Buckeye butterfly and Mangrove Tree Crab may cause other changes to marsh faunal assemblages. Abundance of canopy insects is greater, but diversity is lower in cordgrass habitats than in the mangrove-dominated habitats of South Texas (Loveless and Smee 2019). Similarly, shrimp abundance is significantly higher in monospecific cordgrass marshes, while Blue Crabs and fish are more abundant when mangroves are present (Smee et al. 2017).

Populations of many invertebrate species that currently inhabit Laguna Madre salt marsh and salt marsh–mangrove habitats, such as the Thinstripe Hermit Crab, Laguna Madre Fiddler Crab, Plicate Horn Snail, Coffee Bean Snail, and Marsh Periwinkle, may be altered as mangroves become dominant. Many are common prey items of fish and wading birds and as such link salt marsh–mangrove systems to estuarine and terrestrial systems. Horn snails are an important prey item for the Texas Diamond-backed Terrapin (Alleman 2015). Wading birds forage on the gastropods, crabs, and infauna beneath the mangroves, and songbirds forage for insects in the canopy. Coyotes hunt for fiddler crabs along the shoreline at dawn and dusk.

Many invertebrates common in the Laguna Madre play important roles in sediment reworking, organic matter decomposition, or recycling. Deep fiddler crab burrows increase oxygenation of the sediment and augment growth of salt marsh and mangrove vegetation (Smith et al. 2009; fig. 19.9). However, bioturbation of the sediment by fiddler crabs often produces mounds and holes that bury and reduce survival of salt marsh seeds by 79%–94% (Smith and Tyrrell 2012). The survival of Black Mangrove propagules in Oso Bay was reduced by 40% as the result of fiddler crab activities (D. J. Devlin, unpub. data). High fiddler crab burrow densities also reduce belowground production and accelerate the decomposition of Smooth Cordgrass, resulting in a net loss of 3 g C m^{-2} yr^{-1} (Thomas and Blum 2010). The Coffee Bean Snail consumes and digests the cellulose of mangrove leaves and in high densities increases leaf decomposition rates by orders of magnitude (Proffitt and Devlin 2005a). These and other taxa, along with the microbial community, strongly influence ecological services and functions such as the production and storage of blue carbon. Effects of an ecological regime shift (Guo et al. 2017) from marsh to mangrove dominance on these species is unknown, but the consequences need to be more fully understood.

Figure 19.9. *Fiddler Crabs burrow deeply into the substrate, as demonstrated by the exposed cast of a burrow, and provide oxygen to the sediment. The aeration enhances survival of mangroves and other salt marsh vegetation.*

20 The Laguna Madre

A Conservation Framework

JORGE BRENNER AND JAMES F. BERGAN

Bounded by the Mississippi Delta in Louisiana and the Río Soto la Marina at the southern terminus of the Laguna Madre de Tamaulipas in Mexico is the Gulf Coast Prairies and Marshes (GCPM) ecoregion (Britton and Morton 1989). Parts of the ecoregion lie in two states within two countries, and the whole of it encompasses 22 primary bays, 19 major rivers, and nearly 1,600 km (1,000 mi) of shoreline.

Ecoregional Context

Estimates of total surface area of the GCPM ecoregion range upward of 6.4 million ha (15.8 million ac) (Britton and Morton 1989). As biologically vast and complex as this ecoregion is, the threats to the ecological processes that drive the productivity of the region are equal to or greater in complexity. Humans have always been drawn to the Gulf of Mexico. Nomadic native peoples took advantage of the bounty of food resources—oysters, shrimp, fish, alligators, birds—available in nearshore waters and on the coastal prairies (Ricklis 1996; La Vere 2004; Owen 2003). Today, the attraction is fueled by energy production, industrial development and distribution, business infrastructure, agricultural activities, and the appeal of a coastal lifestyle with associated recreational and aesthetic attributes. Although the Deepwater Horizon oil spill in April 2010 was considered the largest global environmental disaster ever, the event and subsequent restoration work created national awareness of the pollution and other impacts caused by large industrial spills. The spill also generated a multitude of recommendations for industry business models and the governmental role in environmental protection (National Commission on the BP Deepwater Horizon Oil Spill and Offshore Drilling 2011).

The Nature Conservancy (TNC; the Conservancy) recognizes the complexity of the GCPM ecoregion not only in a biological context but also in a socioeconomic setting. Just as there are unique species of animals, plants, and plant communities within the region, so too are there unique population, economic, cultural, and social attributes. To meet TNC's mission, which is to conserve the lands and waters on which all life depends, the Conservancy must envision a GCPM ecoregion where the biotic diversity thrives and people act to conserve nature for its own sake and for its ability to fulfill their needs and enrich their lives. All site-based actions must be dictated by the best available conservation science information. Where data are lacking or inconclusive, the Conservancy must be proactive in filling information gaps to focus our actions better. If the Conservancy is to be successful within the GCPM ecoregion, the TNC must facilitate the means by which humans can live productively and sustainably while conserving biological diversity. This is especially pertinent in the Laguna Madre portion of the GCPM ecoregion within both Texas and Tamaulipas.

Historically the Laguna Madre region was sparsely populated; however, in recent years it has experienced rapid and pervasive development—Cameron County alone grew 26% between 1997 and 2016 (US Census Bureau 2020a). Conversely, the human safety crisis (the drug cartel dominance in that region and associated risks) that has developed in Tamaulipas during the past decade has limited access to some habitats and resources along the Laguna Madre, thus promoting de facto conservation of the area (Alfonso Banda and Leonardo Arellano, pers. comm., 2018).

Before European settlement, the GCPM ecoregion was characterized by a mosaic of tallgrass coastal prairie, riparian bottomland hardwood forests, ephemeral freshwater wetlands, canebrake swamps, extensive coastal forests, chenier woodlands, freshwater tidal wetlands, brush mottes and corridors, barrier islands, estuaries, saltwater marshes, hypersaline lagoons, *lomas* (in South Texas *lomas* typically refers to clay dunes although it translates from Spanish as "hills"), and associated Tamaulipan thornscrub habitats. This integrated matrix of habitat types combined to form one of the most productive ecosystems in the world (Briggs 1974; Smeins et al. 1991). The ecoregion has been transformed dramatically since the early 1900s. Freshwater wetlands have been reduced by 50% (Moulton et al. 1997), coastal forests have been cleared and fragmented (M. Lange 1997), the chenier woodlands of the upper Texas coast are essentially gone (Gosselink et al. 1979), and less than 1% of the tallgrass coastal prairie remains intact (Smeins et al. 1991; TNC 2002). The remaining fragments of most habitat types are generally small and degraded by exotic plants, disrupted hydrology, overgrazing, channelization, or other disruptive factors (see chapter 22 for more information on threats to the Laguna Madre).

Large landholdings elsewhere in Texas are also becoming less common due to development pressures and the consequences of inheritance taxes. However, in the Laguna Madre portion of the GCPM ecoregion, large-scale landownership is still the norm. Many of the development pressures that have impacted other portions of the ecoregion have not occurred along the Laguna Madre. This, along with a history of sound private-land stewardship, provides TNC with a unique opportunity to develop well-leveraged conservation actions in this portion of the ecoregion. In other words, instead of being reactive to a rapid decline in biodiversity, the Conservancy can be proactive.

Nonetheless, in conservation, the only constant is change. The Laguna Madre was established as a high conservation priority for the Conservancy in 1998, with mounting recognition that it is an internationally important resource for shorebirds and sea turtles and an overall hotspot for biodiversity. The Laguna Atascosa National Wildlife Refuge was established in 1946 to protect habitat for migratory songbirds and shorebirds and wintering waterfowl. At present, the refuge hosts the largest population of endangered Ocelots in the United States. Along the central coast of the lagoon, Padre Island National Seashore, established in 1962, protects what is generally considered the longest undeveloped barrier island in the world. This national park contributes significantly to the conservation and recovery of the endangered Kemp's Ridley Sea Turtle. In 1979, Boca Chica and associated inland areas at the southernmost tip of Texas where the Rio Grande meets the Gulf of Mexico was declared the Lower Rio Grande Valley National Wildlife Refuge to protect wildlife corridors and unique plant and animal species. In 2004, the Laguna Madre de Tamaulipas was designated a RAMSAR site (wetlands considered to be of international importance under the RAMSAR Convention). In 2005, the lagoon and Río Bravo Delta areas in Mexico were protected as a Flora and Fauna Protection Area by the National Commission of Natural Protected Areas in Mexico (CONANP 2015). A year later, the protected area was also designated a Biosphere Reserve by the Man and Biosphere Programme of UNESCO.

The TNC's counterpart in Mexico, Pronatura Noreste (a conservation nongovernmental organization), has been leading conservation planning efforts for the Laguna Madre de Tamaulipas. Pronatura Noreste has adopted and adapted the Conservancy's ecoregional planning methodology and in 2014 signed a collaboration agreement with CONANP to support the conservation of the Laguna Madre de Tamaulipas. It is presently working in parallel with our efforts in Texas, prioritizing freshwater wetlands and natural islands with a special focus on migratory birds and working together with local fishing communities in developing sustainable conservation strategies (see chapter 12 for more information on conservation activities in the Laguna Madre de Tamaulipas). By being proactive and working in partnership with other nongovernmental organizations, private landowners, and other entities, the Conservancy hopes to develop long-lasting and sustainable conservation strategies that will benefit the binational resources of the Laguna Madre and its residents and visitors alike.

The Nature Conservancy Approach to Conservation

Every journey starts with a single step. We thought our first step must be as big as possible. Thus, early in our conservation planning for the Laguna Madre, we realized that the published ecological information about the Laguna Madre of Texas and Tamaulipas was disjointed, and many vital portions of the literature were not readily accessible. Substantial data were embedded within unpublished reports and in-house documents. We considered it imperative to develop a "one-stop shopping source" for biological information, not only to assist our own conservation planning efforts but also to support the efforts of all private landowners, researchers, federal and state resource managers, and all those with a deep interest in the Laguna Madre. This revised edition of the original book (Tunnell and Judd 2002) attempts to describe the state of knowledge regarding the Laguna Madre. We hope that the conservation strategies of the Conservancy and other institutions interested in conserving the Laguna Madre will be based, in large part, on the contents of this monumental and revised work.

To successfully accomplish its mission to conserve the lands and waters on which all life depends, in 2001 the Conservancy launched an ambitious and innovative approach. It focused on identifying the best examples of native species and plant communities to conserve. The effort resulted in the publication of the Gulf Coast Prairies and Marshes Ecoregional Conservation Plan (TNC 2002). However, prior to the completion of the ecoregional plan, the Conservancy developed the Conservation Plan for the Texas Portion of the Laguna Madre (also called the Conservation Action Plan, or CAP; TNC 2001). In essence, for biodiversity conservation to be successful, the TNC centers its efforts on larger scales and along ecological boundaries instead of geopolitical lines. This approach, termed "Conservation by Design," emphasizes the goal of securing the long-term survival of all viable native species and community types through the design and conservation of site portfolios within ecoregions (Groves et al. 2000). This goal is achieved through a focused process of conservation target delineation and identification on the landscape (TNC 2000). Each site in a portfolio is an area within which the TNC, its partners, or anyone interested can work at sustainable levels to conserve, or where necessary restore, the structure and ecological processes that maintain the conservation elements for which that site was selected.

The Gulf Coast Prairies and Marshes ecoregion is one of 63 ecoregions the Conservancy has delineated in the continental United States. Boundaries are based primarily on patterns and distribution of species, communities, and associated biotic (plants and animals) and abiotic factors, such as soil type and climate (R. Bailey 1998). While several sites were selected for the ecoregional portfolio (TNC 2002), the Laguna Madre megasite was formally designated as a Phase 1 Ecoregional Portfolio Site in 1996 (TNC 1996).

How the Conservancy carries out ecoregional planning is charted in *Designing a Geography of Hope: A Practitioner's Handbook for Ecoregional Conservation Planning* (Groves et al. 2000). Essentially, the process involves 13 steps, from mapping the ecoregions to taking conservation action (fig. 20.1). At the outset, conservation targets (species and plant communities) are selected, and site-specific ecological processes critical to the long-term viability of those conservation targets are evaluated. The next step involves formulating an assessment of conservation target vulnerability (threats, stresses, and their sources) and then evaluating the leverage and opportunity for conservation success, based in part on location on the landscape, genetic diversity, and flow in relation to existing conservation areas possessing identified targets. Of utmost importance is the need to recognize that biodiversity exists at many levels of ecological scale, such as species, communities, ecosystems, and landscapes. Geographic scale can vary from several square meters to millions of hectares (Poiani and Richter 1999).

Historically, the Conservancy usually worked at local scales (<800 ha [1,977 ac]) with a few exceptions. Such sites have included small-patch communities and ecological systems such as bogs or individual wetland basins. Local-scale species (plants, invertebrates, desert springs fish species) generally have limited ranges. However, in recent years the Conservancy has been working at much larger scales, from intermediate to regional (also called "whole systems"). Intermediate-scale species rely on large-patch communities and ecological systems, such as a coastal salt marsh, that are discrete and easily defined. Intermediate-scale species could include wintering Redheads or Whooping Cranes. Depending on the specific ecological system and regional context, intermediate-scale biodiversity occurs at 400 to 20,000 ha (988 to 49,421 ac). Coarse-scale sites generally range from 8,000 to 400,000 ha (19,768 to 988,422 ac) and consist of terrestrial matrix communities. Examples include Longleaf Pine forests in the southeastern United States and the sky island forest matrix found in mountain ranges within the Chihuahuan Desert ecoregion. The highest order of ecological scale is defined as regional, with 400,000 ha the minimum area required for sustainability for regional-scale species. Migratory animals and high-end predators (bats, fishes, bears) are examples of regional-scale or whole-system animal groups or species (Poiani and Richter 1999).

The Laguna Madre megasite is defined as a functional landscape. This means that multiple ecological systems (barrier island, coastal sand plains, hypersaline lagoon), multiple communities (Tamaulipan thornscrub, freshwater wetlands), and multiple species (Ocelot, Reddish Egret) occur at all scales below regional—that is, coarse, intermediate, and local. The total area of the Laguna Madre region far exceeds 400,000 ha, and indeed, the Conservancy's focus is on the various scales at which conservation strategies will be applied. Functional landscapes possess a high degree of intactness and maintain (or can have restored) most or all of their key ecological patterns, processes, and components. Conservation targets selected within functional landscapes often represent other biodiversity at various scales. For example, selection of a shorebird guild as a conservation target would also capture ecological systems and communities such as wind-tidal flats, beach habitats, washover passes, and some freshwater wetland systems.

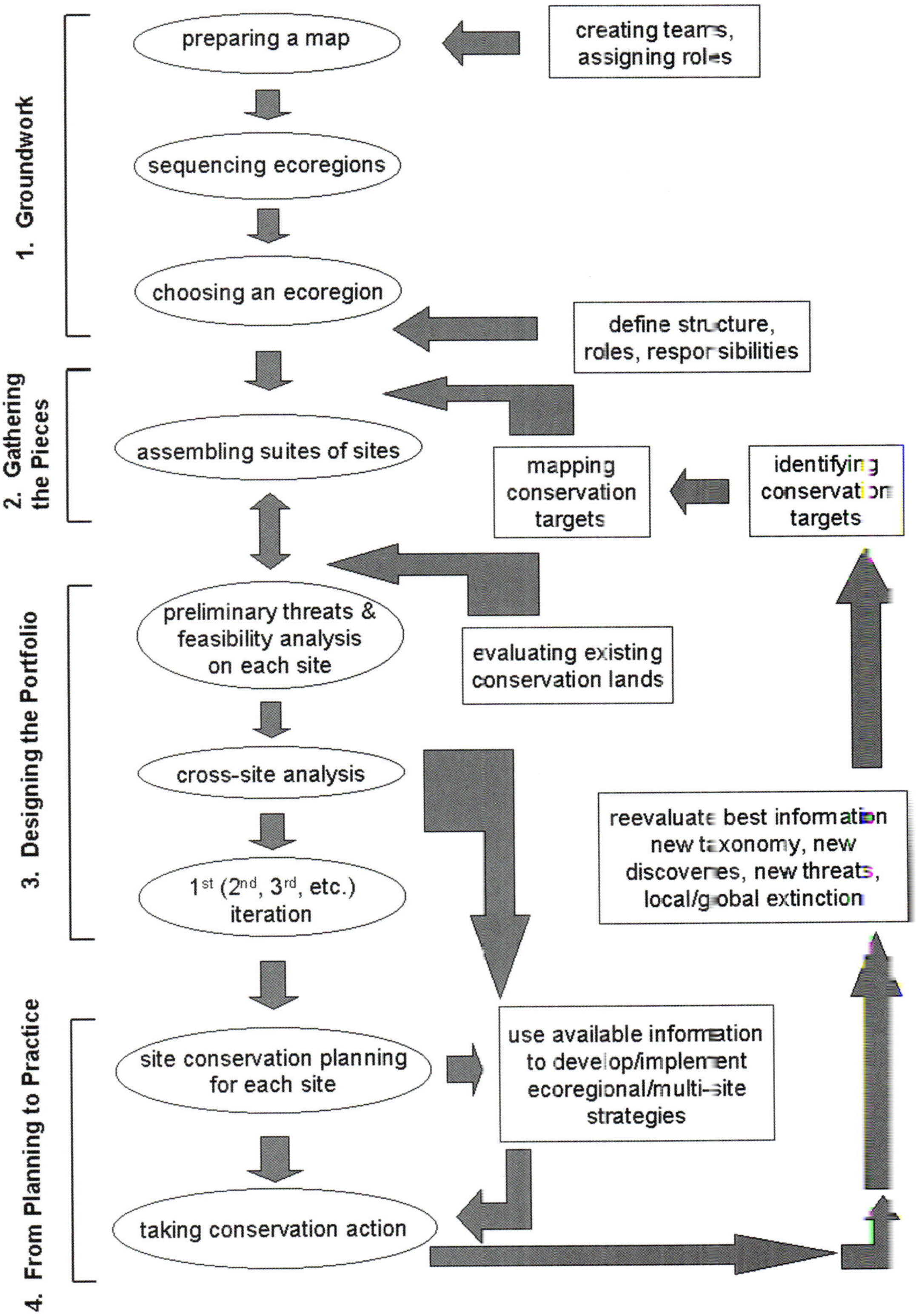

Figure 20.1. *The Ecoregional Planning Process as defined by The Nature Conservancy*

Site Conservation Planning for a Binational Resource

Conservation Target Selection With a conservation vision in which the Conservancy and its partners will work to ensure that the Laguna Madre conservation area remains an ecologically functional landscape, supporting a wide array of native plants and animals as well as the many migratory species that depend on it, the Conservancy's CAP identified 40 zoological targets (36 species and 4 guilds) and 45 botanical targets that are documented from or thought to exist in the Laguna Madre region (table 20.1). Additionally, 13 nested targets constituting communities were identified: resacas, *lomas*, oak brush, coastal prairie, wild-tidal flats, fresh to saline permanent to ephemeral wetlands, upland prairies, nesting and foraging areas on barrier islands and the mainland, South Texas Sand Sheet oak mottes, Black Mangrove–Saltwort shrubland, coquina beach rock, serpulid worm reef, and rookery islands (see TNC 2001, table 3). Similarly, Pronatura Noreste identified 84 zoological targets, 30 botanical targets, and 8 community targets for the Mexican portion of the Laguna Madre. Of these targets, 17 zoological species and guilds, 5 botanical targets, and 5 communities are common to both the Texas and Mexico portions of the region (TNC 2002).

While conservation strategies highlight species of plants and animals and vegetation communities, a selected list of ecological systems and species are also targets that capture multiple scales of biodiversity (see table 20.1). The Conservancy's methodology allows selection of conservation elements at various scales (species, guild, community) but also limits the number of elements in a conservation plan to eight (Groves et al. 2000; TNC 2000). Consequently, care was taken to choose conservation elements at a scale coarse enough to encompass the diverse guilds and individual species of conservation concern. A list of seven focal targets was chosen, primarily based on ecological communities or representatives thereof, including the Tamaulipan thornscrub matrix; Coastal Sand Plains, clay dune, and prairie complex; hypersaline lagoon—seagrasses; barrier islands, Reddish Egret; shorebird guild; and the Ocelot (TNC 2001). Essentially, by maintaining a focus on this suite of systems and species, all of the biodiversity represented within the Laguna Madre region will be captured. For example, by selecting the Tamaulipan thornscrub matrix, numerous other botanical, zoological, and community targets embedded within the matrix can be conserved.

Stresses and Sources Numerous stresses are acting on the Laguna Madre. From incompatible development to nonpoint source pollution, many factors are involved in impacting the ecological functions and processes operating within the Laguna Madre, including more frequent events of hypoxia, eutrophication, and Texas Brown Tide in sub-bays, such as Baffin Bay (Wetz et al. 2017; see also chapters 8 and 22 for more information on threats). However, one can take solace in knowing that the level at which most of these stressors function is less than that documented for other coastal systems, such as Galveston Bay in Texas. In recent times, the most concerning threats to the Laguna Madre area are the multiple impacts that may result from climate change, especially the potential ways valuable conservation targets could be affected. For example, barrier island erosion and storm-surge flooding due to sea-level rise would also likely result in chronic coastal flooding by high-tidal events (UCS 2013, 2018). Another impact might be "tropicalization"—water temperature increases would almost certainly cause shifts in species distributions that would potentially impact trophic relationships and biotic community composition (Scheffel et al. 2018).

The key to conservation success is the ability to correctly identify all the threats to the various targets and then formulate strategies to abate those threats. Table 20.2 lists the various stresses, with a qualitative assessment of how severe each one is for the focal target (system or species) concerned. The Conservancy has also attempted

Table 20.1. *Focal conservation targets (communities, guilds, or species) and associated nested conservation targets (communities, guilds, or species) within the Laguna Madre of Texas*

Conservation Targets	**Focal Targets**						
	Tamaulipan Thornscrub Matrix	**Coastal Sand Plains, Clay Dune, and Prairie Complex**	**Hypersaline Lagoon–Seagrasses**	**Barrier Island**	**Reddish Egret**	**Shorebird Guild**	**Ocelot**
Plants							
Amelia's Sand Verbena		X					
Bailey's Ball Moss		X					
Bigflower Bladderpod		X					
Black Mangrove				X			
Black-laced Cactus	X						
Bristle-free Least Daisy		X					
Clover Grass			X				
Coastal Bend Green-Thread		X					
Cory's Croton		X					
Crestless Wild Onion		X					
Elmendorf's Onion		X					
False Crown Tickseed		X					
Hairy Cutleaf Evening Primrose		X					
Hooker's Palafoxia		X					
Jones' Nailwort		X					
Kleberg Knotweed		X					
Kleberg Saltbush			X				
Large Selenia		X					
Leafless Alkali Weed			X				
Lemon Bee Balm		X					
Lila de los Llanos		X					
Lindheimer's Tephrosia		X					
Lundell's Nailwort		X					
Manatee Grass			X				
Padre Island Dropseed		X		X			
Plains Gumweed		X					
Roughseed Sea-purslane		X					
Sand Brazoria		X					
Sand-belt Bluet		X					
Seaside Beebalm		X					

Table 20.1. Continued

Conservation Targets	Focal Targets						
	Tamaulipan Thornscrub Matrix	**Coastal Sand Plains, Clay Dune, and Prairie Complex**	**Hypersaline Lagoon–Seagrasses**	**Barrier Island**	**Reddish Egret**	**Shorebird Guild**	**Ocelot**
Shinner's Rocket		X					
Shoal Grass			X				
Shortcrown Milkvine		X					
Slender Rushpea		X					
South Texas Clammyweed		X					
South Texas Phacelia		X					
Stinking Rushpea		X					
Texas Stonecrop		X					
Texas Willkommia		X					
Texas Windmill Grass		X					
Texasgrass		X					
Tharp's Rhododon		X					
Turtlegrass			X				
Velvet Spurge		X					
Welder Machaeranthera		X					
Invertebrates							
Blue Land Crab			X				
Laguna Madre Fiddler Crab				X			
Mangrove Crab				X			
Mangrove Periwinkle				X			
Fishes							
Blackfin Goby			X				
Texas Pipefish			X				
Amphibians							
Black-spotted Newt	X	X					
Rio Grande Lesser Siren	X	X					
Sheep Frog	X	X					
Mexican White-lipped Frog	X	X					
Reptiles							
Green Sea Turtle			X				
Hawksbill Sea Turtle			X				
Keeled Earless Lizard		X					

Table 20.1. Continued

Conservation Targets	Focal Targets						
	Tamaulipan Thornscrub Matrix	**Coastal Sand Plains, Clay Dune, and Prairie Complex**	**Hypersaline Lagoon–Seagrasses**	**Barrier Island**	**Reddish Egret**	**Shorebird Guild**	**Ocelot**
Kemp's Ridley Sea Turtle			X				
Loggerhead Sea Turtle			X				
Mexican Black-headed Snake	X						
Regal Black-striped Snake	X						
Speckled Racer	X						
Texas Horned Lizard	X						
Texas Indigo Snake	X						
Texas Tortoise	X						
Birds							
Aplomado Falcon		X					
Audubon's Oriole	X						
Botteri's Sparrow		X					
Brown Pelican			X				
Cactus Ferruginous Pygmy-Owl	X						
Northern Beardless Tyrannulet	X						
Peregrine Falcon				X			
Piping Plover				X		X	
Reddish Egret					X		
Snowy Plover				X		X	
Mammals							
Coues's Rice Rat	X						
Jaguarundi	X						
Maritime Pocket Gopher		X					
Ocelot							X
Southern Yellow Bat	X						
Guild or Community							
Neotropical Songbird				X			
Nesting Sea Turtle				X			
Texas Ebony–Snake-eyes Shrubland	X						
Wintering Waterfowl			X				

Table 20.2. *Identified threats to focal conservation targets within the Laguna Madre of Texas with associated threat rankings by threat for each target*

	Conservation Targets						
	Tamaulipan Thornscrub Matrix	**Coastal Sand Plains Matrix**	**Hypersaline Lagoon—Seagrasses**	**Barrier Island**	**Reddish Egret**	**Shorebird Guild**	**Ocelot**
Incompatible development	Medium	Low	Low	Medium	Medium		Very high
Conversion to agriculture		Low					Very high
Roads							Very high
Brush clearing	High						Low
Texas Brown Tide			High				
Dredging			Medium				
Agricultural runoff			Medium				
Unsuitable grazing practices	Low	Low					Low
Off-road vehicles (ORVs)				Low		Low	
RVs on beaches and mudflats				Low		Low	
Propeller scarring			Low				
Industrial discharge			Low				
Urban runoff			Low				
Wastewater and septic discharges			Low				
Fire suppression		Low					
Threat status for targets and sites	Medium	Low	Medium	Low	Low	Low	Very high

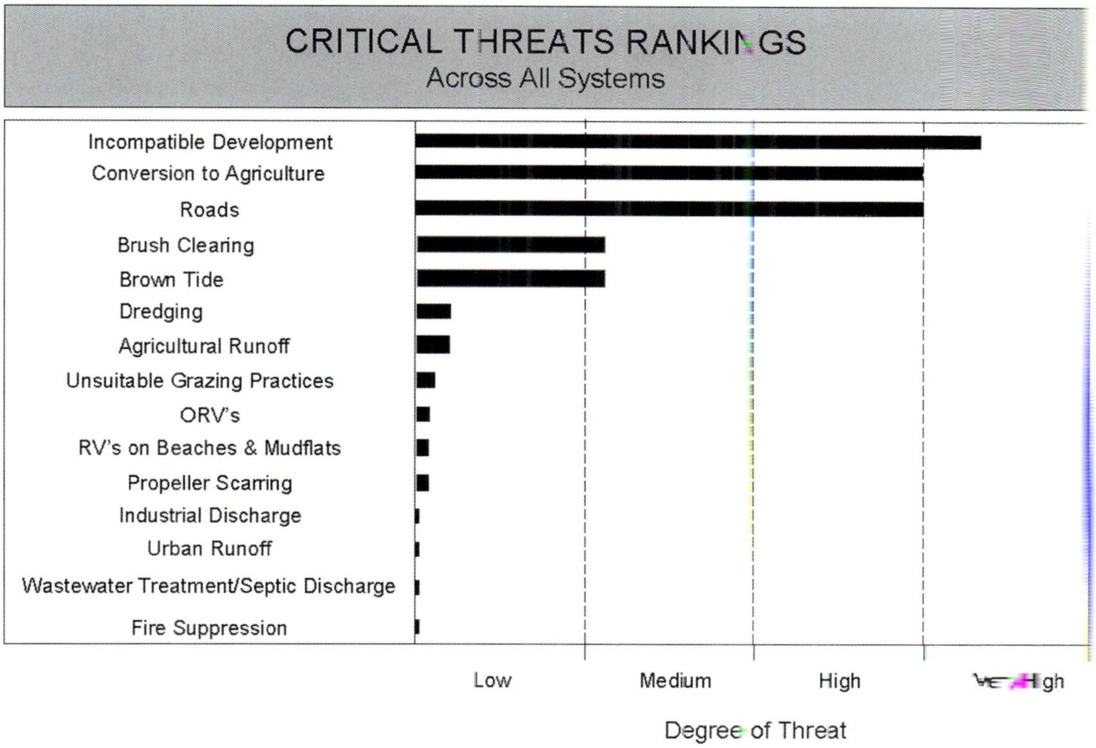

Figure 20.2. *Critical threats rankings across all systems within the Laguna Madre. Degree of threat is a qualitative assessment based on existing data and expert opinions.*

to quantify overall critical threats across all systems (fig. 20.2). Incompatible development, agricultural conversion, and roads obviously are overarching threats for all systems within the Laguna Madre area. It should be clear that the Conservancy does not propose curtailing road construction in the region; however, there probably are ways to engineer and build roads to lessen their impact on specific conservation targets, such as the Ocelot.

When evaluating critical threats for the seven focal conservation targets within Laguna Madre, overall scale must be considered. For example, off-road vehicles (ORVs) cause serious damage to dune systems on South Padre Island. However, when the entire barrier island system fronting the Laguna Madre of Texas and Tamaulipas is considered, ORVs exert damage on a relatively small portion of the system (see fig. 20.2). The National Park Service maintains a high level of control over ORVs within the boundaries of Padre Island National Seashore, thus abating the ORV threat there. This is not to say that ORV-caused dune destruction should not be addressed by means of abatement strategies. Site-specific plans for the South Padre Island barrier island system include specific strategies for addressing problems associated with ORV use.

The conservation targets for the Laguna Madre de Tamaulipas are the same as for the lagoon in Texas, but the threats differ somewhat in type and severity (see chapter 12). For example, overfishing is a much more significant threat in Mexico due to the lack of regulation enforcement, while dredging is much less of a current concern. Overfishing has caused the deterioration or destruction of more than 50% of seagrass habitat (approximately 144 km^2 [56 mi^2]) since the first version of this compendium was published in 2002 (Alfonso Banda and Leonardo Arellano, pers. comm., 20[illegible]).

Although development pressures are much less in Mexico than on South Padre Island and the mainland fronting the Lower Laguna Madre in Cameron County, grazing on the barrier island areas in Tamaulipas is more severe than in Texas. Expansion of the development and aquaculture industry and potential dredging projects could pose future threats to the Laguna Madre de Tamaulipas by also increasing the associated risk of pollution. The plans to develop a new port near Matamoros are especially worrisome.

Strategies for Conservation

The Laguna Madre has not experienced the level of development (residential and industrial), shoreline hardening, and other impacts readily observed elsewhere on the Texas coast. Evaluation of the viability of conservation targets in the Laguna Madre region in the CAP illustrates this well (fig. 20.3). The Ocelot possesses by far the lowest viability ranking with fewer than 39 known individuals left in Texas (in Willacy and Kennedy Counties; USFWS 2016). The Tamaulipan thornscrub matrix has experienced extensive clearing on both sides of the border. Clearing continues today, especially for expanding residential and *colonia* development. However, despite the status of the aforementioned conservation targets, the remaining five are assessed as having good viability. Thus, the Conservancy has a unique opportunity to develop strategies in a more proactive manner to abate threats to conservation targets, rather than react to a declared conservation crisis.

The top priority strategies, which are site based, must be developed to implement landscape-level conservation in the Laguna Madre region: (1) conserve critical barrier island systems, focusing on South Padre Island and dunes; (2) establish a focused private-lands conservation initiative: cooperative projects designed to conserve habitat; and (3) work with partners to identify and protect important habitats. All of the work must be developed through community-based conservation efforts that incorporate socioeconomic concerns, the desires and interests of local community members, and sound application of ecological knowledge (fig. 20.4). Essentially, the CAP identifies several key conservation zones based on best science available and local

Figure 20.3. *Viability rankings for seven conservation targets within the Laguna Madre of Texas based on size, condition, and landscape context*

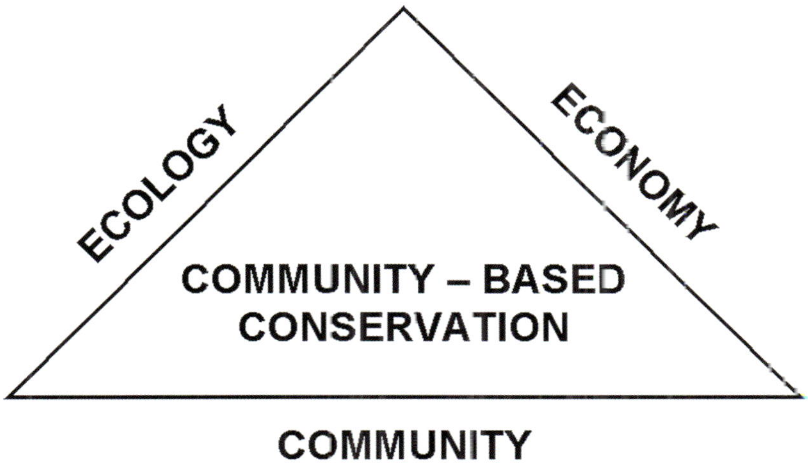

Figure 20.4. *Paradigm for the concept of community-based conservation. The success of community-based conservation rests on a holistic approach bringing together ecological, economic, and community interests.*

input from landowners, resource agencies, and community leaders. These zones are Baffin Bay watershed, South Texas Sand Plains, the Laguna Madre marine ecosystem, South Padre Island, Arroyo Colorado, the Boca Chica–South Bay complex, and the Bahia Grande–Redhead Ridge complex. Pronatura Noreste focuses its work in the Laguna Madre de Tamaulipas on Laguna la Nacha, Laguna el Barril, Laguna Mar Muerto, rookery islands, barrier islands, the Reddish Egret, and community-based development on private and public lands. Additionally, Pronatura Noreste has signed a collaboration agreement with the CONANP management office of the Flora and Fauna Protection Areas of Laguna Madre and Río Bravo Delta to facilitate the collaboration in wetland restoration with the federal agency and support other strategies identified in the management plan (DOF 2015). Each zone will have its own set of discrete strategies addressing the most serious threats. Highest consideration will be given to strategies that can be applied at multiple sites to abate a common threat. For example, if meaningful partnerships could be developed with state and federal agencies, local agricultural interests, and other conservation groups to reduce nonpoint source pollution in the Arroyo Colorado watershed, that aquatic system would benefit along with the Laguna and the variety of conservation targets embedded within these two zones (Flowers et al. 1998).

An ongoing strategy includes site-based acquisition of habitats and promotion of conservation easement applications to assist with the implementation of large-scale Tamaulipan thornscrub restoration efforts on private and public lands. The Conservancy has helped protect 20,475 ha (50,595 ac) in the Laguna since 1980. This effort has assisted federal- and state-management-area networks in acquiring 19,024 ha (47,454 ac) to create and expand state parks, wildlife management areas, and national wildlife refuges. The Conservancy owns and manages three preserves in the Laguna Madre area covering 643 ha (1,589 ac), and it manages a conservation easement on private land covering 808 ha (1,997 ac). This easement directly protects some of the remaining prime habitat for the Ocelot.

Incorporating science and emerging issues into conservation strategies is crucial. For two decades, the Conservancy's work has been guided by the Conservation by Design framework. However, to meet the socioecological challenges of the twenty-first century, such as ever-changing climate, TNC must find better ways to meet conservation goals in times of increasing resource needs for people. The Conservancy now has a new edition of Conservation by Design (2.0) that honors and builds on TNC's successful history and evolves the approach and methodology to meet the challenges of today (TNC 2016). As the strategies are implemented, the new framework will continue to guide where work is focused and how action is taken in building and restoring coastal resilience along the Laguna Madre region.

At present, several other initiatives relevant to the conservation of the Laguna Madre represent regional collaborative efforts. Each contributes in different ways to the strategies that resulted from the CAP. The Coastal Bend Bays and Estuaries Program in Corpus Christi covers the Upper Laguna Madre area and identifies specific goals in its new plan—for example, conserving bird rookery islands to reverse the decline of colonial nesting waterbirds (CBBEP 2018). Although not yet part of the National Estuary Program of the US Environmental Protection Agency, the Lower Laguna Madre Estuary Program managed jointly by Cameron County, University of Texas Rio Grande Valley, and Texas A&M University–Kingsville, is currently developing a plan to balance nature conservation and economically sustainable development goals in the southern region. The Gulf Coast Prairie Landscape Conservation Cooperative, covering the two Laguna Madre areas, was established to provide science capacity and technical expertise for meeting shared natural and cultural resource priorities, such as integrating climate impact assessments into species and habitat conservation goals (Powell et al. 2017). The Laguna Madre Initiative Area of the Gulf Coast Joint Venture is a partnership among federal and state agencies, nonprofit organizations, and private landowners dedicated to the conservation of priority bird habitat along the Gulf of Mexico coast, using innovative models, tools, and partnerships to reverse the decline in their populations (Esslinger and Wilson 2002; Vermillion et al. 2012).

In Tamaulipas the CONANP management office of the Flora and Fauna Protection Areas of Laguna Madre and Río Bravo Delta works in engaging local communities in sustainable development projects by allocating federal funds to support a variety of activities, such as biodiversity monitoring, to reduce the mortality of sea turtles in bycatch by relocating semipermanent fishing gear, as well as to promote reforestation, island protection, and community outreach. The state government also created the Parks and Biodiversity Commission in 2017 to develop a statewide biodiversity strategy that will identify specific goals and actions to promote the sustainable use and conservation of species and systems. One of the main knowledge gaps in the state, and therefore a focus of the strategy, will be to study and set goals for conserving marine biodiversity (Arnulfo Moreno, pers. comm., 2018). It is also important to mention that the narco trafficking and drug cartel crisis in Tamaulipas state has reduced international collaboration in conservation during the past decade by limiting the capacity to access several areas around the Laguna Madre de Tamaulipas (David Lerma, pers. comm., 2018).

People who have experienced the aura of the Laguna Madre may not know the full biological and geographic significance of this unique ecosystem, but they do know that this is a special place in and of itself and for all of the species of wildlife that depend on it to survive. The Conservancy has long recognized these values and is embarking on a plan of action to protect into perpetuity the unique biological richness and mostly intact landscape and marine functions and processes of the Laguna Madre. Moreover, by protecting the ecological functions using the Conservation by Design 2.0 framework, the Conservancy is also conserving the benefits that people obtain from this functional landscape. Hopefully, the Laguna Madre will continue providing communities and visitors with essential goods and services to fulfill their lives, including valuable experiences to enrich them.

Selecting the right places to work is critical in conserving biodiversity. As this revised edition of the compendium attests, the Laguna Madre is obviously such a place. Even more important is implementing a sound plan of action and connecting stakeholders to it. For example, a significant challenge is to connect people and businesses to nature and the benefits that it provides. Although valuable for some and possibly in the short term, conservation actions taken outside the context of an overall ecoregion-based plan likely will not make lasting contributions to sustaining biodiversity. There is still much to be learned about the Laguna Madre, this vital component of the Gulf of Mexico.

21 *Information Gaps and Needs*

JOHN W. TUNNELL JR.

From the earliest writings in the 1920s and 1940s, scientific knowledge about the Laguna Madre of Texas and Tamaulipas focused on the relationships between hypersaline conditions and fisheries productivity. Subsequently, the interest in Texas then turned to the vast seagrass beds (almost 80% of the statewide distribution) and the importance of this submerged vegetation as essential habitat. Most recently, concern over the occurrence of Texas Brown Tide and the impacts of maintenance dredging of the Gulf Intracoastal Waterway (GIWW) have stimulated extensive new research. In Tamaulipas, the recent idea of extending the GIWW from Brownsville, Texas, through the Mexican lagoon has stimulated both concern over the lagoon's future environmental health (and protection) and much research. Unfortunately, much of the work in Mexico has not been publicly published and is available only through the federal government with permission.

During the compilation of literature for this compendium and the accompanying comprehensive bibliography, over 1,250 citations were listed for the Laguna Madre and surrounding area (Padre Island, Rio Grande Valley, adjacent mainland, etc.; table 21.1). About 86% of these citations concerned Texas. Of the nearly 350 citations that were specifically on the Laguna Madre only, about 75% concerned the Texas lagoon and 25% the Tamaulipas lagoon. Most references described projects focused on a specific topic, and most lasted only one or two years, which can be a problem when characterizing such a dynamic system. The only long-term biological data set available is from the Coastal Monitoring Program conducted by Texas Parks and Wildlife Department for fishery species management. The Conrad Blucher Institute for Surveying and Science at Texas A&M University–Corpus Christi began collecting water-level data at eight stations within the Laguna Madre of Texas in the early 1990s. At selected stations data are also collected on other physical parameters, such as temperature, salinity, pH, and dissolved oxygen. No published, long-term data sets were found for the Tamaulipas lagoon, but the Fisheries Institute of Mexico does keep records of landings of selected commercial species.

No comprehensive synthesis of all Laguna Madre information has been compiled. However, extensive topical syntheses and characterizations have been recently completed for the northernmost part of the Upper Laguna Madre system in the Corpus Christi Bay National Estuary Program (now Coastal Bend Bays and Estuaries Program). Reports include numerous topics (living resources, water quality, sediment quality, point source and nonpoint source discharges, seagrasses, wetlands, habitat distribution, marine debris, freshwater inflow, and more) within the Texas Coastal Bend geographic study area.

Table 21.1. *Summary of literature listed in Laguna Madre Comprehensive Bibliography concerning the Laguna Madre (Texas and Tamaulipas) and surrounding areas by geographic area and type of literature*

Location	Journal Article	Proceedings	Report	Thesis	Dissertation	Book	Magazine	Total
United States								
Laguna Madre of Texas	119	23	119	32	5	3	23	324
Baffin Bay	16	3	5	9	2	0	1	36
Padre Island	42	14	32	5	2	1	[illegible]	102
Rio Grande Valley and Arroyo Colorado	40	1	75	7	2	7	[illegible]	132
South Texas coast	44	9	40	3	2	1	3	102
South Texas	82	5	21	21	2	9	[illegible]	[illegible]
Texas coast	105	9	83	5	3	8	19	[illegible]
Texas	22	1	27	2	0	18	2	72
United States	73	6	9	1	1	8	[illegible]	[illegible]
Total United States	543	71	411	85	19	55	55	[illegible]
Mexico								
Laguna Madre de Tamaulipas	42	38	40	4	1	4	5	135
Northeastern Mexico	18	1	2	1	0	0	0	22
Tamaulipas	17	1	0	2	0	0	0	20
Coastal lagoons—Mexico	13	6	1	2	1	1	0	24
Mexico	12	1	1	0	0	3	1	18
Total Mexico	102	47	44	9	2	8	7	[illegible]
Combination papers[a]	91	30	36	3	2	16	0	[illegible]
Miscellaneous papers[b]	175	10	30	3	0	24	3	[illegible]
Grand total	911	158	521	100	23	103	[illegible]	[illegible]

[a] Includes papers concerning both sides of the border (e.g., South Texas and northeastern Mexico, Rio Grande, Gulf of Mexico).

[b] Includes papers primarily on taxonomic groups or events (e.g., hurricanes) within the Laguna Madre area. Most concern Texas species or events.

Table 21.2. *Information gap matrix for selected aspects of knowledge concerning the Laguna Madre of Texas and Tamaulipas*

	ULMTex	**LLMTex**	**NLMTam**	**SLMTam**
Climate	****	****	**	**
Hydrography	***	***	**	**
Geology	****	****	**	**
Tamaulipan Biotic Province	***	***	**	**
Ranching heritage	****	***	**	**
Environment	***	***	**	**
Seagrass meadows	***	***	**	**
Open bay	**	**	**	**
Wind-tidal flats	***	**	**	**
Barrier islands	****	****	**	**
Redheads and other waterfowl	****	***	**	**
Colonial waterbirds and rookery islands	***	***	**	*
Shorebirds and wading birds	***	**	**	*
Sea turtles	**	**	*	*
Fish and fisheries	***	**	**	**
Brown tide	****	**	**	*
Red Tide	**	**	*	*

Note: ULMTex = Upper Laguna Madre of Texas; LLMTex = Lower Laguna Madre of Texas; NLMTam = Northern Laguna Madre de Tamaulipas; SLMTam = Southern Laguna Madre de Tamaulipas.
Rankings based on the amount of literature:
* No information located or available
** Little information available
*** A moderate amount of information available
**** An extensive amount of information available

Information and knowledge gaps discovered during this compendium synthesis are presented in a matrix format for quick review (table 21.2). Overall, geological information seems to be more thorough and complete than biological. A considerable amount of seagrass distribution and ecology information is available for Texas, but much is still unknown about community dynamics and linkage to other ecosystem habitats. Little is known about seagrasses in Tamaulipas.

Extensive multiyear data are available on fish and some invertebrates via the Texas Parks and Wildlife Department's Coastal Fisheries Monitoring Program, but most of it is unanalyzed, except for regulated species. Wind-tidal flats, waterfowl, colonial waterbirds, and shorebirds all have a moderate amount of information available, and the volume of literature concerning Texas Brown Tide is currently expanding.

The following itemization lists some specific information or knowledge gaps that were discovered during compilation of this compendium:

1. Accurate determination of areal coverage of Laguna Madre of Texas and its major habitats is needed.
2. A comprehensive monitoring program is needed to track status and trends of major habitats, biota, and environmental quality.
3. Linkages between habitats (tidal flats and seagrass beds, Laguna Madre and freshwater ponds for waterfowl, etc.) are lacking.
4. Geological question: Why are there so many natural islands in the Laguna Madre de Tamaulipas (approximately 200) and fewer than 10 in the Laguna Madre of Texas?
5. The ecology of Rio Grande Delta ecosystems is practically unknown.
6. Climate data are difficult to access for Tamaulipas; long-term, continuous monitoring stations adjacent to the Laguna are inadequate.
7. Tidal characteristics under today's conditions are not known for the Lower Laguna Madre in Texas or the entire system in Tamaulipas.
8. Little is known of the biology and ecology of certain habitats in Texas (Bahia Grande, Laguna Larga, San Martin Lake).
9. Artificial hard substrates have, for the most part, not been studied.
10. No information is available on jetty habitat in Tamaulipas, and little is available for Texas.
11. Are there serpulid reefs in Tamaulipas (Laguna el Catán), like those in Baffin Bay, Texas?
12. There is no biological or ecological information on coquina outcrops in Texas or Tamaulipas.
13. There is no ecological or genetic information on oysters for the Laguna Madre de Tamaulipas (Laguna el Catán).
14. There are no comprehensive biodiversity surveys or checklists for the Laguna Madre systems.
15. Information about the distribution or abundance of annelids or epibenthic crustaceans in seagrass beds is scarce. Invertebrate nekton is almost completely ignored.
16. The effect of Manatee Grass replacement by Turtle grass as succession proceeds is unknown.

17. There is little information about seagrass biomass decline and invertebrate communities associated with seagrasses in the Laguna Madre de Tamaulipas.
18. Cumulative impacts of tidal flat losses are unknown.
19. No information is available on the role of bacteria and fungi in barrier island vegetation dynamics.
20. Importance of invertebrates as part of the food web has not been quantified for Texas and Tamaulipas barrier island chains.
21. No study has been conducted to identify and quantify the importance of dunes to resident and migrant herbivorous birds.
22. More study is needed on the role of barrier island grasslands in the wintering and migratory ecology of Neotropical migrant songbirds.
23. There are no quantitative studies on invertebrate community structure on barrier islands.
24. With the exception of Redheads, limited information exists on population dynamics of wintering waterfowl in the Laguna Madre of Texas, T amaulipas, Rio Grande Delta, and coastal depressional wetlands.
25. Information concerning ecosystem linkages between the Laguna Madre, mainland, and Gulf of Mexico is needed, along with an ecosystem model (although much research will be needed before an accurate model can be developed).
26. There is little information on population dynamics for most species except commercial fish and shellfish.
27. The distribution and importance of mangroves in Laguna Madre de Tamaulipas need study.
28. The ecological impacts of Black Mangrove expansion in the Laguna Madre of Texas should be monitored, especially where mangroves replace salt marsh and wind-tidal flat habitats.
29. Little is known of wind-tidal flat and seagrass ecology in Laguna Madre de Tamaulipas.
30. The dynamics of invertebrate recruitment to and persistence on wind-tidal flats and its relationship to habitat quality for shorebirds are poorly understood.
31. Shorebird wintering populations and movements within and between Texas and Tamaulipas wind-tidal flats and adjacent habitats are not known well enough to allow management of either the birds or the habitats they frequent.
32. Knowledge of wading-bird wintering ecology is inadequate in both lagoons.
33. Knowledge of shorebird nesting ecology is insufficient in both lagoons.
34. There is no information concerning the effects of droughts and wet cycles on shorebird, wading-bird, and waterfowl use of either lagoon.
35. Relationships between colonial waterbird nesting-guild dynamics and rookery island habitat availability are unknown.

22 Conservation Issues and Recommendations

JOHN W. TUNNELL JR., KIM WITHERS
ELIZABETH H. SMITH, SCOTT MURRAY,
MICHAEL WETZ, AND BART M. BALLARD

Numerous environmental and conservation issues have been raised in the preceding discussion of the Laguna Madre of Texas and Tamaulipas. These difficulties and others are grouped here for a regional overview. Most issues involve anthropogenic impacts due to increasing human population levels and consequent environmental stresses to natural resources and systems. Where appropriate, after discussion of an issue recommendations follow.

Most activities that alter and impact land and water are now governed by various regulatory agencies. For the sake of readers concerned about issues and their consequences but unfamiliar with administrative and regulatory structures, the respective roles and responsibilities of each major state agency involved in the Laguna Madre are listed here:

- Texas Commission on Environmental Quality (TCEQ), air and water quality
- Texas General Land Office (TGLO), submerged lands and tidal shorelines
- Texas Parks and Wildlife Department (TPWD), living natural resources
- Texas Department of Health (TDH), aquatic public health

Following are some federal regulatory agencies and their roles:

- US Fish and Wildlife Service (USFWS), living natural resources
- US Environmental Protection Agency (EPA), environmental protection and regulation
- US Army Corps of Engineers (USACE), dredging and filling activity
- National Marine Fisheries Service (NMFS), federal fisheries and habitat utilization of fishery species

All fishing, ecological, and environmental regulatory issues in Mexico fall within the purview of the Secretaría de Medio Ambiente y Recursos Naturales (SEMARNAT).

The Texas state legislature established the Texas Coastal Management Program (CMP) to improve management of coastal natural resources. This program helps ensure the long-term environmental and economic health of the Texas coast and integrates efforts on associated issues among state, federal, and local agencies that have jurisdiction in coastal areas. The CMP receives funding through the Federal Coastal Zone Management Act and Texas Coastal Coordination Act and strives to achieve consistency between state- and federally funded projects. Approximately $2.2 million per year is available for coastal projects and planning efforts. The TGLO, in conjunction with the Coastal Coordination Advisory Committee, oversees CMP activities and reviews the grant proposals. The Coastal Division of the TGLO serves as administrative staff and oversees the funded CMP projects. The CMP focuses on five primary issues of concern to coastal communities: coastal erosion, wetlands protection, water supply and water quality, dune protection, and shoreline access.

The environmental health of the Laguna Madre of Texas and Tamaulipas is important to the economic health of the entire region, providing up to $400 million per year to the local economy (Cameron, Willacy, Kenedy, and Kleberg Counties) through commercial fishing, sportfishing, birding, recreation, and tourism (K. Chapman, pers. comm., 2018). Fishing trips alone (not counting purchases of durable goods, i.e., fishing gear and boats) contribute approximately $90 million annually to the economy of the Lower Laguna Madre region (Lovell et al. 2011; Ropicki et al. 2016). Visitors to refuges and nature reserves in Cameron County contribute an additional $100 million annually to the local economy. Increasingly, human activity and growth of coastal communities impact the Laguna Madre system.

In 1999–2000 the Texas Center for Policy Studies and Pronatura Noreste set out to define clearly what the Laguna Madre actually means to Texas and Tamaulipas in terms of economic and natural resources and to discuss what processes local communities might use to best protect both the ecology of the area and the livelihood of the people who use this resource. The two organizations jointly conducted the effort in three phases: (1) a binational assessment of the current economic and environmental state of the region; (2) research on local public perception of the links between the economy and the natural environment of the Laguna Madre; and (3) a binational symposium to present the assessment and findings of the research. The overall goal of the project was to foster binational discussion about opportunities for promoting economic activities that encourage conservation and enhancement of natural resources and maintenance of a climate for providing living-wage jobs. The methodology for achieving this goal was educational in nature, and the information produced was intended to assist in planning for the future of the Lower Rio Grande Valley and northeastern Tamaulipas (K. Chapman, pers. comm., 2018). Separate draft reports on the binational project on the Laguna Madre of Texas (K. Chapman 2000) and Tamaulipas (Cruz-Nieto 2000) were made available at "The Future of the Binational Laguna Madre Economic and Natural Resources Symposium," 13–14 April 2000, in South Padre Island, Texas.

Conservation Issues

The remoteness of the Laguna Madre of Texas and Tamaulipas from human population and industrial centers has acted as a substantial means of protection and conservation of natural resources. In Texas, Laguna Atascosa National Wildlife Refuge, large ranches along the mainland shoreline, and Padre Island National Seashore on the barrier island have limited access and development of the Laguna shoreline. Consequently, the nodes of development are primarily to the north near Corpus Christi and to the south near Port Isabel and the community of South Padre Island. Some access and development also occur in western Baffin Bay and at Port Mansfield. The absence of paved highways and utility infrastructure in Tamaulipas has limited access and development to small fishing villages. In the 1970s, paved highways were completed to Mezquital in the north and La Pesca in the south. In the 1990s, two additional highways were completed to the western shoreline, first from San Fernando to Carboneras and Carbajal in the Southern Laguna Madre de Tamaulipas, and most recently from Highway 180 to El Barancón in the Northern Laguna. Rancho Rincón de Anacahuitas, along the northwestern lagoon shoreline, is a model for land-conservation management and protection. For the past 60 years, the Reyes and Martinez families, who own the 12,510 ha (30,000 ac) property, have maintained the thornscrub habitat and its many small ponds and seasonally flooded grasslands as both a working ranch and a prime habitat for wildlife. It is one of the few areas along the Mexican coast that has not been converted to agriculture. Comprehensive conservation and protection of the entire Laguna Madre de Tamaulipas was proposed in the 1990s (DUMAC 1996), and fortunately now the entire lagoon system along with much of the surrounding area is a flora and fauna protected area called the "Laguna Madre y Delta del Río Bravo."

Dredging Impacts During the early years of dredging (1940s–60s) in the Laguna Madre of Texas, large quantities of dredged material were indiscriminately placed in both open water (fig. 22.1) and along shorelines, covering vital seagrass and wind-tidal flat habitats. When the Upper Laguna Madre was surveyed in the 1970s, 109 dredged-material islands existed along the GIWW and an additional 40 were located along privately maintained channels (Chaney et al. 1978). In the entire Laguna Madre of Texas, dredged material covered approximately 78 km^2 (30 mi^2) of lagoon bottom and 114 km^2 (44 mi^2) of adjacent land above sea level (Brown et al. 1976; Brown et al. 1977; Brown et al. 1980). During the 1970s, disposal of dredged material resulting from the creation and maintenance of navigation waterways and harbors became a matter of national interest and concern. One result was establishment of the Dredged Material Research Program at the US Army Corps of Engineers Waterways Experiment Station in Vicksburg, Mississippi (Chaney et al. 1978). Through this program, much scientific information was generated about the flora and fauna in the Laguna Madre, especially concerning the colonial waterbirds (see chapter 14).

Heightened concern about the impacts of maintenance dredging on vital seagrass habitat and how this affected the wintering population of Redheads led to the establishment in 1987 of the Texas Gulf Coast Field Station of the US Geological Survey, Biological Resources Division, in Corpus Christi, to conduct research on the issue. Further concern was raised about placement of dredged material on lands adjacent to the Laguna Madre (King Ranch and Kenedy Trust 1993) and whether further dredging was a wise proposition considering all economic and environmental implications (Diaz and Kelly 1994). Subsequently, the Corps of Engineers established the Laguna Madre Interagency Coordination Team (ICT) to help prepare a 50-year Dredged Material Management Plan for the GIWW in the Laguna Madre. Members of the ICT represented state and federal natural resource and transportation agencies: TCEQ, TGLO, TPWD, TWDB, TXDOT, USFWS, USACE, EPA, and NMFS. Numerous studies completed under the oversight of the ICT covered all aspects of the Laguna Madre dredging operations and impacts: physical oceanography (Brown and Kraus 1997); Piping Plover (EHA 1997; Zonick et al. 1998); containment (EHA and BVA 1998); benthic macrofauna (EHA and BVA 1998); sediment budget (Morton et al. 1998); habitat utilization (Sheridan and Minello 2003; Sheridan 2004), and seagrass model (Dunton 1994; Czerny and Dunton 1995; Dunton et al. 2011). Although absolute answers regarding the best sites and manner of dredged-material placement are still not available, the ICT used extensive science-based information when preparing the Dredged Material Management Plan for the Laguna Madre (USACE and ICT 2002).

a

b

c

Figure 22.1. *Dredging activities along the GIWW in the Laguna Madre: (a) dredge in the lagoon; (b) dredged-material disposal adjacent to an established dredged-material island into the Upper Laguna Madre; (c) dredged-material outflow into a diked contain containment area (Photos by Brian R. Chapman)*

Careful consideration should be given to dredged-material placement, using all available information. Open-water disposal should be avoided, especially in areas of seagrass habitat. Best available technologies should be used for containment of dredged material to reduce or eliminate resuspension of sediments in the water column, which can cause reduced light problems for highly productive seagrass habitat.

Endangered and Threatened Species For a complete listing of endangered, threatened, and rare species in counties adjacent to the Texas Laguna Madre see table 4.7. The following species found in or near the Texas Laguna Madre (LM) and/or Laguna Madre de Tamaulipas (LMT; DUMAC 1993) are listed as endangered or threatened by both the US Department of Interior and TPWD and/or Mexico's bird conservation authority, Sección Mexicana del Consejo Internacional para la Conservación de las Aves (CIPAMEX):

Endangered

- Slender Rushpea (LM)
- Texas Ayenia (LM)
- Black-laced Cactus (LM)
- Star Cactus (LM, LMT)
- South Texas Ambrosia (LM)
- Hawksbill Sea Turtle (LM, LMT)
- Kemp's Ridley Sea Turtle (LM, LMT)
- Leatherback Sea Turtle (LM, LMT)
- Peregrine Falcon (LM, LMT; delisted by US Department of Interior; still considered threatened by TPWD)
- Eskimo Curlew (LM)
- Aplomado Falcon (LM, LMT)
- Yellow-headed Amazon (LMT)
- Interior Least Tern (LM, LMT)
- Jaguarundi (LM, LMT)
- Ocelot (LM, LMT)
- Jaguar (LMT)
- West Indian Manatee (LM, LMT)

Threatened

- Green Sea Turtle (LM, LMT)
- Loggerhead Sea Turtle (LM, LMT)
- Piping Plover (LM, LMT)

Three protected animal species that may be found in either lagoon are the Bottlenose Dolphin, Green Sea Turtle, and Opossum Pipefish. Baseline information on population dynamics and habitat use is needed for most species within the system. Most species recovery efforts use a single-species approach, but habitat conservation and multispecies management are more practical and effective strategies. Standardization of regulatory efforts between the United States and Mexico, especially in regard to conservation of marine mammals and sea turtles, would be desirable. Incentives and education to encourage conservation at the grassroots level could be more effective than legislating conservation.

Nutrient Pollution

Arroyo Colorado The Arroyo Colorado, part of a natural drainage system for the Lower Rio Grande Valley (Bryan 1971; Flores et al. 2017), lies on the northern extent of the Rio Grande Delta (Brown et al. 1980). The arroyo, which stretches 149 km (90 mi), is part of a floodway system that begins south of Mission, Texas, and drains eastward, dividing into the North Floodway and the arroyo at Mercedes, Texas. The arroyo watershed has a drainage area of 1,829 km^2 (706 mi^2). In earlier decades, most of the watershed was used for agricultural production, including row crops, sugar cane, and citrus fruit, but rapid urbanization and population growth are transforming the area into an urbanized metroplex (Flores et al. 2017). The system, which sustains little natural flow except during rainfall runoff, receives treated wastewater from several towns across the valley and provides essentially the only "freshwater" inflow into the Lower Laguna Madre. In the late 1940s the lower 40 km (25 mi) were dredged to a depth of 4.3 m (14 ft) to accommodate barge traffic to the Port of Harlingen (Bryan 1971).

Environmental concerns for the Arroyo Colorado include low dissolved oxygen (DO), high levels of bacteria, and agricultural contaminants. Low DO is a long-standing problem (Bryan 1971; J. R. Davis 1983, 1984, 1985; Baker 1997). For assessment purposes, the Arroyo Colorado was classified by the TCEQ into two segments: Segment 2201 was named the "tidal segment"; and Segment 2202, the "above-tidal segment." The tidal segment has been included on the Texas list of impaired water bodies since 1978 due to periods of low DO that usually occur during the warmer months of May through October (Flores et al. 2017). Fish kills have been reported as a result of these low oxygen levels (Breuer 1962; Bryan 1971).

In 2003, TCEQ completed a total maximum daily load (TMDL) study for the tidal segment, indicating that low DO levels were related as much to the physical setting and geomorphology of the arroyo as they were to the loading of nutrients and oxygen-demanding substances from the watershed (Flores et al. 2017). The study spawned the development of a partnership among local, state, and federal stakeholders called the Arroyo Colorado Watershed Partnership, a group tasked with developing a community-based watershed protection plan to improve water quality in the Arroyo Colorado. After pollution causes and sources were determined, best management practices (BMPs) were selected to be addressed by the local stakeholders. The 10-year watershed protection plan was published in 2007 (Arroyo Colorado Watershed Partnership 2007).

Since implementation, watershed stakeholders have achieved considerable success implementing the plan, and many of their milestones have been accomplished (Flores et al. 2017). A major accomplishment included significant upgrades to approximately eight wastewater treatment facilities and the completion of two new ones that discharge into the arroyo. These actions have significantly decreased pollutant loadings and in-stream ammonia concentrations in the arroyo. Other accomplishments include BMPs for irrigated cropland, centralized wastewater service for many *colonias*, tertiary wetland treatment for some wastewater treatment facilities, watershed models used for education of residents, signage for the watershed, and mapped-out storm drains.

Baffin Bay The reverence that anglers and recreational boaters have for Baffin Bay is best expressed by quotations:

> I love fishing and I grew up fishing the ULM and Baffin. It is a totally unique bay to Texas and all the US. It is a closed system but very beautiful and special.
>
> —Paul Wimberly
> (began fishing in Baffin Bay over 50 years ago)

> I grew up and lived my whole life on the bay; it has been our family's good friend.
>
> —Richard Unterbrink, commercial fisherman
> (began fishing in Baffin Bay 54 years ago at age [illegible])

> Baffin is a true treasure and very unique because it is hypersaline and also has rocks and reefs. If we don't continue to work toward solutions, this bay will eventually die.
>
> —Jim Scoggin
> (began fishing in Baffin Bay 48 years ago)

Since 1989, Baffin Bay and the adjacent Upper Laguna Madre have experienced large, prolonged blooms of the phytoplankton known as Texas Brown Tide (Buskey et al. 1998, 2001). These blooms have caused seagrass die-off (Onuf 1996a, 1996b, [illegible]) and reduced fishing opportunities.

> The brown tide was so bad during the summer of 1990, that on July 27, we moved our Corpus Christi–based operation to the Lower Laguna Madre.
>
> —Bill Sheka, fishing guide
> (began fishing in Baffin Bay in 19[illegible])

Episodic hypoxia coinciding with large fish kills has also been documented on occasion over the past decade (TPWD, unpub. reports). In addition, a large fish kill occurred in late 2010 coincident with hypoxia likely caused by a bloom of the dinoflagellate *Pyrodinium bahamense* (TPWD, unpub. reports). This organism produces toxins under certain conditions (Phlips et al. 2015) and has been implicated in low DO events that led to fish kills elsewhere (Morrison and Greening 2011). The dinoflagellate is typically most competitive when nutrient inputs are relatively high (Phlips et al. 2015).

Despite these troubling signs, until recently little was known about the overall health of Baffin Bay or if the seemingly negative conditions for fish were a result of natural variability or symptoms of human impacts on the watershed and bay. This changed in 2012, when a group of citizens shared their concerns about the health of Baffin Bay. Their persistence brought new attention to Baffin Bay, and by 2013 several new scientific studies were undertaken to address these concerns. One of these was a water-quality sampling program that greatly expanded on previous sampling efforts.

The researchers relied on a dedicated group of "citizen-scientists," including Jim Atkins, Frank Baumann, Jerry Bjork, Jan Cannamore, Al Gordon, Michael Henry, Carl Hooker, Thomas Mullenix, Scott Murray, Bobby Richter, David Rowsey, Ron Ryon, Jim Scoggins, Charles Sellers, Diane Sellers, John Sutton, and Paul Wimberly. These volunteers provided manpower and boats on a monthly basis for four years so that TAMU–CC scientists could collect water-quality data necessary to assess the health of Baffin Bay. The following quotes exemplify the motivation for their efforts:

> I want a healthy bay for my kids and grandkids now and after I'm gone.
>
> —John Sutton (began fishing in Baffin Bay 60 years ago at 13 years old)

> I wanted to be part of an effort to save and recuperate Baffin Bay.
>
> —Jim Scoggins

This study initially received funding from Kleberg County, the Coastal Conservation Association, Saltwater Fisheries Enhancement Association, and Coastal Bend Bays and Estuaries Program. Later, the Celanese Corporation, Texas Sea Grant, TGLO's Coastal Management Program, and NOAA provided "boosts" to continue the sampling program.

A key finding of the volunteer-based water-quality study is that nitrogen concentrations are consistently higher in Baffin Bay than in other Texas estuaries and estuaries worldwide (see Wetz et al. 2017; Felix and Campbell 2019). Furthermore, the nitrogen is primarily in an organic form that is conducive to supporting Texas Brown Tide growth but not that of competing phytoplankton considered ecologically beneficial. Prior studies ignored organic nitrogen, which explains why Texas Brown Tide blooms originally were thought to be naturally occurring and not a result of nutrient enrichment.

It is now clear that reducing nitrogen loads to Baffin Bay is the key to ultimately improving water-quality conditions. The good news is that this has been accomplished in many other locations. In early 2018, a group of stakeholders was formed, representing a cross section of private citizens, agencies, academics, and nonprofits. The goal of this group is to reduce the frequency and extent of Texas Brown Tide blooms by reducing external nitrogen loads to Baffin Bay. At a broader level, the group essentially wants what many other longtime users of Baffin Bay have expressed:

> I want to see the overall decline turned around. Baffin Bay has provided a good economy locally from both commercial and sport fishing and duck hunting. We want to have that back and sustain it.
>
> —Richard Unterbrink

> I want future generations to enjoy and appreciate what I have experienced in my lifetime.
>
> —Mike Blackwood
> (began fishing in Baffin Bay with his Dad in 1955)

> I want to see the general public have a better understanding of what is going on in Baffin Bay. I want to see local, regional, and state involvement in finding and implementing solutions.
>
> —Jim Scoggins

This group has begun to the lay the groundwork for watershed restoration activities while also developing a watershed protection plan (e.g., Arroyo Colorado) that will offer a menu of options and watershed management activities aimed at ensuring the viability of Baffin Bay for decades to come.

Island Inhabitation: Tamaulipas

The Laguna Madre de Tamaulipas contains nearly 200 natural islands (INEGI 1994). Of the 107 inhabited islands in Mexico, 22 exist in Tamaulipas, more than in any other state, and all of the inhabited islands in Tamaulipas are located in the Laguna Madre. Their total population in the early 1990s was about 1,400 (Mendoza et al. 20[illegible]). Population size on individual islands ranged from single-family dwellings of several people to small villages of about 237 people on Isla las Malvinas. Most of the people inhabiting the islands are participants in the fishing industry (INEGI 1994). There are concerns that human cohabitation on islands with nesting colonial waterbirds is detrimental to the birds. Removal of vegetation for housing and firewood destroys critical habitat, and domestic dogs may severely impact nesting success. Vegetated islands also stabilize substrates, so destruction or removal of vegetation can lead to increased erosion. Other destructive activities include clearing for housing development construction, constructing roads, and creating a proliferation of off-road trails (Brown et al. 1976; Brown et al. 1977; Brown et al. 1980; Judd et al. 1989). Devegetation of vegetated barrier island flats and foredunes renders these habitats highly susceptible to erosion by wind and water, especially during tropical storms and hurricanes which can carry large volumes of sediment into the Laguna Madre.

Island management plans need to be developed that enable fishermen to utilize non-nesting colony islands and to stay clear of rookeries during nesting season. Some government assistance may be necessary to help transfer them from rookery to nonrookery islands. Destruction of natural coastal vegetation should be avoided, especially on barrier islands and areas adjacent to the Laguna Madre. Since vegetation stabilizes coastal land and minimizes land loss by island or shoreline erosion, vegetation should be restored when destroyed by construction or other activities (Brown et al. 1976; Brown et al. 1977; Brown et al. 1980).

Island Cabins and Houseboats

Of the approximately 400 permitted cabins registered under the Texas Cabin Program managed by the TGLO, 287 are located on islands in the Laguna Madre. The presence of cabins on dredged-material islands may have impacts on the lagoon's ecosystem (fig. 22.2). New cabins cannot be constructed, so most TGLO regulations deal with repairs, modifications, or approved construction on existing structures. When cabins are located over water and where docks and piers extend over water, seagrass meadows can be shaded, affecting plant productivity. Some illegal dredging of channels occurs when cabin lessees "prop-wash" a shallow area for better access to piers, thus impacting seagrass meadows. Most toilet facilities are primitive, ranging from pit toilets on the islands to dock facilities over the water. The TGLO does not have jurisdiction over

a

b

Figure 22.2. *Fishing cabins located on dredged-material islands in the northern parts of the Land-Cut between the GIWW and The Hole (a) and the Upper Laguna Madre, Texas (b) (Photographs by J. W. Tunnell, October 1995 [a] and November 1983 [b])*

these practices, and permits generally require that the lessee conform to local county health codes. Weekend household trash and debris from cabin or dock modifications often accumulate. TGLO usually notifies the lessee by letter of the problem, requesting immediate removal. Abandoned cabins are usually destroyed.

After commercial and recreational fishermen began bringing floating cabins into Texas bays and lagoons, the state legislature passed a law requiring all owners of floating houseboats to either apply for a permit or remove the cabin. The TPWD was charged with oversight of the Texas Floating Cabin Permit Program as of 1 September 2001. The TPWD approved a total of 151 applications for floating cabin permits in all Texas bays, then ceased to issue permits for new floating structures. Permitted owners of floating cabins can sell their houseboat to a new owner, but a fee is required to transfer the permit.

Most floating cabins are not self-propelled, so the state issued specific regulations concerning the location, lighting, and sanitation devices. Each floating cabin permit is issued annually, and failure to renew within 90 days of the permit expiration results in the loss of the permit. Permit holders may rebuild, repair, or replace a cabin as long as the size does not exceed the specifications of the original permit. Should a cabin be found in violation of the permit provisions, the permit holder may be required to remove the structure from coastal waters.

Shoreline Construction

Construction of groins, piers (fig. 22.3), jetties, and oil and gas wellsite peninsulas has altered water circulation and sediment transport patterns along Laguna Madre and Gulf shorelines in Texas. In Tamaulipas, jetties along Gulf beaches and *charangas* (fish traps; see fig. 17.11) in the Laguna Madre have similar effects. *Charangas*, usually placed specifically to catch fish or shrimp on flood or ebb tide, have caused considerable shoaling and erosion of the lagoon bottom (J. W. Tunnell, pers. observ.). The state of a shoreline, whether erosional, depositional, or in equilibrium, is primarily caused by natural processes, such as availability of source sediments and wave intensity (Brown et al. 1976; Brown et al. 1977; Brown et al. 1980). Shoreline construction alters the natural dynamics, and each alteration in one place is compensated for in another. For example, construction of a jetty or groin along an erosional shoreline will often trap sediment on one side while potentially speeding up erosion on the other.

In Texas, the Coastal Erosion Planning and Response Act, implemented in 1999, is being managed by the TGLO with scientific guidance from the Bureau of Economic Geology of the University of Texas at Austin.

Proper management and use of shorelines along the open Gulf and Laguna Madre require recognition of the natural characteristics and processes governing each area. Any modification of a particular shoreline should take full account of these characteristics and processes and avoid disrupting them (Brown et al. 1976; Brown et al. 1977; Brown et al. 1980).

Figure 22.3. *Numerous long fishing piers extend from the shoreline at Port Mansfield, Texas, on the Lower Laguna Madre. (Photograph by J. W. Tunnell, July 1987)*

Land Reclamation and Filling

Although not a serious concern in the Laguna Madre, the artificial filling of shallow lagoonal waters adjacent to land may create potentially valuable shorefront property for housing development or industrial expansion. The process also permanently destroys important coastal habitats and alters shoreline configuration, water circulation, and sediment transport (Brown et al. 1976; Brown et al. 1977; Brown et al. 1980). Filling and land reclamation projects should be considered not only in terms of economic value but also in terms of environmental impact to natural systems.

Artificial Passes

Hypersalinity due to limited circulation between the Gulf of Mexico and Laguna Madre combined with low freshwater inflow and high evaporation has been identified as a "problem" since at least the early 1920s (Higgins and Lord 1926). Despite overall high fisheries productivity, frequent summer fish kills, particularly in Baffin Bay, in the early part of the twentieth century resulted in cries to "do something," beginning nearly a century of engineering projects designed to decrease salinities in the Laguna Madre. One popular "solution" to hypersalinity has been dredging passes through the barrier islands. Passes opened by hurricanes combined with torrential rainfall have a short-term impact on salinities in the Laguna Madre because they usually close within a short period after storm passage (Gunter 1946; Behrens 1969; Hildebrand 1969, 1980; Behrens et al. 1977). Permanent, jettied passes have ameliorated salinities within the Texas Laguna Madre (Quammen and Onuf 1993) and Laguna Madre de Tamaulipas to a certain extent. However, permanent passes also interrupt longshore sand transport, resulting in erosion downcurrent from jetties. Closed, natural passes or washover passes are often chosen for stabilization, but these habitats are vital to many wintering birds. Jettied passes may completely destroy these habitats when channels are dredged, or their value may be changed because of alteration of hydrology or other natural processes.

Because of the low tidal range along the Texas coast, generally only one pass per bay or lagoon system can be maintained by natural processes. Additional passes reduce tidal exchange through existing ones, necessitating increased dredging to maintain them (Brown et al. 1976; Brown et al. 1977; Brown et al. 1980). Artificial passes also alter natural circulation patterns and subject protected lagoonal waters to greater effects from storm surges. Only Mansfield Pass has remained open as an artificial pass, and it requires considerable maintenance dredging at a cost of approximately $800,000 per year. Yarborough Pass and Fish Pass (Corpus Christi Water Exchange Pass) are examples of failed artificial passes (Gunter 1946; Behrens et al. 1977). In Tamaulipas, all four currently dredged passes were artificially opened, although the one at Boca Ciega was the site of a natural washover pass. All four were strategically placed for access and water exchange: two in the northern and two in the southern Laguna Madre (see figs. 2.1, 2.4, and 2.5).

Before construction, the economic benefit of artificial passes should be weighed against the cost of additional maintenance dredging and increased potential damage from storm surges (Brown et al. 1976; Brown et al. 1977; Brown et al. 1980). Because passes are often dredged through habitats used by birds, there is a need to study local effects of passes on hydrology and ecology of tidal flats and other bay margin habitats.

Oil Spills

Exploration for oil and gas and subsequent production are major revenue-generating activities in South Texas. Impacts from oil and gas navigation channelization, seismic activity, and other issues are addressed elsewhere in this chapter. Oil spills from

Figure 22.4. *Sheets of oil accumulated on Padre Island beaches following the Ixtoc I oil spill in 1979. (Photograph by Brian R. Chapman)*

exploration and production occasionally occur and can potentially cause major environmental damage, especially in sensitive coastal environments. No major oil spills have occurred in the Laguna Madre in Texas or Tamaulipas. However, the major IXTOC I oil spill in 1979 in the Bay of Campeche heavily impacted Texas Gulf of Mexico beaches (fig. 22.4), and some oil extended through Brazos Santiago Pass and Mansfield Pass, impacting adjacent lagoonal habitats (Hooper 1981; Tunnell et al. [illegible]; Rabalais and Flint 1983). Environmental Sensitivity Index (ESI) maps were created as a result of that spill (Hayes et al. 1980), and oil spill prevention and preparedness became a priority in Texas. Today the Oil Spill Prevention and Response Division of the TGLO maintains a ready force to prevent and combat oil spills. The division also maintains a constantly updated geographic information system set of ESI maps for the entire Texas coast, including the Laguna Madre, and its Oil Spill Toolkit brings all of this information, and more, together and ready for the next spill (TGLO [illegible]; Buschang 2018).

TGLO is providing excellent oversight, preventive measures, and cleanup of Texas oil spills. Support for this system should be continued.

Contaminants

The Lower Rio Grande Valley was once one of the most intensively farmed areas in the United States with over 425,000 ha (1,050,175 ac) of cultivated land that received repeated pesticide applications throughout the year (Custer and Mitchell 1987a, [illegible]; Jahrsdoerfer and Leslie 1988). Runoff was transported to the Lower Laguna Madre via the Arroyo Colorado and other agricultural drains. Elevated chlordane and total DDT concentrations (Dick 1982), DDE (White, Mitchell, Kennedy, et al. 1983), and arsenic (Wells et al. 1988) accumulated within the water, sediment, fish and fish-eating birds, and water associated with this drainage system. At or near the outlets of these drainages into the Lower Laguna Madre, DDE and arsenic were found in tissues of wintering shorebirds (White, Mitchell, Kennedy, et al. 1983), breeding Willets (Custer and Mitchell 1991), nesting Black Skimmers, and Laughing Gulls (White, Mitchell,

Kennedy, et al. 1983; White et al. 1984; Custer and Mitchell 1987b). Within the Lower Laguna Madre, concentrations of mercury, arsenic, selenium, lead, cadmium, and organochlorine compounds were generally at background levels in sediments and biota and did not suggest a threat to wildlife (Custer and Mitchell 1993). Higher concentrations of mercury and DDE recorded in some samples in the Lower Laguna Madre probably originated from agricultural runoff (Custer and Mitchell 1993). Fortunately, TCEQ initiated a program to monitor for "legacy pollutants," those substances whose use has been banned or severely restricted by the EPA, and former elevated levels of these substances in sediments and fish tissues have continued to decline (TCEQ 2003).

Accumulations of Debris

Padre Island National Seashore has implemented a long-term effort in debris cleanup and evaluation of sources (origins). Most studies of marine debris, including the increased accumulation of nurdles (tiny plastic pellets used in the manufacture of plastic products), are targeted toward Gulf shorelines; bay debris data are more difficult to locate and quantify. However, a report on the incidence of bay debris in the Texas Coastal Bend identified potential areas where bay debris may originate or accumulate (Amos et al. 1997). Locations such as boat ramps, harbors, industrial boat yards, and marinas were designated as point sources, while roadways adjacent to the shoreline and storm runoff discharges were designated nonpoint sources. Eighteen boat ramps were identified in the Upper Laguna Madre as potential point source areas. Few roads are actually located near the shoreline of either the Upper or Lower Laguna Madre, so trash accumulation at boat ramps may be a significant source. Other sources of trash and debris may include illegal dumping at remote areas; trash left at campsites or cabins or blown out of boats; and debris that has floated in through Gulf passes. Nurdles, which are about the size of a caper, enter Gulf waters and float into bays and lagoons after accidental release from plastics manufacturing plants (Sikes 2018). These tiny, bead-like pellets often harbor toxic chemicals and can cause physical damage in addition to poisoning after being ingested by fish and other marine organisms (Tanaka et al. 2013; Wright et al. 2013).

The following recommendations have been made to understand debris issues better in the Laguna Madre of Texas: (1) properly design bay debris surveys that result in good-quality information; (2) train volunteers to collect and categorize trash and debris at regular intervals; (3) use and maintain the database to assess status and trends of bay debris; and (4) identify sources and implement a public educational program to reduce bay trash and debris (Amos et al. 1997).

Seagrass Scarring

Although the most obvious cause of current seagrass loss in the Laguna Madre is light attenuation, other factors may also be important. Many seagrass meadows in the Upper Laguna Madre, especially near the JFK Causeway, have been scarred by boat propellers, although damage is currently not classified as severe (Dunton, Montagna, et al. 1998). There has been no study of the extent of seagrass scarring in the Lower Laguna Madre. Depending on the length and depth of scars, as well as the species of seagrass affected, damage can result in permanent loss. Deep, long scars tend to deepen and erode along the edges, causing greater local turbidity as well as increased water flow, often resulting in continuing seagrass loss after the initial damage. Some species, particularly Shoal Grass, may begin recolonizing scars that are small and/or shallow almost immediately. Damage to Turtlegrass meadows with their extensive, often massive, rhizome systems is typically more long-lasting than damage to Shoal Grass meadows, an annual species. Many seagrass meadows are accidentally scarred when boaters misjudge depths or the location of channels, but some damage is deliberate, as some boaters try to maintain illegally marked, propwashed channels into otherwise inaccessible shallow waters or take shortcuts between channels or other areas

Because of the significant role that seagrasses play in the marine ecosystem, the Texas state legislature enacted a new law in 2013 to protect these vital habitats. The law prohibits the uprooting of seagrasses with an outboard motor propeller. Boaters who uproot seagrasses can be charged with a Class C misdemeanor and can be fined up to $500 for each offense. Electric trolling motors are exempt from this regulation. Consequently, TPWD encourages boaters who are in an area where there is a potential for seagrass destruction to "lift, drift, pole, and troll" (Salinas 2013).

Accidental losses of seagrasses due to scarring could be avoided through greater education of boaters and better channel markers, especially in privately maintained channels. In fact, TPWD encourages all boaters to take a boater education course that includes information about seagrasses. Boaters should also familiarize themselves with their equipment and the area in which they will be boating before heading out onto the water. In addition, preservation of seagrass meadows in shallow areas might be accomplished through establishment of "no boat zones." This concept has merit and has been suggested by the Coastal Conservation Association. No boat zones are currently under consideration by TPWD for several areas in the Texas Laguna Madre.

Sportfishing Pressure

Fishing pressure along the Texas coast, particularly in the Laguna Madre and Baffin Bay, has reached an all-time high. While fishery managers continue to grapple with data and trends, the numbers of saltwater fishermen continue to grow at an amazing pace. Our state's population is expected to increase to about 34 million by 2030. According to a recent survey, Texas saltwater license sales are growing at the rate of about 2000 new fishermen per month. How many more fishermen will be on the Laguna Madre and Baffin Bay by 2030? The number will be staggering by today's standards, and the demands on fishery resources will be unprecedented.

Perhaps no other example better illustrates the megatrend in fishing pressure than the Upper Laguna Madre. Between 1984 and 2001, the Upper Laguna Madre and Baffin Bay experienced a 243% increase in total combined fishing pressure with an incredible 600,000 man-hours of combined fishing time in 2000–2001.

It is noteworthy that beginning in 2000, guide boats in the Upper Laguna Madre and Baffin Bay began landing more trout than private boats did. For example, during 2001, guide boats landed almost 30% more trout than private boats did. Considering the fact that the number of private anglers far outweighs the number of fishing guides, this is an amazing statistic, which illustrates the harvest effectiveness of fishing guides. A big difference between the private fishermen and fishing guides, in terms of fishing pressure, is the comparable time spent fishing. Successful guides fish perhaps 200 days per year or more, while most private anglers fish a fraction of that time.

Licensed saltwater fishing guides are at an all-time high in Texas. As of 2008, there were 936 licensed fishing guides in Texas. There has been a 360% increase in fishing guides since 1980. We simply cannot continue this trend in unchecked fishing guide growth and mass harvest and expect to have quality sport fisheries.

Limited entry of fishing guides is not a new concept in fisheries management and is currently practiced in a number of states. Texas should consider placing a cap on the current number of guides with term limits and nontransferability of licenses. We should also consider developing an accurate science-based approach for determining the number of fishing guides that should be allowed to fish in the Laguna Madre and Baffin Bay, or any other Texas bay system.

Reports of "double trips" during the day for some fishing guides have become commonplace, particularly during "croaker bait" season. Currently, Texas has no regulations governing the number of trips that a fishing guide can charter during a 24-hour period, which obviously needs immediate regulatory attention.

The rapidly growing number of saltwater fishing tournaments along the Texas coast is also contributing to fishing pressure, and this is certainly true for the Laguna Madre and Baffin Bay. Although TPWD performed an interim study of 183 saltwater fishing tournaments in 2005–6, it is difficult to identify the many hundreds of sponsored and nonsponsored events that occur each year. The interim study concluded that the overall impact from tournaments on fish populations is unknown and that tournaments in Texas are on the rise. What we do know is that fishing tournaments that kill fish either in small or large aggregates do have an accumulative impact on the resources. Live "weigh-ins" are practiced for bass fishing tournaments and have been for years. This conservation practice should be required and adopted for all saltwater tournaments as well.

It is recommended that TPWD and/or academia perform a more comprehensive study of both sponsored and nonsponsored inshore fishing tournaments. There is a potential need for some form of regulation on tournaments and, quite possibly, some combination of permitting and associated registration fees, record-keeping requirements to determine harvest impacts, and consideration of the timing and number of large tournaments impacting other users on our crowded bays.

Finally, catch and release is a valuable conservation tool to hedge against rising fishing pressure. First and foremost, releasing fish allows the fish to be caught more than once and provides more fish in the pool to be caught. Second, releasing fish allows the fish to spawn more than once or twice. For example, most trout in the Laguna Madre and Baffin Bay are caught before their second year of age and, therefore, have only one or possibly two spawning seasons before they are harvested. Third, catch and release allows a sport fish to reach its maximum trophy potential.

Although catch and release is being practiced by some fishing guides and private anglers, we are still a long way from maximizing the use of this valuable conservation tool. It is recommended that TPWD, Coastal Conservation Association of Texas (CCA), Saltwater-Fisheries Enhancement Association (SEA), and academia develop and implement outreach educational programs emphasizing and promoting catch and release as an important conservation tool to hedge against present and future fishing pressure.

Although TPWD has traditional fisheries management tools, such as size and bag limit restrictions to offset fishing pressure, the individual attitudes and actions of fishing guides, private fishermen, and tournament fishermen will ultimately determine the amount of fishing pressure and its impact on sport fisheries in the Laguna Madre and Baffin Bay.

Impacts of Wind Farms

The recent demand for alternative sources of energy has resulted in wind power becoming the fastest-growing energy source in the world, particularly in the United States (Wiser and Bolinger 2017). This push for wind-energy development has come with little knowledge of its impacts on wildlife and their habitats (Kuvlesky et al. 2007). Because the footprint of wind energy is much larger than that for other forms of energy, its potential impacts on habitats and certain species can be significant. As wind farm development expands across South Texas, many effects remain unknown due to limited physical access allowed to wind farms on private lands and information stemming from on-the-ground investigations of a proprietary nature preventing transfer of knowledge about current impacts. Given these limitations, considerable information has been gained over the last decade about the potential impacts of wind farms on wildlife and their habitats.

Bird and bat mortality from collisions with wind turbines is the most publicized impact of wind farms on wildlife. This is especially of concern along the lower Texas coast because this region supports some of the highest passage rates of migratory birds in the world (Contreras Walsh et al. 2017). Additionally, during migratory movements many birds fly at altitudes consistent with the rotor-swept height of wind turbines erected in the region. During spring, weather conditions that attenuate visibility are relatively common along the lower Texas coast, and birds tend to fly in these poor-visibility conditions during spring, making them more susceptible to collision with turbines (Contreras Walsh et al. 2017). Indirect effects of wind farms on wildlife have also been observed for wind farms adjacent to the Laguna Madre. For instance, disturbance displacement of Redheads from the proximity of the wind farm, as well as impacts on local hydrology caused by wind farm infrastructure, have impacted use of coastal ponds by Redheads in their core wintering area (C. Lange 2014; Lange et al. 2018).

A less-studied impact of wind farms on wildlife is the effects wind farms can have on local climate. Wind turbines modify surface-atmosphere exchanges caused by vertical mixing of air masses (Baidya Roy et al. 2004; Baidya Roy and Traiteur 2010; Baidya Roy 2011; Zhou et al. 2012). Effects of wind farms, particularly large wind farms, on modifying local meteorological conditions can be significant (Adams and Keith 2007; Baidya Roy et al. 2004). Regions such as the lower Texas coast that exhibit relatively small ambient turbulent kinetic energy and relatively stratified atmospheric layers are impacted more by the warming effects of the increased surface-atmosphere exchange (Baidya Roy and Traiteur 2010; Fitch et al. 2013). This is especially of concern in the semiarid climate of the lower Texas coast, where some species (Northern Bobwhite) appear to already be close to their thermal limit (Guthery 2000). Even small increases in temperature (<1 °C) are projected to have detrimental effects on ground-dwelling species in these semiarid environments (Carroll et al. 2015).

The need for alternative sources of energy is clear; however, proper siting of energy development with consideration of direct and indirect impacts to wildlife and their habitats is critical. This is not only important for the sustainability of the many other natural resources in the region but also for support of these alternative energy forms by the public.

Afterword

Cabins, Floaters, and Lagunatics

DAVID A. MCKEE, AARON BAXTER,
AND JIM ATKINS

It may at first seem strange that a chapter on the island-situated and floating cabins (houseboats) is included in this book when most chapters are exclusively dedicated to the biology, ecology, and conservation of the rich and diverse living resources of the Laguna Madre. In reality, however, the anglers, birders, duck hunters, and sightseers are the ones who enjoy all the lagoon has to offer (fig. 23.1). Among the people most appreciative of its natural beauty and exquisite solitude are the cabin and floating-cabin owners who have long regarded the lagoon as their "hallowed ground."

The abundance of shorebirds, waterfowl, and fish has long attracted anglers and nature lovers to the Laguna Madre. Some simply escape to the lagoon to enjoy the pleasure of horizon-wide vistas, relaxing serenity, and total darkness at night. No one knows who built the first cabin on an island in the Laguna Madre, but it began a tradition that has established friendships and bonded families for generations. Each of the following sections is written by a person who has benefited from his own special relationships with the lagoon, a cabin, or a houseboat and those individuals with whom

Figure 23.1. *Anglers often enjoy beautiful mornings on the Laguna Madre. (Watercolor painting by Dr. John Pettigrove)*

his experiences were shared. Their accounts provide a more personal gauge of how the "Mother Lagoon" has value well beyond the biological, ecological, and commercial attributes described in this book.

Introduction

David A. McKee

About 10 years ago, Aaron Baxter and I decided it might be timely to invite all the permitted cabin owners in the Texas Laguna Madre to share their stories and photographs for a book we would publish. We were granted permission from the Texas General Land Office (TGLO) and obtained contact information for each cabin "owner" from that agency. Packets were mailed to over 400 permit holders, stating what we needed from them if they chose to participate. Surprisingly, only 13 returned their acceptance forms along with stories and photographs. Thus, this project was dead in the water. Only after the invitation by Wes Tunnell to include a chapter on cabins was this project jolted back into action. Aaron Baxter was back on board, and I invited Jim Atkins (owner of the floating cabin called "Church" and a longtime fellow "lagunatic") to join us. Jim also provided the firsthand narrative about the storied "Old 97A" cabin. Late in the game, John Pettigrove joined our ranks by agreeing to provide paintings of some of the structures.

With hundreds of cabins and floaters present in the lagoon, it is not possible to include all the stories and histories associated with them. Thus, the stories surrounding three cabins located on dredged-material islands, a famous cabin located on a low hummock at the edge of the lagoon, and one floating cabin were selected for inclusion. Several of these are well-known to anglers, and some are used as landmarks. Two cabins, the "Summer House" and "Old 97A," are iconic and legendary. I have been a partner for nearly 40 years in a third cabin, "Hewit South," that I regard as my personal lodge. Texas A&M University–Corpus Christi conducts various types of marine research and education at its Laguna Madre Field Station, and it is included as well. Although the floating-cabin program is much more recent, the first and most storied of all of them is the "Church." It is included because it was in place decades before the floating-cabin program was initiated.

Some of the first-ever marine conservation efforts on the Texas coast were "hatched" by Laguna Madre cabin and floating-cabin owners back in the 1970s. The Gulf Coast Conservation Association (GCCA), precursor to the Coastal Conservation Association (CCA), was formed in the late 1970s, and some of the organization's first directors "owned" cabins and floating cabins in the Laguna Madre. Results of these early brainstorming sessions "down south" generated a myriad of successful marine conservation activities up and down the Texas coast.

History, Rules, and Regulations

Aaron Baxter

Since the early 1900s, cabins on Texas bays and lagoons have been constructed and used by fishermen and duck hunters. Known initially as "fishing shacks," the earliest cabins were established as base camps, allowing access to remote areas within the bays. These buildings were often positioned on one of the few naturally occurring small islands or shell outcroppings. The dearth of suitable building sites restricted the number of cabins that could be built within the bays. In 1949, the United States Army Corp of Engineers (USACE) completed the dredging of the Gulf Intracoastal Waterway (GIWW) to provide a shipping lane from Corpus Christi to Brownsville. As the channel was excavated, the slurry of dredged material was poured along the edges of the GIWW. Eventually, the mounds of spoil removed from the channel dried and consolidated, forming hundreds of new islands. Soon, other channels dredged to reach oil and gas platforms followed, and the number of islands grew. With the increase in available acreage came a proliferation of newly built cabins. Some were nothing more than ramshackle huts, but others were more substantial. In 1973, the Texas legislature enacted Chapter 33 of the Texas Natural Resources Code. This new legislation charged the School Land Board, acting through the TGLO, with the management of these cabins. Someone could register any cabin in existence on or before 27 August 1973 under the new management system. If a cabin was not registered, the cabin "owner" would lose any claim to the structure. At that time, 619 cabin permits were registered with the State.

The TGLO administers a set of rules and regulations pertaining to the permitting of a cabin in Texas. Permits are issued for five years, and cabins can be used only for recreational purposes. Every five years, a permit holder is required to renew the permit and has some assurance that the permit will be continued if the cabin and the permit holder are in compliance. Cabin permits may be amended or transferred as long as these actions are approved by the managing agency. Periodically, the TGLO offers the public an opportunity to bid for a permit for the construction of new cabins on carefully predetermined sites. The sites are selected using criteria designed to minimize impacts to natural resources. Between 2005 and 2011, 10 permit bids were offered. A bid winner is responsible for assuming all costs related to the construction and maintenance of the cabin, but the State of Texas retains ownership of both the cabin and the property. Each cabin permittee must install a self-contained, portable waste system and remove all debris from the cabin site. Additionally, the permit holder must comply with all guidelines for the construction of new or rebuilt piers.

Today, 401 permitted cabins are located in nine counties along the Texas coast. Most of these cabins (287) are located in the Upper Laguna Madre. The cabins average just over 93 m^2 (1,000 ft^2) in size, although recently many larger cabins have been constructed. It has always been a common practice for permit holders to take on partners who help cover the expenses associated with maintaining the structure. These costs are often high due to the remote location and constant exposure of these cabins. This trend continues today, and many permit holders have multiple partners to assist in efforts to keep the structure in compliance with state regulations.

In addition to the land-based Texas Cabin Program, there is also a Coastal Floating Cabin Permit Program, which was initiated after numerous commercial and recreational fishermen began placing floating structures within the state's public waters. During the 77th Texas legislature, Senate Bill 1573 charged the Texas Parks and Wildlife Department (TPWD) with the management of floating cabins effective 1 September 2001. Those with existing cabins were invited to either apply for a permit or remove the structure. The TPWD received a total of 151 applications and issued permits to all applicants, but from that time on, no new permits for floating cabins have been issued.

To maintain a permit for a "floater," as they are often called, a permit holder must comply with a set of rules and regulations that are somewhat similar to those established for land-based cabins. However, some special regulations apply to the location of a floating cabin. For example, a floater cannot be moored in the waters of a state park, state wildlife refuge, or state coastal preserve or within 150 m (500 ft) of another TPWD permitted cabin or TGLO permitted structure. Floating cabins must also be located more than 75 m (250 ft) from a navigational obstruction and must not damage oyster reefs, serpulid reefs, or seagrass beds. Other rules include use of a marine sanitation device and the display of both the cabin permit number and a white 360° light visible from sunset to sunrise. The permits are issued on an annual basis, and failure to renew within 90 days of the permit expiration results in the loss of that permit. Permit holders may relocate a cabin up to twice a year and may rebuild, repair, or replace a cabin as long as the size does not exceed that which was stated on the original permit. If a cabin is found to be in violation of the provisions of the code, the permit holder may be required to remove the structure from coastal waters.

While floating cabins occur in all Texas bays, the largest concentration of coastal floating cabins is found in the area known as the Land-Cut. This portion of the GIWW connects the Upper and Lower Laguna Madre through an area in Kenedy and Kleberg Counties. Because accessible boat ramps are scarce in the area, the floaters provide stationary camps that allow access to the many fishing and hunting opportunities found in this part of the Laguna Madre.

Since no new permits are issued for floating cabins in Texas, the only way to obtain a permit is through the transfer of an existing one. The TPWD requires a $300 transfer fee for the transaction, but the exchange from one permit holder to another often involves the payment of large cash sums. Many permit holders take on partners to help cover the cost of purchasing the permit and the expense of maintaining the structure in its harsh, remote location.

The Summer House

David A. McKee

Veteran Upper Laguna Madre regulars refer to the Summer House as a landmark for a great fishing area along the Kenedy Ranch shoreline. However, few know the history of this structure. The ranch hands called it the "Tide Gauge Cabin" because the government built a tide gauge pier there in the 1940s to monitor water levels in the Upper Laguna Madre. Most anglers are familiar only with the other tide gauge pier that still stands behind the northern end of the Tide Gauge Bar in Baffin Bay.

All that remains today of this one-time favorite retreat of rancher Sarita K. East is the rounded rooftop, which can still be seen extending above the tree line on the western shore at the northern end of the Land-Cut. Construction of the Summer House began in 1950 under the supervision of two longtime Kenedy Ranch employees, Fred Yaklin Sr. and Louis Turcott. It was built primarily as a getaway for Sarita, but it was also used by the "coast rider" cowboys during cattle roundups, ranch security riders, and game wardens when they were working the eastern sections of the 81,000 ha (200,000 ac) ranch. The structure rested on a massive slab of concrete and consisted of a kitchen, a screen porch that encircled the cabin, and a carport. A butane light, refrigerator, and cooking stove were in place along with a wood-burning stove. A central, two-sided fireplace and chimney was used to heat the cabin during the winter. Drinking water was hauled in by trailer in a 1,983 L (500 gal) cypress tank. The grounds surrounding the encampment were planted with native vegetation and enclosed with a barbed-wire fence in an attempt to keep deer, feral hogs, and javelinas from consuming Sarita's yard plants and the produce in her garden. The 2 m (6 ft) high knoll on which the Summer House was built provided a magnificent panoramic view of the surrounding tidal flats, Laguna Madre, and the ranch.

Located only a half-day drive from the ranch headquarters and county seat in Sarita, Texas (Sarita East's namesake), the cabin was frequently occupied. It was the only place for anyone to stay on the eastern section of the ranch. A sand, clay, and caliche road was constructed across the mudflats so vehicles could reach the cabin. Sarita would stay there for weeks on end, and she loved to fish from the tide gauge pier. She often waded the worm reef flats in front of her beloved Summer House to catch Redfish, Spotted Seatrout, flounder (likely the Southern Flounder), and Black Drum. She used

Figure 23.2. *The "Summer House," an early cabin built on the Kenedy Ranch shoreline, was built for Sarita Kenedy East, the ranch owner and an avid angler. The school of fish swimming beneath the Laughing Gulls illustrates why the cabin was located here. (Watercolor painting by Dr. John Pettigrove)*

the cabin until her death in 1961. Thereafter, it was used less and less, and by the mid-1970s it was beyond repair. Illegal commercial fishermen occasionally hid their nets inside the structure in later years to avoid capture by TPWD game wardens.

The unique beach rock along the shoreline that begins at Penascal Point and continues south in front of the cabin to the Land-Cut is reminiscent of an old Gulf of Mexico coastline. The worm reefs that dot the flats in front of the Summer House and extend out to the Intracoastal Canal provide excellent fish habitat. The area is well-known by anglers because fish moving south along that shoreline stack up there before entering the Land-Cut and fish leaving the Land-Cut fan out on the shoreline as they travel north.

No photographs of the Summer House in its prime were available when I first wrote this piece in 1995, nor were any available for this writing. John Pettigrove discovered a print by renowned wildlife artist John Dearman titled *Summer House* and used watercolors to paint the image used here (fig. 23.2). The Summer House and surroundings are private property, so attempts to visit the structural remains are discouraged. Descriptions of this famous structure and its history were provided by Louis Turcotte (deceased), Paul Wimberly, Mike Blackwood, and Scott Murray.

Old 97A

Jim Atkins

Soon after the Gulf Intracoastal Waterway was completed through the Laguna Madre, a few resourceful individuals began constructing fishing cabins on the islands created from the disposal of dredged material. One of the earliest of these cabins was built in 1952 and 1953 by an individual whose name no one seems to recall. He arranged to haul the building materials through the Laureles Division of the King Ranch. He stockpiled them on the shoreline for later transport by boat to a dredged-material

Figure 23.3. *Aerial photograph of "Old 97A" on a busy weekend. The "scooter" built by Ford "Docky" Allen can be seen to the upper right side of the pier. (Photograph contributed by David McKee; photographer unknown)*

island approximately a mile north of Baffin Bay. Shortly after completing the cabin, the new owner suffered a serious heart attack. His doctors recommended that he not risk having heart issues 40 km (25 mi) south of Corpus Christi "in the middle of nowhere." Following his doctor's advice, he sold the cabin to Red Tyler, who owned the boat ramp and bait house located at what is now Bluff's Landing. Red began renting the cabin to fishermen, and a group including Chatter Allen, Smiley Davis, Ford Allen, and Charlie Haas leased the cabin for the summer months for two consecutive years in the mid-1950s. After the second summer they approached Red about buying the cabin and made a deal. No one is sure about the selling price, but it is thought to be between $500 and $600 (fig. 23.3). They adopted the "97A" name for the cabin because of its proximity to a can buoy on the Intracoastal Waterway numbered "97A." Other partners joined and left this group over the years, but this original group remained partners, fishing buddies, and friends for the next 35 years.

Chatter and Smiley were both coaches in the Corpus Christi Independent School District (CCISD); Chatter ultimately became the CCISD athletic director, and Smiley became his assistant. Their jobs provided them with free time in the summer, which they often spent at the cabin. Several times they stayed in excess of 30 days without returning to town. Other partners would often join them on weekends, and one friend, Bond Critzberg from Georgetown, was always a fixture on extended stays. They welcomed friends who stopped by for afternoon visits and possibly a card game. They never took crushed ice to the cabin but carried blocks of ice in large coolers and chipped shards to chill their beverages. When friends came down to fish for a couple of days, the price of admission was several coolers full of *block* ice, food other than fish, and a couple of extra 23 L (6 gal) cans of gasoline. Large Igloo ice chests full of huge Redfish and Spotted Seatrout lined the wall of the screened-in back porch, and departing guests were encouraged to take as many fish as they desired. A few of these fish headed to market in the ice boxes of their commercial fishing friends, who also frequently stopped by for visits. In exchange, the commercial fishermen acted as "security guards" when no one was using the cabin. Consequently, Old 97A was one of the only cabins never broken into or vandalized.

All of the partners frequented the cabin on a regular basis with family and friends. Many of their children grew up fishing the Laguna Madre. Many of their fondest memories are the cabin trips, the assortment of "characters" they encountered, and the enduring friendships they made.

A Fourth of July cabin trip was an annual affair with all of the owners and their guests, who spent several days fishing, cooking, and socializing. Someone always seemed to have a guitar or mandolin, and after a hard day of fishing, the afternoon gathering usually evolved into a group seated around the kitchen table playing music, singing, recalling old fishing tales, and sharing the latest jokes. The early morning of the Fourth always saw the US flag ceremoniously raised at the end of the pier, where everyone assembled to sing the "Star-Spangled Banner." I specifically recall one Fourth of July morning in the mid-1970s as I sipped coffee and visited with Chatter on the pier while observing boats headed south in the Intracoastal Waterway. After counting 25 boats, Chatter turned to me and said, shaking his head, "The whole damn town is coming to Baffin Bay." Obviously, the recreational pressure on the fishery at that time was nothing compared to that today.

These original owners evolved into some of the most respected fishermen on the Laguna Madre and were likely the first shallow-water sight casters on the Texas coast. They became exclusively artificial-lure fishermen and were the first in this area to use polarized sunglasses. They also evolved from wading in tennis shoes to booties. They lived by the "run deep and fish shallow" rule and always respected other fishermen by giving them a wide berth. They were, however, highly competitive, and each always tried to outfish the other. Chatter was arguably the best of the group, often stating before he left in the morning that he would fish only for trout that day. Later, he would return with an amazing stringer of only trout. The following day, he would do the same with Redfish. Their initial boat was a 5 m (16 ft) HiggsCraft with a walk-through windshield, although Chatter later acquired a 16 ft Vannoy skiff. Ford, or "Dooley" as he was called, built a small scooter steered by leaning to one side or the other.

They were some of the first to fish the Graveyard/Nine Mile Hole when the only access was either beaching a boat along the Intracoastal Waterway and walking across the dredged-material islands or driving down the beach and hiking through the dunes. The real challenge when accessing from the Intracoastal Waterway was getting the fish back to the boats. Although they limited themselves to 10 fish each, they each had to portage approximately 45 kg (100 lb) of fish. They tried various ways to lessen the efforts involved, finally settling on strapping their fish to old World War II–era stretchers. The four corners of the stretcher were each toted by one participant while the other fishermen lugged all the rods, reels, and tackle. After one of these exhausting trips the group was relaxing and recuperating while enjoying the southeast breeze on the east porch of the cabin when the screen door on the west side slammed. One of the group questioned who might be coming in, whereupon Bond Critzberg calmly replied, "Don't worry, that's just my ass dragging in from the Graveyard."

Having observed the thousands of fish doomed to death each year after being trapped in the landlocked body of water due to seasonally low tides, the Old 97A group discussed what could be done to save these fish. After several years of developing various ideas, the group requested the help of Senator John Tower. Eventually, the US Army Corps of Engineers dredged channels connecting the Graveyard to the Intracoastal Waterway, thus providing escape routes for the previously doomed fish.

When efforts began in the mid-1970s to increase regulations on the commercial fishing industry, the Old 97A partners were somewhat reluctant to get involved due to their years-long friendship with many commercial fishermen and their concerns about vandalism to their cabin. However, the youngsters they had mentored became extremely involved, and this group composed a large percentage of the original Corpus Christi Chapter Board of the GCCA, now CCA. Several Old 97A partners attended GCCA fund-raising dinners, and Chatter joined us on a chartered bus bound for Austin in 2001 and testified in favor of legislation declaring Redfish and Spotted Seatrout as game fish.

The original group of owners sold 97A in 1990 to a group that ultimately included Chatter's son Butch, and they continue to use it regularly. Amazingly, the cabin has endured everything Mother Nature has thrown at it, including hurricanes, high tides, howling northers, and the erosion of the island. Although the composition of partners evolved during the past 65 years, the cabin still stands looking much like it did the first time I visited in the early 1970s.

Hewit South

David A. McKee

The cabin got its name following the divorce of one of my original partners who lived in the exclusive Hewit Drive neighborhood. Following the divorce, his ex-wife got the Hewit property and he got the fishing cabin down south, which he tongue in cheek named "Hewit South."

The cabin, located on an island we named "Twin Palms Island," was probably built in the early 1960s. It was purchased by my partners for $200 from two commercial

Figure 23.4. *Two cabins at "Hewit South" were originally connected by a narrow plank but were later joined when the space between them was enclosed and made into a breezeway. The pier was constructed of recycled lumber. (Watercolor painting by Dr. John Pettigrove)*

fishermen in the early 1970s. At the time, two separate buildings, a cook house and a bunk house, were connected by a narrow board. The many hand and knee prints in the sand below the board were evidence that the walk across the plank was not always successful. Later the area between the two structures was enclosed and made into a breezeway (fig. 23.4).

I became a partner in 1980 but was very familiar with the cabin as I had spent quite a few nights there as a guest over the previous three or four years. About 1985, we reroofed and repainted the structure and repaired the pier. We also jacked up the entire structure and set it on telephone poles that we had hauled through the King Ranch and floated across to the cabin. Over the years, several new partners were involved with rebuilding the cabin and pier. The pier is constructed entirely of recycled lumber, which makes it unique because it is the only one on the lagoon made using this long-lasting material and the only one that is impervious to boring marine worms that destroy wooden piers. The most recent renovation was about 10 years ago when the wall studs were replaced with 2×4 treated lumber (the original studs were only 2×2s), and the exterior plywood walls were replaced with HardiePlank. More recently, a generator shed was added.

One story worth telling involves the troublesome brother of a high-ranking naval officer at Naval Station Corpus Christi in the early 1960s. The brother's behavior around the city was continually embarrassing the officer, so he relocated the brother to Twin Palms Island, where he could do no harm. The brother lived on the island for quite some time and raised goats and rabbits to supplement the fish he caught. My understanding is that he learned his lesson and was eventually brought back to civilization. In the 1980s, the remnants of the rabbit pens were still visible. At least one very large domestic rabbit was still on the island in the late 1980s.

Breaking and entering was a problem for all cabin owners back during the so-called Redfish Wars in the 1980s. One break-in involved an intruder cutting into the side of the cabin with a chainsaw and taking an outboard engine. Back then, few people were down the lagoon, and it was quite easy for anyone to take what they wanted, undetected. Such is no longer the case.

Pet burials were common on the islands, especially for dogs that spent much time down there. Our first Golden Retriever, Hydro, loved the water (thus his name), and when he passed, the family took him down there at slow speed wrapped in a blanket. We laid him to rest near the cabin as we laughed and cried, remembering all the good times we had there with him.

No one characterized the Laguna Madre in recent times more than the "Baron of Baffin," Louis Peetz. A large group of friends spread his ashes on Twin Palms Island and made a weekend into an Irish Wake, where we laughed and cried, telling stories and recalling all the good times we had there with Peetzy.

One thing all cabin "owners" (actually, we do not own the cabins—the TGLO is the owner) share in common is the treasured memories old and new that are created on the islands. Many like me have children that have grown up down there, roaming and exploring the islands. Socializing with other cabin owners after a day of fishing or at night sharing meals and stories is a universal activity enjoyed by all cabin and floater owners. Nighttime fishing under pier lights can be productive and allows an angler to avoid the blistering summer sun. Back in the day, everyone knew everyone down there, as most had had their cabins from the outset. Most owners flew flags so all knew who was there as they came south down the GIWW. All knew where another would be fishing and avoided interfering with that spot. The Laguna Madre remained an isolated bay system until the late 1980s. Before the development of tunnel-hull boats, few could navigate the shallow waters of Baffin Bay. There was always the fear of hitting one of the many serpulid worm reefs that lurked just below the surface. Of course, the cabin owners knew the bay well and knew where to run and what to avoid. Although the Laguna Madre is now known and frequented by an ever-increasing number of anglers and sightseers, it is still a magical place. The fishing experience is always great, but when the catching is sporadic, bird-watching is a pleasant alternative. Good times are always guaranteed!

Laguna Madre Field Station

Aaron Baxter

The Laguna Madre Field Station (LMFS) is situated on a dredged-material island in the Upper Laguna Madre, approximately 5 mi (8 km) south of the JFK Causeway, which connects the mainland to Padre Island. Leased in the mid-1970s from the TGLO under an Educational Facilities Lease, the LMFS was constructed through volunteer efforts by Texas A&M University–Corpus Christi (TAMUCC) faculty, staff, and students. The original structure was small, serving as a site to provide field training and education, while also supporting research. As time passed, the harsh coastal environment caused the cabin to become weathered beyond repair, and a new facility was constructed. Under a lease amendment approved by the TGLO, construction began on a new structure in the winter of 1999–2000. The current complex, managed by the Center for Coastal Studies (CCS) at TAMUCC, measures over 158 m^2 (1,700 ft^2) and consists of a laboratory/classroom, two dormitories, a kitchen, and two bathrooms (fig. 23.5). The LMFS has propane-, gasoline-, and solar-powered utilities. A rainwater collection system capable of storing up to 9,500 L (2,500 gal) provides water to the cabin.

The location of the LMFS makes it an opportune staging area for sample collections, environmental monitoring, and continued outreach/education. Undergraduate- and

Figure 23.5. *The Laguna Madre Field Station, located in the Upper Laguna Madre, serves as a base for field studies by Texas A&M University–Corpus Christi faculty, students, and visiting artists. (Watercolor painting by Dr. John Pettigrove)*

graduate-level classes utilize the LMFS, as do various field camps. Past and current research has been conducted by TAMUCC faculty and students, as well as researchers from other organizations, including the Coastal Bend Bays and Estuaries Program. Recent areas of research have focused on harmful algal blooms, Black Mangrove expansion, avian movements, and hydrological modeling.

While research and education at the LMFS still remain at the forefront, the CCS has recently introduced a new Artist in Residence (AIR) program whose purpose is to blend the arts with biological sciences to foster collaboration within the local, national, and international communities. Applications are accepted from artists in all mediums, and chosen applicants are immersed in a week-long stay at the LMFS, where they create a portfolio inspired by their visit to the Laguna Madre. Three artists, all painters, have participated in the program. It is hoped that artists in other mediums will also take advantage of this unique opportunity to work in such a secluded, yet beautiful setting. The sale of artwork created during the residency goes toward maintaining the LMFS so that students, researchers, school groups, and artists can enjoy this unique facility well into the future.

The Church

Jim Atkins

After graduating from college and moving back to Corpus Christi, I was fortunate to fish with a group of guys who owned a cabin, Old 97A, located north of Baffin Bay. After a couple of years of being their guest, Paul Wimberly and I decided we would like to buy or build a cabin in the same general area. We contacted the General Land Office to inquire about the process to obtain a permit, but we were told that no additional permits would be issued and even attempting to buy an existing permit was problematic. Hearing this news, Paul contacted a friend with Ottis Engineering, who had a small floating structure west of the Old 97A cabin and asked if they were interested in selling. After learning they were interested, we began contacting various agencies that we thought might have some regulatory authority. Finding none, we bought the only floating cabin on the Laguna Madre in 1975. It was a small cabin on an approximately 12 m (40 ft) deck with bunks for four, a small butane range, and table and chairs for dining.

Soon after purchasing the cabin, Paul and I were returning from a quail hunt and discussing the need for some type of reflector on the cabin so we could locate it at night with a spotlight from the Intracoastal Waterway. As we talked, suddenly our headlights hit this yellow reflective sign on the side of the highway near Petronila, and we said, “Problem solved.” We decided to “borrow” the sign to serve as our reflector, and it happened to be a “Church” sign. Now, over 40 years later, Church number three sits in about the same location with a yellow “Church” sign, this one acquired by ordering and paying for it.

After using the cabin extensively, we decided to tow it to town in 1980 with the intention of enlarging it and doing other maintenance. Carl Perry tied his shrimp boat, the *Blue Lady*, onto the floating cabin, intending to tow it to his family’s bait shop and marina, located at what is now Bluff’s Landing. Paul and I released the four anchors holding the cabin in place as Carl applied power to his boat, but the cabin did not move. Thinking it might be aground, Carl allowed slack in his line and jerked the line taunt, thinking the jolt might free the cabin. After this effort failed several times, Carl backed down beside the cabin, feeding rope off his boat. From several feet behind the cabin, Carl applied full throttle, and with his diesel engine bellowing black smoke, the *Blue Lady* roared by the cabin, the rope skipping over the water. When he hit the end of the rope the stern of the boat lifted out of the water and shifted to one side, while the cabin shook and creaked. As the boat settled back into the water, the slack was again jerked out of the rope and the cabin began to move. The tow to Flour Bluff was under way. Not until we pulled the cabin out of the water at the marina did we discover the fifth anchor attached to the center of the substructure. Evidently, we had plowed the lagoon for 40 km (25 mi) and in the process had crossed two natural gas pipelines.

Richard Haas became a partner, and we completed the cabin enlargement and other maintenance and launched the cabin in preparation for the trip back to Baffin Bay. Unfortunately, at the same time Hurricane Allen developed in the Gulf of Mexico and hit Corpus Christi in August 1980. We did all we could to secure the cabin prior to the storm, but high tides and wind broke it loose. It made several excursions through the marina, destroying all the piers and sinking a couple of shrimp boats before eventually coming to rest, undamaged, in what is now the parking lot of Laguna Reef Restaurant. When the parking lot dried enough to support a crane, we picked the cabin up, set it back in the marina, and much to the delight of Buddy Perry, the marina owner, returned it to its mooring north of Baffin Bay.

The mid-1970s and early 1980s saw a huge increase in illegal commercial fishing, primarily the utilization of gill nets in the closed waters of the Laguna Madre. Paul, Richard, and I were early members of GCCA (now CCA) and heavily involved in developing legislation and regulations addressing these illegal activities. Through this involvement we developed friendships with several game wardens, including Frank Dickerson, who was the regional director of enforcement for an area covering most of South Texas, including all state waters from north of Rockport to Brownsville. Conversations led to the idea of the game wardens using our floater as a base. This allowed them to remain in the Baffin Bay area for multiple days. The wardens started using the cabin in 1980 and continued until a Monday in 1982 when Paul telephoned saying the Church had burned. When I had left the cabin late Sunday afternoon, everything was fine. Within an hour of his phone call, Paul and I were in my boat headed south, where we found the cabin destroyed and still smoldering. It was obvious that an accelerant had been used, and we assumed the arson was retaliation for our involvement in GCCA and/or allowing the use of the cabin by the wardens. When we returned to town, we telephoned Frank Dickerson and informed him of the fire. Frank was beyond mad and felt certain commercial interests were involved. He moved wardens from throughout the state to the Laguna Madre and contacted the Texas Rangers to investigate. The Laguna was effectively closed to commercial fishing for the next three to four months, and though the responsible parties were not identified, it was confirmed to have been arson. Several years later we got a confirmation of the responsible parties.

Fortunately, the cabin was insured and we towed the pontoons to town. Erik Doughty joined our partnership, and we purchased new pontoons and started construction on a new, slightly larger Church. We donated the old pontoons to TPWD and arranged for GCCA to purchase materials for TPWD to construct both a floating cabin and a land-based cabin. They sold the land-based cabin several years later, but they still maintain a floater located in the Baffin Bay area.

The second Church served as a fishing base and afternoon gathering spot for family and friends until 2000, when we brought it to Flour Bluff for repairs and maintenance. After getting it into town and closely examining all that needed to be done as a result of 20 years on the Laguna Madre, the decision was made to junk the second Church and construct a new one. The third Church is currently sitting within a few feet of where

Figure 23.6. *The original floating cabin named "The Church" was the first "floater" in the Laguna Madre. The watercolor painting by Dr. John Pettigrove depicts the third floater occupying the same location used for 40 years.*

the previous cabins were moored, and it continues to provide a great base for fishing and enjoying the unique qualities of the Laguna Madre and Baffin Bay (fig. 23.[illegible])

The number of floating cabins grew substantially over the years, raising concerns about identifying ownership and liability in case of sinking or abandonment. We put together a group including several floating-cabin owners and approached TPWD about the possibility of the department becoming the regulatory agency since these structures are floating and could be registered much like a boat. They agreed, and with the assistance of our local legislative delegation, especially Representative Judy Hawley, a bill passed in the 2001 Legislative Session, becoming effective 1 September 2001, establishing rules and regulations for floating-cabin ownership.

Through 40-plus years, the Church has seen plenty of fish caught, numerous afternoon gatherings welcoming nearby cabin owners, some great meals, lots of politics and legislation discussed, fisheries management tools conceived, friendships enhanced and new ones made, children maturing into young adults, and grandchildren accompanying their fathers to the Laguna Madre. I just hope the cycle continues!

Appendix

Common and Scientific Names

Scientific names of organisms mentioned in the text appear in this appendix. Alternative common names and scientific names no longer in use but occurring in older sources, such as the first edition of this book, are also included in brackets; some accounts include informative notations. Federally listed threatened and endangered species are denoted by superscript E (^{E}) following the scientific name. Introduced (i.e., exotic) species are indicated by an asterisk (*) following the scientific name. The abbreviation SPN indicates Spanish common names mentioned in reference to the Laguna Madre de Tamaulipas.

Bacteria, Algae, Fungi, Lichens, and Protozoa

Cyanobacterium, phylum Cyanobacteria [photosynthetic bacteria; plural, Cyanobacteria]
Mermaid's Wine Glass, *Acetabularia* spp.
Red algae, division Rhodophyta [Several of the 6,000 species occur in the Laguna Madre.]
Red Tide, caused by either of two dinoflagellates: *Karenia brevis* [*Gymnodinium breve*] or *Alexandrium monilita* [*Gonyaulax monilita*]
Sargassum, *Sargassum fluitans* and *S. natans* [floating brown algae that wash ashore on Gulf beaches]
Texas Brown Tide, *Aureoumbra lagunensis*
Texas Fever, *Babesia bigemina* or *B. bovis* [These protozans transmitted by ticks cause the disease in cattle.]

Green Plants

Agarito, *Mahonia trifoliata* [Agarita, *Berberis trifoliata*]
Allthorn, *Castela erecta* [Goatbush, *C. texana*]
Amelia's Sand Verbena, *Abronia amelia*
Anacahuita, *Cordia boissieri* [Mexican Olive, Texas Wild Olive]
Anacua, *Ehretia anacua*
Annual Seepweed, *Suaeda linearis*
Arizona White Oak, *Quercus arizonica*
Australian Pine, *Casuarina equisetifolia**
Aztec Pine, *Pinus teocote*
Bailey's Ball Moss, *Tillandsia baileyi*
Bald Cypress, *Taxodium distichum*
Beach Carpet, *Blutaparon vermiculare*
Beach Evening Primrose, *Oenothera drummondii*
Beach Morning Glory, *Ipomoea imperati*
Beach Tea, *Croton punctatus*

Beebrush, *Aloysia gratissima*
Berlandier's Wolfberry, *Lycium berlandieri*
Bermuda Grass, *Cynodon dactylon*
Bigflower Bladderpod, *Lesquerella grandiflora*
Bitter Panicum, *Panicum amarum*
Blackbrush, *Vachellia rigidula* [*Acacia rigidula*]
Black-laced Cactus, *Echinocereus reichenbachii* var. *alberti*[E]
Black Mangrove, *Avicennia germinans* [SPN: *Mangle Negro*]
Black Mimosa, *Mimosa pigra* [Giant Sensitive Tree]
Black Willow, *Salix nigra*
Bluewood, *Condalia hookeri*
Border Paloverde, *Parkinsonia texana* var. *macra*
Brazilian Peppertree, *Schinus terebinthifolius**
Bristle-free Least Daisy, *Cheatopappa imberbis*
Brownseed Paspalum, *Paspalum plicatulum*
Buffalograss, *Bouteloua dactyloides*
Buffelgrass, *Cenchrus cilaris* [*Pennisetum ciliare*]*
bulrush, *Scirpus* spp.
Bush Seepweed, *Suaeda nigra*
Bushy Bluestem, *Andropogon glomeratus*
Buttonwood, *Conocarpus erectus* [SPN: Mangle Botoncillo]
Cabbage Palm, *Sabal palmetto*
Camphor Daisy, *Rayjacksonia phyllocephala* [*Haplopappus phyllocephalus*]
Camphorweed, *Heterotheca subaxillaris*
Cane Bluestem, *Bothriochloa barbinodis*
Carolina Wolfberry, *Lycium carolinianum* [Wolfberry]
Cedar Elm, *Ulmus crassifolia*
Chandler's Craiglily, *Echeandia chandleri* [Chandler's Amberlily, *Anthericum chandleri*] [E]
Clover Grass, see Star Grass
Coastal Bend Greenthread, *Thelesperma nuecense*
coldenia, *Tiquilia* spp.
Common Buttonbush, *Cephalanthus occidentalis*
Common Ragweed, *Ambrosia artemisiifolia*
Common Reed, *Phragmites australis* [River Cane]
cordgrass, *Spartina* spp.
Corpus Christi Fleabane, *Erigeron procumbens*
Cory's Croton, *Croton coryi*
Coyotillo, *Karwinskia humboltiana* [SPN: Tullidora]
Creosote Bush, *Larrea tridentata*
Crestless Wild Onion, *Allium canadense* var. *mobilense*
Crinkleawn, *Trachypogon secundus*
croton, *Croton* spp.
dayflowers, *Commelina* spp.
Denseflower Cordgrass, *Sporobolus montevidensis* [*Spartina densiflora*]
Desert Yaupon, *Schaefferia cuneifolia*
Dogweed, *Thymophylla pentachaeta* [Common Dogweed]
Dune Sandbur, *Cenchrus tribuloides*
Elmendorf's Onion, *Allium elmendorfii*
False Crown Tickseed, *Coreopsis nuecensoides*

Fiddleleaf Morning Glory, *Ipomoea imperati*
Fringed Signalgrass, *Urochloa ciliatissima*
Fringeleaf Paspalum, *Paspalum setaceum* var. *ciliatifolium* [Thin Paspalum]
Frogfruit, *Phyla nodiflora*
Giant Ball Moss, *Tillandsia baileyi* [Bailey's Ballmoss]
glasswort, *Salicornia* spp. [*S. bigelovii* and *S. virginica* occur in the Texas Laguna Madre.]
Goatfoot Morning Glory, *Ipomoea pes-caprae* [Railroad Vine]
Granjeno, *Celtis pallida* [Spiny Hackberry]
greenbriar, *Smilax* spp.
Groundsel Bush, *Baccharis halmifolia*
Guajillo, *Senegalia berlandieri* [Roundflower Acacia, *Acacia berlandieri*]
Guayacan, *Guaiacum angustifolium*
Guinea Grass, *Megathyrsus maximus**
Gulf Cordgrass, *Sporobolus spartinus* [*Spartina spartinae*; SPN: Zacahusitle]
Gulfdune Paspalum, *Paspalum monostachyum*
Gumbo-limbo, *Bursera simaruba* [SPN: Chaca]
Gutta-percha, *Maytenus texana*
Hairy Cutleaf Evening Primrose, *Oenothera mexicana*
Hairy Grama, *Bouteloua hirsuta*
Hall's Panic Grass, *Panicum hallii*
Hedgehog Cactus, *Thelocactus setispinus* [*Ferocactus setispinus*]
Honey Mesquite, *Prosopis glandulosa*
Hooded Windmillgrass, *Chloris cucullata*
Hooker's Palafoxia, *Palafoxia hookeriana*
Huisache, *Acacia farnesiana* [*A. smallii, Vachellia farnesiana, Mimosa farnesiana*]
Indian blanket, *Gaillardia* spp.
Jones' Nailwort, *Paronychia jonesii*
Kleberg Knotweed, *Polygonium striatulum*
Kleberg Saltbush, *Atriplex klebergorum*
Knotgrass, *Paspalum distichum*
Knotweed Leafflower, *Phyllanthus polygonoides*
La Coma, *Sideroxylon celastrinum* [*Bumelia celastrina*]
Lace Echinocereus, *Echinocereus reichenbachii*
Large Selenia, *Selenia grandis*
Largeleaf Marsh Pennywort, *Hydrocotyle bonariensis*
Leafless Alkali Weed, *Cressa nudicaulis*
Leatherstem, *Jatropha dioica* [Rubber Plant]
Lemon Bee Balm, *Monarda citriodora*
Lila de los Llanos, *Echeandia chandleri*
Lime Prickly Ash, *Zanthoxylum fagara* [Lime Pricklyash]
Lindheimer's Tephrosia, *Tephrosia lindheimeri*
Little Nipple Cactus, *Mammillaria heyderi* var. *hemisphaerica* [Pichilinga]
Live Oak, *Quercus virginiana* [Southern Live Oak]
Longleaf Pine, *Pinus palustris*
Lotebush, *Ziziphus obtusifolia*
Lundell's Nailwort, *Paronychia lundellorum*
Manatee Grass, *Cymodocea filiforme* [*Syringodium filiforme*]
Marsh-Elder, *Iva frutescens*
Marsh Fimbry, *Fimbristylis castanea*

Matted Sandmat, *Euphorbia serpens* [Mat Euphorbia]*
mesquite, *Prosopis* spp.
Mexican Ash, *Fraxinus berlandieriana*
Mexican Mud Plantain, *Heteranthera mexicana*
Mexican Pinyon, *Pinus cembroides*
Narrow-leaf Forestiera, *Forestiera angustifolia* [Desert Olive]
Narrow-leaved Cat-tail, *Typha domingensis*
Nelson's Pinyon, *Pinus nelsonii* [Nelson's Pine]
Olney's Three-square Bulrush, *Schoenoplectus americanus*
Padre Island Dropseed, *Sporobolus tharpii* [Tharp's Dropseed]
Paloverde, *Parkinsonia aculeata* [Retama, Palo Verde]
Pan American Balsamscale, *Elionurus tripsacoides*
Partridge Pea, *Chamaecrista fasciculata*
Pencil Cactus, *Cylindropuntia leptocaulis* [Tasajillo, *Opuntia leptocaulis*]
Perennial Glasswort, *Salicornia ambigua*
Pink Pappusgrass, *Pappophorum bicolor*
Pitaya, *Echinocereus enneacanthus*
Plains Gumweed, *Grindelia oolepis*
pondweed, *Potomogeton* spp.
Post Oak, *Quercus stellata*
prickly pear, *Opuntia* spp.
Purple Threeawn, *Aristida purpurea* var. *purpurea*
Railroad Vine, *Ipomea pes-caprae* [SPN: Riñonina]
Rattlebush, *Sesbania drummondii*
Red Grama, *Bouteloua trifida*
Red Lovegrass, *Eragrostis secundiflora*
Red Mangrove, *Rhizophora mangle* [SPN: Mangle Rojo]
Red Threeawn, *Aristida purpurea* var. *longiseta*
Rio Grande Dewberry, *Rubus riograndis*
Rio Grande Ragweed, *Ambrosia cheiranthifolia* [South Texas Ambrosia][E]
Roughseed Sea-purslane, *Sesuvium triathemoides*
Runyon's Beehive Cactus, *Coryphanta macromeris* var. *runyonii* [Runyon's Cory Cactus]
Runyon's Huaco, *Manfreda longiflora*
Runyon's Water-willow, *Justicia runyonii* [Runyon's Justicia]
rushes, *Juncus* spp.
saltbush, *Atriplex* spp.
Saltgrass, *Distichlis spicata*
Saltmeadow Cordgrass, *Sporobolus pumilus* [Marshhay Cordgrass, *Spartina patens*]
Saltwort, *Batis maritima* [SPN: Verdolaga de Playa]
San Patricio Tansyaster, *Psilactus heterocarpa* [Welder Machaeranthera, *Machaeranthera heterocarpa*]
Sand Brazoria, *Brazoria arenaria*
Sand-belt Bluet, *Houstonia subviscosa*
Sea Blite, *Suaeda linearis*
Sea Lavender, *Limonium carolinianum*
Sea Oats, *Uniola paniculata* [SPN: Avena de Mar]
Sea Oxeye, *Borrichia frutescens* [Sea Ox-eye, Sea Ox-eye Daisy]
Sea Purslane, *Sesuvium portulacastrum*

Seacoast Beardgrass, *Schizachyrium scoparium* var. *littoralis* [Seacoast Bluestem, *Andropogon scoparius*]
Seacoast Goldenrod, *Solidago sempervirens*
Seashore Dropseed, *Sporobolus virginicus*
Seashore Paspalum, *Paspalum vaginatum*
Seaside Beebalm, *Monarda maritima*
Seaside Heliotrope, *Heliotropium curassavicum* var. *curassavicum*
sedge, *Cyperus* spp.
Seep Willow, *Baccharis salicifolia*
Seepweed, *Suaeda nigrescens*
sensitive briar, *Schrankia* spp.
Shinner's Rocket, *Thelypodiopis shinnersii*
Shoal Grass, *Halodule beaudettei* [*H. wrightii, Diplantheria wrightii*]
Shoregrass, *Distichlis littoralis* [*Monanthochloe littoralis*; SPN: Saladilla]
Shortcrown Milkvine, *Matelea brevicoronata*
Shrubby Wood-sorrel, *Oxalis frutescens* [Shrubby Oxalis, *O. berlandieri*]
Silver Bluestem, *Bothriochloa laguroides* ssp. *torreyana*
Silverleaf Nightshade, *Solanum elaeagnifolium*
Silverleaf Sunflower, *Helianthus argophyllus*
Slender Bluestem, *Schizachyrium tenerum*
Slender Dwarf Morning-glory, *Evolvulus alsinoides*
Slender Rushpea, *Hoffmannseggia tenella*[E]
Slim Tridens, *Tridens muticus* var. *muticus*
smartweeds, *Polygonum* spp.
Smooth Cordgrass, *Sporobolus alterniflora* [*Spartina alterniflora*]
Smut Grass, *Sporobolus indicus*
Snake-eyes, *Phaulothamnus spinescens*
snakewood, *Condalia* spp.
Soapbush, *Guaiacum angustifolium* [SPN: Guyacán]
South Texas Clammyweed, *Polanisia erosa breviglandulosa*
South Texas Phacelia, *Phacelia patuliflora*
South Texas Rushpea, *Caesalpinia phyllanthoides*
Southern Cattail, *Typha domingensis*
Spanish Dagger, *Yucca treculeana* [SPN: Palma Chocha]
spike rushes, *Eleocharis* spp.
Spiny Allthorn, *Koeberlinia spinosa*
Spiny Hackberry, *Celtis ehrenbergiana* [SPN: Palo Blanco]
Star Cactus, *Astrophytum asterias* [Star Peyote][E]
Star Grass, *Halophila engelmannii* [Clover Grass, Peanut Grass]
Stinking Rushpea, *Pomaria austrotexana*
Sugar Hackberry, *Celtis laevigata*
sunflowers, *Helianthus* spp.
Switchgrass, *Panicum virgatum*
Tanglehead, *Heteropogon contortus*
Tenaza, *Harvardia pallens* [Mimosa Bush, *Pithecellobium pallens*]
Texas Ayenia, *Ayenia limitaris*[E]
Texas Bullnettle, *Cnidoscolus texanus*
Texas Colubrina, *Colubrina texensis*
Texas Ebony, *Ebenopsis ebano* [SPN: *Ébano*]

Texas Kidneywood, *Eysenhardtia texana*
Texas Lantana, *Lantana urticoides* [Common Lantana, *L. horrida*]
Texas Mountain Laurel, *Sophora secundiflora*
Texas Palmetto, *Sabal mexicana* [Texas Sabal Palm, *S. texana*]
Texas Persimmon, *Diospyros texana*
Texas Prickly Pear, *Opuntia engelmannii*
Texas Sage, *Leucophyllum frutescens* [Cenizo, Texas Silverleaf; SPN: Cenizo]
Texas Sandmint, *Rhododon ciliatus* [Tharp's Rhododon]
Texas Sea Purslane, *Sesuvium trianthemoides* [Roughseed Sea Purslane]
Texas Stonecrop, *Lenophyllum texanum*
Texas Willkommia, *Willkommia texana*
Texas Windmill Grass, *Chloris texensis*
Texasgrass, *Vaseyochloa multinervosa*
Tharp's Rhododon, see Texas Sandmint
threeawn, *Aristida* spp.
Trichloris, *Trichloris pluriflora*
tridens, *Tridens* spp.
Tropical Royalblue Waterlily, *Nymphaea elegans* [SPN: Lirio Acuático]
Tumble Lovegrass, *Eragrostis sessilispica*
Turtlegrass, *Thalassia testudinum*
Twisted Acacia, *Vachella bravoensis* [*Acacia shaffneri* var. *bravoensis*]
Twoleaf Senna, *Senna roemeriana*
Vasey's Wild-lime, *Adelia vaseyi* [Vasey's Adelia]
Velvet Spurge, *Euphorbia innocua*
Water Oak, *Quercus nigra*
Water Tupelo, *Nyssa aquatica*
Waternymph, *Najas guadalupensis*
Welder Machaeranthera, *Psilactis tenuis* [*Machaeranthera tenuis*]
White Leadtree, *Leucaena leucocephala*
White Mangrove, *Laguncularia racemosa* [SPN: Mangle Blanco]
White Poppymallow, *Callirhoe involucrata* var. *lineariloba* [Slimelobe Poppymallow]
White Spikerush, *Eleocharis albida*
Widgeon Grass, *Ruppia maritima*
windmillgrass, *Chloris* spp.
Woolly Stemodia, *Stemodia lanata* [*S. tomentosa*]
yucca, *Yucca* spp.

Invertebrate Animals

Acorn Barnacle, *Balanus balanus*
Adams Miniature Ark, *Arcopsis adamsi*
Alternate Tellin, *Eurytellina alternata alternata*
American Carrier Shell, *Xenophora conchyliophora*
American Horse Mussel, *Modiolus americanus*
amphipod, *Gammarus* spp. among others
Angelwing, *Cyrtopleura costata*
Angular Triton, *Cymatium femorale*
Angulate Tellin, *Eurytellina angulosa*
Arrow Shrimp, *Tozeuma carolinense*
Asian Tiger Shrimp, *Penaeus monodon** [SPN: Camarón Tigre Gigante]
Atlantic Brief Squid, *Lolliguncula brevis*

Atlantic Calico Scallop, *Argopecten gibbus*
Atlantic Cyclinella, *Cyclinella tenuis*
Atlantic Deer Cowrie, *Macrocypraea cervus*
Atlantic Ghost Crab, *Ocypode quadrata*
Atlantic Gould Clam, *Gouldia cerina*
Atlantic Jewelbox, *Pseudochama radians*
Atlantic Kitten's Paw, *Plicatula gibbosa*
Atlantic Nut Clam, *Nucula proxima*
Atlantic Papermussel, *Amygdalum papyrium*
Atlantic Rangia, *Rangia cuneata*
Atlantic Sand Crab, *Emerita talpoida*
Atlantic Thorny Oyster, *Spondylus americanus*
Atlantic Trumpet Triton, *Charonia variegata*
Atlantic Winged Oyster, *Pteria colymbus*
Banded Tulip, *Fasciolaria lilium*
Bandeye Hermit, *Paguristes tortugae*
Barratt Corbula, *Corbula barratiana*
Baughman's Ark, *Anadara baughmani*
Bay Scallop, *Argopecten irradians amplicostatus*
Bay Squid, *Lolliguncula brevis*
Beautiful Caecum, *Caecum pulchellum*
Bicolor Purse-oyster, *Isognomon bicolor*
Bigclaw Snapping Shrimp, *Alpheus heterochaelis*
Black Date Mussel, *Lithophaga nigra*
Blackback Land Crab, *Gecarcinus lateralis*
Blood Ark, *Lunarca ovalis* [*Anadara ovalis*]
Blotched Swimming Crab, *Portunus spinimanus*
Blue Crab, *Callinectes sapidus* [SPN: Jaiba Azul]
Blue Land Crab, *Cardisoma guanhumi*
Brackish Grass Shrimp, *Palaemonetes intermedius*
Branham's Tulip, *Cinctura bramhamae* [*Fasciolaria branhamae*]
brine fly, *Ephydra* spp.
brine shrimp, *Artemia* spp.
Broad-ribbed Carditid, *Carditamera floridana*
Brown Mussel, *Perna perna*
Brown Rangia, *Rangia flexuosa*
Brown Shrimp, *Farfantepenaeus aztecus* [*Penaeus aztecus*; SPN: Camarón Cafe]
Bruised Nassa, *Phrontis vibex* [*Nassarius vibex*]
Bush's Keeled Pandora, *Pandora bushiana*
Buttercup Lucine, *Anodontia alba*
Buttonsnail, *Modulus modulus*
Calico Box Crab, *Hepatus epheliticus*
Calico Clam, *Macrocallista maculata*
Campeche Angelwing, *Pholas campechiensis*
Cancellate Semele, *Semele bellastriata*
Caribbean Armstripe Octopus, *Amphioctopus burryi*
Caribbean Corbula, *Caryocorbula caribaea*
Caribbean Olive, *Oliva scripta*
Caribbean Rock-boring Urchin, *Echinometra lucunter*
Caribbean Spiny Lobster, *Panulirus argus*

Carolina Marsh Clam, *Polymesoda caroliniana*
Catesby's Risso, *Schwartziella catesbyana*
Cayenne Keyhole Limpet, *Diadora cayensis*
Chalky Lucine, *Pegophysema philippiana* [*Anodontia philippiana*]
Channel Barrel-bubble, *Acteocina canaliculata*
Channeled Duck Clam, *Raeta plicatella*
Channelnose Spider Crab, *Coelocerus spinosus*
Chemnitz Ark, *Anadara chemnitzii*
Cloudy Periwinkle, *Litoraria nebulosa*
Coffee Melampus, *Melampus coffea*
Colorful Sea Whip, *Leptogorgia virgulata*
Common American Auger, *Terebra dislocata*
Common Atlantic Abra, *Abra aequalis*
Common Atlantic Slippersnail, *Crepidula fornicata*
Common Jingle, *Anomia simplex*
Common Octopus, *Octopus vulgaris*
Concentric Nut Clam, *Nuculana concentrica*
Conrad's Paperbubble, *Haminoea succinea*
Constricted Macoma, *Macoma constricta*
Contracted Corbula, *Caryocorbula contracta*
Convex Slippersnail, *Crepedula convexa*
Corregated Razor Clam, *Solecurtus cumingianus*
Crested Oyster, *Ostrea equestris*
Cross-barred Venus, *Chione elevata*
Cut-ribbed Ark, *Anadara floridana* [*A. secticostata*]
Daggerblade Grass Shrimp, *Palaemonetes pugio*
Dana Swimming Crab, *Callinectes danae*
Dark Cerith, *Cerithium atratum*
Dark Shore Crab, *Pachygrapsus gracilis*
Disk Dosinia, *Dosinia discus*
Dwarf Oyster, *Ostrea stentina*
Dwarf Surf Clam, *Mulinia lateralis*
Dyed Micro-cockle, *Nemocardium tinctum*
Eared Ark, *Anadara notabilis*
Eastern Melampus, *Melampus bidentatus*
Eastern Oyster, *Crassostrea virginica* [American, Atlantic, or Virginia Oyster; SPN: Ostión]
Eastern White Slippersnail, *Crepidula depressa*
Egg Cockle, *Laevicardium serratum*
Elegant Dosinia, *Dosinia elegans*
Elongate Macoma, *Macoma tenta*
Emerald Nerite, *Smaragdia viridis*
False Angelwing, *Petricolaria pholadiformis*
False Limpet, *Siphonaria pectinata*
False Shark Eye, *Neverita delessertiana*
Fat-clawed Hermit Crab, *Pagurus pollicaris* [Flatclaw Hermit]
Fat Dovesnail, *Parvanachis obesa*
fiddler crabs, *Uca* spp. [Seven species of fiddler crabs occur in the Laguna Madre.]
Flame Box Crab, *Calappa flammea*
Flamingo Tongue, *Cyphoma gibbosum* [Flamingo Tongue Snail]

Flat Tree Oyster, *Isognomon alatus*
Florida Cross-barred Venus, *Chione elevata* [formerly misidentified in Texas as *C. cancellata*]
Florida Fighting Conch, *Strombus alatus*
Florida Prickly Cockle, *Trachycardium egmontianum*
Florida Rocksnail, *Stramonita haemastoma*
Florida Spiny Jewelbox, *Arcinella cornuta*
Fragile Barnacle, *Chthamalus fragilis*
Fragile Surf Clam, *Macrotoma fragilis*
Furcate Spider Crab, *Stenocionops furcatus coleata*
Giant American Bittersweet, *Glycymeris americana*
Giant Atlantic Cockle, *Dinocardium robustum*
Giant Eastern Murex, *Hexaplex fulvescens*
Giant False Coquina, *Iphigenia brasiliana*
Giant Hermit, *Petrochirus diogenes*
Giant Triton, *Cymatium parthenopeum*
Giant Tun, *Tonna galea*
Glassy Lyonsia, *Lyonsia hyalina*
Grass Cerith, *Bittiolum varium*
grass shrimp, *Palaemonetes* spp.
Gray Pygmy Venus, *Timoclea grus*
Greedy Dovesnail, *Costoanachis avara*
Green Porcelain Crab, *Petrolisthes armatus*
Grooved Odostome, *Eulimastoma canaliculata* [*Odostomia canaliculata*]
Gulf Dovesnail, *Costoanachis semiplicata*
Gulf Frog Crab, *Raninoides louisianensis*
Gulf Grassflat Crab, *Dyspanopeus texanus*
Gulf Stone Crab, *Menippe adina*
Hairback Fiddler, *Uca vocator*
Hairy Sponge Crab, *Cryptodromiopsus antillensis*
Half-naked Pen Shell, *Atrina seminuda*
Harris Mud Crab, *Rhithropanopeus harrisii*
Hawkwing Conch, *Tricornis raninus* [*Strombus raninus*]
Hooked Mussel, *Iphigenia brasiliana*
Horse Conch, *Triplofusus giganteus* [*Pleuroploca giganteus*]
Humic Marsh Crab, *Sesarma ricordi*
Imperial Venus, *Lirophora latilirata*
Impressed Odostome, *Boonea impressa* [*Odostomia impressa*]
Incongruous Ark, *Anadara brasiliana*
Interrupted Periwinkle, *Echinolittorina interrupta* [*Littorina ziczac*]
Interrupted Turbonille, *Turbonilla interrupta*
Iridescent Swimming Crab, *Portunus gibbesi*
Ivory Bush Coral, *Occulina diffusa*
Ivory Cerith, *Cerithium eburneum*
Lady-in-Waiting Venus, *Puberella intapurpurea*
Laguna Madre Fiddler Crab, *Uca subcylindrica*
land hermit crab, family Coenobitidae
Lesser Blue Crab, *Callinectes similis* [Dwarf Crab, Swimming Crab]
Lightning Whelk, *Busycon contrarium*
Longfin Inshore Squid, *Loligo pealeii*

Longnose Spider Crab, *Libinia dubia*
Los Olmos Tiger Beetle, *Cicindela nevadica olmosa*
Lunar Dovesnail, *Astyris lunata*
Lunate Crassinella, *Crassinella lunulata*
Manfreda Giant Skipper, *Stallingsia maculosus* [Maculated Manfreda Skipper]
Mangrove Crab, *Ucides cordatus*
Mangrove Oyster, *Crassostrea rhizophoraea*
Mangrove Periwinkle, *Littorina angulifera*
Marsh Grass Shrimp, *Palaemonetes vulgaris*
Marsh Periwinkle, *Littoraria irrorata* [*Littorina irrorata*]
Matagorda Macoma, *Ameritella mitchelli* [*Macoma mitchelli*]
Mexican Hermit, *Clibanarius digueti*
Miniature Moonsnail, *Tectonatica pusilla*
Minor Jackknife Clam, *Ensis minor*
Morton Eggcockle, *Laevicardium mortoni*
Mossy Ark, *Arca imbricata*
Mottled Purse Crab, *Persephona mediterranea*
Mottled Shore Crab, *Pachygrapsus transversus*
Mudflat Fiddler, *Uca rapax*
Northern Lion's Paw, *Nodipecten fragosus* [*Lyropecten fragosus*]
Olive Nerite, *Neritina usnea*
Olmec Fiddler, *Uca marguerita*
One-toothed Simnia, *Simnialena uniplacata*
Orangeclaw Hermit, *Calcinus tibicen*
Ornate Cardiomya, *Cardiomya ornatissima*
Ornate Verticord, *Trigonulina ornata*
Pallas's Sea Whip, *Leptogorgia setacea*
Paper Scallop, *Amusium papyraceum*
Paphia Imperial Clam, *Lirophora paphia* [*Chione paphia*]
penaeid shrimp, family Penaeidae, which includes Brown, Pink, and White Shrimp
Pigmy Octopus, *Octopus joubini*
Pink Shrimp, *Farfantepenaeus duorarum* [SPN: Camarón Rosado]
Pitted Baby-bubble, *Japonacteon punctostriatus* [*Rictaxis punctostriatus*]
Plicate Horn Shell, *Cerithideopsis pliculosa* [Plicate Hornsnail, *Cerithidea pliculosa*]
Plicate Mangelia, *Pyrgocythara plicosa*
Pointed Nut Clam, *Nuculana acuta*
Pointed Venus, *Anomalocardia auberiana*
polychaetes, worms (phylum Annelida)
Ponderous Ark, *Noetia ponderosa*
Portly Spider Crab, *Libinia emarginata*
Prickly Cockle, *Trachycardium isocardia*
Purple Sea Urchin, *Arbacia punctulata*
Purplish Semele, *Semele purpurascens*
Purplish Tagelus, *Tagelus divisus*
Red Imported Fire Ant, *Solenopsis invicta**
Redfinger Rubble Crab, *Eriphia gonagra*
Regal Sea Fan, *Leptogorgia hebes*
Ridged Slipper Lobster, *Scyllarides nodifer*
Ridged Venus, *Globivenus rigida*
Rostrate Cardiomya, *Cardiomya perrostrata*

Rough Scallop, *Aequipecten muscosus*
Salle's Auger, *Hastula salleana*
Sally Lightfoot Crab, *Grapsus grapsus*
Sand Snapping Shrimp, *Alpheus floridanus*
Saw-toothed Pen Shell, *Atrina serrata*
Scaled Slipper Lobster, *Scyllarus depressus*
Scissor Date Mussel, *Lithophaga aristata*
Scorched Mussel, *Brachidontes exustus*
Scotch Bonnet, *Phalium granulatum*
Scuttle, *Octopus vulgaris* [SPN: Pulpo de Roca]
Sea Roach, *Ligia exotica*
Seawhip Barnacle, *Conopea galeata*
Seawhip Shrimp, *Neopontonides beaufortensis*
Semiplicate Dovesnail, *Costoanachis semiplicata* [*Anachis semiplicata*]
serpulid polychaetes, phylum Annelida, marine segmented worms
Shark Eye, *Neverita duplicata*
Sharp Nassa, *Nassarius acutus*
Sharptoothed Swimming Crab, *Callinectes rathbunae*
Shoulder Pearwhelk, *Fulguropsis plagosus* [*Busycotypus plagosus*]
Slender Sargassum Shrimp, *Latreutes fucorum*
slippershell, *Crepidula* spp.
Smooth Duck Clam, *Anatina anatina*
Smyth's Tiger Beetle, *Cincidela chlorocephala smythi*
Southern Quahog, *Mercenaria campechiensis*
Southern Ribbed Mussel, *Geukensia granosissima*
Speckled Swimming Crab, *Arenaeus cribrarius*
Spiny Jewelbox, *Arcinella arcinella*
Spongy Decorator Crab, *Macrocoeloma trispinosum*
Spotted Porcelain Crab, *Porcellana sayana*
Spotted Slippersnail, *Crepidula maculosa*
Stareye Hermit, *Dardanus venosus*
Stiff Pen Shell, *Atrina rigida*
Stout Tagelus, *Tagelus plebeius*
Striate Bubble, *Bulla striata*
Striate Piddock, *Martesia striata*
Striate Tellin, *Merista aequistriata*
Striped Porcelain Crab, *Porcellana sigsbeiana*
Sunray Venus, *Macrocallista nimbosa*
Sunrise Tellin, *Tellina radiata*
Superb Grasshopper, *Eximacris superbum*
Surf Hermit, *Isocheles wurdemanni*
Tagelus-like Macoma, *Macoma tageliformis*
Tampa Tellin, *Tampaella tampaensis* [*Angulus tampaensis*, *Tellina tampaensis*]
Taylor's Alternate Tellin, *Eurytellina alternata tayloriana*
Tenspine Spider Crab, *Stenocionops spinosissimus*
Texas Aligena, *Kelliopsis texasiana* [*Aligena texasiana*]
Texas Coquina, *Donax texasianus*
Texas Mud Crab, *Dyspanopeus texanus* [*Neopanope texana*]
Texas Tellin, *Tellina texana*
Texas Venus, *Agriopoma texasianum*

Thick Lucine, *Phacoides pectinata* [*Lucina pectinata*]
Thinstripe Hermit Crab, *Clibanarius vittatus*
Tidal Spray Crab, *Plagusia depressa*
Tiger Lucine, *Codakia orbicularis*
Titan Acorn Barnacle, *Megabalanus coccopoma**
Transverse Ark, *Anadara transversa*
True Tulip, *Fasciolaria tulipa*
Turkey Wing, *Arca zebra*
Turnip Whelk, *Busycoarctum coarctatum*
Two-toned Tree Oyster, *Isognomon bicolor*
Unequal Spoon Clam, *Periploma margaritaceum*
Variable Bittium, *Bittium varium*
Variable Cerith, *Cerithium lutosum*
Variable Coquina, *Donax variabilis*
Virgin Nerite, *Neritina virginea*
Warty Anemone, *Bunodosoma cavernata*
Wedge Piddock, *Martesia cuneiformes*
West Indian Cardita, *Carditamera gracilis*
West Indian Chank, *Turbinella angulata*
West Indian Dovesnail, *Columbella mercatoria*
West Indian False Cerith, *Batillaria minima*
Western Bay Scallop, *Argopecten irradians amplicostatus*
Wharf Roach, *Ligia exotica*
White Miniature Ark, *Acar dominguensis* [*Barbatia dominguensis*]
White Shrimp, *Litopenaeus setiferus* [SPN: Camarón Blanco]
White Spotwrist Hermit Crab, *Pagurus criniticornis*
White-spotted Octopus, *Callistoctopus macropus*
Yellow Box Crab, *Calappa sulcata*
Yellow Egg Cockle, *Laevicardium mortoni*
Yellow Prickly Cockle, *Dallocardia muricata* [*Trachycardium muricatum*]
Yellowline Arrow Crab, *Stenorhynchus seticornis*
Zostera Shrimp, *Hippolyte zostericola*

Fishes

American Eel, *Anguilla rostrata*
American Gizzard Shad, *Dorosoma cepedianum*
American Halfbeak, *Hyporhampus meeki* [*H. unifasciatus*]
Atlantic Bumper, *Chloroscombrus chrysurus*
Atlantic cichlids, *Cichlasoma* spp.
Atlantic Croaker, *Micropogonias undulates*
Atlantic Cutlassfish, *Trichiurus lepturus*
Atlantic Devilray, *Microgobius gulosus*
Atlantic Guitarfish, *Rhinobatos lentiginosus*
Atlantic Moonfish, *Selene setapinnis*
Atlantic Needlefish, *Strongylura marina*
Atlantic Spadefish, *Chaetodipterus faber*
Atlantic Stingray, *Dasyatis sabina*
Atlantic Thread Herring, *Opisthonema oglinum* [SPN: Sardina de Hebra]
Atlantic Threadfin, *Polydactylus octonemus*
Banded Astyanax, *Astyanax fasciatus*

Bank Cusk-eel, *Ophidion holbrookii*
Barred Grunt, *Conodon nobilis*
Bay Anchovy, *Anchoa mitchilli* [SPN: Anchoa]
Bay Whiff, *Citharichthys macrops*
Bigeye Searobin, *Prionutus longispinosus*
Bighead Searobin, *Prionotus tribulus*
Black Drum, *Pogonias cromis*
Black Grouper, *Mycteroperca bonaci*
Black Margate, *Anisotremus surinamensis*
Blackcheek Tonguefish, *Symphurus plagiusa*
Blackfin Goby, *Gobionellus atripinnis*
Blue Catfish, *Ictalurus furcatus*
Bull Pipefish, *Syngnathus springeri*
Bull Shark, *Carcharhinus leucas*
Channel Catfish, *Ictalurus punctatus*
Clown Goby, *Microgobius gulosus*
Code Goby, *Gobiosoma robustum* [SPN: Gobio Clave]
Common Carp, *Cyprinus carpio*
Common Snook, *Centropomus undecimalis*
Cowfish, *Acanthostracion quadricornis* [SPN: Torito Cornudo]
Cownose Ray, *Rhinoptera quadriloba*
Crested Blenny, *Hypleurochilus geminatus*
Crevalle Jack, *Caranx hippos*
Dark Seacatfish, *Cathorops melanopus*
Darter Goby, *Ctenogobius boleosoma*
Dog Snapper, *Lutjanus jocu*
Dotterel Filefish, *Aluterus heudelotii*
Double-spot Tonguefish, *Symphurus ocellatus*
Dusky Anchovy, *Anchoa lyolepis*
Dusky Pipefish, *Syngnathus floridae*
Dwarf Seahorse, *Hippocampus zosterae*
Estuarine Seacatfish, *Cathorops aguadulce*
Fat Sleeper, *Dormitator maculatus* [SPN: Pujul]
Finescale Menhaden, *Brevoortia gunteri* [SPN: Lacha]
Flagfin Mojarra, *Eucinostomus melanopterus*
Florida Pompano, *Trachinotus carolinus*
Freshwater Goby, *Ctenogobius shufeldti*
Frilled Goby, *Bathygobius soporator*
Fringed Flounder, *Etropus crossotus*
Fringed Sole, *Gymnachirus texae*
Gafftopsail Catfish, *Bagre marinus*
Gizzard Shad, *Dorosoma cepedianum*
Goliath Grouper, *Epinephalus itajara*
gray sharks, *Carcharhinus* spp.
Gray Snapper, *Lutjanus griseus*
Great Barracuda, *Sphyraena barracuda*
Green Goby, *Microgobius thalassinus*
Ground Drummer, *Bairdiella ronchus*
Gulf Flounder, *Paralichthyes albigutta*
Gulf Killifish, *Fundulus grandis*

Gulf Kingfish, *Menticirrhus littoralis*
Gulf Menhaden, *Brevoortia patronus*
Gulf Pipefish, *Syngnathus scovelli* [SPN: Pez Pipa del Golfo]
Gulf Stream Flounder, *Citharichthys arctifrons* [SPN: Lenguado Golfino]
Gulf Toadfish, *Opsanus beta*
Hairy Blenny, *Labrisomus nuchipinnis*
Halfbeak, *Hyporhamphus unifasciatus*
Hardhead Catfish, *Ariopsis felis* [*Arius felis*]
Highfin Blenny, *Lupinoblennius nicholsi* [SPN: Borracho Aletón]
Highfin Goby, *Gobionellus oceanicus*
Horse-eye Jack, *Caranx latus*
Inland Silversides, *Menidia beryllina*
Inshore Lizardfish, *Synodus foetens*
Irish Pompano, *Diapterus auratus*
Jack Crevalle, *Caranx hippos*
Ladyfish, *Elops saurus*
Lane Snapper, *Lutjanus synagris*
Least Puffer, *Sphoeroides parvus*
Leatherjack, *Oligoplites saurus*
Lined Sole, *Achirus lineatus*
Liza, *Mugil liza*
Longnose Gar, *Lepisosteus osseus*
Longnose Killifish, *Fundulus similis*
Lookdown, *Selene vomer*
Lyre Goby, *Evorthodus lyricus*
Man-of-War Fish, *Nomeus gronovii*
Mexican Goby, *Ctenogobius claytoni* [Blackfin Goby, *Gobionellus atripinnus*]
Mexican Mojarra, *Eucinostomus melanopterus* [SPN: Mojarra del Ley]
Molly Miller, *Scartella cristata* [*Blennius cristatus*]
Mosquitofish, see Western Mosquitofish
Mottled Mojarra, *Eucinostomus lefroyi*
Naked Goby, *Gobiosoma bosc* [SPN: Gobio Desnudo]
Northern Harvestfish, *Peprilus paru*
Northern Kingfish, *Menticirrhus saxatilis*
Northern Puffer, *Sphoeroides maculatus*
Offshore Tonguefish, *Symphurus civitatium*
Opossum Pipefish, *Microphis lineatus* [*M. brachyurus lineatus*]
Permit, *Trachinotus falcatus*
Pigfish, *Orthopristis chrysoptera*
Pinfish, *Lagodon rhomboides*
Planehead Filefish, *Stephanolepis hispidus*
Pygmy Filefish, *Stephanolepis setifer*
Rainwater Killifish, *Lucania parva*
Red Drum, *Sciaenops ocellatus* [Redfish, Channel Bass]
Red Lionfish, *Pterois volitans** [SPN: Pez León]
Redfish, see Red Drum
Rhomboid Mojarra, *Diapterus rhombeus*
Rio Grande Shiner, *Notropis jemezanus*
River Goby, *Awaous banana* [incorrectly listed as *A. tajasica* in first edition]
Rough Silversides, *Membras martinica* [SPN: Pejerrey Rasposo]

Roundel Skate, *Raja texana*
Sailfin Molly, *Poecilia latipinna*
Sand Seatrout, *Cynoscion arenarius*
Sargassum Frogfish, *Histrio histrio*
Sargassum Pipefish, *Syngnathus pelagicus*
Sash Flounder, *Trichopsetta ventralis*
Scaled Sardine, *Harengula jaguana*
Scrawled Cowfish, *Acanthostracion quadricornis*
Sergeant Major Damselfish, *Abudefduf saxitilis*
Sharksucker, *Echeneis naucrates*
Sheepshead, *Archosargus probatocephalus*
Sheepshead Minnow, *Cyprinidon variegatus*
Silver Jenny, *Eucinostomus gula* [SPN: Mojarra Española]
Silver Perch, *Bairdiella chrysoura*
Silver Seatrout, *Cynoscion nothus*
silversides, *Menidia* spp.
Skilletfish, *Gobiesox strumosus* [SPN: Cazoleta]
Smallscale Fat Snook, *Centropomus parallelus*
Smooth Puffer, *Lagocephalus laevigatus*
Snook, *Centropomus undecimalis*
Southern Flounder, *Paralichthys lethostigma*
Southern Hake, *Urophycis floridana*
Southern Kingfish, *Menticirrhus americanus*
Southern Stargazer, *Astroscopus y-graecum*
speckled tilapia, *Tilapia* spp.
Speckled Worm Eel, *Myrophis punctatus*
Spot, *Leiostomus xanthurus*
Spotfin Butterflyfish, *Chaetodon ocellatus*
Spotfin Mojarra, *Eucinostomus argenteus*
Spotted Eagle Ray, *Aetobatus narinari*
Spotted Scorpionfish, *Scorpaena plumieri*
Spotted Seatrout, *Cynoscion nebulosus*
Spotted Whiff, *Citharichthys macrops*
Star Drum, *Stellifer lanceolatus*
Striped Anchovy, *Anchoa hepsetus* [SPN: Anchoa Legítima]
Striped Burrfish, *Chilomycterus schoepfii*
Striped Mullet, *Mugil cephalus* [SPN: Lisa]
Tarpon, *Megalops atlanticus*
Texas Pipefish, *Syngnathus affinis*
Tidewater Silversides, *Menidia peninsulae* [*M. beryllina*]
Western Mosquitofish, *Gambusia affinis*
White Mullet, *Mugil curema* [SPN: Lebrancha]
Yellowfin Mojarra, *Gerres cinereus*

Amphibians

Black-spotted Newt, *Notophthalmus meridionalis*
Cane Toad, *Rhinella marina* [Marine Toad, Giant Toad, *Bufo marinus*]
Couch's Spadefoot, *Scaphiopus couchii*
Eastern Narrow-mouthed Toad, *Gastrophryne carolinensi*
Gulf Coast Toad, *Incillius nebulifer* [Coastal Plain Toad, *Bufo cognatus*]

Great Plains Narrow-mouthed Toad, *Gastrophryne olivacea*
Greater Siren, *Siren lacertina*
Green Treefrog, *Hyla cinerea*
Hurter's Spadefoot, *Scaphiopus hurteri*
Mexican Burrowing Toad, *Rhinophrynus dorsalis*
Mexican Treefrog, *Smilisca baudinii*
Mexican White-lipped Frog, *Leptodactylus fragilis* [White-lipped Frog, *L. labialis*]
Rio Grande Chirping Frog, *Eleutherodactylus cystignathoides*
Rio Grande Leopard Frog, *Lithobates berlandieri* [*Rana berlandieri*]
Rio Grande Lesser Siren, *Siren intermedia* [South Texas Siren, large form]
Sheep Frog, *Hypopachus variolosus*
Spotted Chorus Frog, *Pseudacris clarkia*
Texas Toad, *Anaxyrus speciosus* [*Bufo speciosus*]
Woodhouse's Toad, *Anaxyrus woodhousii* [*Bufo woodhousii*]

Reptiles

American Alligator, *Alligator mississippiensis*
Blue Spiny Lizard, *Sceloporus cyanogenys*
Bullsnake, *Pituophis cantifer sayi*
Cat-eyed Snake, *Leptodeira septentrionalis* [Northern Cat-eyed Snake]
Checkered Gartersnake, *Thamnophis marcianus*
Desert Kingsnake, *Lampropeltis getula splendida*
Desert Massasaugua, *Sisturus catenatus edwardsii*
Diamond-backed Watersnake, *Nerodia rhombifera*
Eastern Hog-nosed Snake, *Heterodon platirhinos*
Eastern Massasaugua, *Sisturus cantenatus*
Eastern Snapping Turtle, *Chelydra serpentina* [Common Snapping Turtle]
Eastern Yellow-bellied Racer, *Coluber constrictor flaviventris*
Flat-headed Snake, *Tantilla gracilis*
Great Plains Ratsnake, *Pantherophis emoryi emoryi* [*Elaphe guttata emoryi*]
Great Plains Skink, *Plestiodon obsoletus* [*Eumeces obsoletus*]
Green Anole, *Anolis carolinensis*
Green Iguana, *Iguana iguana*
Green Sea Turtle, *Chelonia mydas*[E]
Gulf Coast Ribbonsnake, *Thamnophis proximus orarius*
Hawksbill Sea Turtle, *Eretochelys imbricata*[E]
Keeled Earless Lizard, *Holbrookia propinqua*
Kemp's Ridley Sea Turtle, *Lepidochelys kempii*[E] [SPN: Tortuga Lora]
Leatherback Sea Turtle, *Dermochelys coriacea*[E]
Little Brown Skink, *Scincella lateralis* [Ground Skink]
Loggerhead Sea Turtle, *Caretta caretta*[E]
Marsh Brownsnake, *Storeria dekayi limnetes*
Mediterranean Gecko, *Hemidactylus turcicus**
Mexican Black-headed Snake, *Tantilla atriceps*
Mexican Hook-nosed Snake, *Ficimia streckeri*
Mexican Milksnake, *Lampropeltis triangulum annulata*
Ornate Box Turtle, *Terrapene ornata*
Red-eared Slider, *Trachemys scripta* [SPN: Tortuga de Oreja Roja]
Regal Black-striped Snake, *Coniophanes imperialis* [Black-striped Snake]
Reticulate Collared Lizard, *Crotaphytus reticularis*

Rose-bellied Lizard, *Sceloporus variabilis*
Rough Earthsnake, *Virginia striatula*
Saltmarsh Watersnake, *Nerodia clarkii* [Gulf Saltmarsh Snake]
Scarlet Snake, *Cemophora coccinea*
Six-lined Racerunner, *Aspidocelis sexlineatus* [*Cnemidophorus sexlineatus*]
Slender Glass Lizard, *Ophisaurus attenuatus*
Southern Texas Groundsnake, *Sonora semiannulata taylori*
Speckled Racer, *Drymobius margaritiferus*
Spiny Softshell, *Apalone spinifera* [SPN: Tortuga de Concha Blanda]
Spot-tailed Earless Lizard, *Holbrookia lacerata*
Texas Brownsnake, *Storeria dekayi texana*
Texas Coralsnake, *Micrurus tener*
Texas Diamond-backed Terrapin, *Malaclemys terrapin littoralis*
Texas Glossy Snake, *Arizona elegans arenicola*
Texas Horned Lizard, *Phrynosoma cornutum*[E] [SPN: Lagartija Cornuda]
Texas Indigo Snake, *Drymarchon melanurus erebennius* [Indigo Snake, Central American Indigo Snake, *D. corais*]
Texas Patch-nosed Snake, *Salvadora rahamiae lineata*
Texas Ratsnake, *Pantherophis obsoleta* [*Elaphe obsoleta*]
Texas Spiny Lizard, *Sceloporus olivaceus*
Texas Spotted Whiptail, *Aspidocelis gularis* [*Cnemidophorus gularis*]
Texas Threadsnake, *Leptotyphlops dulcis* [Texas Blind Snake]
Texas Tortoise, *Gopherus berlandieri* [SPN: Tortuga del Desierto]
Yellow Mud Turtle, *Kinosternon flavescens*

Birds

American Avocet, *Recuvirostra americana*
American Oystercatcher, *Haematopus palliatus*
American Peregrine Falcon, *Falco peregrinus anatum*
American White Pelican, *Pelecanus erythrorhynchos* [SPN: Pelícano Blanco]
American Wigeon, *Mareca americana* [*Anas americana*]
Aplomado Falcon, see Northern Aplomado Falcon
Audubon's Oriole, *Icterus graduacauda audubonii*
Baird's Sandpiper, *Calidris bairdii*
banded plovers, *Charadrius* spp.
Black Skimmer, *Rhynchops niger*
Black-bellied Plover, *Pluvialis squatarola*
Black-bellied Whistling-Duck, *Dendrocygna autumnalis*
Black-crowned Night-Heron, *Nycticorax nycticorax*
Black-necked Stilt, *Himantopus mexicanus*
Blue-winged Teal, *Spatula discors* [*Anas discors*]
Botteri's Sparrow, *Peucaea botterii*
Brown Pelican, *Pelecanus occidentalis*
Brownsville Common Yellowthroat, *Geothlypis trichas insperata*
Buff-breasted Sandpiper, *Calidris subruficollis* [*Tryngites subruficollis*]
Bufflehead, *Bucephala albeola*
Canvasback, *Aythya valisineria*
Caspian Tern, *Hydroprogne caspia* [*Sterna caspia*]
Cattle Egret, *Bubulcus ibis*
Common Black Hawk, *Buteogallus anthracinus* [SPN: Aguililla Negra Menor]

Common Goldeneye, *Bucephala clangula*
cormorants, *Phalacrocorax* spp.
dowitchers, *Limnodromous* spp.
Dunlin, *Calidris alpina*
Eastern Meadowlark, *Sturnella magna*
Eskimo Curlew, *Numenius borealis*[E]
Ferruginous Pygmy-Owl, *Glaucidium brasilianum* [including the "cactus" subspecies, *G. b. cactorum*]
Forster's Tern, *Sterna fosteri*
Fulvous Whistling-Duck, *Dendrocygna bicolor*
Gadwall, *Mareca strepera* [*Anas strepera*]
Great Blue Heron, *Ardea herodias*
Great Egret, *Ardea alba*
Greater Scaup, *Aythya marila*
Greater White-fronted Goose, *Anser albifrons*
Greater Yellowlegs, *Tringa melanoleuca*
Green Heron, *Butorides virescens*
Green-winged Teal, *Anas crecca*
Grey-lined Hawk, *Buteo nitidus* [SPN: Aguililla Gris]
Gull-billed Tern, *Gelochelidon nilotica* [*Sterna nilotica*]
Henslow's Sparrow, *Ammodramus henslowii*
Hudsonian Godwit, *Limosa haemastica*
Interior Least Tern, *Sterna antillarum athalassos*[E]
Killdeer, *Charadius vociferous* [SPN: Chorlo Culirojo]
Laughing Gull, *Leucophaeus atricilla* [*Larus atricilla*]
Least Sandpiper, *Calidris minutilla*
Least Tern, *Sternula antillarum* [*Sterna antillarum*]
Lesser Canada Goose, *Branta canadensis parvipes*
Lesser Golden Plover, *Pluvialis dominica*
Lesser Scaup, *Aythya affinis*
Lesser Yellowlegs, *Tringa flavipes*
Little Blue Heron, *Egretta caerulea*
Loggerhead Shrike, *Lanius ludovicianus*
Long-billed Curlew, *Numenius americanus*
Mallard, *Anas platyrhynchos*
Marbled Godwit, *Limosa fedoa*
Mottled Duck, *Anas fulvigula*
Neotropic Cormorant, *Phalacrocorax brasilianus*
Northern Aplomado Falcon, *Falco femoralis septentrionalis*[E]
Northern Beardless Tyrannulet, *Campostoma imberbe*
Northern Bobwhite, *Colinus virginianus*
Northern Cardinal, *Cardinalis cardinalis*
Northern Mockingbird, *Mimus polyglottos*
Northern Pintail, *Anas acuta* [SPN: Pato Golondrino]
Northern Shoveler, *Spatula clypeata*
Osprey, *Pandion haliaetus*
Pectoral Sandpiper, *Calidris melanotos*
Peregrine Falcon, *Falco peregrinus anatum*
Piping Plover, *Charadrius melodus* [SPN: Chorlo Melódico]
Red Knot, *Calidris canutus*

Red-breasted Merganser, *Mergus serrator*
Reddish Egret, *Egretta rufescens* [SPN: Garza Melenuda]
Redhead, *Aythya americana* [SPN: Cabeza Rojo]
Ring-necked Duck, *Aythya collaris*
Roseate Spoonbill, *Platalea ajaja* [*Ajaia ajaja*]
Rose-throated Becard, *Pachyramphus aglaiae*
Ross's Goose, *Chen rossii*
Royal Tern, *Thalasseus maximus* [*Sterna maxima*]
Ruddy Duck, *Oxyura jamaicensis*
Ruddy Turnstone, *Arenaria interpres*
Sanderling, *Calidris alba* [SPN: Playero Blanco]
sandpipers, *Calidris* spp.
Sandwich Tern, *Thalasseus sandvicensis* [*Sterna sandvicensis*]
Semipalmated Plover, *Charadius semipalmatus* [SPN: Chorlo Semipalmeado]
Semipalmated Sandpiper, *Calidris semipalmatus*
Sennett's Hooded Oriole, *Icterus cucullatus senetti*
Short-billed Dowitcher, *Limnodromus griseus* [SPN: Costurero Pico Corto]
Snow Goose, *Chen caerulesens*
Snowy Egret, *Egretta thula*
Snowy Plover, *Charadius alexandrinus* [SPN: Chorlo Patinegro]
Solitary Sandpiper, *Tringa solitaria*
Sooty Tern, *Onychoprion fuscatus* [*Sterna fuscata*]
Spotted Sandpiper, *Actitis macularius*
Stilt Sandpiper, *Calidris himantopus*
Texas Botteri's Sparrow, *Aimophila botteri texana*
Tricolored Heron, *Egretta tricolor* [Louisiana Heron]
Tropical Parula, *Parula pitiayumi*
Upland Sandpiper, *Bartramia longicauda*
Western Sandpiper, *Calidris mauri*
Whimbrel, *Numenius phaeopus* [SPN: Zarapico Trinador]
White Ibis, *Eudocimus albus*
White Pelican, *Pelecanus erythrorhynchos*
White-bellied Wren, *Uropsila leucogastra* [SPN: Chivirín Vientre Blanco]
White-faced Ibis, *Plegadis chihi*
White-rumped Sandpiper, *Calidris fusicollis*
White-tailed Hawk, *Buteo albicaudatus*
Whooping Crane, *Grus americana*[E]
Willet, *Catoptrophorus semipalmatus*
Wilson's Phalarope, *Phalaropus tricolor*
Wilson's Plover, *Charadius wilsonia*
Wood Stork, *Mycteria americana*
Yellow-crowned Night-Heron, *Nyctanassa violacea*
Yellow-headed Amazon, *Amazona oratrix* [SPN: Loro Cabeza Amarilla]
Zone-tailed Hawk, *Buteo albonotatus*

Mammals

American Badger, *Taxidea taxus*
American Perimyotis, *Perimyotis subflavus* [Georgia Bat, *Pipistrellus subflavus*]
Black-eared Rice Rat, *Handleomys melanotis* [*Oryzomys melanotis*]
Black-tailed Jackrabbit, *Lepus californicus*

Bobcat, *Lynx rufus* [SPN: Gato Rabón]
Bottlenose Dolphin, *Tursiops truncatus*
Brazilian Free-tailed Bat, *Tadarida brasiliensis* [Mexican Free-tailed Bat]
Brown Rat, *Rattus norvegicus**
Brush Deermouse, *Peromyscus boylii*
Collared Peccary, *Pecari tajacu* [Javelina, *Tayassu tajacu*]
Common Bottlenose Dolphin, *Tursiops truncatus* [SPN: Delfín Nariz de Botella]
Common Gray Fox, *Urocyon cinereoargenteus*
Common Muskrat, *Ondatra zibethicus*
Coues's Rice Rat, *Oryzomys couesi*
Coyote, *Canis latrans*
Crawford's Desert Shrew, *Notiosorex crawfordi*
Desert Cottontail, *Sylvilagus audubonii*
Domestic Dog, *Canis familiaris*
Eastern Cottontail, *Sylvilagus floridanus*
Eastern Fox Squirrel, *Sciurus niger*
Eastern Mole, *Scalopus aquaticus*
Eastern Spotted Skunk, *Spilogale putorius interrupta*
Feral Pig, *Sus scrofa**
Fulvous Harvest Mouse, *Reithrodontomys fulvescens*
Ghost-faced Bat, *Mormoops megalophylla* [Leaf-chinned Bat]
Gulf Coast Kangaroo Rat, *Dipodomys compactus* [SPN: Rata Canguro del Padre]
Hispid Cotton Rat, *Sigmodon hispidus*
Hispid Pocket Mouse, *Chaetodipus hispidus*
Hog-nosed Skunk, *Conepatus leuconotus*
House Mouse, *Mus musculus**
Jaguar, *Panthera onca* [*Felis onca*][E]
Jaguarundi, *Herpailurus yagouaroundi* [*Felis yagouaroundi*][E]
Least Shrew, *Cryptotis parva*
Margay, *Leopardus wiedii*
Maritime Pocket Gopher, *Geomys personatus maritimus*
Marsh Rice Rat, *Oryzomys palustris*
Merriam's Pocket Mouse, *Perognathus merriami*
Mexican Ground Squirrel, *Ictidomys mexicanus* [*Spermophilus mexicanus*]
Mexican Long-tongued Bat, *Choeronycteris mexicana*
Mexican Spiny Pocket Mouse, *Heteromys irroratus* [*Liomys irroratus*]
Mountain Lion, *Puma concolor* [Cougar, Panther, Puma, *Felis concolor*]
Nilgai, *Boselaphus tragocamelus**
Nine-banded Armadillo, *Dasypus novemcinctis*
North American Deermouse, *Peromyscus maniculatus*
Northern Grasshopper Mouse, *Onychomys leucogaster*
Northern Pygmy Mouse, *Baiomys taylori*
Northern Yellow Bat, *Dasypterus intermedius* [*Lasiurus intermedius*]
Nutria, *Myocastor coypus**
Ocelot, *Leopardus pardalis* [SPN: Ocelote][E]
Ord's Kangaroo Rat, *Dipodomys ordii* [SPN: Rata Canguro]
Raccoon, *Procyon lotor*
Roof Rat, *Rattus rattus**
Southern Plains Woodrat, *Neotoma micropus*
Southern Yellow Bat, *Dasypterus ega* [*Lasiurus ega*]

Spotted Ground Squirrel, *Xerospermophilus spilosoma* [*Spermophilus spilosoma*; SPN: Ardilla de Tierra]
Striped Skunk, *Mephitis mephitis*
Tamaulipan Mole, *Scalopus aquaticus inflatus*
Texas Pocket Gopher, *Geomys personatus*
Tropical Pocket Gopher, *Geomys tropicalis*
Virginia Opossum, *Didelphis virginiana*
West Indian Manatee, *Trichechus manatus*[E]
White-faced Hog-nosed Skunk, *Conepatus leuconotus* [Eastern Hog-nosed Skunk]
White-footed Deermouse, *Peromyscus leucopus*
White-nosed Coati, *Nasua narica*
White-tailed Deer, *Odocoileus virginianus*
Yellow-faced Pocket Gopher, *Cratogeomys castanops*
Yuma Myotis, *Myotis yumanensis*

Literature Cited

Adair, S. E. 1990. Factors influencing wintering diving duck use of coastal ponds in South Texas. MS thesis, Texas A&M University, College Station.

Adair, S. E., J. L. Moore, and W. H. Kiel. 1990. *Winter ecology of Redhead ducks in the Gulf Coast Region*. Cooperative Agreement 14-16-009-87-909. College Station: Texas A&M University and US Fish and Wildlife Service Wetlands Research Center.

Adair, S. E., J. L. Moore, and W. H. Kiel Jr. 1996. Winter diving duck use of coastal ponds: An analysis of alternative hypotheses. *Journal of Wildlife Management* 60:83–93.

Adair, S. E., J. L. Moore, and C. P. Onuf. 1994. Distribution and status of submerged vegetation in estuaries of the upper Texas coast. *Wetlands* 14:110–121.

Adams, A. S., and D. W. Keith. 2007. Wind energy and climate: Modeling the atmospheric impacts of wind energy turbines. *Transactions of the American Geophysical Union* 88 (Fall Meeting Supplement): Abstract B44B-08.

Adcock, M. E. 2016. Soil and vegetative associations of heteromyid rodents in Central and South Texas with comments on trapping techniques. MS thesis, Texas State University, San Marcos.

Agostoni, M., and D. L. Erdner. 2011. Analysis of ammonium transporter and urease gene expression in *Aureoumbra lagunensis*. *Harmful Algae* 10:549–556.

Aguilar-Perera, A., and A. Tuz-Sulub. 2010. Non-native, invasive Red Lionfish (*Pterois volitans* [Linnaeus, 1758]: Scorpaenidae), is first recorded in the southern Gulf of Mexico, off the northern Yucatan Peninsula, Mexico. *Aquatic Invasions* 5 (Suppl. 1): S9–S12.

Aguirre-Guzmán, G., J. G. Sánchez-Martínez, R. Pérez-Castañeda, and R. Orta-Rodríguez. 2010. Detection of necrotizing hepatopancreatitis (NHP) in wild shrimp from Laguna Madre, Mexico by a multiplex polymerase chain reaction. *Thai Journal of Veterinary Medicine* 40:337–341.

Ajemian, M. J., K. S. Mendenhall, J. B. Pollack, M. S. Wetz, and G. W. Stuntz. 2018. Moving forward in a reverse estuary: Habitat use and movement patterns of Black Drum (*Pogonias cromis*) under distinct hydrological regimes. *Estuaries and Coasts* 41:1410–1421.

Alber, M. 2002. A conceptual model of estuarine freshwater inflow management. *Estuaries* 25:1246–1261.

Albins, M. A. 2013. Effects of invasive Pacific Red Lionfish *Pterois volitans* versus a native predator on Bahamian coral-reef fish communities. *Biological Invasions* 15:29–43.

Aldrich, D. V., and W. B. Wilson. 1960. The effect of salinity on growth of *Gymnodinium breve* Davis. *Biological Bulletin* 119:57–64.

Alleman, B. J. 2015. Diet analysis and assessment of habitat and prey availability associated with Texas Diamond-backed Terrapin (*Malaclemys terrapin littoralis*). MS thesis, University of Houston–Clear Lake.

Alleman, L., and M. Hester. 2011. Reproductive ecology of Black Mangrove (*Avicennia germinans*) along the Louisiana coast: Propagule production cycles, dispersal limitations and establishment elevations. *Estuaries and Coasts* 34:1068–1077.

Allen, J. A. 1892. The geographical distribution of North American mammals. *American Museum of Natural History Bulletin* 4: 199–243.

Allen, J. A. 1893. The geographical origin of North American birds, considered in relation to the faunal areas of North America. *Auk* 10: 97–150.

Allen, R. P. 1954. The Reddish Egret: Bird of colors and contrasts. *Audubon* 55:252–255.

Allen, R. P. 1955. The Reddish Egret: Bird of colors and contrast. *Audubon* 57:24–27.

Aller, R. C. 1978. The effects of animal-sediment interactions on geochemical processes near the sediment-water interface. Pp. 157–172 in M. Wiley, ed., *Estuarine Interactions*. New York: Academic Press.

Alvarado, S. A. 1996. Hard substrate. Pp. 111–150 in J. W. Tunnell Jr., Q. R. Dokken, E. H. Smith, and K. Withers, eds., *Current Status and Historical Trends of the Estuarine Living Resources within the Corpus Christi Bay National Estuary Program Study Area*. Corpus Christi National Estuary Program, CCBNEP-06A. Corpus Christi: Texas Natural Resources Conservation Commission.

Alvarez, T. 1963. The recent mammals of Tamaulipas, Mexico. *University of Kansas Publication of the Museum of Natural History* 14:363–473.

Amdurer, M. 1978. Geochemistry, hydrology, and mineralogy of the Laguna Madre flats, South Texas. MS thesis, University of Texas at Austin.

Amdurer, M., and L. S. Land. 1982. Geochemistry, hydrology, and mineralogy of the sand bulge area, Laguna Madre flats, South Texas. *Journal of Sedimentary Petrology* 52:703–716.

Amos, A. F., A. R. Wickham, and K. C. Keplar. 1997. *Current Status and Historical Trends in the Incidence of Marine/Bay Debris in the Corpus Christi Bay National Estuary Program Study Area*. Corpus Christi Bay National Estuary Program CCBNEP-12. Corpus Christi: Texas Natural Resource Conservation Commission.

Anderson, B., R. S. Scalon, E. W. Behrens, and P. L. Parker. 1992. Stable carbon isotope variations in sediment from Baffin Bay, Texas, USA: Evidence for cyclic changes in organic matter source. *Chemical Geology* 101:227–233.

Anderson, J. D., W. J. Karel, C. E. Mace, B. L. Bartram, and M. P. Hare. 2014. Spatial genetic features of Eastern Oysters (*Crassotrea virginica* Gmelin) in the Gulf of Mexico: Northward movement of a secondary contact zone. *Ecology and Evolution* 4:1671–1685.

Anderson, J. D., D. J. Shaver, and W. J. Karel. 2013. Genetic diversity and natal origins of Green Turtles (*Chelonia mydas*) in the western Gulf of Mexico. *Journal of Herpetology* 47:251–257.

Andrews, P. S. 1964. Serpulid reefs, Baffin Bay, Southeast Texas. Pp. 101–120 in A. J. Scott, ed., *Depositional Environments, South Central Texas Coast: Field Trip Guidebook*. Houston: Gulf Coast Association of Geological Societies.

Anonymous. 1967. *Estudio preliminar para la rehabilitacion de la Laguna Madre de Tamaulipas, México*. Mexico City: Secretaría de Marina.

Anonymous. 1993. Introduction. In *Environmental Impacts of the Plan to Dump Dredged Spoil from the Gulf Intracoastal Waterway on Lands Bordering the Laguna Madre*. Kingsville, TX: King Ranch, Inc. and John G. Kenedy Charitable Trust.

Anonymous. 2005. *Carta Nacional Pesquera*. Diario Oficial de la Federación, 25 de agosto de 2006 (1ª sección). Mexico City.

Arellano-Méndez, L. U. 2011. Estructura de los pastos marinos y su relación con la variabilidad ambiental a diferentes escalas espaciales. Tesis doctoral, Centro de Investigación y de Estudios Avanzados-Instituto Politécnico Nacional, Unidad Mérida, Mexico.

Arellano-Méndez, L. U., A. Mora-Olivo, C. Zamora-Tovar, and E. de la Rosa-Manzano. 2017. First report of the invasive Red Lionfish *Pterois volitans* (Linnaeus, 1758) (Scorpaenidae) on the coast of Tamaulipas, Mexico. *BioInvasions Records* 6:255–258.

Arellano-Méndez, L. U., S. Morales-Ojeda, and J. Herrera-Silveira. 2015. Carbono orgánico de las praderas de *Thalassia testudinum* en Bahía de la Ascensión (Quintana Roo, México): Una primera estimación del contenido de carbono azul en una laguna costera del Caribe Mexicano. In F. Paz and J. Wong, eds., *Estado actual del conocimiento del ciclo del carbono y sus interacciones en México: Síntesis a 2014*. Mérida, Mexico: Centro de Investigación y Estudios Avanzados del Instituto Politécnico Nacional.

Armitage, A. R., W. E. Highfield, S. D. Brody, and P. Louchouarn. 2015. The contribution of mangrove expansion to salt marsh loss on the Texas Gulf Coast. *PLoS ONE* 10(5):e0125404.

Arms, S. A. 1996. Overwintering behavior and movement of immature Green Sea Turtles in South Texas waters. MS thesis, Texas A&M University, College Station.

Armstrong, N. E. 1987. *The Ecology of Open-Bay Bottoms of Texas: A Community Profile*. US Fish and Wildlife Service Biological Report 85(7.12). Washington, DC.

Armstrong, N. E., and H. T. Odum. 1964. Photoelectric system. *Science* 143:256–258.

Arroyo Colorado Watershed Partnership. 2007. *A Watershed Protection Plan for the Arroyo Colorado Phase I.* Arroyo Colorado Watershed Partnership and Texas Sea Grant. College Station: Texas Water Resources Institute, Texas AgriLife Extension Service.

Arvin, J. 2010. *A Survey of Upper Texas Coast Critical Habitats for Migratory and Wintering Piping Plover and Associated "Sand Plovers."* Final report to Texas Parks and Wildlife Department Section 6, Grant No. TX E-95-R, Austin.

Auffenberg, W., and W. G. Weaver Jr. 1969. *Gopherus berlandieri* in southeastern Texas. *Bulletin of the Florida State Museum* 13:141–203.

Avalos, C. G. 2015. Presencia y biometria del camarón tigre, *Penaeus monodon* (Fabricius 1798), en la bahía de Amatique, Izabal. Tesis de licenciatura, Centro de Estudios del Mar y Acuicultura, Universidad de San Carlos de Guatemala.

Avens, L., L. R. Goshe, L. Coggins, B. Higgins, A. M. Landry Jr., and R. Bailey. 2017. Variability in age and size at maturation, reproductive longevity, and long-term growth dynamics for Kemp's Ridley Sea Turtles in the Gulf of Mexico. *PLoS ONE* 12 (3): e0173999. https://doi.org/10.1371/journal.pone.0173999.

Avens, L., L. R. Goshe, C. A. Harms, E. T. Anderson, A. G. Hall, W. M. Cluse, M. H. Godfrey, et al. 2012. Population characteristics, age structure, and growth dynamics of neritic juvenile Green Turtles in the northeastern Gulf of Mexico. *Marine Ecology Progress Series* 458:213–229.

Bacak-Clements, P. 1988. A survey of avian use and the aquatic fauna of three ponds, in Willacy County, Texas. MS thesis, Texas A&I University, Kingsville.

Baccus, J. T., J. K. Horton, and P. D. Carangelo. 1977. *A Study of Beach and Dunes, Floral and Faunal Interrelations as Influenced by Recreational and Use Impact on Padre Island National Seashore.* Santa Fe, NM: Office of Natural Resources, Southwest Region, National Park Service.

Baidya Roy, S. 2011. Simulating impacts of wind farms on local hydrometeorology. *Journal of Wind Engineering and Industrial Aerodynamics* 99:491–498.

Baidya Roy, S., S. W. Pacala, and R. L. Walko. 2004. Can large scale wind farms affect local meteorology? *Journal of Geophysical Research* 109 (D19). https://doi.org/10.1029/2004JD004763.

Baidya Roy, S., and J. J. Traiteur. 2010. Impacts of wind farms on surface air temperatures. *Proceedings of the National Academy of Science* 107:17899–17904.

Bailey, R. G. 1998. *Ecoregions: The Ecosystem Geography of the Oceans and Continents.* New York: Springer-Verlag.

Bailey, T. W. 1905. Biological survey of Texas. *North American Fauna* 25:1–222.

Baker, K., and N. Rabalais. 1978. Manuscript for the biological survey of Padre Island. Unpublished manuscript on file at Padre Island National Seashore, Corpus Christi, TX.

Baker, M. C., and A. E. M. Baker. 1973. Niche relationships among six species of shorebirds on their wintering and breeding range. *Ecological Monographs* 43:193–212.

Baker, P. 1997. Coastal shrimp farming in Texas. Pp. 117–130 in R. Goldburg and T. Triplett, eds., *Murky Waters: Environmental Effects of Aquaculture in the United States.* Washington, DC: Environmental Defense Fund.

Baker, R. H., and D. W. Lay. 1938. Notes on the mammals of Galveston and Mustang Islands. *Journal of Mammalogy* 19:505.

Baldassarre, G. A. 2014. *Ducks, Geese, and Swans of North America.* Baltimore: Johns Hopkins University Press.

Baldassare, G. A., A. R. Brazda, and D. R. Woodyard. 1989. The east coast of Mexico. Pp. 407–425 in L. M. Smith, R. L. Pederson, and R. M. Kaminski, eds., *Habitat Management for Migrating and Wintering Waterfowl in North America.* Lubbock: Texas Tech University Press.

Ball, M. C. 1987. Salinity-induced potassium deficiency causes loss of functional photosystem II in leaves of the Grey Mangrove, *Avicennia marina*, through depletion of the atrazine-binding polypeptide. *Australian Journal of Plant Physiology* 14:351–361.

Ball, M. C., J. R. Cowan, and G. D. Farquhar. 1988. Maintenance of leaf temperature and the optimization of carbon gain in relation to water loss in a tropical mangrove forest. *Australian Journal of Plant Physiology* 15:263–276.

Ballard, B. M. 2002. Parasitism of a Laughing Gull nest by Black-bellied Whistling Ducks. *Wilson Bulletin* 113:339–340.

Ballard, B. M., J. D. James, R. L. Bingham, M. J. Petrie, and B. C. Wilson. 2010. Coastal pond use by Redheads wintering in the Laguna Madre, Texas. *Wetlands* 30:669–674.

Ballard, B. M., J. E. Thompson, M. J. Petrie, M. Chekett, and D. G. Hewitt. 2004. Diet and nutrition of Northern Pintails wintering along the southern coast of Texas. *Journal of Wildlife Management* 68:371–382.

Barba, E., A. Raz-Guzmán, and A. J. Sánchez. 2005. Distribution patterns of estuarine caridean shrimps in the southwestern Gulf of Mexico. *Crustaceana* 78:709–726.

Barba, E., and A. J. Sánchez. 2005. Peracarid crustaceans of central Laguna Madre Tamaulipas in the southwestern Gulf of Mexico. *Gulf of Mexico Science* 2:241–247.

Barbier, E. B. 2013. Valuing ecosystem services for coastal wetland protection and restoration: Progress and challenges. *Resources* 2:213–230.

Barnes, D. 1971. Anatomy of a spoil island. MS thesis, Texas A&I University, Kingsville.

Bates, E. M., and B. M. Ballard. 2014. Factors influencing behavior and success of foraging Reddish Egrets. *Waterbirds* 37:191–202.

Bates, E. M., R. W. DeYoung, and B. M. Ballard. 2009. Genetic diversity and population structure of Reddish Egrets along the Texas coast. *Waterbirds* 32:430–436.

Bates, E. M., L. M. Koczur, and B. M. Ballard. 2015. Post-fledging survival of juvenile Reddish Egrets (*Egretta rufescens*). *Waterbirds* 38:401–406.

Bates, E. M., L. M. Koczur, A. Krainyk, B. M. Ballard, and A. C. Kasner. 2016. Spatial and temporal dynamics of foraging habitat availability for Reddish Egrets in the Laguna Madre, Texas. *International Journal of Biodiversity and Conservation* 8:251–258.

Baudouin, M., B. De Thoisy, P. Chambault, R. Berzins, M. Entraygues, L. Kelle, A. Turny, Y. L. Maho, and D. Chevallier. 2015. Identification of key marine areas for conservation based on satellite tracking of post-nesting migrating Green Turtles (*Chelonia mydas*). *Biological Conservation* 184:36–41.

Bauer, D. W. 1993. Avian use of wetlands in southern Texas during nonbreeding seasons. MS thesis, Texas A&M University–Kingsville.

Beasom, S. L., and H. H. Haucke. 1975. A comparison of four distance sampling techniques in South Texas Live Oak mottes. *Journal of Range Management* 28:142–144.

Behrens, E. W. 1963. Buried Pleistocene river valleys in Aransas and Baffin Bays, Texas. *Publications of the Institute of Marine Science, University of Texas* 9:7–18.

Behrens, E. W. 1966. Recent emerged beach in eastern Mexico. *Science* 152:642–643.

Behrens, E. W. 1969. Hurricane effects on a hypersaline bay. Pp. 301–308 in A. A. Ayala Castañares and F. B. Phleger, eds., *Coastal Lagoons: A Symposium*. Mexico City: Universidad Nacional Autónoma de México.

Behrens, E. W. 1974. Holocene sea level rise effect on the development of an estuarine carbonate depositional environment: Baffin Bay, Texas. *Memorias Institucion Géologia Bassin Aquitaine* 7:337–341.

Behrens, E. W., and R. L. Watson. 1973. *Corpus Christi Water Exchange Pass: A Case History of Sedimentation and Hydraulics during Its First Year*. Contract DACW 72-72-C-002. Springfield, VA: US Army Corps of Engineers, Coastal Research Center.

Behrens, E. W., R. L. Watson, and C. C. Mason. 1977. *Hydraulics and Dynamics of New Corpus Christi Pass, Texas: A Case History, 1972–1973*. Report No. DACW-72-72-C-0026 and DACW-72-72-C-0027. Austin: Bureau of Economic Geology, University of Texas.

Bello Pineda, J., L. Gómez Mendoza, V. Magaña Rueda, Boris Graizbord, P. H. Rodríguez Herrero, and M. Cervantes Ábrego. 2009. Sitio piloto Río San Fernando–Laguna la Nacha, descripción física y relevancia ecológica. Pp. 391–413 in F. J. Buenfil Friedman, ed., *Adaptación a los impactos del cambio climático en los humedales costeros del Golfo de México*, vol. 2. Mexico City: INE, SEMARNAT.

Bellrose, F. C. 1980. *Ducks, Geese, and Swans of North America*. Harrisburg, PA: Stackpole Books.

Bengtson, S. A., A. Nilson, S. Nordstrom, and S. Rundgren. 1976. Effect of bird predation on lumbricid populations. *Oikos* 27:9–12.

Bennett, D. C., and M. R. Hughes. 2003. Comparison of renal and salt gland function in three species of wild ducks. *Journal of Experimental Biology* 206:3273–3284.

Benson, L. H. 1979. *Plant Classification*. 2nd ed. Boston: D. C. Heath.

Bent, A. C. 1921. *Life Histories of North American Gulls and Terns*. New York: Dover Publications.

Bent, A. C. 1926. *Life Histories of North American Marsh Birds* US National Museum Bulletin No. 135. Washington, DC: US National Museum.

Bevan, E., T. Wibbels, B. M. Z. Najera, L. Sarti, F. I. Martinez, J. M. Cuevas, B. J. Gallaway, L. J. Pena, and P. M. Burchfield. 2016. Estimating the historic size and current status of the Kemp's Ridley Sea Turtle (*Lepidochelys kempii*) population. *Ecosphere* 7:1–15.

Bezanson, D. 2000. Natural vegetation types of Texas and their representation in conservation areas. MS thesis, University of Texas at Austin.

Bianchi, T. S., A. A. Mead, J. Zhao, X. Li, R. S. Comeaux, R. A. Feagin, and R. W. Kulawardhana. 2013. Historical reconstruction of mangrove expansion in the Gulf of Mexico: Linking climate change with carbon sequestration in coastal wetlands. *Estuarine, Coastal and Shelf Science* 119:7–16.

Bielfelt, B. J., and A. R. Litt. 2016. Effects of increased *Heteropogon contortus* (Tanglehead) on rangelands: The tangled issue of native invasive species. *Rangeland Ecology and Management* 69:508–512.

Bildstein, K. 1999. The forced migration of the Broad-Winged Hawk. Pp. 79–102 in K. P. Able, ed. *Gatherings of Angels: Migrating Birds and Their Ecology*. Ithaca, NY: Comstock Books.

Bjorndal, K. A. 1980. Nutrition and grazing behavior of the Green Turtle, *Chelonia mydas*. *Marine Biology* 56:147–154.

Bjorndal, K. A. 1982. The consequences of herbivory for the life history pattern of the Caribbean Green Turtle. Pp. 111–116 in K. Bjorndal, ed., *Biology and Conservation of Sea Turtles*. Washington, DC: Smithsonian Institution Press.

Bjorndal, K. A. 1997. Foraging ecology and nutrition of sea turtles. Pp. 199–232 in P. L. Lutz and J. A. Musick, eds., *The Biology of Sea Turtles*. Boca Raton, FL: CRC Press.

Bjorndal, K. A. 1999. Priorities for research in foraging habitats. Pp. 12–14 in K. L. Eckert, K. A. Bjorndal, F. A. Abreu-Grobois, and M. Donnelly, eds., *Research and Management Techniques for the Conservation of Sea Turtles*. Washington, DC: IUCN/SSC Marine Turtle Specialist Group.

Blair, W. F. 1950. The biotic provinces of Texas. *Texas Journal of Science* 2:93–117.

Blair, W. F. 1952. Mammals of the Tamaulipan Biotic Province in Texas. *Texas Journal of Science* 4:230–250.

Blanchard, G. F., and P. A. Montagna. 1992. Photosynthetic response of natural assemblages of marine benthic microalgae to short- and long-term variation of incident irradiance in Baffin Bay, Texas. *Journal of Phycology* 28:7–14.

Boesch, D., and R. E. Turner. 1984. Dependence of fishery species on salt marshes: The role of food and refuge. *Estuaries* 7:460–468.

Bogusch, E. R. 1952. Brush invasion in the Rio Grande Plain of Texas. *Texas Journal of Science* 4:85–91.

Bolam, S. G., P. Whomersley, and M. Schratzberger. 2004. Macrofaunal recolonization on intertidal mudflats: Effect of sediment organic and sand content. *Journal of Experimental Marine Biology and Ecology* 306:157–180.

Bolen, E. G., and B. R. Chapman. 1981. Estimating winter sex ratios for Buffleheads. *Southwestern Naturalist* 26:49–52.

Bouillon, S., A. V. Borges, E. Castañeda-Moya, K. Diele, T. Dittmar, N. C. Duke, E. Kristensen, et al. 2008. Mangrove production and carbon sinks: A revision of global budget estimates. *Global Biogeochemical Cycles* 22:1–12.

Bowen, B. W., A. L. Bass, S.-M. Chow, M. Bostrom, K. A. Bjorndal, A. B. Bolten, T. Okuyama et al. 2004. Natal homing in juvenile Loggerhead Turtles (*Caretta caretta*). *Molecular Ecology* 13:3797–3808.

Bowen, B. W., W. S. Grant, Z. Hillis-Starr, D. J. Shaver, K. A. Bjorndal, A. B. Bolten, and A. L. Bass. 2007. Mixed-stock analysis reveals migration of juvenile Hawksbill Turtles (*Eretmochelys imbricata*) in the Caribbean Sea. *Molecular Ecology* 16:49–60.

Bowen, B. W., A. B. Meylan, and J. C. Avise. 1991. Evolutionary distinctiveness of the endangered Kemp's Ridley Sea Turtle. *Nature* 352:709–711.

Bowen, G. J., L. L. Wassenaar, and K. A. Hobson. 2005. Global application of stable hydrogen and oxygen isotopes to wildlife forensics. *Oecologia* 143:337–348.

Brager, S. 1993. Diurnal and seasonal behavior patterns of Bottlenose Dolphins (*Tursiops truncatus*). *Marine Mammal Science* 9:438–440.

Brandvold, D. K., C. J. Popp, and J. A. Brierley. 1976. Waterfowl refuge effect on water quality: II. Chemical and physical properties. *Journal (Water Pollution and Control Federation)* 48:680–687.

Breaux, N. J. 2017. Ecosystem resilience following salinity change in a hypersaline estuary. MS thesis, Texas A&M University–Corpus Christi.

Brenner, J. 1997. Estimación de batimetra datos multispectrales caso de estudio Laguna la Nacho, Tamaulipas. MS tesis, Instituto Tecnológico y de Estudios Superiores de Monterrey, Monterrey, Nuevo León, Mexico.

Brennan, L. A., and W. P. Kuvlesky Jr. 2005. North American grassland birds: An unfolding conservation crisis? *Journal of Wildlife Management* 69:1–13.

Breuer, J. P. 1957. An ecological survey of Baffin and Alazan Bays, Texas. *Publications of the Institute of Marine Science, University of Texas* 4:134–155.

Breuer, J. P. 1962. An ecological survey of the Lower Laguna Madre of Texas, 1953–1959. *Publications of the Institute of Marine Science, University of Texas* 8:153–183.

Bricker, S. B., C. G. Clement, D. E. Pirhalla, S. P. Orlando, and D. R. G. Farrow. 1999. *National Estuarine Eutrophication Assessment: Effects of Nutrient Enrichment in the Nation's Estuaries.* Silver Spring, MD: NOAA, National Ocean Service, Special Projects Office, and the National Centers for Ocean Science.

Briggs, J. C. 1974. *Marine Zoogeography*. New York: McGraw-Hill.

Briggs, R. J., and D. D. Everett. 1983. Avian use of small aquatic habitats in South Texas. *Proceedings of the Annual Conference of the Southeast Association of Fish and Wildlife Agencies* 37:86–94.

Britton, J. C., and B. Morton. 1989. *Shore Ecology of the Gulf of Mexico*. Austin: University of Texas Press.

Brogden, W. B., C. H. Oppenheimer, and D. Bowman. 1977. Selected biological data, Mustang and North Padre Islands. Chapter IV in R. S. Kier and E. G. Fruh, eds., *Environmental and Economic Impacts of Recreational Community Development, Mustang Island and North Padre Island*, vol. 2. Austin: University of Texas.

Broussard, A., and N. Martin, eds. 1986. *Texas Coast Hurricanes*. Texas A&M University, Sea Grant College Program TAMU-SG-86-505, College Station.

Brown, C. A., and N. C. Kraus. 1997. *Environmental Monitoring of Dredging and Processes in Lower Laguna Madre, Texas*. Technical Report TAMU-CC-CBI-96-01. Corpus Christi: Texas A&M University–Corpus Christi Conrad Blucher Institute for Surveying and Science.

Brown, L. F., J. L. Brewton, T. J. Evans, J. H. McGowen, W. A. White, C. G. Groat, and W. L. Fisher. 1980. *Environmental Geologic Atlas of the Texas Coastal Zone: Brownsville-Harlingen Area*. Austin: Bureau of Economic Geology, University of Texas.

Brown, L. F., J. L. Brewton, J. H. McGowen, T. J. Evans, W. L. Fisher, and C. G. Groat. 1976. *Environmental Geologic Atlas of the Texas Coastal Zone: Corpus Christi Area*. Austin: Bureau of Economic Geology, University of Texas.

Brown, L. F., J. H. McGowen, T. J. Evans, C. G. Groat, and W. L. Fisher. 1977. *Environmental Geologic Atlas of the Texas Coastal Zone: Kingsville Area*. Austin: Bureau of Economic Geology, University of Texas.

Brune, G. 2002. *Springs of Texas*. Vol. 1. College Station: Texas A&M University Press.

Bruno, J. F., and M. D. Bertness. 2001. Habitat modification and facilitation in benthic marine communities. Chapter 8 in M. D. Bertness, S. D. Gaines, and M. E. Hay, eds., *Marine Community Ecology*. Sunderland, MA: Sinauer Associates.

Brusati, E. D., P. J. DuBowy, and T. E. Lacher. 2001. Comparing ecological functions of natural and created wetlands for shorebirds in Texas. *Waterbirds* 24:371–380.

Brush, T. 1995. Habitat use by wintering shorebirds along the Lower Laguna Madre of South Texas. *Texas Journal of Science* 47:179–190.

Brush, T. 2005. Nesting birds of a tropical frontier: The Lower Rio Grande Valley of Texas. College Station: Texas A&M University Press.

Bryan, C. E. 1971. *An Ecological Survey of the Arroyo Colorado, Texas 1966–1969*. Technical Series No. 10. Austin: Texas Parks and Wildlife Department.

Bryant, D. M. 1979. Effects of prey density and site character on estuary usage by overwintering waders (Charadrii). *Estuarine Coastal Marine Science* 9:369–384.

Buckley, P. A., and F. G. Buckley. 1977. Hexagonal packing of Royal Tern nests. *Auk* 94:36–43.

Bugica, K., D. Sterba-Boatwright, and M. S. Wetz. 2020. Water quality trends in Texas estuaries. *Marine Pollution Bulletin* 152. https://doi.org/10.1016/j.marpolbul.2020.110903.

Bumguardner, B. W., R. L. Colura, E. Young, D. Westbrook, and R. Buckley. 1996. *Black Drum Life History in Texas Bays with Emphasis on the Upper Laguna Madre*. Final report for US Department of Interior, Fish and Wildlife Service, Grant F-36-R, Project 18. Washington, DC: US Department of Interior.

Burger, J. 2018. *Birdlife of the Gulf of Mexico*. College Station: Texas A&M University Press.

Burger, J., and M. Gochfeld. 1990. *The Black Skimmer: Social Dynamics of a Colonial Species*. New York: Columbia University Press.

Burger, J., and B. L. Olla, eds. 1984. *Shorebirds: Migration and Foraging Behavior*. Behavior of Marine Animals, vol. 6. New York: Plenum Press.

Burke, V. J., and E. A. Standora. 1991. Factors affecting stranding of cold-stunned juvenile Kemp's Ridley and Loggerhead Sea Turtles in Long Island, New York. *Copeia* 1991:1136–1138.

Buroker, N. E. 1983. Population genetics of the American Oyster *Crassostrea virginica* along the Atlantic Coast and the Gulf of Mexico. *Marine Biology* 75:99–112.

Burr, J. G. 1930. A sail down the Laguna. *Year Book on Texas Conservation of Wildlife* 1929–1930:54–58.

Burr, J. G. 1945a. Adventure on Laguna Madre. *Texas Game and Fish* 4 (4): 8–9, 17–20.

Burr, J. G. 1945b. Nature wins. *Texas Game and Fish* 4 (9): 4, 16–17, 25–26.

Burr, J. G. 1947. The Laguna Madre Sphinx. *Texas Game and Fish* 6 (5): 4, 21–22.

Burr, J. G. 1950. Pass cutting on Padre Island. *Texas Game and Fish* 9 (3): 22–23.

Buschang, S. 2018. *Oil Spill Planning and Response Toolkit*. Texas General Land Office. https://glo.texas.gov/ost/download/index.html.

Buskey, E. J. 2008. How does eutrophication affect the role of grazers in harmful algal bloom dynamics? *Harmful Algae* 8:152–157.

Buskey, E. J., H. DeYoe, F. J. Jochem, and T. A. Villareal. 2003. Effects of mesozooplankton removal and ammonium addition on planktonic trophic structure during a bloom of the Texas "Brown Tide": A mesocosm study. *Journal of Plankton Research* 25:215–228.

Buskey, E. J., and C. J. Hyatt. 1995. Effects of the Texas (USA) "Brown Tide" alga on planktonic grazers. *Marine Ecology Progress Series* 126:285–292.

Buskey, E. J., H. Liu, C. Collumb, and J. G. F. Bersano. 2001. The decline and recovery of a persistent Texas brown tidal algal bloom in the Laguna Madre (Texas, USA). *Estuaries* 24:337–346.

Buskey, E. J., P. A. Montagna, A. F. Amos, and T. E. Whitledge. 1997. Disruption of grazer populations as a contributing factor to the initiation of the Texas Brown Tide algal bloom. *Limnology and Oceanography* 42:1216–1222.

Buskey, E. J., S. Stewart, J. Peterson, and C. Collumb. 1996. *Current Status and Historical Trends of Brown Tide and Red Tide Phytoplankton Blooms in the Corpus Christi Bay National Estuary Program Study Area*. Publication CCBNEP-07. Corpus Christi: Texas National Resource Conservation Commission.

Buskey, E. J., and D. A. Stockwell. 1993. Effects of a persistent "brown tide" on zooplankton populations in the Laguna Madre of South Texas. Pp. 659–666 in T. J. Smayda and Y. Shimizu, eds., *Toxic Phytoplankton Blooms in the Sea: Proceedings of the Fifth International Conference on Toxic Marine Phytoplankton*. Amsterdam: Elsevier.

Buskey, E. J., T. A. Villareal, and T. Lopez-Barreiro. 1999. Reconstructing the initiation of the Texas Brown Tide bloom of *Aureoumbra lagunensis* from archived samples using an immunofluorescence assay. *Planktonic Biology and Ecology* 46:159–161.

Buskey, E. J., B. Wysor, and C. Hyatt. 1998. The role of hypersalinity in the persistence of the Texas "Brown Tide" in the Laguna Madre. *Journal of Plankton Research* 20:1553–1565.

Bustard, H. R. 1972. *Sea Turtles: Their Natural History and Conservation*. London: Collins.

Buttke, D. E., A. Walker, I. S. Huang, L. Flewelling, J. Lankton, A. E. Ballmann, T. Clapp, J. Lindsay, and P. V. Zimba. 2018. Green Tree Frog (*Hyla cinerea*) and Ground Squirrel (*Xerospermophilus spilosoma*) mortality attributed to inland brevetoxin transport at Padre Island National Seashore, Texas, USA, 2015. *Journal of Wildlife Diseases* 54:142–146.

Cahn, A. R. 1923. Louisiana Herons and Reddish Egrets at home. *Natural History* 23:471–479.

Cahoon, D. R., P. Hensel, J. Rybczyk, K. L. McKee, C. E. Proffitt, and B. C. Perez. 2003. Mass tree mortality leads to mangrove peat collapse at Bay Islands, Honduras after Hurricane Mitch. *Journal of Ecology* 91:1093–1105.

Caillouet, C. W., Jr. 2014. Interruption of the Kemp's Ridley population's pre-2010 exponential growth in the Gulf of Mexico and its aftermath: One hypothesis. *Marine Turtle Newsletter* 143:1–7.

Caillouet, C. W., Jr., M. J. Duronslet, A. M. Landry, and D. J. Shaver. 1991. Sea turtle strandings and shrimp fishing effort in the northwestern Gulf of Mexico, 1986–1989. *Fishery Bulletin* 89:712–718.

Caillouet, C. W., Jr., B. J. Gallaway, and N. Putman. 2016. Kemp's Ridley Sea Turtle saga and setback: Novel analyses of cumulative hatchlings released and time-lagged annual nests in Tamaulipas, Mexico. *Chelonian Conservation and Biology* 15:115–131.

Caillouet, C. W., Jr., D. J. Shaver, W. G. Teas, J. M. Nance, and D. B. Revera. 1996. Relationship between sea turtle stranding rates and shrimp fishing intensities in the northwestern Gulf of Mexico: 1986–1989 versus 1990–1993. *Fishery Bulletin* 94:237–249.

Calabrese, A. 1969. Individual and combined effects of salinity and temperature on embryos and larvae of the Coot Clam, *Mulinia lateralis* (Say). *Biological Bulletin* 137:417–428.

Calvert, R. A., and A. De León. 1990. *The History of Texas*. Arlington Heights, IL: Harlan Davidson.

Cannicci, S., D. Burrows, S. Fratini, T. J. Smith III, J. Offenberg, and F. Dahdouh-Guebas. 2008. Faunal impact on vegetation structure and ecosystem function in mangrove forests: A review. *Aquatic Botany* 89:186–200.

Capone, D. G., and B. F. Taylor. 1980. Microbial nitrogen cycling in a seagrass community. Pp. 153–161 in V. S. Kennedy, ed., *Estuarine Perspectives*. New York: Academic Press.

Carangelo, P., R. Heiser, M. Knight, J. Snodgrass, K. Allen, C. Hughes, and G. Penn. 1975. Final report: Plant and animal study group. Chapter 6 in C. P. McRoy, ed., *Biology of Seagrass Ecosystems: Report of a Summer Field Course*. Seagrass Bulletin No. 1. Fairbanks: Institute of Marine Science, University of Alaska.

Carls, E. G., D. B. Fenn, and S. A. Chaffey. 1995. Soil contamination by oil and gas drilling and production operations in Padre Island National Seashore, Texas, USA. *Journal of Environmental Management* 45:273–286.

Carls, E. G., R. I. Lonard, and D. B. Fenn. 1990. Impact of oil and gas operations on the vegetation of Padre Island National Seashore, Texas, USA. *Ocean and Shoreline Management* 14:85–104.

Carney, K. M., and W. J. Sydeman. 1999. A review of human disturbance effects on nesting colonial waterbirds. *Waterbirds* 22:68–79.

Carpelan, L. H. 1967. Invertebrates in relation to hypersaline habitats. *Contributions in Marine Science* 12:219–229.

Carr, A. F. 1986. Rips, FADS, and little Loggerheads. *Bioscience* 36:92–100.

Carr, E., N. Damude, and M. Lindsay. 1993. *Endangered, Threatened, and Watch List of Texas Plants*. Publication No. 9, 3rd rev. Austin: Texas Organization for Endangered Species.

Carr, J. T., Jr. 1967. *The Climate and Physiography of Texas*. Report 53. Austin: Texas Water Development Board.

Carr, W. E. S., and C. A. Adams. 1973. Food habits of juvenile marine fishes occupying seagrass beds in the estuarine zone near Crystal River, Florida. *Transactions of the American Fisheries Society* 102:511–540.

Carr, W. R. n.d. Some plants of the South Texas Sand Sheet. Accessed 23 February 2018. http://w3.biosci.utexas.edu/prc/DigFlora/WRC/Carr-SandSheet.html.

Carrera, E., and G. de la Fuente. 1996. Laguna Madre. *Revista DUMAC* 16 (1): 28–31.

Carroll, J. J. 1927. Down Bird Island way. *Wilson Bulletin* 39:195–208.

Carroll, J. J. 1930a. Breeding of the American White Pelican on the Texas coast. *Condor* 32:202–204.

Carroll, J. J. 1930b. More about the White Pelican on the Texas coast. *Condor* 32:304.

Carroll, J. M., C. A. Davis, R. D. Elmore, S. D. Fuhlendorf, and E. T. Thacker. 2015. Thermal patterns constrain diurnal behavior of a ground-dwelling bird. *Ecosphere* 6 (11). https://doi.org/10.1890?ES15–00163.1.

Carroll, S. P., and K. L. Cramer. 1985. Age differences in kleptoparasitism by Laughing Gulls (*Larus atricilla*) on adult and juvenile Brown Pelicans (*Pelecanus occidentalis*). *Animal Behaviour* 33:201–205.

Castle, K. T., L. J. Flewelling, J. Bryan, A. Kramer, J. Lindsay, C. Nevada, W. Stablein, D. Wong, and J. H. Landsberg. 2013. Coyote (*Canis latrans*) and Domestic Dog (*Canis familiaris*) mortality and morbidity due to a *Karenia brevis* Red Tide in the Gulf of Mexico. *Journal of Wildlife Diseases* 49:955–964.

Caudle, C. S. 1992. Population dynamics of the fish fauna found in three earthen ponds on North Padre Island, Texas. MS thesis, Southwest Texas State University, San Marcos.

Cavanaugh K. C., J. R. Kellner, A. J. Forde, D. S. Gruner, J. D. Parker, W. Rodriguez, and I. C. Feller. 2014. Poleward expansion of mangroves is a threshold response to decreased frequency of extreme cold events. *Proceedings of the National Academy of Sciences* 111:723–727.

Caveny, R. J., S. J. Voelkel, W. T. Brademan, J. B. Hardin, M. J. Peterson, and B. A. Collier. 2011. Distribution, fidelity, and abundance of Rio Grande wild turkey roosts in the Texas Coastal Sand Plains. *Proceedings of the Annual Conference of the Southeastern Association of Fish and Wildlife Agencies* 65:45–50.

Cayan, D. R., T. Das, D. W. Pierce, T. P. Barnett, M. Tyree, and A. Gershunov. 2010. Future dryness in the southwest US and the hydrology of the early 21st century drought. *PNAS* 107 (50): 21271–21276. doi: 10.1073/pnas.0912391107.

CBBEP (Coastal Bend Bays and Estuaries Program). 2018. *Coastal Bend Bays Plan: Protecting the Coastal Bend Bays and Estuaries.* 2nd ed. Corpus Christi, TX: CBBEP.

Celis, A., G. Rodríguez-Almaraz, and F. Álvarez. 2007. Los cirripedios torácicos (Crustacea) de aguas someras de Tamaulipas, México. *Revista Mexicana de Biodiversidad* 78:325–337.

Chabreck, R. H., and A. W. Palmisano. 1973. The effects of Hurricane Camille on the marshes of the Mississippi River Delta. *Ecology* 54:1118–1123.

Chaney, A. H. 1981. *A Study of the Bird Use of the Wetlands in the Middle Rio Grande Valley*. Report to US Fish and Wildlife Service Ecological Services Division, Agreement 20181-0885 FY 80, Corpus Christi.

Chaney, A. H. 1988. *An Analysis of the Nekton and Plankton around a Shoal Grass Bed in the Laguna Madre of Texas*. Report to Padre Island National Seashore Contract No. PX7490-7-0009, Corpus Christi.

Chaney, A. H., G. W. Blacklock, and S. G. Bartels. 1996. *Current Status and Historical Trends of Avian Resources in the Corpus Christi Bay National Estuary Program Study Area*. Corpus Christi Bay National Estuary Program CCBNEP-06B. Corpus Christi: Texas Natural Resource Conservation Commission.

Chaney, A. H., B. R. Chapman, J. P. Karges, D. A. Nelson, and R. R. Schmidt. 1978. *The Use of Dredged Material Islands by Colonial Seabirds and Wading Birds in Texas*. Waterways Experiment Station Report No. D-78-8. Vicksburg, MS: US Army Corps of Engineers.

Chapman, B. R. 1984. *Seasonal Abundance and Habitat-Use Patterns of Coastal Bird Populations on Padre and Mustang Island Barrier Beaches (following the Ixtoc I Oil Spill)*. Publication FWS-OBS-83/31. Washington, DC: US Fish and Wildlife Service.

Chapman, B. R. 1988. History of American White Pelican colonies in South Texas and northern Tamaulipas. *Colonial Waterbirds* 11:275–283.

Chapman, B. R., and E. G. Bolen. 2018. *The Natural History of Texas*. College Station: Texas A&M University Press.

Chapman, B. R., and R. L. Packard. 1974. An ecological study of Merriam's Pocket Mouse in southeastern Texas. *Southwestern Naturalist* 21:275–280.

Chapman, K. 2000. *The Binational Laguna Madre—Texas*. Draft report. Austin: Texas Center for Policy Studies.

Chapman, S. S., and B. R. Chapman. 1990. Bats from the coastal region of southern Texas. *Texas Journal of Science* 42:13–22.

Cheeseman, B. S. 1994. Richard King: Pioneering market capitalism on the frontier. Pp. 86–91 in J. S. Graham, ed., *Ranching in South Texas: A Symposium*. Kingsville: Texas A&M University–Kingsville.

Cheeseman, B. S. 1997. "Let us have 500 good determined Texans": Richard King's account of the Union invasion of South Texas, November 12, 1863, to January 20, 1864. *Southwestern Historical Quarterly* 101:76–95.

Chin, J. L. 1978. *Distribution of Marine Grasses in Southern Laguna Madre*. Report to Texas Parks and Wildlife Department on file at Texas A&M University–Corpus Christi, Center for Coastal Studies Library.

Church, J. A., and N. J. White. 2006. A 20th century acceleration in global sea-level rise. *Geophysical Research Letters* 33, L01602. doi:10.1029/2005GL024826.

Churchwell, B. 2017. Fireworks spark a 320-acre grass fire near the National Seashore. KIII TV. Corpus Christi. https://www.kiiitv.com/article/news/local/fireworks-spark-320-acre-grass-fire/453949303.

Cintron, G., A. E. Lugo, D. J. Pool, and G. Morris. 1978. Mangroves of arid environments in Puerto Rico and adjacent islands. *Biotropica* 10:110–121.

Clark, D. A., S. Brown, D. W. Kicklighter, J. Q. Chambers, J. R. Thomlinson, J. Ni, and E. A. Holland. 2001. Net primary production in tropical forests: An evaluation and synthesis of existing field data. *Ecological Applications* 11:371–384.

Clark, T. L. 1961. *Coastal Waterfowl Project: Ecology of Wintering Waterfowl in Lower Laguna Madre*. Texas Parks and Wildlife Federal Aid Project, W-29-R-14, Job 16. Austin: Texas Parks and Wildlife Department.

Clarke, P. J., R. A. Kerrigan, and C. J. Westpal. 2001. Dispersal potential and early growth in 14 tropical mangroves: Do early life history traits correlate with patterns of adult distribution? *Journal of Ecology* 89:648–659.

Claydon, J. A. B., M. C. Calosso, and S. B. Traiger. 2012. Progression of invasive lionfish in seagrass, mangrove and reef habitats. *Marine Ecology Progress Series* 448:119–129.

Clover, E. U. 1937. Vegetational survey of the Lower Rio Grande Valley, Texas. *Madroño* 4 (2–3): 41–66, 77–100.

Coffey, G. N. 1909. Clay dunes. *Journal of Geology* 17:754–755.

Coldren, G. A., C. Barreto, D. Wykoff, E. M. Morrissey, J. A. Langley, I. C. Feller, and S. K. Chapman. 2016. Chronic warming stimulates growth of marsh grasses more than mangroves in a coastal wetland ecotone. *Ecology* 97:3167–3175.

Coldren, G. A., and C. E. Proffitt. 2017. Mangrove seedling freeze tolerance depends on salt marsh presence, species, salinity, and age. *Hydrobiologia* 803:159–171.

Cole, R. M. 1981. The serpulid reefs of Baffin Bay, Texas. Pp. 63–74 in J. L. Russell and R. W. Shum, eds., *Geology of Clay Dunes, Baffin Bay, and the South Texas Sand Sheet*. Field Trip Guidebook, 83rd Annual Meeting, Texas Academy of Science. Austin: Texas Academy of Science.

Collias, N. E., and E. C. Collias. 1984. *Nest Building and Bird Behavior*. Princeton, NJ: Princeton University Press.

Collier, A., and J. W. Hedgpeth. 1950. An introduction to the hydrography of tidal waters of Texas. *Publications of the Institute of Marine Science, University of Texas* 1:121–194.

Colwell, M. A. 2010. *Shorebird Ecology, Conservation, and Management*. Berkeley: University of California Press.

Colwell, M. A., and S. L. Landrum. 1993. Nonrandom shorebird distribution and fine-scale variation in prey abundance. *Condor* 95:94–103.

Comeaux, R. S., A. A. Mead, and T. S. Bianchi. 2012. Mangrove expansion in the Gulf of Mexico with climate change: Implications for wetland health and resistance to rising sea levels. *Estuarine, Coastal and Shelf Science* 96:81–95.

CONABIO (Comisión Nacional para el Conocimiento y Uso de la Biodiversidad). 2009. *Sitios de manglar con relevancia biológica y con necesidades de rehabilitación ecológica*. https://www.biodiversidad.gob.mx/ecosistemas/manglares.

CONANP (Comisión Nacional de Áreas Naturales Protegidas). 2014. *Procedimos: Primer Simposio de Investigación en el APFF Laguna Madre y Delta del Río Bravo "Biodiversidad y Cambio Climático."* Ciudad Victoria, Tamaulipas: CONANP.

CONANP. 2015. *Programa de Manejo,* Área de Protección de Flora y Fauna Laguna Madre y Delta del Río Bravo. Mexico City: Secretaría de Medio Ambiente y Recursos Naturales. https://simec.conanp.gob.mx/pdf_libro_pm/122_libro_pm.pdf.

Conant, R. 1975. *A Field Guide to Reptiles and Amphibians of Eastern/Central North America*. Boston: Houghton Mifflin.

Congdon, V. M., and K. H. Dunton. 2016. *Tracking Long-Term Trends in Seagrass Cover and Condition in Texas Coastal Waters*. Austin: Texas General Land Office.

Conover, J. T. 1964. The ecology, seasonal periodicity, and distribution of benthic plants in some Texas lagoons. *Marine Botany* 7:4–41.

Contreras, E. F., and O. Castañeda. 2004. Las lagunas costeras y estuarios del Golfo de México: Hacia el establecimiento de índices ecológicos. Pp. 373–416 in M. Caso, I. Pisanty, and E. Excurra, eds., *Diagnóstico ambiental del Golfo de México*, vol. 1. Mexico City: INE, SEMARNAT.

Contreras Walsh, S., M. B. Ballard, D. B. Wester, W. P. Kuvlesky Jr., L. A. Brennan, M. L. Morrison, and K. Boydston. 2017. High passage rates and different seasonal migration strategies of birds along the lower Texas coast. *International Journal of Biodiversity and Conservation* 9:183–199.

Contreras-Balderas, A. J., J. A. Garcia-Salas, and J. I. Gonzalez-Rojas. 1990. Aves acuaticas y semiacuaticas de la Laguna Madre, Tamaulipas, México: Otono-invierno 1988–1989, su aprovechamiento cinegetico. *BIOTAM* 2 (2): 23–30.

Contreras-Lozano, J. A., D. Lazcano, J. A. García-Salas, and A. J. Contreras-Balderas. 2013. Herpetofauna in two coastal sites in the municipality of Soto la Marina, Tamaulipas, Mexico. *Acta Zoológica Mexicana*, n.s., 29 (1): 144–152.

Cooke, W. W. 1910. *US Bureau of Biological Survey Bulletin 35*. Washington, DC: US Bureau of Biological Survey.

Cooper, J. A. G., D. A. Lewis, and O. H. Pilkey. 2007. Fetch limited barrier islands: Overlooked coastal landforms. *GSA Today* 17:4–9.

Coover, J. R., and C. A. Rechenthin. 1965. *Report of a Preliminary Investigation of the Wind Erosion and Salt Problem on the Salt Flats of Laguna Madre*. Temple, TX: US Department of Agriculture, Soil Conservation Service.

Copeland, B. J. 1967. Environmental characteristics of hypersaline lagoons. *Contributions in Marine Science* 12:207–218.

Copeland, B. J., and R. S. Jones. 1965. Community metabolism in some hypersaline waters. *Texas Journal of Science* 17:188–205.

Copeland, B. J., J. H. Thompson Jr., and W. Ogletree. 1968. Effects of wind on water levels in the Texas Laguna Madre. *Texas Journal of Science* 20:196–199.

Cornejo, E. M. 2009. Fish assemblage dynamics in the re-flooded Bahia Grande. MS thesis. University of Texas at Brownsville.

Cornelius, S. E. 1975. Food choice of wintering Redhead ducks (*Aythya americana*) and utilization of available resources in Lower Laguna Madre, Texas. MS thesis, Texas A&M University, College Station.

Cornelius, S. E. 1977. Food and resource utilization by wintering Redheads on Lower Laguna Madre. *Journal of Wildlife Management* 41:374–385.

Cornelius, S. E. 1984. *An Ecological Survey of Alazan Bay, Texas*. Technical Bulletin No. 5. Kingsville, TX: Caesar Kleberg Wildlife Research Institute.

Correa-Sandoval, A., and J. H. Rodríguez-Castro. 2013. Zoogeografía de los bivalvos marinos de la costa de Tamaulipas, México. *Revista de Biología Marina y Oceanografía* 48:565–584.

Coste, R. L., and M. K. Skoruppa. 1989. *Colonial Waterbird Rookery Island Management Plan for the South Texas Coast*. Technical Report CCSU-1089-CCS. Corpus Christi: Texas A&M University–Corpus Christi, Center for Coastal Studies.

Côté, I. M., S. J. Green, and M. Hixon. 2013. Predatory fish invaders: Insights from Indo-Pacific lionfish in the western Atlantic and Caribbean. *Biological Invasions* 164:50–61.

Cotner, J. B., M. W. Suplee, N. W. Chen, and D. E. Shormann. 2004. Nutrient, sulfur and carbon dynamics in a hypersaline lagoon. *Estuarine, Coastal and Shelf Science* 59:639–652.

Cowardin, L. M., V. Carter, F. C. Golet, and E. T. LaRoe. 1979. *Classification of Wetlands and Deep-Water Habitats of the United States*. Publication FWS/OBS-79/31. Washington, DC: US Fish and Wildlife Service, Office of Biological Services.

Cowper, S. W. 1978. The drift algae community of seagrass beds in Redfish Bay, Texas. *Contributions in Marine Science* 21:125–132.

Coyne, M. S. 1994. Feeding ecology of subadult Green Sea Turtles in South Texas waters. MS thesis, Texas A&M University, College Station.

Craighead, F. C., and V. C. Gilbert. 1962. The effects of Hurricane Donna on the vegetation of southern Florida. *Quarterly Journal of the Florida Academy of Sciences* 25:1–28.

Crimm, A. C. C. 1997. Introduction. Pp. xv–xxxvii in J. Monday and B. Colley, *Voices from the Wild Horse Desert: The Vaquero Families of the King and Kenedy Ranches*. Austin: University of Texas Press.

Cronin, L. E., R. B. Biggs, D. A. Flemer, G. T. Pitzmeyer, F. Goodwin Jr., W. L. Dovel, and D. E. Richie Jr. 1970. *Gross Physical and Biological Effects of Overboard Spoil Disposal in Upper Chesapeake Bay*. Special Report No. 3. College Park, MD: University of Maryland Chesapeake Biological Laboratory, Natural Resources Institute.

Cruz, G. R. 1988. *Let There Be Towns: Spanish Municipal Origins in the American Southwest, 1610–1810*. College Station: Texas A&M University Press.

Cruz, J. A. 2003. Parámetros de crecimiento y mortalidad del ostión (*Crassostrea virginica*) en la Laguna Madre, Tamaulipas durante la temporada 2002. Pp. 11–12 in B. E. Espino, M. A. Carrasco, E. G. Cabrals, and M. Puente, eds., *Memorias del II Foro Científico de Pesca Ribereña*. Colima, Mexico: SAGARPA (Secretaría de Agricultura, Ganadería, Desarrollo Rural, Pesca y Alimentación), INP (Instituto Nacional de Pesca), CRIP (Centro Regional de Investigación Pesquera)-Manzanillo.

Cruz-Nieto, M. A. 2000. *The Binational Laguna Madre—Tamaulipas*. Draft report. Austin: Texas Center for Policy Studies.

Cuddy, M. R. 2018. Pattern in iscoscapes and N:P stoichioscapes of the dominant seagrasses (*Halodule wrightii* and *Thalassia testudinum*) in the western Gulf of Mexico. MS thesis, University of Texas at Austin.

Cunjak, R. A., J.-M. Roussel, M. A. Gray, J. P. Dietrich, D. F. Cartwright, K. R. Munkittrick, and T. D. Jardine. 2005. Using stable isotope analysis with telemetry or mark-recapture data to identify fish movement and foraging. *Oecologia* 144:636–646.

Curray, J. R. 1960. Sediments and history of Holocene transgression, continental shelf, northwest Gulf of Mexico. Pp. 221–266 in F. P. Shepard, F. B. Phleger, and T. H. van Andel, eds., *Recent Sediments, Northwest Gulf of México*. Tulsa, OK: American Association of Petroleum Geologists.

Custer, C. M., T. W. Custer, and P. J. Zwank. 1997. Migration chronology and distribution of Redheads on the Lower Laguna Madre, Texas. *Southwestern Naturalist* 42:40–51.

Custer, T. W., and C. A. Mitchell. 1987a. Exposure to insecticides of brushland wildlife within the Lower Rio Grande Valley, Texas. *Environmental Pollution* 45:207–220.

Custer, T. W., and C. A. Mitchell. 1987b. Organochlorine contaminants and reproductive success of Black Skimmers in South Texas, 1984. *Journal of Field Ornithology* 58:480–489.

Custer, T. W., and C. A. Mitchell. 1991. Contaminant exposure of Willets feeding in agricultural drainages of the Lower Rio Grande Valley of South Texas. *Texas Environmental Monitoring and Assessment* 16:189–220.

Custer, T. W., and C. A. Mitchell. 1993. Trace elements and organochlorines in the Shoal Grass community of the Lower Laguna Madre, Texas. *Texas Environmental Monitoring and Assessment* 25:235–246.

Czerny, A. B. 1994. Growth and photosynthetic responses of two tropical seagrasses, *Thalassia testudinum* and *Halodule wrightii*, to in situ manipulations of irradiance. MS thesis, University of Texas at Austin.

Czerny, A. B., and K. H. Dunton. 1995. The effects of in situ light reduction on the growth of two subtropical seagrasses, *Thalassia testudinum* and *Halodule wrightii*. *Estuaries* 18:418–427.

Dalquest, W. W. 1968. Mammals of North-Central Texas. *Southwestern Naturalist* 13:13–21.

Dalrymple, D. W. 1964. Sedimentary facies in Baffin Bay. PhD diss., Rice University, Houston.

Dangremond, E. M., L. T. Simpson, T. Z. Osborne, and I. C. Feller. 2019. Nitrogen enrichment accelerates mangrove range expansion in the temperate-tropical ecotone. *Ecosystems* 23:703–714. DOI: doi.org/10.1007/s10021–019–00441–2.

Daniell, L. E. 1892. *Personnel of the Texas State Government, with Sketches of Representative Men of Texas*. San Antonio: Maverick Printing House.

Darnell, R. M. 1958. Food habits of fishes and larger invertebrates of Lake Pontchartrain, Louisiana, an estuarine community. *Publications of the Institute of Marine Science, University of Texas* 5:353–416.

Davenport, J. 1997. Temperature and the life-history strategies of sea turtles. *Journal of Thermal Biology* 22:479–488.

Davis, A. M. 1942. A Study of Boscaje de la Palma in Cameron County, Texas and of *Sabal texana*. MA thesis, University of Texas at Austin.

Davis, J. H. 1940. The ecology and geologic role of mangroves in Florida. *Publications of the Carnegie Institution of Washington* 517:303–412.

Davis, J. R. 1983. *Intensive Survey of the Arroyo Colorado Segment 2201*. Publication No. IS-49. Austin: Texas Department of Water Resources.

Davis, J. R. 1984. *Intensive Survey of the Arroyo Colorado Segment 2201*. Publication No. IS-61. Austin: Texas Department of Water Resources.

Davis, J. R. 1985. *Intensive Survey of the Arroyo Colorado Segment 2201, August 22–25, 1983*. Publication No. IS-69. Austin: Texas Department of Water Resources.

Davis, R. A., Jr. 2011. *Sea Level Change in the Gulf of Mexico*. College Station: Texas A&M University Press.

Davis, W. B., and D. J. Schmidly. 1994. *The Mammals of Texas*. Austin: Texas Parks and Wildlife Press.

den Hartog, C. 1977. Structure, function, and classification in seagrass communities. Pp. 89–122 in C. P. McRoy and C. Helfferich, eds., *Seagrass Ecosystems: A Scientific Perspective*. New York: Marcel Dekker.

Denton, W., and C. Contreras. 2004. *The Red Tide (*Karenia brevis*) Bloom of 2000*. Austin: Texas Parks and Wildlife Department.

DePue, J. 1974. Nesting and reproduction of the Black Skimmer (*Rhynchops niger* Linnaeus) on four spoil islands in the Laguna Madre, Texas. MS thesis, Texas A&I University, Kingsville.

Devlin, D. J. 2004. Analysis of the relationship between parasitic beetle *Coccotrypes rhizophorae*, and a host plant, the Red Mangrove (*Rhizophora mangle*). PhD diss., University of Louisiana–Lafayette.

DeYoe, H. R., A. M. Chan, and C. A. Suttle. 1995. Phylogeny of *Aureococcus anophagefferens* and a morphologically similar bloom-forming alga from Texas as determined by 18S ribosomal RNA sequence analysis. *Journal of Phycology* 31:413–418.

DeYoe, H., W. Pulich, and N. Fernando. 2016. *Modeled Inflow Validation and Nutrient Loading Estimation in Two Subwatersheds of the Lower Laguna Madre*. Coastal Management Program Report, CMP contract number 14-085-000-7953. https://glo.texas.gov/coastal-grants/_documents/grant-project/14-085-final-rpt.pdf.

DeYoe, H. R., D. A. Stockwell, R. R. Bidagare, M. Latasa, P. W. Johnson, P. E. Hargraves, and C. A. Suttle. 1997. Description of the algal species *Aureoumbra lagunensis* gen. et sp. nov. and referral of *Aureoumbra* and *Aureococcus* to the Pelagophyceae. *Journal of Phycology* 33:1042–1048.

DeYoe, H. R., and C. A. Suttle. 1994. The inability of the Texas "Brown Tide" alga to use nitrate and the role of nitrogen in the initiation of a persistent bloom of this organism. *Journal of Phycology* 30:800–806.

Diamond, D. D., and T. E. Fulbright. 1990. Contemporary plant communities of upland grasslands of the Coastal Sand Plain, Texas. *Southwestern Naturalist* 35:385–392.

Diamond, D. D., D. H. Riskind, and S. L. Orzell. 1987. A framework for plant community classification and conservation in Texas. *Texas Journal of Science* 39:203–221.

Diaz, A., and M. Kelly. 1994. *Subsidized Destruction: The Gulf Intracoastal Waterway and the Laguna Madre*. Austin: Texas Center for Policy Studies.

Dice, L. R. 1937. Mammals of the San Carlos Mountains and vicinity. *University of Michigan Studies and Scientific Series* 12:245–268.

Dice, L. R. 1943. *The Biotic Provinces of North America*. Ann Arbor: University of Michigan Press.

Dick, M. 1982. *Pesticide and PCB Concentrations in Texas—Water, Sediment and Fish Tissues*. Report No. 264. Austin: Texas Department of Water Resources.

Diener, R. A. 1975. *Cooperative Gulf of Mexico Estuarine Inventory and Study—Texas: Area Description*. NOAA-Technical Report NMFS Circ-393. Seattle, WA: NOAA.

Dixon, J. R. 2000. *Amphibians and Reptiles of Texas*. College Station: Texas A&M University Press.

Dobie, J. L., L. H. Ogren, and J. F. Fitzpatrick Jr. 1961. Footnotes and records of the Atlantic Ridley Turtle (*Lepidochelys kempii*) from Louisiana. *Copeia* 1961:109–110.

Dodd, C. K., Jr. 1988. *Synopsis of the Biological Data on the Loggerhead Sea Turtle* Caretta caretta *(Linnaeus 1758)*. US Fish and Wildlife Service Biological Report 88(14), Washington, DC.

DOF (Diario Oficial de la Federación). 1994a. *Norma Oficial Mexicana NOM-015-PESC-1994, para regular la extracción de las existencias naturales de ostión en los sistemas lagunarios estuarinos del Estado de Tabasco*. Mexico City: SEMARNAT.

DOF. 1994b. *Norma Oficial Mexicana NOM-016-PESC-1994, para regular la pesca de lisa y liseta o lebrancha en aguas de jurisdicción federal del Golfo de México y Mar Caribe, así como del Océano Pacífico, incluyendo el Golfo de California*. Mexico City: SEMARNAT.

DOF. 2005. *Decreto por el que se declara área natural protegida, con el carácter de área de protección de flora y fauna, la región conocida como Laguna Madre y Delta del Río Bravo, ubicada en los municipios de Matamoros, San Fernando y Soto La Marina, en el Estado de Tamaulipas, con una superficie total de 572,808–60–94.22 hectáreas*. Mexico City: SEMARNAT.

DOF. 2015. *Acuerdo por el que se da a conocer el resumen del programa de manejo del Área de Protección de Flora y Fauna Laguna Madre y Delta del Río Bravo*. Mexico City: SEMARNAT.

Dokken, Q. R. 1981. Spatial and temporal distribution and species composition of larval fish populations within Alazan Bay, Texas. MS thesis, Corpus Christi State University, Corpus Christi.

Dokken, Q. R., G. C. Matlock, and S. Cornelius. 1984. Distribution and composition of larval fish populations within Alazan Bay, Texas. *Contributions in Marine Science* 27:205–222.

Doughty, R. W. 1984. Sea turtles in Texas: A forgotten commerce. *Southwestern Historical Quarterly* 88:43–70.

Douglas, C. L. 1968. *Cattle Kings of Texas*. Fort Worth: Branch-Smith.

Drake, K. R., J. E. Thompson, and K. L. Drake. 2001. Movements, habitat use, and survival of nonbreeding Piping Plovers. *Condor* 103:259–267.

Drawe, D. L., and K. R. Kattner. 1978. Effect of burning and mowing on vegetation of Padre Island. *Southwestern Naturalist* 23:273–278.

Drawe, D. L., K. R. Kattner, W. H. McFarland, and D. D. Neher. 1981. Vegetation and soil properties of five habitat types on North Padre Island. *Texas Journal of Science* 33:145–157.

Drawe, D. L., and I. M. Ortega. 1996. Impacts of geophysical seismic survey vehicles on Padre Island National Seashore vegetation. *Texas Journal of Science* 48:107–118.

Duarte, C. M. 1991. Seagrass depth limits. *Aquatic Botany* 40:363–377.

DUMAC (Ducks Unlimited de México, A.C.). 1990a. Areas lacustres de México: Plan Maestro, Año 2000. *Revista DUMAC* 12 (3): 4–12.

DUMAC. 1990b. Laguna Madre y Laguna de Tamaulipas. *Revista, DUMAC* 12 (3): 15–16.

DUMAC. 1993. *Diagnostico eco-grafico para el manejo y conservacion de la Laguna Madre, Tamaulipas y Zonas de Influencia: Informe final*. Technical Report. Monterrey, Nuevo León, Mexico: Ducks Unlimited of Mexico, A.C.

DUMAC. 1996. *Vegetacion acuatica de la Laguna Madre de Tamaulipas (Pastos Marinos): Informe final de proyecto*. Monterrey, Nuevo León, Mexico: Ducks Unlimited of México A.C. and Instituto de Ecologia y Alimentos de la Universidad Autónoma de Tamaulipas.

Dunton, K. H. 1994. Seasonal growth and biomass of the subtropical *Halodule wrightii* in relation to continuous measurements of underwater irradiance. *Marine Biology* 120:479–489.

Dunton, K. H. 1996. Photosynthetic production and biomass of the subtropical seagrass *Halodule wrightii* along an estuarine gradient. *Estuaries* 19:436–447.

Dunton, K. H., A. Burd, L. Cifuentes, P. J. Eldridge, and J. W. Morse. 1998. *The Effect of Dredge Deposits on the Distribution and Productivity of Seagrasses: An Integrative Model for Laguna Madre*. Draft final report. Galveston: Interagency Coordination Team, US Army Corps of Engineers, Galveston District.

Dunton, K. H., A. Burd, L. Cifuentes, P. M. Eldridge, and J. W. Morse. 2003. *Effects of Dredge Deposits on Seagrasses: An Integrative Model for Laguna Madre*. Vol. 2, *Executive Summary*. Concluding report. Galveston: US Army Corps of Engineers, Galveston District.

Dunton, K. H., P. A. Montagna, and S. A. Holt. 1998. *Characterization of Anthropogenic and Natural Disturbance on Vegetated and Unvegetated Bay Bottom Habitats in the Corpus Christi Bay National Estuary Program Study Area*. Vol. 2, *Assessment of Scarring in Seagrass Beds*. Corpus Christi Bay National Estuary Program CCBNEP-25B. Corpus Christi: Texas Natural Resource Conservation Commission.

Dunton, K. H., W. Pulich Jr., and T. Mutchler. 2011. *A Seagrass Monitoring Program for Texas Coastal Waters*. Coastal Bend Bays and Estuaries Program Report, Contract No. 0627. http://www.texasseagrass.org/doc/A%20Seagrass%20Monitoring%20Program%20for%20Texas%20 1-10-11.pdf.

Duran, M. C. 2004. *An Inventory of Reptiles and Amphibians of Padre Island National Seashore*. San Antonio: Texas Conservation Data Center, The Nature Conservancy.

Dyer, K. R. 1973. *Estuaries: A Physical Introduction*. London: Wiley and Sons.

Dykstra, R. F. 1966. Investigation of some algae of the Texas Gulf Coast. MS thesis, University of Texas at Austin.

Ecoservices. 1993a. *Bird Use of the Padre Island National Seashore Gulf Beach from September 1992–August 1993*. Report to Padre Island National Seashore, Contract No. 1443PX749092188, Corpus Christi.

Ecoservices. 1993b. *Laguna Madre Bird Project from Yarborough Pass to Mansfield Channel during July 1992 through April 1993*. Report to Padre Island National Seashore, Contract No. 1443PX7000092582, Corpus Christi.

Edwards, P., and D. F. Kapraun. 1973. Benthic marine algal ecology in the Port Aransas, Texas area. *Contributions in Marine Science* 17:15–52.

EHA (Espey, Huston and Associates, Inc.). 1977. *Ecological Baseline Inventory (Appendix B) and Environmental Effects Assessment, Hidalgo and Willacy County Drainage Districts, Permit Application No. 11374*. Report to Galveston District Corps of Engineers, Galveston.

EHA. 1997. *Piping Plover Habitat Survey of Dredged Material Placement Areas along the Gulf Intracoastal Waterway from Port Isabel Bay to the Mud Flats*. Report to US Army Corps of Engineers, Galveston District, EH&A Job No. 18383, Document No. 970507, Austin.

EHA. 1998. *Laguna Madre, Texas Contaminant Assessment*. Report to US Army Corps of Engineers, Galveston District, EH&A Job No. 18798, Document No. 971410, Austin.

EHA and BVA (Barry A. Vittor and Associates). 1998. *Benthic Macroinfaunal Analysis of Dredged Material Placement Areas in the Laguna Madre, Texas, Spring and Fall 1996 Surveys*. Report to US Army Corps of Engineers, Galveston District, EH&A Job No. 15650-53, Document No. 970740, Austin.

Ehrhart, L. M., W. Redfoot, D. Bagley, and K. Mansfield. 2014. Long-term trends in Loggerhead (*Caretta caretta*) nesting and reproductive success at an important western Atlantic rookery. *Chelonian Conservation and Biology* 13:173–181.

Eleuterius, L. N. 1987. Seagrass ecology along the coasts of Alabama, Louisiana, and Mississippi. Pp. 11–24 in M. J. Durako, R. C. Phillips, and R. R. Lewis III, eds., *Proceedings of the Symposium on Subtropical-Tropical Seagrasses of the Southeastern United States*. Florida Department of Natural Resources Marine Research Publication 42. St. Petersburg: Florida Department of Natural Resources.

Elliott-Smith, E., M. Bidwell, A. E. Holland, and S. Haig. 2015. *Data from the 2011 International Piping Plover Census*. US Geological Survey Data Series 922. Reston, VA: US Geological Survey.

Ellis, M. J. 1986. *The Corpus Christi Hurricane Almanac, Texas*. Corpus Christi: Hurricane Publications.

Ellison, J. C. 1993. Mangrove retreat with rising sea-level, Bermuda. *Estuarine, Coastal and Shelf Science* 37:75–87.

Epperly, S. P., J. Braun, A. J. Chester, F. Cross, J. V. Merriner, P. A. Tester, and J. H. Churchill. 1996. Beach strandings as an indicator of at-sea mortality of sea turtles. *Bulletin of Marine Science* 59:289–297.

Esslinger, C. G., and B. C. Wilson. 2002. *North American Waterfowl Management Plan*. Gulf Coast Joint Venture: Laguna Madre Initiative. Albuquerque, NM: US Fish and Wildlife Service.

Estrada, E. C., J. A. Villarreal Q., D. S. Alfonso, P. M. Marisela, S. M. Laura, E. Jurado, and Y. M. Carmen. 2006. Diversity and distributional patterns of legumes in southern Nuevo León, Mexico. *Southwestern Naturalist* 51:1–10.

Evans, P. R., D. M. Herdson, P. J. Knights, and M. W. Pienkowski. 1979. Short-term effects of reclamation of part of Seal Sands, Teesmouth, on wintering waders and shelduck. *Oecologia* 41:83–206.

Everitt, J. H., and D. L. Drawe. 1993. *Trees, Shrubs and Cacti of South Texas*. Lubbock: Texas Tech University Press.

Everitt, J. H., F. W. Judd, D. E. Escobar, M. A. Alaniz, M. R. Davis, and W. MacWhorter. 1996. Using remote sensing and spatial information technologies to map Texas Palmetto in the Lower Rio Grande Valley of Texas. *Southwestern Naturalist* 41:218–226.

Everitt, J. H., C. Yang, F. W. Judd, and K. R. Summy. 2010. Use of archive aerial photography for monitoring Black Mangrove populations. *Journal of Coastal Research* 26:649–653.

Everitt, J. H., C. Yang, S. Sriharan, and F. W. Judd. 2009. Using high resolution satellite imagery to map Black Mangrove on the Texas Gulf Coast. *Journal of Coastal Research* 24:1582–1586.

Everitt, J. H., C. Yang, K. R. Summy, F. Judd, and M. R. Davis. 2007. Evaluation of color-infrared photography and digital imagery to map Black Mangrove on the Texas Gulf Coast. *Journal of Coastal Research* 23:230–235.

Ewe, S. M. L., E. E. Gaiser, D. L. Childers, D. Iwaniec, V. H. Rivera-Monroy, and R. R. Twilley. 2006. Spatial and temporal patterns of aboveground net primary productivity (ANPP) along two freshwater-estuarine transects in the Florida coastal Everglades. *Hydrobiologia* 569:459–474.

Fall, B. A. 1973. Noteworthy bird records from South Texas (Kenedy County). *Southwestern Naturalist* 18:244–247.

Farmer, M. 1991. *Reddish Egrets of the Lower Laguna Madre, Texas*. Report to US Fish and Wildlife Service, Lower Rio Grande National Wildlife Refuge, Alamo, TX.

Farmer, M. 1996. *Reddish Egrets along the western Gulf of Mexico*. Report to US Fish and Wildlife Service, Ecological Services, Clear Lake, TX.

Farmer, M., and E. Carrera Gonzalez. 1993. *Mid-winter Surveys of Reddish Egret Foraging Habitat in the Laguna Madre, Mexico*. Report to US-Mexico Joint Committee of US Fish and Wildlife Service and Secretaría de Desarrollo Social, Albuquerque, NM.

Faust, D. R., M. J. Hooper, G. P. Cobb, M. Barnes, D. Shaver, S. Ertolacci, and P. N. Smith. 2014. Inorganic elements in Green Sea Turtles (*Chelonia mydas*): Relationships among external and internal tissues. *Environmental Toxicology and Chemistry* 1–8. DOI 10.1002/etc.2650.

Felix, J. D., and J. Campbell. 2019. *Investigating Reactive Nitrogen Sources That Stimulate Algal Blooms in Baffin Bay*. Final report to the Coastal Bend Bays and Estuaries Program, Publication CBBEP-129, Project Number 1818, Corpus Christi.

Feller, I. C. 1995. Effects of nutrient enrichment on growth and herbivory of dwarf Red Mangrove (*Rhizophora mangle*). *Ecological Monographs* 65:477–505.

Feller, I. C., E. M. Dangremond, D. J. Devlin, C. E. Lovelock, C. E. Proffitt, and W. Rodriguez. 2015. Nutrient enrichment intensifies hurricane impact in scrub mangrove ecosystems in the Indian River Lagoon, Florida, USA. *Ecology* 96:2960–2972.

Feller, I. C., C. E. Lovelock, U. Berger, K. L. McKee, S. B. Joye, and M. C. Ball. 2010. Biocomplexity in mangrove ecosystems. *Annual Review of Marine Science* 2:395–417.

Ferguson-Wood, E. J. 1963. A study of the diatom flora of fresh sediments of the South Texas bays and adjacent waters. *Publications of the Institute of Marine Science, University of Texas* 9:237–310.

Fikes, R. L., and R. L. Lehman. 2008. Small-scale recruitment of flora to a newly developed tidal inlet in the northwest Gulf of Mexico. *Gulf of Mexico Science* 26:130–132.

Fikes, R. L., and R. L. Lehman. 2010. Recruitment and colonization of macroalgae to a newly constructed rocky intertidal habitat in the northwest Gulf of Mexico. *Gulf and Caribbean Research* 22:9–20.

Fikes, R. L., R. L. Lehman, and K. V. Klootwyk. 2010. An update on the benthic algae of Mansfield Pass, Texas. *Texas Journal of Science* 62:183–194.

Finkbeiner, M., J. D. Simons, C. Robinson, and A. Summers. 2009. *Atlas of Shallow Water Benthic Habitats of Coastal Texas: Espiritu Santo Bay to Lower Laguna Madre, 2004 and 2007*. Charleston, SC: NOAA Coastal Services Center.

Fisk, H. N. 1949. *Geological Investigation of the Laguna Madre Flats*. Report to Humble Oil and Refining Company, Houston.

Fisk, H. N. 1959. Padre Island and the Laguna Madre flats: Coastal South Texas. Pp. 103–152 in R. J. Russell, ed., *Second Coastal Geography Conference*. Baton Rouge, LA: Coastal Studies Institute, Louisiana State University.

Fitch, A. C., J. B. Olson, and J. K. Lundquist. 2013. Parameterization of wind farms in climate models. *Journal of Climatology* 26:6439–6458.

Flint, R. W., S. Rabalais, and R. D. Kalke. 1982. Estuarine benthos and ecosystem functioning. Pp. 185–201 in J. R. Davis, ed., *Proceedings of the Symposium on Recent Benthological Investigations in Texas and Adjacent States*. Austin: Texas Academy of Science.

Flint, R. W., and J. A. Younk. 1983. Estuarine benthos: Long-term community structure variations, Corpus Christi Bay, Texas. *Estuaries and Coasts* 6:126–141.

Flores, J., K. Wagner, and L. Gregory. 2017. *Update to the Arroyo Colorado Watershed Protection Plan*. Technical Report 504. College Station: Arroyo Colorado Watershed Partnership, Texas Water Resources Institute.

Flowers, J., N. Easterling, and L. Hauck. 1998. *Prediction of Effects of Best Management Practices on Agricultural Nonpoint Source Pollution in the Arroyo Colorado Watershed*. Report No. PR97-05. Stephenville: Tarleton State University, Texas Institute for Applied Environmental Research.

Foley, A. M., J. A. Goolsby, A. Ortega-S. Jr., A. Ortega-S., A. Pérez de León, N. K. Singh, A. Schwartz, D. Ellis, D. G. Hewitt, and T. A. Campbell. 2017. Movement patterns of Nilgai antelope in South Texas: Implications for Cattle Fever Tick management. *Preventive Veterinary Medicine* 146:166–172.

Foley, A. M., K. Singel, P. H. Dutton, T. M. Summers, A. E. Redlow, and J. Lessman. 2007. Characteristics of a Green Turtle (*Chelonia mydas*) assemblage in northwestern Florida determined during a hypothermic stunning event. *Gulf of Mexico Science* 2007:131–143.

Fonseca, M. S. 1994. *Seagrasses: A Guide to Planting Seagrasses in the Gulf of Mexico*. Technical Report TAMU-SG-94–601. College Station: Texas A&M University Sea Grant College Program.

Ford, G. 1998. "Tio" Kleberg leaving King Ranch. *Corpus Christi Caller-Times* 116 (125C): 5, 10.

Forgason, C. A., and T. E. Fulbright. 2003. Cattle, wildlife, and range management on the King Ranch over the years. Pp. 9–21 in C. A. Forgason, F. C. Bryant, and P. Genho, eds., *Range Management: Integrating Cattle, Wildlife, and Range*. Kingsville, TX: King Ranch.

Forman, S. L., L. Nordt, J. Gomez, and J. Pierson. 2009. Late Holocene dune migration on the South Texas Sand Sheet. *Geomorphology* 108:159–170.

Foroughbakhch, R. 1992. Establishment and growth potential of fuelwood species in northeastern Mexico. *Agroforestry Systems* 19:95–108.

Foroughbakhch, R., L. A. Hauad, A. E. Cespedes, E. E. Pence, and N. González. 2001. Evaluation of 15 indigenous and introduced species for reforestation and agroforestry in northeastern Mexico. *Agroforestry Systems* 51:213–221.

Foroughbakhch, R., J . L. Hernandez-Pinero, M. A. Alvarado-Vazquez, E. Céspedes-Cabriales, A. Rocha-Estrada, and M. L. Cárdenas-Avila. 2009. Leaf biomass determination on woody shrub species in semiarid zones. *Agroforestry Systems* 77:181–192.

Franco-Pizana, J., T. E. Fulbright, and D. T. Gardiner. 1995. Spatial relations between shrubs and *Prosopis glandulosa* canopies. *Journal of Vegetation Science* 6:73–78.

Franco-Pizana, J., T. E. Fulbright, D. T. Gardiner, and A. R. Tipton. 1996. Shrub emergence and seedling growth in microenvironments created by *Prosopis glandulosa. Journal of Vegetation Science* 7:257–264.

Fredericq, S., T. O. Cho, S. A. Earle, C. F. Gurgel, D. M. Krayesky, L. E. Mateo-Cid, A. C. Mendoza-González, J. N. Norris, and A. M. Suárez. 2009. Seaweeds of the Gulf of Mexico. Pp. 187–259 in D. L. Felder and D. K. Camp, eds., *Gulf of Mexico Origins, Waters, and Biota,* vol. 1, *Biodiversity*. College Station: Texas A&M University Press.

Freeman, T. 1962. Quiet water oölites from Laguna Madre, Texas. *Journal of Sedimentary Petrology* 32:475–483.

Friess, D. A. 2016. Ecosystem services and disservices of mangrove forests: Insights from historical colonial observations. *Forests* 7 (183): 1–16. doi:10.3390/f7090183.

Frost, C. C. 1998. Presettlement fire frequency regimes of the United States: A first approximation. Pp. 70–81 in T. L. Pruden and L. A. Brennan, eds., *Proceedings of the 20th Tall Timbers Fire Ecology Conference: Fire in Ecosystem Management: Shifting the Paradigm from Suppression to Prescription*. Tallahassee, FL: Tall Timbers Research Station.

Fry, B. 2006. *Stable Isotope Ecology*. New York: Springer.

Fry, B., and P. L. Parker. 1979. Animal diet in Texas seagrass meadows: $\delta^{13}C$ evidence for the importance of benthic plants. *Estuarine and Coastal Marine Science* 8:499–509.

Fry, B., and E. B. Sherr. 1984. $\delta^{13}C$ measurements as indicators of carbon flow in marine and freshwater ecosystems. *Contributions in Marine Science* 27:13–47.

Fulbright, T. E. 1994. Mesquite–Liveoak–Seacoast Bluestem, SRM 719. P. 97 in T. N. Shiflet, ed., *Rangeland Cover Types of the United States*. Denver, CO: Society for Range Management.

Fulbright, T. E., and F. C. Bryant. 2002. *The Last Great Habitat*. Special Publication 1. Kingsville, TX: Caesar Kleberg Wildlife Research Institute.

Fulbright, T. E., D. D. Diamond, J. Rappole, and J. Norwine. 1990. The Coastal Sand Plain of southern Texas. *Rangelands* 12:337–340.

Fulbright, T. E., J. O. Kuti, and A. R. Tipton. 1995. Effects of nurse-plant canopy temperature on shrub seed germination and seedling growth. *Acta Oecologica* 16:621–632.

Fulbright, T. E., E. A. Lozano-Cavazos, D. C. Ruthven III, and A. R. Litt. 2013. Plant and small vertebrate composition and diversity 36–39 years after root plowing. *Rangeland Ecology and Management* 66:19–25.

Fulbright, T. E., J. A. Ortega-S., D. G. Hewitt, T. W. Boutton, and A. Ortega-S Jr. 2017. White-tailed Deer, cattle and Nilgai: Sharing the landscape. *Texas Wildlife*, December, 24–25.

Fuls, B. E. 1974. Further ecological studies on the macroichthyofauna of the Laguna Salada, Texas. MS thesis, Texas A&I University, Kingsville.

Fulton, K. J. 1976. Subsurface stratigraphy, depositional environments, and aspects of reservoir continuity—Rio Grande Delta, Texas. PhD diss., University of Cincinnati, Ohio.

Gabler, C. A., M. J. Osland, J. B. Grace, C. L. Stagg, R. H. Day, S. B. Hartley, N. M. Enwright, A. S. From, M. L. McCoy, and J. L. McLeod. 2017. Macroclimate change expected to transform coastal wetland ecosystems this century. *Nature Climate Change* 7:142–147.

Galan, A. P., and J. E. Light. 2017. Reptiles and amphibians associated with Texas Pocket Gopher (*Geomys personatus*) burrow systems across the Texas Sand Sheet. *Herpetological Review* 48:517–521.

Gallaway, B. J., W. J. Gazey, C. W. Caillouet Jr., P. T. Plotkin, F. A. A. Grobois, A. F. Amos, P. M. Burchfield, et al. 2016. Development of a Kemp's Ridley Sea Turtle stock assessment model. *Gulf of Mexico Science* 33:138–157.

Gallaway, B. J., W. J. Gazey, T. Wibbels, E. Bevan, D. J. Shaver, and J. George. 2016. Evaluation of the status of Kemp's Ridley Sea Turtle following the 2010 Deepwater Horizon Oil Spill. *Gulf of Mexico Science* 33:192–205.

Garcia, C. P. 1979. *Padre Island and Padre Jose Nicolas Balli: Application for Historical Marker.* Corpus Christi, TX: Grunwald Publications.

García, E. 2004. *Modificaciones al sistema de clasificación climática de Köppen.* Mexico City: Instituto de Geografía, Universidad Nacional Autónoma de México.

Garcia, J. F. 2011. Can environmental variation affect seedling survival of plants in northeastern Mexico? *Archives of Biological Science Belgrade* 63:731–737.

García-Cubas, A., Jr. 1968. *Ecologia y distribucion de los micromoluscos recientes de la Laguna Madre Tamaulipas, México.* Instituto de Geologia Boletín Numero 86. Mexico City: Universidad Nacional Autónoma de México.

Garcia-Gil, G., J. Rendón-von Osten, J. Garcia Guzman, E. Carrera González, C. Tejeda Cruz, F. E. Galán Amaro, and B. Ortiz Espejel. 1993. Diagnostico ambiental de Laguna Madre, Tamaulipas. Pp. 535–552 in S. I. Salazar-Vallejo and N. E. Gonzalez, eds., *Biodiversidad marina y costera de México.* Mexico City: Comisión Nacional para el Conocimiento y Aprovechamiento de la Biodiversidad y Centro de Investigaciones de Quintana Roo.

Gardner, L. R., W. K. Michener, E. R. Blood, T. M. Williams, D. J. Lipscomb, and W. H. Jefferson. 1991. Ecological impact of Hurricane Hugo—salinization of a coastal forest. *Journal of Coastal Research* 8:301–317.

Gardner, L. R., W. K. Michener, T. M. Williams, E. R. Blood, B. Kjerve, L. A. Smock, D. J. Lipscomb, and C. Gresham. 1992. Disturbance effects of Hurricane Hugo on a pristine coastal landscape: North Inlet, South Carolina, USA. *Netherlands Journal of Sea Research* 30:249–26[illegible].

Garrett, J. M., and D. G. Barker. 1987. *A Field Guide to Reptiles and Amphibians of Texas.* Houston: Gulf Publishing.

Garza, H., Jr. 1997. Habitat use and activities of the Piping Plover, *Charadrius melodus*, wintering on South Padre Island, Texas. MS thesis, University of Texas–Pan American, Edinburg.

Garza-Torres, H. A., and A. G. Navarro-Sigüenza. 2001. *Informe final: Proyecto S085 avifauna de la Laguna Madre de Tamaulipas.* Universidad Autónoma de Tamaulipas Technical Report. Cd. Victoria, Tamaulipas: Comisión Nacional para el Conocimiento y Uso de la Biodiversidad. http://www.conabio.gob.mx/institucion/proyectos/resultados/InS085.pdf.

Gaston, A. J. 2004. *Seabirds: A Natural History.* New Haven, CT: Yale University Press.

Gawlik, D. E. 2002. The effects of prey availability on the numerical response of wading birds. *Ecological Monographs* 72:329–346.

Geary, B., M. C. Green, and B. M. Ballard. 2015. Movements and survival of juvenile Reddish Egrets (*Egretta rufescens*) on the Gulf of Mexico coast. *Endangered Species Research* 28:123–133.

Gobler, C. J., F. Koch, Y. Kang, D. L. Berry, Y. Z. Tang, M. Lasi, L. Walters, L. Hall, and J. D. Miller. 2013. Expansion of harmful brown tides caused by the pelagophyte, *Aureoumbra lagunensis* to the US East Coast. *Harmful Algae* 27:29–41.

Godfrey, M. 2013. Turtles left out in the cold? *Current Conservation* 5 (2): 31–33.

Goetze, J. R., A. Nelson, and V. French. 1999. Community relationships of some small mammals on Padre Island National Seashore, Texas. *Occasional Papers, Museum of Texas Tech University* 189:1–8.

Golder, W., D. Allen, S. Cameron, and T. Wilder. 2008. Dredged material as a tool for management of tern and skimmer nesting habitats. DOER Technical Notes Collection (ERDC TN-DOER-E24). Vicksburg, MS: US Army Engineer Research and Development Center. https://apps.dtic.mil/sti/pdfs/ADA485771.pdf.

Gómez Ortiz, M. G. 2012. *La pesca de lisa* Mugil cephalus *en Laguna Madre, Tamaulipas.* Mexico City: Instituto Nacional de la Pesca.

Gomez Soto, A., and S. Contreras Balderas. 1987. Ictiofauna de la Laguna Madre, Tamaulipas, México. Pp. 8–17 in *Memorias del Congreso, IX Congreso Nacional de Zoologia*, vol. 2. Villahermosa, Tabasco, Mexico.

Gonzalez, J. I., R. K. Skowronek, and B. I. Lovett. 2015. Deflation troughs, water, and prehistoric occupation on the margins of the South Texas Sand Sheet. *Journal of Texas Archaeology and History* 1:70–93.

Gonzáles-Rodríguez, H., I. Cantú-Silva, M. V. Gómez-Meza, and R. G. Ramírez-Lozano. 2004. Plant-water relations of thornscrub shrub species, northeastern Mexico. *Journal of Arid Environments* 58:483–503.

Gorga, C. C. T. 2010. Foraging ecology of Green Turtles (*Chelonia mydas*) on the Texas coast, determined by stable isotope analysis. MS thesis, Texas A&M University, College Station.

Gorman, L. R., and S. M. Haig. 2002. Distribution and abundance of Snowy Plovers in eastern North America, the Caribbean, and the Bahamas. *Journal of Field Ornithology* 73:38–52.

Goss-Custard, J. D. 1970. The responses of Redshank (*Tringa totanus* (L.)) to spatial variations in the density of their prey. *Journal of Animal Ecology* 39:1–13.

Goss-Custard, J. D. 1979. Role of winter food supplies in the population ecology of common British wading birds. *Verhandlungen Ornithologische Gesellschaft Bayern* 23:125–146.

Goss-Custard, J. D. 1980. Competition for food and interference among waders. *Ardea* 68:31–52.

Goss-Custard, J. D. 1983. Spatial and seasonal variations in the food supply of waders Charadrii wintering in the British Isles. Pp. 85–96 in J. Fjeldså and H. Meltofte, eds., *Proceedings of the 3rd Nordic Congress of Ornithology, 1981*. Copenhagen: Dansk Ornitologisk Forening—Zoologisk Museum.

Goss-Custard, J. D. 1985. Foraging behaviour of wading birds and the carrying capacity of estuaries. Pp. 169–188 in R. M. Sibly and R. H. Smith, eds., *Behavioural Ecology*. Oxford, UK: Blackwell Scientific.

Goss-Custard, J. D., and K. Charman. 1976. Predicting how many wintering waterfowl an area can support. *Wildfowl* 27:157–158.

Goss-Custard, J. D., and S. E. A. le V. ditDurrell. 1990. Bird behaviour and environmental planning: Approaches in the study of wader populations. *Ibis* 132:273–289.

Goss-Custard, J. D., D. G. Kay, and R. M. Blindell. 1977. The density of migratory and overwintering Redshank *Tringa totanus* (L.) and Curlew *Numenius arquata* (L.) in relation to the density of their prey in south-east England. *Estuarine Coastal Marine Science* 15:491–510.

Gosselink, J. D., C. L. Cordes, and J. W. Parsons. 1979. *An Ecological Characterization Study of the Chenier Plain Coastal Ecosystem of Louisiana and Texas*. 3 vols. FWS/OBS-78/9, 10, 11. Washington, DC: US Fish and Wildlife Service, Office of Biological Services.

Gotto, J. W., F. R. Tabita, and C. Van Baalen. 1981. Nitrogen fixation in intertidal environments of the Texas Gulf Coast. *Estuarine Coastal Shelf Science* 12:231–235.

Gould, F. W. 1969. *Texas Plants—a Checklist and Ecological Summary*. Texas Agricultural Experiment Station MP-585. College Station: Texas A&M University.

Gould, F. W. 1975. *The Grasses of Texas*. College Station: Texas A&M University Press.

Gould, F. W., and T. W. Box. 1965. *Grasses of the Texas Coastal Bend (Calhoun, Refugio, Aransas, San Patricio, and Northern Kleberg Counties)*. College Station: Texas A&M University Press.

Grant, M. L. 1998. A corporate end to a cowboy era. *Corpus Christi Caller-Times* 116 (227A): 1, 6–8.

Gray, J. D., T. L. King, and R. L. Colura. 1991. Effect of temperature and hypersalinity on hatching success of Spotted Seatrout eggs. *Progressive Fish Culturist* 53:81–84.

Green, L. M., and R. P. Campbell. 2008. *Trends in Finfish Landings of Sport Boat Anglers in Texas Marine Waters, May 1974–May 2008*. Texas Parks and Wildlife Department-Coastal Fisheries Management Data Series 257. Austin: Texas Parks and Wildlife Department.

Green, S. J., J. L. Akins, A. Maljkovic, and I. M. Côté. 2012. Invasive lionfish drive Atlantic coral reef fish declines. *PLoS ONE* 7:1–3.

Griffith, G., S. Bryce, J. Omerlink, and A. Rogers. 2007. *Ecoregions of Texas*. Project report to Texas Commission on Environmental Quality, Austin.

Groue, K. J., and L. J. Lester. 1982. A morphological and genetic analysis of geographic variation among oysters in the Gulf of Mexico. *Veliger* 24:331–341.

Groves, C., L. Valutis, D. Vosick, B. Neely, K. Wheaton, J. Touval, and B. Runnels. 2000. *Designing a Geography of Hope: A Practitioner's Handbook for Ecoregional Planning*. 2 vols. Arlington, VA: The Nature Conservancy.

Grubbs, F., A. Morris, A. Nunez, Z. Olsen, and J. Tolan. 2013. *Emaciated Black Drum Event Baffin Bay and the Upper Laguna Madre*. Corpus Christi: Texas Parks and Wildlife Department, Coastal Fisheries Division.

Guilfoyle, M. P., R. A. Fischer, D. N. Pashley, and C. A. Lott, eds. 2006. *Summary of First Regional Workshop on Dredging, Beach Nourishment, and Birds on the South Atlantic Coast*. EDRC/EL TR-06-10. Vicksburg, MS: US Army Engineer Research and Development Center.

Gulf of Mexico Fishery Management Council. 1987. *Red Drum Amendment 1: Environmental Assessment and Supplemental Regulatory Impact Review and Initial Regulatory Flexibility Analysis to the Secretarial Fisheries Management Plan for Red Drum Fishery of the Gulf of Mexico. Mar.* Tampa, FL: Gulf of Mexico Fishery Management Council.

Gulf of Mexico Fishery Management Council. 2015. *Status Determination Criteria for Penaeid Shrimp and Adjustments to the Shrimp Framework Procedure: Amendment 15 to the Fishery Management Plan for the Shrimp Fishery of the Gulf of Mexico, U.S. Waters*. Tampa, FL: Gulf of Mexico Fishery Management Council.

Gunter G. 1945a. Some characteristics of ocean waters and Laguna Madre. *Texas Game and Fish* 3 (10): 7, 19–22.

Gunter, G. 1945b. Studies on marine fishes of Texas. *Publications of the Institute of Marine Science University of Texas* 1 (1): 1–90.

Gunter, G. 1946. Problems of the Texas coast. *Texas Game and Fish* 3 (12): 9, 25–28.

Gunter, G. 1961. Some relations of estuarine organisms to salinity. *Limnology and Oceanography* 6:182–190.

Gunter, G. 1967. Vertebrates in hypersaline waters. *Contributions in Marine Science* 12:230–241.

Gunter, G., and H. H. Hildebrand. 1951. Destruction of fishes and other organisms on the South Texas coast by the cold wave of January 28–February 3, 1951. *Ecology* 32:731–736.

Guo, H., C. Weaver, S. P. Charles, A. Whitee, J. Dastidar, P. D'Odorico, J. D. Fuentes, J. S. Kominoski, A. R. Armitage, and S. C. Pennings. 2017. Coastal regime shifts: Rapid responses of coastal wetlands to changes in mangrove cover. *Ecology* 98:762–772.

Guo, H., Y. Zhang, Z. Lan, and S. C. Pennings. 2013. Biotic interactions mediate the expansion of Black Mangrove (*Avicennia germinans*) into salt marshes under climate change. *Global Change Biology* 19:2765–2774.

Guthery, F. S. 2000. *On Bobwhites*. College Station: Texas A&M University Press.

Guzmán-Sáenz, F. M., Z. J. Molina-Garza, R. Pérez-Castañeda, José C. Ibarra-Gámez, and Lucio Galaviz-Silva. 2009. Virus de la necrosis hipodérmica y hematopoyética infecciosa (IHHNV) y virus del síndrome de Taura (TSV) en camarón silvestre (*Farfantepenaeus aztecus* Ives, 1891 y *Litopenaeus setiferus* Linnaeus, 1767) de la Laguna Madre Golfo de México. *Revista de Biología Marina y Oceanografía* 44:663–672.

Hackney, A. D. 2016. *Texas Coastal Rookery Conservation Plan*. Final report, FWS Cooperative Agreement F11AC00242. New York: National Audubon Society.

Haig, S. M., C. L. Ferland, F. J. Cuthbert, J. Dingledine, A. Hecht, and N. McPhillips. 2005. A complete species census and evidence for regional declines in Piping Plovers. *Journal of Wildlife Management* 69:160–173.

Haig, S. M., and J. H. Plissner. 1993. Distribution and abundance of Piping Plovers: Results and implications of the 1991 international census. *Condor* 95:145–156.

Haines, E. B., and C. L. Montague. 1979. Food sources of estuarine invertebrates analyzed using $^{13}C/^{12}C$ ratios. *Ecology* 60:48–56.

Hall, E. R. 1951. Mammals obtained by Dr. Curt von Wedel from the barrier beach of Tamaulipas, Mexico. *University of Kansas Museum of Natural History Publications* 5:33–47.

Hallam, A. 1965. Environmental causes of stunting in living and fossil marine benthonic invertebrates. *Palaeontology* 8:132–155.

Hammer, B. T., M. L. Fogel, and T. C. Hoering. 1998. Stable carbon isotope ratios of fatty acids in seagrass and Redhead ducks. *Chemical Geology* 152:29–41.

Hanselka, C. W. n.d. The Coastal Prairie: Southern extensions and associated contacts. Unpublished manuscript held by T. E. Fulbright.

Hansen, B., P. K. Bjornsen, and P. J. Hansen. 1994. The size ratio between planktonic predators and their prey. *Limnology and Oceanography* 39:395–403.

Hardegree, B. 1997. Biological productivity associated with the serpulid reefs of Baffin Bay, Texas. MS thesis, Texas A&M University–Corpus Christi.

Harmon, B. G. 1962. Mollusks as food of Lesser Scaup along the Louisiana coast. *Transactions of the 27th North American Wildlife Conference* 27:132–137.

Harrington, B. A., and R. I. G. Morrison. 1979. Semipalmated Sandpiper migration in North America. *Studies in Avian Biology* 2:83–99.

Harrington, R. W., and E. S. Harrington. 1972. Food of female Marsh Killifish, *Fundulus confluentus* Goode and Bean, in Florida. *American Midland Naturalist* 87:492–502.

Harris, R. V. 1988. The mammals of Padre Island, Texas. Unpublished manuscript on file, Texas A&M University–Corpus Christi, Center for Coastal Studies Library.

Harris, V. T., and R. H. Chabreck. 1958. Some effects of Hurricane Audrey on the marsh at Marsh Island, Louisiana. *Louisiana Academy of Science* 21:47–50.

Hart, R. A. 2013. *Review of the Status and Health of the Gulf of Mexico Shrimp Stocks for 2012*. Tampa, FL: Gulf of Mexico Fisheries Management Council.

Hayes, M. O. 1965. Sedimentation on a semi-arid, wave dominated coast (South Texas) with emphasis on hurricane effects. PhD diss., University of Texas at Austin.

Hayes, M. O. 1967. *Hurricanes as Geological Agents: Case Studies of Hurricanes Carla, 1961, and Cindy, 1963*. Report Investigation 61. Austin: Bureau of Economic Geology, University of Texas at Austin.

Hayes, M. O., D. D. Domeracki, C. D. Getter, T. W. Kana, and G. I. Scott. 1980. *Sensitivity of Coastal Environments to Spilled Oil, South Texas Coast*. Research Planning Institute, Inc., Report No. RPI/R/80/4/11-12. Columbia, SC: NOAA Office of Marine Pollution Assessment.

Hayes, M. O., and A. J. Scott. 1964. Environmental complexes, South Texas coast. *Transactions of the Gulf Coast Association of Geology Societies* 14:237–240.

Hedgpeth, J. W. 1947. The Laguna Madre of Texas. *North American Wildlife Conference* 12:364–380.

Hedgpeth, J. W. 1953. An introduction to the zoogeography of the northwestern Gulf of Mexico with reference to the invertebrate fauna. *Publications of the Institute of Marine Science, University of Texas* 3:111–224.

Hedgpeth, J. W. 1954. Bottom communities of the Gulf of Mexico. Pp. 203–222 in P. S. Galtsoff, ed., *Gulf of Mexico: Its Origin, Waters, and Marine Life*, Fishery Bulletin 89. Washington, DC: US Government Printing Office.

Hedgpeth, J. W. 1967. Ecological aspects of the Laguna Madre, a hypersaline estuary. Pp. 408–419 in G. H. Lauff, ed., *Estuaries*. Publication 83. Washington, DC: American Association for Advancement of Science.

Hellier, T. R., Jr. 1962. Fish production and biomass studies in relation to photosynthesis in the Laguna Madre of Texas. *Publications of the Institute of Marine Science, University of Texas* 8:1–22.

Henry, W. K., D. M. Driscoll, and J. P. McCormack. 1980. Hurricanes on the Texas Coast. Report TAMU-SG-75-504. College Station: Texas A&M University Sea Grant Program, Center for Applied Geosciences.

Hensley, R. A., and B. E. Fuls. 1998. *Trends in Relative Abundance and Size of Selected Finfishes and Shellfishes along the Texas Coast: November 1975–December 1996*. Management Data Series No. 159. Austin: Texas Parks and Wildlife Department.

Herber, J. P. 1981. Holocene sediments under Laguna Madre, Cameron County, Texas. MA thesis, University of Texas at Austin.

Hermand, R., C. Salen-Picard, E. Alliot, and C. Degiovanni. 2008. Macrofaunal density, biomass and composition of estuarine sediments and their relationship to the river plume of the Rhone River (NW Mediterranean). *Estuarine, Coastal and Shelf Science* 79:367–376.

Hernandez, E. A., and V. Uddameri. 2014. Standardized precipitation evaporation index (SPEI)-based drought assessment in semi-arid South Texas. *Environmental Earth Sciences* 71:2491–2501.

Herzka, S. Z., and K. H. Dunton. 1997. Seasonal photosynthetic patterns of the seagrass *Thalassia testudinum* in the western Gulf of Mexico. *Marine Ecology Progress Series* 153:103–117.

Hester, T. R., S. L. Black, D. G. Steele, B. W. Olive, A. A. Fox, K. Reinhard, and L. C. Bement. 1989. From the Gulf to the Rio Grande: Human adaptation in Central, South, and lower Pecos Texas. *Karl Reinhard Papers/Publications* 59. http://digitalcommons.unl.edu/natresreinhard/59.

Hice, D. L., and D. J. Schmidly. 2002. The mammals of coastal Texas: A comparison between mainland and barrier island faunas. *Southwestern Naturalist* 47:244–256.

Hicklin, P. W. 1987. The migration of shorebirds in the Bay of Fundy. *Wilson Bulletin* 99:540–570.

Hicklin, P. W., and P. C. Smith. 1984. Selection of foraging sites and invertebrate prey by migrant Semipalmated Sandpipers, *Calidris pusilla* (Pallas), in Minas Basin, Bay of Fundy. *Canadian Journal of Zoology* 62:2201–2210.

Hicks, D. W. 2010. Molluscan ecology and habitats. Pp. 28–75 in J. W. Tunnell Jr., J. Andrews, N. C. Barrera, and F. Moretzsohn, *Encyclopedia of Texas Seashells*. College Station: Texas A&M University Press.

Hicks, D. W., and J. W. Tunnell Jr. 1993. Invasion of the South Texas coast by the edible Brown Mussel *Perna perna* (Linnaeus, 1758). *Veliger* 36:92–94.

Hicks, D. W., and J. W. Tunnell Jr. 1994. *The Invasive Biology of the Edible Brown Mussel,* Perna perna *(Linnaeus, 1758), on the South Texas Coast: Year 1.* Technical Report TAMUCC-9403-CCS. Corpus Christi: Texas A&M University–Corpus Christi, Center for Coastal Studies.

Higgins, E., and R. Lord. 1926. *Preliminary Report on the Marine Fisheries of Texas*. Document No. 1009. Washington, DC: Bureau of Fisheries.

Hildebrand, H. 1958. Estudios biologicos preliminares sobre la Laguna Madre de Tamaulipas. *Ciencia* (Mexico) 17:151–173.

Hildebrand, H. 1963. Hallazgo del area de anidacion de la tortuga "lora" *Lepidochelys kempii* (Garman), en la costa occidental del Golfo de México (Reptila, Chelonia). *Ciencia* (Mexico) 22:105–112.

Hildebrand, H. 1969. Laguna Madre de Tamaulipas: Observations on its hydrography and fisheries. Pp. 679–686 in A. A. Castañares and F. B. Phleger, eds., *Coastal Lagoons: A Symposium*. Mexico City: Universidad Nacional Autónoma de Mexico.

Hildebrand, H. 1980. The Laguna Madre de Tamaulipas: Its hydrography and shrimp fishery. Unpublished manuscript, submitted to National Marine Fisheries Service, on file at Texas A&M University–Corpus Christi, Center for Coastal Studies Library.

Hildebrand, H. 1981. A historical review of the status of sea turtle populations in the western Gulf of Mexico. Pp. 447–453 in K. A. Bjorndal, ed., *Biology and Conservation of Sea Turtles: Proceedings of the World Conference of Sea Turtle Conservation*. Washington, DC: Smithsonian Institution Press and World Wildlife Fund.

Hildebrand, H. H. 1982. A historical review of the sea turtle populations in the western Gulf of Mexico. Pp. 447–453 in K. Bjorndal, ed., *Biology and Conservation of Sea Turtles*. Washington, DC: Smithsonian Institution Press.

Hildebrand, H. H. 1983. Random notes on sea turtles in the western Gulf of Mexico. Pp. 34–40 in D. Owens, D. Crowell, G. Dienberg, M. Grassman, S. McCain, Y. Morris, N. Schwantes, and T. Wibbels., *Western Gulf of Mexico Sea Turtle Workshop Proceedings*, TAMUSG84105, Sea Grant College Program. College Station: Texas A&M University.

Hildebrand, H., and D. King. 1978. *A Biological Study of the Cayo del Oso and Pita Island Area of the Laguna Madre*. Final report to Central Power and Light Company, vol. 1, Corpus Christi.

Hillestad, H. O., J. I. Richardson, and G. K. Williamson. 1978. *Incidental Capture of Sea Turtles by Shrimp Trawler Men in Georgia*. Athens, GA: Southeastern Wildlife Service.

Hines, S. L. 2016. Cattle, deer, and Nilgai interactions. PhD diss., Texas A&M University–Kingsville.

Hirth, H. H., L. C. Klikorr, and K. T. Harper. 1973. Sea grasses at Khor Unaira, People's Democratic Republic of Yemen, with reference to their role in the diet of the Green Turtle, *Chelonia mydas*. *US Fisheries Bulletin* 71:1093–1097.

Hodnett, C. 1999. Grass fire near Park Road 22 burns across Padre: Corpus Christi, Flour Bluff firefighters hoped half-mile blaze would burn itself out. *Corpus Christi Caller-Times*, 3 December.

Hoese, H. D., B. J. Copeland, F. N. Mosely, and E. D. Lane. 1968. Fauna of the Aransas Pass Inlet, Texas III: Diel and seasonal variation in trawlable organisms of the adjacent area. *Texas Journal of Science* 20:33–60.

Hollandsworth, S. 1998. When we were kings. *Texas Monthly* 26 (8): 112–117, 140–144.

Hood, S. L., and S. J. Dinsmore. 2007a. Abundance of Snowy and Wilson's Plovers in the Lower Laguna Madre of Texas. *Journal of Field Ornithology* 78:362–368.

Hood, S. L., and S. J. Dinsmore. 2007b. The influence of habitat on nest survival of Snowy and Wilson's Plovers in the Lower Laguna Madre region of Texas. *Studies in Avian Biology* 34:124–159.

Hook, J. H. 1991. Seasonal variation in relative abundance and species diversity of fishes in South Bay. *Contributions in Marine Science* 32:127–141.

Hooper, C. H., ed. 1981. *The IXTOC I Oil Spill: The Federal Scientific Response*. Boulder, CO: NOAA Hazardous Materials Response Project.

Hoover, C. A., and P. M. Gaffney. 2005. Geographic variation in nuclear genes of the Eastern Oyster, *Crassostrea virginica* Gmelin. *Journal of Shellfish Research* 24:103–112.

Horton, R., C. Herweijer, C. Rosenzweig, J. Liu, V. Gornitz, and A. C. Ruane. 2008. Sea level rise projections for current generation CGCMs based on the semi-empirical method. *Geophysical Research Letters* 35 (2). https://doi.org/10.1029/2007GL032486.

Howell, A. H. 1914. Revision of the American harvest mice (genus *Reithrodontomys*). *North American Fauna* 36:1–97.

Howell, L. N., K. J. Reich, D. J. Shaver, A. M. Landry Jr., and C. C. Gorga. 2016. Ontogenetic shifts in diet and habitat of juvenile Green Sea Turtles in the northwestern Gulf of Mexico. *Marine Ecology Progress Series* 559:217–229.

Hu, X., J. B. Pollack, M. R. McCutcheon, P. A. Montagna, and Z. Ouyang. 2015. Long-term alkalinity decrease and acidification of estuaries in the northwestern Gulf of Mexico. *Environmental Science and Technology* 49:3401–3409.

Huang, I. S., L. J. Pinnell, J. W. Turner, H. Abdulla, L. Boyd, E. W. Linton, and P. V. Zimba. 2020. Preliminary assessment of cyanobacterial community structure of wind-tidal algal flats in the Laguna Madre, Texas, USA. *Biology* (Basel) 9 (8). https://doi.org/10.3390/biology9080183.

Huffman, G. G., and W. A. Price. 1949. Clay dune formation near Corpus Christi, Texas. *Journal of Sedimentary Petrology* 19:118–127.

Humm, H. J., and H. H. Hildebrand. 1962. Marine algae from the Gulf Coast of Texas and Mexico. *Publications of the Institute of Marine Science, University of Texas* 8:227–268.

Hunt, W. G., R. R. Rogers, and D. L. Slowe. 1975. Migratory and foraging behavior of Peregrine Falcons on the Texas Coast. *Canadian Field Naturalist* 89:111–123.

Hunt, W. G., and F. P. Ward. 1988. Habitat selection by spring migrant peregrines at Padre Island, Texas. Pp. 527–535 in T. J. Cade, J. H. Enderson, C. G. Thelander, and C. M. White, eds., *Peregrine Falcon Populations: Their Management and Recovery*. Boise, ID: Peregrine Fund.

Hunter, R. E., R. L. Watson, G. W. Hill, and K. A. Dickinson. 1972. Modern depositional environments and processes, northern and central Padre Island, Texas. Pp. 1–17 in Corpus Christi Geological Society, ed., *Padre Island National Seashore Field Guide*. Corpus Christi, TX: Golden Banner Press.

Hussong, D., J. M. Damare, R. J. Limpert, W. J. Sladen, R. M. Weiner, and R. R. Colwell. 1979. Microbial impact of Canada Geese (*Branta canadensis*) and Whistling Swans (*Cygnus columbianus columbianus*) on aquatic systems. *Applied and Environmental Microbiology* 37:14–20.

Ibáñez-Aguirre, A. L., and M. Gallardo-Cabello. 2004. Reproduction of *Mugil cephalus* and *M. curema* (Pisces: Mugilidae) from a coastal lagoon in the Gulf of Mexico. *Bulletin of Marine Science* 75:37–49.

Igl, L. D., and B. M. Ballard. 1999. Habitat associations of migrating and overwintering grassland birds in southern Texas. *Condor* 101:771–782.

INEGI (Instituto Nacional de Estadística, Geografía y Informática). 1980. *Carta topográfica 1:250,000, Matamoros G14-6-9-12 Tamaulipas y Texas*. Aguascalientes, Mexico: Instituto Nacional de Estadística, Geografía y Informática. [one sheet]

INEGI. 1994. *Atlas del territorio insular habitado de los Estados Unidos Mexicanos, 1990 anexo cartográgrafico*. Aguascalientes, Mexico: INEGI.

Inglis, J. M. 1964. *A History of Vegetation on the Rio Grande Plain*. Bulletin no. 45. Austin: Texas Parks and Wildlife Department.

IPCC (Intergovernmental Panel on Climate Change). 2014. *Climate Change 2014: Synthesis Report. Contribution of Working Groups I, II and III to the Fifth Assessment Report of the Intergovernmental Panel on Climate Change*. (Core Writing Team, R. K. Pachauri and L. A. Meyer [eds.]). Geneva, Switzerland: IPCC. https://www.ipcc.ch/site/assets/uploads/2018/02/SYR_AR5_FINAL_full_wcover.pdf.

Jackson, J. 1986. *Los Mesteños: Spanish Ranching in Texas, 1721–1821*. College Station: Texas A&M University Press.

Jahrsdoerfer, S. E., and D. M. Leslie Jr. 1988. *Tamaulipan Brushland of the Lower Rio Grande Valley of South Texas: Description, Human Impacts, and Management Options*. US Fish and Wildlife Service Biological Report 88(36), Washington, DC.

Jakobsen, H. H., C. Hyatt, and E. J. Buskey. 2001. Growth and grazing on the Texas Brown Tide alga *Aureoumbra lagunensis* by the tintinnid *Amphorides quadrilineata*. *Aquatic Microbial Ecology* 23:245–252.

James, J. D. 2006. Utilization of Shoal Grass resources and nutritional ecology of wintering Redheads in the Laguna Madre of Texas. PhD diss., Texas A&M University–Kingsville.

Javor, B. 1989. *Hypersaline Environments*. New York: Springer-Verlag.

Jevrejeva, S., J. C. Moore, and A. Grinsted. 2010. How will sea level respond to changes in natural and anthropogenic forcings by 2100? *Geophysical Research Letters* 37 (7). https://doi.org/10.1029/2010GL042947.

Johnson, C. M., and G. S. Baldassarre. 1988. Aspects of the wintering ecology of Piping Plovers in coastal Alabama. *Wilson Bulletin* 100:214–215.

Johnson, R. G. 1970. Variations in diversity within benthic marine communities. *American Naturalist* 104:285–300.

Johnston, M. C. 1955. Vegetation of the Eolian Plain and associated coastal features of southern Texas. PhD diss., University of Texas at Austin.

Johnston, M. C. 1963. Past and present grasslands of southern Texas and northeastern Mexico. *Ecology* 44:456–466.

Jones, F. B. 1975. *Flora of the Texas Coastal Bend*. Welder Wildlife Foundation Contribution B-6. Corpus Christi: Mission Press.

Jones, G. D., and J. K. Frey. 2008. First records of Gray Fox (*Urocyon cinereoargenteus*) on Texas barrier islands. *Texas Journal of Science* 60:225–227.

Jones, G. D., and J. Frey. 2013. Mammals of Padre Island National Seashore, Texas. *Special Publications, Museum of Texas Tech University* 61:1–63.

Jud, Z. R., and C. A. Layman. 2012. Site fidelity and movement patterns of invasive lionfish, *Pterois* spp., in a Florida estuary. *Journal of Experimental Marine Biology and Ecology* 414:69–74.

Judd, F. W. 1973. The ecology of the Keeled Earless Lizard, *Holbrookia propinqua*. PhD diss., Texas Tech University, Lubbock.

Judd, F. W. 1976. Food and feeding behavior of the Keeled Earless Lizard, *Holbrookia propinqua*. *Southwestern Naturalist* 21:17–26.

Judd, F. W., and R. I. Lonard. 1985. Effects of perturbations on South Padre Island. Pp. 1855–1869 in O. T. Magoon, H. Converse, D. Miner, D. Clark, and L. Thomas, eds., *Proceedings of the Fourth Symposium on Coastal and Ocean Management, Coastal Zone '85*. Baltimore: American Society of Civil Engineers.

Judd, F. W., and R. I. Lonard. 1987. Disturbance and community development. Pp. 1731–1748 in O. T. Magoon, H. Converse, D. Miner, L. T. Tobin, D. Clark, and G. Domurat, G. (eds.)

Proceedings of the Fifth Symposium on Coastal and Ocean Management, Coastal Zone '87. Seattle, WA: American Society of Civil Engineers.

Judd, F. W., R. I. Lonard, J. H. Everitt, and D. E. Escobar. 1991. Resilience of Seacoast Bluestem barrier island communities. Pp. 3513–3524 in O. T. Magoon, H. Converse, V. Tippie, L. T. Tobin, and D. Clark, eds., *Proceedings of the Seventh Symposium on Coastal and Ocean Management, "Coastal Zone '91."* Long Beach, CA: American Society of Civil Engineers.

Judd, F. W., R. I. Lonard, J. H. Everitt, and R. Villarreal. 1989. Effect of vehicular traffic in secondary dunes and vegetated dunes and vegetated flats of South Padre Island, Texas. Pp. 4634–4635 in O. T. Magoon, ed., *Coastal Zone '89: Proceedings of the Sixth Symposium on Coastal and Ocean Management*. New York: American Society of Civil Engineers.

Judd, F. W., R. I. Lonard, and S. L. Sides. 1977. The vegetation of South Padre Island, Texas in relation to topography. *Southwestern Naturalist* 22:31–48.

Judd, F. W., H. Nieuwendaal, and D. L. Hockaday. 1991. The Leatherback Turtle, *Dermochelys coriacea*, in southernmost Texas. *Texas Journal of Science* 43:101–103.

Judd, F. W., and S. L. Sides. 1983. Effects of Hurricane Allen on the nearshore vegetation of South Padre Island, Texas. *Southwestern Naturalist* 28:365–369.

Jurado, E., A. O. Aguirre, J. Flores, J. Navar, H. Villalón, and D. Wester. 2000. Germination in Tamaulipan thornscrub of north-eastern Mexico. *Journal of Arid Environments* 46:413–424.

Jurado, E., E. Estrada, and A. Moles. 2001. Characterizing plant attributes with emphasis on seeds in Tamaulipan thornscrub in semi-arid Mexico. *Journal of Arid Environments* 48:309–321.

Jurado, E., J. Flores, J. Navar, and J. Jimenez. 1998. Seedling establishment under native Tamaulipan thornscrub and *Leucaena leucocephala* plantation. *Forest Ecology and Management* 105:151–157.

Jurado, E., J. F. Garcia, J. Flores, and E. Estrada. 2006. Leguminous seedling establishment in Tamaulipan thornscrub of northeastern Mexico. *Forest Ecology and Management* 221:133–139.

Jurado, E., J. Navar, H. Villalon, and M. Pando. 2001. Germination associated with season and sunlight for Tamaulipan thornscrub plants in north-eastern Mexico. *Journal of Arid Environments* 49:833–841.

Kaldy, J. E., and K. H. Dunton. 2000. Above- and below-ground production, biomass and reproductive ecology of *Thalassia testudinum* (Turtle Grass) in a subtropical coastal lagoon. *Marine Ecology Progress Series* 193:271–283.

Kaldy, J. E., K. H. Dunton, and A. B. Czerny. 1995. Variation in macroalgae species composition and abundance on a rock jetty in the northwest Gulf of Mexico. *Botanica Marina* 38:519–527.

Kang, Y., F. Koch, and C. J. Gobler. 2015. The interactive roles of nutrient loading and zooplankton grazing in facilitating the expansion of harmful algal blooms caused by the pelagophyte, *Aureoumbra lagunensis*, to the Indian River Lagoon, FL, USA. *Harmful Algae* 49:162–173.

Kantrud, H. A. 1991. *Widgeon Grass (*Ruppia maritima *L.): A Literature Review*. US Fish and Wildlife Service Fish and Wildlife Research Report 10, Washington, DC.

Kapraun, D. F. 1980. Summer aspect of algal zonation on a Texas jetty in relation to wave exposure. *Contributions in Marine Science* 23:101–109.

Karlsson, S., E. Saillant, B. W. Bumguardner, R. R. Vega, and J. R. Gold. 2008. Genetic identification of hatchery-released Red Drum in Texas bays and estuaries. *North American Journal of Fisheries Management* 28:1294–1304.

Kattner, K. R. 1973. Secondary successional vegetation on Padre Island National Seashore. MS thesis, Texas A&I University, Kingsville.

Kauffman, J. B, M. F. Adame, V. Arifainti, L. M. Schile-Beers, A. F. Bernardino, R. K. Bhomia, D. C. Donato, et al. 2020. Total ecosystem carbon stocks of mangroves across broad global environmental and physical gradients. *Ecological Monographs* 90 (2). https://doi.org/10.1002/ecm.1405.

Keim, B. D., and R. A. Muller. 2009. *Hurricanes of the Gulf of Mexico*. Baton Rouge: Louisiana State University Press.

Kelleway, J. J., K. Cavanaugh, K. Rogers, I. C. Feller, E. Ens, C. Doughty, and N. Saintilan. 2017. Review of the ecosystem service implications of mangrove encroachment into salt marshes. *Global Change Biology* 2017:3967–3983.

Kelley, P. 1986. *River of Lost Dreams: Navigation on the Rio Grande*. Lincoln: University of Nebraska Press.

Kelsey, M. G., and M. Hassall. 1989. Patch selection by Dunlin on a heterogeneous mudflat. *Ornis Scandinavia* 20:250–254.

Kennedy, J. P., M. Pil, C. E. Proffitt, W. A. Boeger, A. M. Stanford, and D. J. Devlin. 2016. Postglacial expansion of the Red Mangrove, *Rhizophora mangle* L., in the Caribbean Sea and Florida. *American Journal of Botany* 103:260–276.

Kennedy, M. L., T. L. Best, and D. J. Rookstool. 1973. Activity patterns of the Padre Island Kangaroo Rat, *Dipodomys compactus* True. *Southwestern Naturalist* 18:242–243.

Kenworthy, W. J., and D. E. Haunert. 1991. *The Light Requirements of Seagrasses: Proceedings of a Workshop to Examine the Capability of Water Quality Criteria, Standards and Monitoring Programs to Protect Seagrasses*. NOAA Technical Memorandum NMFS-SEFC-250, Washington, DC.

Kiel, W. H., Jr. 1957. *Coastal Waterfowl Survey: Ecology of Wintering Waterfowl in the Lower Laguna Madre*. Texas Parks and Wildlife Federal Aid Project, W-29-R-10, Job 16, Austin.

Kieslich, J. M. 1977. *A Case History of Port Mansfield Channel, Texas*. US Army Coastal Engineering Research Center and US Army Engineer Waterways Experimental Station Report GITI Report 12, Springfield, VA.

Kikuchi, T. 1980. Faunal relations in temperate seagrass beds. Pp. 153–172 in R. C. Phillips and C. P. McRoy, eds., *Handbook of Seagrass Biology: An Ecosystem Perspective*. New York: Garland STPM Press.

Kikuchi, T., and J. M. Peres. 1977. Consumer ecology of seagrass beds. Pp. 147–191 in C. P. McRoy and C. Helfferich, eds., *Seagrass Ecosystems: A Scientific Perspective*. New York: Marcel Dekker.

Kim, H. C., and P. A. Montagna. 2012. Effects of climate-driven freshwater inflow variability on macrobenthic secondary production in Texas lagoonal estuaries: A modeling study. *Ecological Modelling* 235:67–80.

King Ranch and Kenedy Trust. 1993. *Environmental Impacts of the Plan to Dump Dredged Spoil from the Gulf Intracoastal Waterway on Lands Bordering the Laguna Madre*. Kingsville, TX: King Ranch.

King, K. A., D. R. Blankinship, E. Payne, A. J. Krynitsky, and G. L. Hensler. 1985. Brown Pelican populations and pollutants in Texas 1975–1981. *Wilson Bulletin* 97:201–214.

King, K. A., E. L. Flickinger, and H. H. Hildebrand. 1977. The decline of Brown Pelicans on the Louisiana and Texas Gulf Coast. *Southwestern Naturalist* 21:417–431.

King, T. L., R. Ward, and E. G. Zimmerman. 1994. Population structure of Eastern Oysters (*Crassostrea virginica*) inhabiting the Laguna Madre, Texas, and adjacent bay systems. *Canadian Journal of Fisheries and Aquatic Science* 51:215–222.

Kirkpatrick, B, L. E. Fleming, D. Squicciarini, L. C. Backer, R. Clark, W. Abraham, J. Benson, et al. 2004. Literature review of Florida Red Tide: Implications for human health effects. *Harmful Algae* 3:82–86.

Kitting, C. L., B. Fry, and M. D. Morgan. 1984. Detection of inconspicuous epiphytic algae supporting food webs in seagrass meadows. *Oecologia* 62:145–149.

Klima, E. F., G. R. Gitschlag, and M. L. Renaud. 1988. Impacts of the explosive removal of offshore petroleum platforms on sea turtles and dolphins. *Marine Fisheries Review* 50:33–42.

Klootwyk, K. V. 2006. Seasonal changes in the benthic macroalgae composition and abundance at the Mansfield Pass jetties, Port Mansfield, Texas. MS thesis, Texas A&M University–Corpus Christi.

Koch, F., Y. Kang, T. A. Villareal, D. M. Anderson, and C. J. Gobler. 2014. A novel immunofluorescence flow cytometry technique detects expansion of brown tides caused by *Aureoumbra lagunensis* to the Caribbean Sea. *Applied and Environmental Microbiology* 80:4947–4957.

Koenig, R. L. 1969. A comparison of the winter food habits of three species of waterfowl from the Upper Laguna Madre of Texas. MS thesis, Texas A&I University, Kingsville.

Koepke, J. F. 1969. The mammals of northern Padre Island and Nueces County, Texas. MS thesis, Pittsburg State University, Pittsburg, KS.

Kopecky, A. L., and K. H. Dunton. 2006. Variability in drift macroalgal abundance in relation to biotic and abiotic factors in two seagrass dominated estuaries in the western Gulf of Mexico. *Estuaries and Coasts* 29:617–629.

Kottek, M., J. Grieser, C. Beck, B. Rudolf, and F. Rubel. 2006. World map of the Köppen-Geiger climate classification updated. *Meterologische Zeitschrift* 15:259–263.

Kowalski, J. L., H. R. DeYoe, and T. C. Allison. 2009. Seasonal production and biomass of the seagrass, *Halodule wrightii* Aschers. (Shoal Grass), in a subtropical Texas lagoon. *Estuaries and Coasts* 32:467–482. doi: 10.1007/s12237-009-9146-z.

Kowalski, J. L., D. L. Hockaday, G. H. Boza, and H. R. DeYoe. 2007. Re-occurrence of the tropical green macroalga, *Penicillus capitatus* (Chlorophyta: Bryopsidales), in the Lower Laguna Madre of South Texas. *Texas Journal of Science* 59:1–7.

Krauss, K., and M. Ball. 2013. On the halophytic nature of mangroves. *Trees* 27:7–11.

Kreuz, J. L. 1973. A comparison of the polychaete fauna of two different geographical areas in South Texas. MS thesis, Texas A&I University, Kingsville.

Kritzer, J. P., M. B. DeLucia, E. Greene, C. Shumway, M. F. Topolski, J. Thomas-Blate, L. A. Chiarella, K. B. Davy, and K. Smith. 2016. The importance of benthic habitats for coastal fisheries. *BioScience* 66:274–284.

Krull, R. M. 1976. The small fish fauna of a disturbed hypersaline environment. MS thesis, Texas A&I University, Kingsville.

Kucera, C. J., C. K. Faulk, and G. J. Holt. 2002. The effect of spawning salinity on eggs of Spotted Seatrout (*Cynoscion nebulosus*, Cuvier) from two bays with historically different salinity regimes. *Journal of Experimental Marine Biology and Ecology* 272:147–158.

Kuriyama, Y., N. Mochizuki, and T. Nakashima. 2005. Influence of vegetation on aeolian sand transport rate from a backshore to a foredune at Hasaki, Japan. *Sedimentology* 52:1123–1132.

Kuvlesky, W. P., Jr., L. A. Brennan, M. L. Morrison, K. K. Boydston, B. M. Ballard, and F. C. Bryant. 2007. Wind energy development and wildlife conservation: Challenges and opportunities. *Journal of Wildlife Management* 71:2487–2498.

La Vere, D. 2004. *The Texas Indians*. College Station: Texas A&M University Press.

Lack, D. 1968. *Ecological Adaptations for Breeding in Birds*. London: Methuen.

Lacson, J. M., and W. Y. Lee. 1997. *Status and Trends of Selected Marine Fauna in the Corpus Christi Bay National Estuary Program Study Area*. Corpus Christi Bay National Estuary Program, CCBNEP-24. Corpus Christi: Texas Natural Resource Conservation Commission.

Land, L. S., E. W. Behrens, and S. A. Frishman. 1979. The ooids of Baffin Bay, Texas. *Journal of Sedimentary Petrology* 49:1269–1278.

Landry, A. M., Jr., J. A. Williams, and C. W. Caillouet Jr. 1992. Sea turtle capture and habitat characterization study. Unpublished report submitted to National Marine Fisheries Service on file in Texas A&M University–Corpus Christi, Center for Coastal Studies Library.

Lange, C. J. 2014. Influence of development and sea-level rise on conservation of Redheads along the lower Texas coast. MS thesis, Texas A&M University–Kingsville.

Lange, C. J., B. M. Ballard, and D. P. Collins. 2018. Impacts of wind turbines on Redheads in the Laguna Madre. *Journal of Wildlife Management* 82:531–537.

Lange, M. A. 1997. *Austin's Woods Land Protection Compliance Document*. Austin: US Fish and Wildlife Service.

Lankford, R. R. 1977. Coastal lagoons of Mexico, their origin and classification. Pp. 182–215 in M. Wiley, ed., *Estuarine Processes*. London: Academic Press.

Lassuy, D. R. 1983. *Species Profiles: Life Histories and Environmental Requirements (Gulf of Mexico)—Spotted Seatrout*. Biological Service Division FWS/OBS-82/11.4, US Army Corps of Engineers Technical Report El-82-4. Washington, DC: US Fish and Wildlife Service.

Lawson, A. L., M. L. Morrison, and R. D. Slack. 2011. Impacts of oil and gas development on wintering grassland birds at Padre Island National Seashore, Texas. *Southeastern Naturalist* 10:303–320.

Lea, T. 1957a. Captain King of Texas. *Atlantic Monthly* 199 (4): 40–46.

Lea, T. 1957b. *The King Ranch*. 2 vols. Kingsville, TX: Carl Hertzog Publishing.

Leach, D. 2017. *Caesar Kleberg and the King Ranch*. Kingsville, TX: Caesar Kleberg Foundation for Wildlife Conservation.

Leary, T. R. 1957. A schooling of Leatherback Turtles, *Dermochelys coriacea*, on the Texas Coast. *Copeia* 1957:232.

Leatherman, S. P., and P. J. Godfrey. 1979. *The Impact of Off-Road Vehicles on Coastal Ecosystems in Cape Cod National Seashore: An Overview*. The Environmental Institute, University of Massachusetts/NPS Cooperative Research Unit UM/NPSCRU Report 34, Amherst.

Leber, K. M. 1985. The influence of predatory decapods, refuge, and microhabitat selection on seagrass communities. *Ecology* 66:1951–1964.

LeBlanc, R. J., and W. D. Hodgson. 1959. Origin and development of the Texas shoreline. *Transactions of the Gulf Coast Association of Geological Societies* 9:197–220.

LeDee, O. E. 2005. Winter distribution of Piping Plovers on the US Gulf of Mexico: An analysis of habitat characteristics at major wintering sites. MS thesis, University of Minnesota, St. Paul.

LeDee, O. E., F. J. Cuthbert, and P. V. Bolstad. 2008. A remote sensing analysis of coastal habitat composition for a threatened shorebird, the Piping Plover (*Charadrius melodus*). *Journal of Coastal Research* 24:718–726.

Lee, M. C. 1995. Factors influencing shorebird habitat use and interhabitat movements on Mustang Island, Texas from September 1990 through September 1991. MS thesis, Texas A&M University–Corpus Christi.

Lehman, R. L. 1999. A checklist of benthic marine macroalgae from the Corpus Christi Bay area. *Texas Journal of Science* 51:241–252.

Lehman, R. L. 2013. *Marine Plants of the Texas Coast*. College Station: Texas A&M University Press.

Lehman, R. L., R. O'Brien, and T. White. 2005. *Plants of the Texas Coastal Bend*. College Station: Texas A&M University Press.

Lehmann, V. W. 1948. Restocking on the King Ranch. *Transactions of the North American Wildlife Conference* 30:236–239.

Lehmann, V. W. 1969. *Forgotten Legions: Sheep in the Rio Grande Plain of Texas*. El Paso: Texas Western Press.

Leija-Tristán, A., A. Contreras-Arquieta, M. E. Garcia-Garza, A. J. Contreras-Balderas, M. de L. Lozano-Vilano, S. Contreras-Balderas, M. E. García-Ramiréz, et al. 2000. Taxonomic, bioecological and biogeographic aspects of selected biota of Laguna Madre, Tamaulipas, Mexico. Pp. 399–435 in M. Munawar, S. G. Lawrence, L. F. Munawar, and D. F. Malley, eds., *Aquatic Ecosystems of Mexico: Status and Scope*. Ecovision World Monograph Series. Leiden Netherlands: Backhuys Publishers.

Leopold, A. S. 1959. *Wildlife of Mexico: The Game Birds and Mammals*. Berkeley: University of California Press.

Levine, J. M., J. S. Brewer, and M. D. Bertness. 1998. Nutrients, competition and plant zonation in a New England salt marsh. *Journal of Ecology* 86:285–292.

Lewison, B., L. Crowder, and D. J. Shaver. 2003. The impact of turtle excluder devices and fisheries closures on Loggerhead and Kemp's Ridley strandings in the western Gulf of Mexico. *Conservation Biology* 17:1089–1097.

Lichlyter, S. A. 2006. An applied paleoecology case study: Bahia Grande, Texas prior to construction of the Brownsville Ship Channel. MS thesis, University of Texas at Brownsville.

Lin, G., and L. S. L. Sternberg. 1992. Effect of growth form, salinity, nutrient and sulfide on photosynthesis, carbon isotope discrimination and growth of Red Mangrove (*Rhizophora mangle* L.). *Functional Plant Biology* 19:509–517.

Lipschultz, M. L. 2016. Effect of drought and grazing on land bird populations in South Texas. MS thesis, Texas A&M University–Kingsville.

Liu, H., and E. J. Buskey. 2000. Hypersalinity enhances the production of extracellular polymeric substance (EPS) in the Texas Brown Tide alga *Aureoumbra lagunensis* (Pelagophyceae). *Journal of Phycology* 35:1–7.

Liu, H., E. A. Laws, T. A. Villareal, and E. J. Buskey. 2001. Nutrient-limited growth of *Aureoumbra lagunensis* (Pelagophyceae), with implications for its capability to outgrow other phytoplankton species in phosphate-limited environments. *Journal of Phycology* 37:500–508.

Lloyd, W. 1891. *Mammals of Padre Island: Aug. 20 to 31, Nov. 1 to 13, 1891*. Smithsonian Institution Archives, Record Unit 7176, Box 94, Folder 18, US Fish and Wildlife Service, 1860–1961 Field Reports, Washington, DC.

Lockett, L., and R. W. Read. 1990. Extension of native range of *Sabal mexicana* (Palmae) in Texas to include central coast. *Sida* 14:79–85.

Lockley, J. C. 1995. Effect of imported fire ant predation on a population of the Least Tern: An endangered species. *Southwestern Entomologist* 20:517–519.

Lohse, E. A. 1958. Geochronology of mud flats through varve analysis. Pp. 1–47 in *Sedimentology of South Texas: Field Trip Guidebook*, Gulf Coast Association of Geological Societies annual meeting, October 27–November 1, 1958, Corpus Christi.

Lonard, R. I. 1993. Personal affidavit, Exhibit 6. In *Environmental Impacts of the Plan to Dump Dredged Spoil from the Gulf Intracoastal Waterway on Lands Bordering the Laguna Madre*. Kingsville, TX: King Ranch and John G. Kenedy Charitable Trust.

Lonard, R. I., J. H. Everitt, and F. W. Judd. 1991. *Woody Plants of the Lower Rio Grande Valley, Texas*. Miscellaneous Papers, Texas Memorial Museum (Miscellaneous Publications No. 7). Austin: University of Texas at Austin.

Lonard, R. I., and F. W. Judd. 1980. Phytogeography of South Padre Island, Texas. *Southwestern Naturalist* 23:497–510.

Lonard, R. I., and F. W. Judd. 1981. *The Terrestrial Flora of South Padre Island, Texas*. Miscellaneous Papers, Texas Memorial Museum (Miscellaneous Publication No. 6). Austin: University of Texas at Austin.

Lonard, R. I., and F. W. Judd. 1985. Effects of a severe freeze on native woody plants in the Lower Rio Grande Valley, Texas. *Southwestern Naturalist* 30:397–403.

Lonard, R. I., and F. W. Judd. 1991. Comparison of the effects of the severe freezes of 1983 and 1989 on native woody plants in the Lower Rio Grande Valley, Texas. *Southwestern Naturalist* 36:213–217.

Lonard, R. I., and F. W. Judd. 1993. Recovery of vegetation of barrier island washover zones. Pp. 2324–2331 in O. T. Magoon, W. S. Wilson, H. Converse, and L. T. Tobin, eds., *Proceedings of the Eighth Symposium on Coastal and Ocean Management, "Coastal Zone '93."* New Orleans: American Society of Civil Engineers.

Lonard, R. I., and F. W. Judd. 2002. Riparian vegetation of the Lower Rio Grande. *Southwestern Naturalist* 47:420–432.

Lonard, R. I., F. W. Judd, J. H. Everitt, and D. E. Escobar. 1991. Roadside associated disturbance on coastal dunes. Pp. 2823–2836 in O. T. Magoon, H. Converse, V. Tippie, L. T. Tobin, and D. Clark, eds., *Proceedings of the Seventh Symposium on Coastal and Ocean Management, "Coastal Zone '91."* Long Beach, CA: American Society of Civil Engineers.

Lonard, R. I., F. W. Judd, and S. L. Sides. 1978. Annotated checklist of the flowering plants of South Padre Island, Texas. *Southwestern Naturalist* 23:497–510.

Lonard, R. I., F. W. Judd, and E. H. Smith. 2003. Recovery of vegetation following a wild fire on the margins of tidal flats, Padre Island National Seashore, Texas. *Texas Journal of Science* 55:347–364.

Lonard, R. I., F. W. Judd, E. H. Smith, and C. Yang. 2004. Recovery of vegetation following a wildfire in a barrier island grassland, Padre Island National Seashore, Texas. *Southwestern Naturalist* 49:173–188.

Long, D. T., and R. Gudramovics. 1983. Major-element geochemistry of brines from the wind tidal flat area, Laguna Madre, Texas. *Journal of Sedimentary Petrology* 53:797–810.

Longley, W. L. 1994. *Freshwater Inflows to Texas Bays and Estuaries: Ecological Relationships and Methods for Determination of Needs*. Austin: Texas Water Development and Texas Parks and Wildlife Department.

Longley, W. L., W. B. Brogden, and S. N. James. 1989. *Texas Barrier Island Characterization Conceptual Models*. US Fish and Wildlife Service, National Wetland Research Center Open File Report 89-05, Lafayette, LA.

Lopez-Barreiro, T. 1998. Detection and distribution of the Texas Brown Tide species *Aureoumbra lagunensis* in the Gulf of Mexico using an immunofluorescence assay. MS thesis, University of Massachusetts, Boston.

Lopez-Barreiro, T., A. Villareal, and S. L. Morton. 1998. Development of an antibody against the Texas Brown Tide (*Aureoumbra lagunensis*). Pp. 263–265 in B. Reguera, J. Blanco, M. L. Fernandez, and T. Wyatt, eds., *Harmful Algae*. Viga, Spain: Xunta de Galicia and Intergovernmental Oceanographic Commission of UNESCO.

López-Martínez, J. 2013. Caracterización de las praderas de pastos marinos en tres localidades de la Laguna Madre de Tamaulipas, México. Tesis, Universidad Autónoma de Tamaulipas, Facultad de Ingeniería y Ciencias.

Loveless, J. B., and D. L. Smee. 2019. Changes in arthropod communities as Black Mangroves *Avicennia germinans* expand into Gulf of Mexico salt marshes. *Arthropod Plant Interactions* 13:465–475.

Lovell, S., S. Steinbeck, and J. Hilger. 2011. *The Economic Contribution of Marine Angler Expenditures in the United States, 2011*. NOAA Technical Memo, NMFS-F/SPO-134. Washington, DC: US Department of Commerce.

Lugo, A. E., G. Cintron, and C. Goenaga. 1981. Mangrove ecosystems under stress. Pp. 129–153 in G. W. Barrett and R. Rosenberg, eds., *Stress Effects on Natural Ecosystems*. London: John Wiley and Sons.

Lugo, A. E., E. Medina, and K. McGinley. 2014. Issues and challenges of mangrove conservation in the Anthropocene. *Madera y Bosques* 20 (special issue): 11–38.

Lugo, A. E., and S. Snedaker. 1974. The ecology of mangroves. *Annual Review of Ecology and Systematics* 5:39–64.

Lund, F. 1974. *Marine Turtle Nesting in the United States*. Report to the US Fish and Wildlife Service, Washington, DC.

Lunt, J., K. McGlaun, and E. M. Robinson. 2013. Effects of Black Mangrove (*Avicennia germinans*) expansion on saltmarsh (*Spartina alterniflora*) benthic communities of the South Texas coast. *Gulf and Caribbean Research* 25:125–129.

Mabee, T. J., J. H. Plissner, S. M. Haig, and J. P. Goosen. 2001. Winter distributions of North American plovers in the Laguna Madre regions of Tamaulipas, Mexico and Texas, USA. *Wader Study Group Bulletin* 94:39–43.

Mackin, J. G. 1971. *A Study of the Effect of Oil Field Brine Effluents on Biotic Communities in Texas Estuaries*. Texas A&M University Research Foundation Project 735, College Station.

Magana, H. A., C. Contreras, and T. A. Villareal. 2003. A historical assessment of *Karenia brevis* in the western Gulf of Mexico. *Harmful Algae* 2:163–171.

Magnuson, J. J., K. A. Bjorndal, W. D. DuPaul, G. L. Graham, D. W. Owens, C. H. Peterson, P. C. H. Pritchard, J. I. Richardson, G. E. Saul, and C. W. West. 1990. *Decline of the Sea Turtles: Causes and Prevention*. Washington, DC: National Academy Press.

Mannino, A., and P. A. Montagna. 1997. Small-scale spatial variation of macrobenthic community structure. *Estuaries* 20:159–173.

Marquez, M. A., A. Fierro-Cabo, and C. E. Cintra-Buenrostro. 2017. Can ecosystem functional recovery be traced to decomposition and nitrogen dynamics in estuaries of the Lower Laguna Madre, Texas? *Restoration Ecology* 25:618–628.

Marquez, M., R. A. Villanueva, and P. M. Burchfield. 1989. Nesting population and production of hatchlings of Kemp's Ridley Sea Turtles at Rancho Nuevo, Mexico. Pp. 16–19 in K. A. Bjorndal, ed., *Biology and Conservation of Sea Turtles*. Washington, DC: Smithsonian Institution Press.

Marsh, S. L. 1979. Factors affecting the distribution, food habits, and lead toxicosis of Redhead ducks in the Laguna Madre, Texas. MS thesis, Texas A&M University, College Station.

Martin, C. O., and M. F. Hehnke. 1981. South Texas potholes: Their status and value as wildlife habitat. *Wetlands* 1:19–46.

Martin, F. D., and M. Cooper. 1981. A comparison of fish fauna found in pure stands of two tropical Atlantic seagrasses, *Thalassia testudinum* and *Syringodium filiforme*. *Northeast Gulf Science* 5:31–37.

Martin, J. H., and L. W. McEachron. 1996. *Historical Annotated Review of Winter Kills of Marine Organisms in Texas Bays*. Management Data Series No. 50. Austin: Texas Parks and Wildlife Department, Coastal Fisheries Branch.

Martin, J. J. 1979. A study of the feeding habits of the Black Drum (*Pogonias cromis* Linnaeus) in Alazan Bay and the Laguna Salada, Texas. MS thesis, Texas A&I University, Kingsville.

Martínez, M., and A. Novelo. 1993. La vegetación aquática del estado de Tamaulipas, México. *Anales del Instituto de Biología Universidad Nacional Autónoma Serie Botánica* 64 (2): 59–86.

Martinez-Andrade, F., P. Campbell, and B. Fuls. 2005. *Trends in Relative Abundance and Size of Selected Finfish and Shellfish along the Texas Coast: November 1975–December 2003*. Management Data Series No. 232. Austin: Texas Parks and Wildlife Department, Coastal Fisheries Branch.

Martínez-Palacios, C. A., A. Campos Mendoza, E. Díaz-Pardo, F. Arreguín Sánchez, R. Rueda Jasso, J. Fonseca Madrigal, A. Gutíerrez-Hernández, et al. 2010. Bagres armados: ¿Erradicación o utilización? *Ciencia y Desarrollo* 36 (247): 28–33.

Mason, A. J. 2014. The simultaneous effects of competition and predation on the foraging behavior of the Gulf Coast Kangaroo Rat (*Dipodomys compactus*) and Merriam's Pocket Mouse (*Perognathus merriami*) in South Texas. MS thesis, Texas A&M University–Kingsville.

Masson, P. H. 1955. An occurrence of gypsum in Southwest Texas. *Journal of Sedimentary Petrology* 25:72–77.

Matlock, G. C. 1987. Maximum total length and age of Red Drum off Texas. *Northeast Gulf Science* 9:49–52.

Matthiessen, P. 1967. *The Shorebirds of North America*. New York: Viking Press.

Mays, J. L. 1996. Population size and distribution of Cactus Ferruginous Pygmy-Owls in Brooks and Kenedy Counties, Texas. MS thesis, Texas A&M University–Kingsville.

McAdams, M. S. 1987. Classification and waterfowl use of ponds in South Texas. MS thesis, Texas A&M University, College Station.

McCampbell, C. 1952. *Texas Seaport: The Story of the Growth of Corpus Christi and the Coastal Bend Area*. New York: Exposition Press.

McEachron, L. W., R. L. Colura, B. Bumguardner, and R. Ward. 1998. Survival of stocked Red Drum in Texas. *Bulletin of Marine Science* 62:359–368.

McGowen, J. H., L. F. Brown Jr., T. R. Calnan, J. L. Chin, J. P. Herber, and C. L. Lewis. 1977. *History and Processes Involved in Development of South Padre Island, Laguna Madre, and Los Bancos de en Medio*. Report to Texas General Land Office. Austin: Bureau of Economic Geology, University of Texas.

McGowen, J. H., C. G. Groat, L. F. Brown Jr., W. L. Fisher, and A. J. Scottl. 1970. *Effects of Hurricane Celia: A Focus on Environmental Geologic Problems of the Texas Coastal Zone*. Circular 70-3. Austin: Bureau of Economic Geology, University of Texas.

McGowen, J. H., C. V. Proctor, L. F. Brown Jr., T. H. Evans, W. L. Fisher, and C. G. Groat. 1976. *Environmental Geologic Atlas of the Texas Coastal Zone: Port Lavaca Area*. Austin: Bureau of Economic Geology, University of Texas.

McGowen, J. H., and A. J. Scott. 1975. Hurricanes as geologic agents on the Texas coast. Pp. 23–43 in L. E. Cronin, ed., *Estuarine Research*, vol. 2, *Geology and Engineering*. New York: Academic Press.

McGraw, A. J. 1984. Archaeology of the South Texas Sand Sheet: A study of Chevron properties in Brooks County. *Index of Texas Archaeology: Open Access Gray Literature from the Lone Star State*, vol. 1984, article 7. https://doi.org/10.21112/ita.1984.1.7.

McKee, K. L. 1995. Seedling recruitment patterns in a Belizean mangrove forest: Effects of establishment ability and physico-chemical factors. *Oecologia* 101:448–460.

McKee, K. L., D. R. Cahoon, and I. C. Feller. 2007. Caribbean mangroves adjust to rising sea level through biotic controls on change in soil elevation. *Global Ecology and Biogeography* 16:545–556.

McKee, K. L., and J. A. Cherry. 2009. Hurricane Katrina sediment slowed elevation loss in subsiding brackish marshes of the Mississippi River Delta. *Wetlands* 29:2–15.

McKee, K. L., I. A. Mendelssohn, and M. K. Hester. 1988. Reexamination of pore water sulfide concentrations and redox potentials near the aerial roots of *Rhizophora mangle* and *Avicennia germinans*. *American Journal of Botany* 75:1352–1359.

McKee, K. L., J. E. Rooth, and I. C. Feller. 2007. Mangrove recruitment after forest disturbance is facilitated by herbaceous species in the Caribbean. *Ecological Applications* 17:1678–1693.

McKee, K., and C. Vervaeke. 2018. Will fluctuations in salt marsh–mangrove dominance alter vulnerability to a subtropical wetland to sea-level rise? *Global Change Biology* 24 (3): 1224–1238. DOI: 10.1111/gcb.13945.

McKenzie, D. F. 1987. Utilization of rootstock and browse by waterfowl in moist-soil impoundments in Missouri. MS thesis, University of Missouri, Columbia.

McLendon, T. 1991. Preliminary description of the vegetation of South Texas exclusive of coastal saline zones. *Texas Journal of Science* 43:13–32.

McMahan, C. A. 1965–1967. *Ecology of Principal Waterfowl Foods of Laguna Madre*. Texas Parks and Wildlife Department Job Completion Reports, Federal Aid Projects W-29-R-18 to 20, Austin.

McMahan, C. A. 1966. *Ecology of Principal Waterfowl Foods of Laguna Madre*. Texas Parks and Wildlife Department Job Completion Report, Federal Aid Project No. W-29-R-19, Austin.

McMahan, C. A. 1967a. *Coastal Waterfowl Survey: Ecology of Principal Wintering Waterfowl in Lower Laguna Madre*. Texas Parks and Wildlife Federal Aid Project, W-29-R-20, Job 6 Austin.

McMahan, C. A. 1967b. *Ecology of Principal Waterfowl Foods of Laguna Madre*. Texas Parks and Wildlife Department Job Completion Report, Federal Aid Project No. W-29-R-20, Austin.

McMahan, C. A. 1968. Biomass and salinity tolerance of Shoalgrass and Manateegrass in the Lower Laguna Madre. *Journal of Wildlife Management* 32:501–506.

McMahan, C. A. 1970. Food habits of ducks wintering in Laguna Madre, Texas. *Journal of Wildlife Management* 34:946–949.

McMahan, C. A., R. G. Frye, and K. L. Brown. 1984. *The Vegetation Types of Texas including Cropland*. Austin: Wildlife Division, Texas Parks and Wildlife Department.

McMillan, C. 1971. Environmental factors affecting seedling establishment of the Black Mangrove on the central Texas coast. *Ecology* 52:927–930.

McMillan, C. 1975a. Adaptive differentiation to chilling in mangrove populations. Pp. 62–68 in G. E. Walsh, S. C. Snedaker, and H. J. Teas, eds., *Proceedings of the International Symposium on the Biology and Management of Mangroves*, vol. 1. Gainesville: University Press of Florida.

McMillan, C. 1975b. Interaction of soil texture with salinity tolerances of Black Mangrove (*Avicennia*) and White Mangrove (*Laguncularia*) from North America. Pp 561–566 in G. E. Walsh, S. C. Snedaker, and H. J. Teas, eds., *Proceedings of the International Symposium on the Biology and Management of Mangroves*. Gainesville: University Press of Florida.

McMillan, C. 1979. Differentiation in response to chilling temperatures among populations of three marine spermatophytes, *Thalassia testudinum*, *Syringodium filiforme* and *Halodule wrightii*. *American Journal of Botany* 66:810–819.

McMillan, C. 1985. The seed reserve for *Halodule wrightii*, *Syringodium filiforme* and *Ruppia maritima* in Laguna Madre, Texas. *Contributions in Marine Science* 28:141–149.

McMillan, C., and F. N. Moseley. 1967. Salinity tolerances of five marine spermatophytes of Redfish Bay, Texas. *Ecology* 48:503–506.

McMurry, S. L. 1971. Nesting and development of the Reddish Egret (*Dichromanassa rufescens* Gmelin) on a spoil bank chain in the Laguna Madre. MS thesis, Texas A&I University, Kingsville.

McNeil, R., and J. Burton. 1977. Southbound migration of shorebirds from the Gulf of St. Lawrence. *Wilson Bulletin* 89:167–171.

McNulty, J. K., W. N. Lindal Jr., and J. E. Sykes. 1972. *Cooperative Gulf of Mexico Estuarine Inventory and Study, Florida Phase I: Area Description*. NOAA Technical Report Circular 368, Washington, DC.

McRoy, C. P., and C. Helfferich. 1977. Preface. Pp. iii–v in C. P. McRoy and C. Helfferich, eds., *Seagrass Ecosystems: A Scientific Perspective*. New York: Marcel Dekker.

McRoy, C. P., and C. McMillan. 1977. Production ecology and physiology of seagrasses. Pp. 53–87 in C. P. McRoy and C. Helfferich, eds., *Seagrass Ecosystems: A Scientific Perspective*. New York: Marcel Dekker.

Medina, E., and M. Francisco. 1997. Osmolality and δ^{13} C of leaf tissues of mangrove species from environments of contrasting rainfall and salinity. *Estuarine, Coastal and Shelf Science* 45:337–344.

Meire, P., and E. Kuyken. 1984. Relations between the distribution of waders and the intertidal benthic fauna of the Oosterschelde, Netherlands. Pp. 55–68 in P. R. Evans, J. D. Goss-Custard, and W. G. Hale, eds., *Coastal Waders and Wildfowl in Winter*. Cambridge: Cambridge University Press.

Mendelssohn, I. A., M. R. Byrnes, R. T. Kneib, and B. A. Vittor. 2017. Coastal habitats of the Gulf of Mexico. Pp. 359–640 in C. Ward, ed., *Habitats and Biota of the Gulf of Mexico: Before the Deepwater Horizon Oil Spill*. New York: Springer.

Mendenhall, K. S. 2015. Diet of Black Drum (*Pogonias cromis*) based on stable isotope and stomach content analyses. MS thesis, Texas A&M University–Corpus Christi.

Mendoza, R., N. Arreaga, J. Hernandez, V. Segovia, I. Jasso, and D. Pérez. 2011. *Aquatic Invasive Species in the Río Bravo / Laguna Madre Ecological Region*. Background Paper 2011-02, Commission for Environmental Cooperation, Montreal, Canada. http://www3.cec.org/islandora/en/item/10259-aquatic-invasive-species-in-rio-bravolaguna-madre-ecological-region-en.pdf.

Merkord, G. W. 1978. The distribution and abundance of seagrasses in Laguna Madre of Texas. MS thesis, Texas A&I University, Kingsville.

Metz, T. L., and A. M. Landry Jr. 2013. An assessment of Green Turtle (*Chelonia mydas*) stocks along the Texas coast, with emphasis on the Lower Laguna Madre. *Chelonian Conservation and Biology* 12:293–302.

Meylan, P. A., A. B. Meylan, and J. A. Gray. 2011. The ecology and migrations of sea turtles 8: Tests of the developmental habitat hypothesis. *Bulletin of the American Museum of Natural History* 357:1–70.

Michot, T. C. 2000. Comparison of wintering Redhead populations in four Gulf of Mexico seagrass beds. Pp. 243–260 in F. A. Comin, J. A. Herrera, and J. Ramirez, eds., *Limnology and Aquatic Birds: Monitoring, Modelling and Management*. Mérida, Mexico: Universidad Autónoma de Yucatán.

Michot, T. C., M. C. Woodin, S. E. Adair, and E. B. Moser. 2006. Diurnal time-activity budgets of Redheads (*Aythya americana*) wintering in seagrass beds and coastal ponds in Louisiana and Texas. *Hydrobiologia* 567:113–128.

Milbrandt, E. C., and M. N. Tinsley. 2006. The role of Saltwort (*Batis maritima* L.) in regeneration of degraded mangrove forests. *Hydrobiologia* 568:369–377.

Miller, J. A. 1975. Facies characteristics of Laguna Madre wind-tidal flats. Pp. 67–73 in R. N. Ginsburg, ed., *Tidal Deposits: A Casebook of Recent Examples and Fossil Counterparts*. New York: Springer-Verlag.

Mini, A. E., E. R. Harrington, E. Rucker, B. D. Ducker, and T. B. Mowbray. 2014. American Wigeon (*Mareca americana*), version 2.0. In A. F. Poole, ed., *The Birds of North America*. Ithaca, NY: Cornell Lab of Ornithology.

Mitchell, C. A. 1991. Ecology of wintering Redheads (*Aythya americana*) on the Lower Laguna Madre, Texas. PhD diss., Louisiana State University, Baton Rouge.

Mitchell, C. A. 1992. Water depth predicts Redhead distribution in the Lower Laguna Madre, Texas. *Wildlife Society Bulletin* 20:420–424.

Mitchell, C. A., T. W. Custer, and P. J. Zwank. 1992. Herbivory on Shoalgrass by wintering Redheads in Texas. *Journal of Wildlife Management* 58:131–141.

Mock, D. W., H. Drummond, and C. H. Stinson. 1990. Avian siblicide. *American Scientist* 78:438–449.

Molina, K. C., R. M. Erwin, E. Palacios, E. Mellink, and N. W. H. Seto. 2010. *Status Review and Conservation Recommendations for the Gull-billed Tern (*Gelochelidon nilotica*) in North America*. US Fish and Wildlife Service, Biological Technical Publication BTP-R1013-2010, Washington, D.C.

Moncreiff, C. A., and M. J. Sullivan. 2001. Trophic importance of epiphytic algae in subtropical seagrass beds: Evidence from multiple stable isotope analyses. *Marine Ecology Progress Series* 215:93–106.

Monday, J., and B. Colley. 1997. *Voices from the Wild Horse Desert: The Vaquero Families of the King and Kenedy Ranches*. Austin: University of Texas Press.

Montagna, P. A. 1992. *Benthic Samples Taken from the Padre Island National Seashore*. University of Texas Marine Science Institute Technical Report TR/92-002, Port Aransas.

Montagna, P. A. 1993. *Comparison of Ecosystem Structure and Functions of Created and Natural Seagrass Habitats in Laguna Madre, Texas*. University of Texas Marine Science Institute Technical Report TR/93-007, Port Aransas.

Montagna, P. A., J. Brenner, J. Gibeaut, and S. Morehead. 2011. Coastal impacts. Pp. 96–123 in J. Schmandt, G. R. North, and J. Clarkson, eds., *The Impact of Global Warming in Texas*, 2nd ed. Austin: University of Texas Press.

Montagna, P., J. Gibeaut, and J. W. Tunnell Jr. 2007. South Texas climate 2100: Coastal impacts. Pp. 57–77 in J. Norwine and K. John, eds., *The Changing Climate of South Texas 1900–2100: Problems and Prospects, Impacts and Implications*. Kingsville, TX: CREST-RESSACA, Texas A&M University–Kingsville.

Montagna, P. A., and R. D. Kalke. 1995. Ecology of infaunal mollusca in South Texas estuaries. *American Malacological Bulletin* 15:307–326.

Montagna, P. A., and T. A. Palmer. 2012. *Water and Sediment Quality Status and Trends in the Coastal Bend. Phase 2: Data Analysis*. Publication CBBEP 73, Project Number 1206, Coastal Bend Bays and Estuaries Program, Corpus Christi.

Montagna, P. A., T. A. Palmer, and J. B. Pollack. 2013. *Hydrological Changes and Estuarine Dynamics*. Springer Briefs in Environmental Science 8. New York: Springer-Verlag. DOI: 10.1007/978-1-4614-5833-3.

Montagna, P. A., D. A. Stockwell, and R. D. Kalke. 1993. Dwarf Surfclam *Mulinia lateralis* (Say, 1822) populations and feeding during the Texas Brown Tide event. *Journal of Shellfish Research* 12:433–442.

Moore, H. B., and B. F. McPherson. 1963. Colonization of the Miami area by the barnacle *Balanus trigonus* Darwin and a note on its occurrence on the test of an echinoid. *Bulletin of Marine Science of the Gulf and Caribbean* 13:418–421.

Moore, J. L. 1991. Habitat-related activities and body mass of wintering Redhead ducks on coastal ponds in South Texas. MS thesis, Texas A&M University, College Station.

Moore, R. H. 1976. Reproductive habits and growth of *Bufo speciosus* on Mustang Island, Texas, with notes on the ecology and reproduction of other anurans. *Texas Journal of Science* 27:173–178.

Moore, S. 2010. Effects of Guineagrass on Northern Bobwhite habitat use. MS thesis, Texas A&M University–Kingsville.

Moorman, A. M., T. E. Moorman, G. A. Baldassarre, and D. M. Richard. 1991. Effects of saline water on growth and survival of Mottled Duck ducklings. *Journal of Wildlife Management* 55:471–476.

Morán-Silva, A., M. L. Jiménez-Badillo, S. Cházaro-Olvera, C. Meiners, G. Galindo-Cortes, J. L. Oviedo Pérez, and A. México. 2014. First record of Tiger Shrimp (*Penaeus monodon* Fabricius 1798), in the south-central coast of Veracruz, Gulf of Mexico. *BIOCYT* 7 (28): 509–514.

Mora-Olivo, A. 1996. *Informe final del proyecto: Vegetación acuática de la Laguna Madre de Tamaulipas (pastos marinos)*. Tamaulipas: Universidad Autónoma de Tamaulipas.

Morgan, D., and C. L. Kitting. 1984. Productivity and utilization of the seagrass *Halodule wrightii* and its attached epiphytes. *Limnology and Oceanography* 29:1066–1076.

Morreale, S. J., A. B. Meylan, S. S. Sadove, and E. A. Standora. 1992. Annual occurrence and winter mortality of marine turtles in New York waters. *Journal of Herpetology* 26:301–308.

Morris, J. A., Jr., and P. E. Whitfield. 2009. Biology, ecology, control and management of the invasive Indo-Pacific lionfish: An updated integrated assessment. *NOAA Technical Memorandum NOS NCCOS* 99:1–57.

Morrison, G., and H. Greening. 2011. Seagrass. Chapter 4 in K. Yates, H. Greening, and G. Morrison, eds., *Integrating Science and Resource Management in Tampa Bay, Florida*, Circular 1348. Reston, VA: US Department of the Interior and US Geological Survey.

Morrison, R. I. G. 1984. Migration systems of some New World shorebirds. Pp. 125–202 in J. Burger and B. L. Olla, eds., *Behavior of Marine Animals*, vol. 6, *Shorebirds: Migration and Foraging Behavior*. New York: Plenum Press.

Morrison, R. I. G., R. K. Ross, J. Guzman P., and A. Estrada. 1993. Aerial surveys of Nearctic shorebirds wintering in Mexico: Preliminary results of surveys on the Gulf of Mexico and Caribbean coasts. *Canadian Wildlife Service Progress Notes* 206:1–14.

Morton, R. A., and L. E. Garner. 1993. *Geological Investigation of the Wind-Tidal Flat Area, Kenedy County, Texas*. No. IAC (92-93) 2170. Austin: Bureau of Economic Geology, University of Texas.

Morton, R. A., and C. W. Holmes. 2009. Geological processes and sedimentation rates of wind-tidal flats, Laguna Madre, Texas. *Gulf Coast Association of Geological Societies Transactions* 59:519–538.

Morton, R. A., and J. H. McGowen. 1980. *Modern Depositional Environments of the Texas Coast*. Guidebook 20. Austin: Bureau of Economic Geology, University of Texas.

Morton, R. A., and M. J. Pieper. 1975. *Shoreline Changes on Brazos Island and South Padre Island (Mansfield Channel to Mouth of the Rio Grande): An Analysis of Historical Changes of the Texas Gulf Shoreline*. Report No. 75-2. Austin: Bureau of Economic Geology, University of Texas.

Morton, R. A., W. A. White, R. S. Kerr, and W. D. Kuenzi. 1977. Geological aspects of barrier island development: Mustang and North Padre Islands, Texas. Chapter 3 in R. S. Kier and E. G. Fruh, eds., *Environmental and Economic Impacts of Recreational Community Development, Mustang Island and North Padre Island*, vol. 2. Austin: Center for Research in Water Resources, University of Texas.

Morton, R. A., W. A. White, and R. C. Nava. 1998. *Sediment Budget Analysis for Laguna Madre, Texas: An Examination of Sediment Characteristics, History, and Recent Transport*. Report DACW64-96-C-0018. Austin: Bureau of Economic Geology, University of Texas.

Moulton, D. F., T. E. Dahl, and D. M. Dall. 1997. *Texas Coastal Wetlands: Status and Trends, Mid-1950s to Early 1990s*. Albuquerque, NM: US Fish and Wildlife Service, Southeastern Region.

Muehl, G. T. 1994. Distribution and abundance of water birds and wetlands in coastal Texas. MS thesis, Texas A&M University–Kingsville.

Muhlstein, H. I., and T. A. Villareal. 2007. Organic and inorganic nutrient effects on growth rate-irradiance relationships in the Texas Brown Tide alga *Aureoumbra lagunensis* (Pelagophyceae). *Journal of Phycology* 43:1223–1226.

Murray, L. S., and T. S. Jinnette. 1978. Survival of dominant estuarine organisms impinged on fine mesh traveling screens at the Barney M. Davis Power Station. Pp. 79–87 in R. K. Sharma and J. B. Palmer, eds., *Larval Exclusion Systems for Power Plant Cooling Water Intakes*. Publication ANVES-66. Lemont, IL: Argonne National Laboratory.

Murray, N. J., S. R. Phinn, M. DeWitt, R. Ferrari, R. Johnston, M. B. Lyons, N. Clinton, D. Thau, and R. A. Fuller. 2019. The global distribution and trajectory of tidal flats. *Nature* 565:222–265.

Musick, J. A., and C. J. Limpus. 1997. Habitat utilization and migration in juvenile sea turtles. Pp. 137–163 in P. L. Lutz and J. A. Musick, eds., *The Biology of Sea Turtles*. Boca Raton, FL: CRC Press.

Myers, J. P., R. I. G. Morrison, P. Z. Antas, B. A. Harrington, T. E. Lovejoy, M. Sallaberry, S. E. Senner, and A. Tarak. 1987. Conservation strategy for migratory species. *American Scientist* 75:19–26.

Myers, J. P., M. Sallaberry, A. E. Ortiz, G. Castro, L. M. Gordon, J. L. Maron, C. T. Schick, E. Tabilo, P. Antas, and T. Below. 1990. Migration routes of New World Sanderlings (*Calidris alba*). *Auk* 107:172–180.

Nagelkerken, I., S. J. M. Blaber, S. Bouillon, P. Green, M. Haywood, L. G. Kirton, J.-O. Meynecke, et al. 2008. The habitat function of mangroves for terrestrial and marine fauna: A review. *Aquatic Botany* 89:155–185.

Naidoo, G. 1987. Effects of salinity and nitrogen on growth and water relations in the mangrove, *Avicennia marina* (Forsk.) vierh. *New Phytologist* 107:317–325.

National Commission on the BP Deepwater Horizon Oil Spill and Offshore Drilling. 2011. *Deep Water: The Gulf Oil Disaster and the Future of Offshore Drilling: Report to the President*. https://www.govinfo.gov/app/details/GPO-OILCOMMISSION.

NatureServe. 2002. *International Classification of Ecological Communities: Terrestrial Vegetation of the United States, Tamaulipan Thornscrub Ecoregion*. Arlington, VA: NatureServe; San Antonio: The Nature Conservancy of Texas.

Návar, J. 2011. Stemflow variation in Mexico's northeastern forest communities: Its contribution to soil moisture content and aquifer recharge. *Journal of Hydrology* 408:35–42.

Návar, J., J. M. González, and H. Gonzalez. 2009. Gross precipitation and throughfall chemistry in legume species planted in northeastern Mexico. *Plant and Soil* 318:15–26.

Návar, J., E. Méndez, and V. Dale. 2002. Estimating stand biomass in the Tamaulipan thornscrub of northeastern Mexico. *Annals of Forest Science* 59:813–821.

Návar, J., E. Méndez, A. Nájera, J. Graciano, V. Dale, and B. Parresol. 2004. Biomass equations for shrub species of Tamaulipan thornscrub of north-eastern Mexico. *Journal of Arid Environments* 59:657–674.

Navar, J., F. de J. Rodriguez-Flores, P. A. Dominguez-Calleros, and G. Perez-Verdin. 2014. Diversity-productivity relationship in the northeastern Tamaulipan thornscrub forest of Mexico. *International Journal of Ecology* 2014:1–8.

Neckles, H. A., B. S. Kopp, B. J. Peterson, and P. S. Pooler. 2012. Integrating scales of seagrass monitoring to meet conservation needs. *Estuaries and Coasts* 35:23–46. doi: 10.1007/s12237-011-9410-x.

Negrete, I. G., A. D. Nelson, J. Goetze, L. Macke, T. Wilburn, and A. Day. 1999. A checklist for the vascular plants of Padre Island National Seashore. *Sida* 18:1241–1259.

Nelson, A., J. Goetze, and A. Lucksinger. 2001. A comparison of the flora of northern Padre Island to that of Matagorda Island, Mustang Island and southern Padre Island. *Occasional Papers, Museum of Texas Tech University* 209 1–23.

Nelson, A. D., J. R. Goetze, I. G. Negrete, V. E. French, M. P. Johnson, and L. M. Macke. 2000. Vegetational analysis and floristics of four communities in the Big Ball Hill region of Padre Island National Seashore. *Southwestern Naturalist* 45:431–442.

Nelson, D. M., ed. 1992. *Distribution and Abundance of Fishes and Invertebrates in Gulf of Mexico Estuaries*. Vol. 1, *Data Summaries*. NOAA/NOS Strategic Environmental Assessments Division, ELMR Report No. 10, Silver Spring, MD.

Nelson, E. W., and E. A. Goldman. 1926. Mexico. Pp. 574–596 in V. E. Shelford, ed., *Naturalist's Guide to the Americas*. Baltimore: Williams and Wilkins.

Newstead, D., and K. Vale. 2014. *Protecting Important Shorebird Habitat Using Piping Plovers as an Indicator Species*. Final report to the US Fish and Wildlife Service, Agreement No. 1448-20181-A-J839, Corpus Christi.

Nichols, J. L., and G. A. Baldassare. 1990. Habitat selection and interspecific associations of Piping Plovers wintering in the United States. *Wilson Bulletin* 102:581–590.

Nichols, M. R., J. Diaz, and L. C. Shaffner. 1990. Effects of hopper dredging and sediment dispersion, Chesapeake Bay. *Environmental Geology and Water Science* 15:31–43.

Nickerson, N. H., and F. R. Thibodeau. 1985. Association between pore water sulfide concentrations and distributions of mangroves. *Biogeochemistry* 1:183–192.

Nixon, J. 1986. *Stewards of a Vision*. Hong Kong: Everbest Printing.

NMFS and USFWS (National Marine Fisheries Service and US Fish and Wildlife Service). 1991. *Recovery Plan for U.S. Population of Atlantic Green Turtle*. Washington, DC: National Marine Fisheries Service.

NMFS and USFWS. 1992. *Recovery Plan for Leatherback Turtles in the U.S. Caribbean, Atlantic and Gulf of Mexico*. Washington, DC: National Marine Fisheries Service.

NMFS and USFWS. 1993. *Recovery Plan for Hawksbill Turtles in the U.S. Caribbean Sea, Atlantic Ocean, and Gulf of Mexico*. St. Petersburg, FL: National Marine Fisheries Service.

NMFS and USFWS. 2007. *Green Sea Turtle (*Chelonia mydas*) 5-Year Review: Summary and Evaluation*. Silver Spring, MD: National Marine Fisheries Service.

NMFS and USFWS. 2008. *Recovery Plan for the Northwest Atlantic Population of the Loggerhead Sea Turtle (*Caretta caretta*), Second Revision.* Silver Spring, MD: National Marine Fisheries Service.

NMFS and USFWS. 2015. *Kemp's Ridley Sea Turtle (*Lepidochelys kempii*) 5-Year Review: Summary and Evaluation.* Silver Spring, MD: National Marine Fisheries Service.

NMFS, USFWS, and SEMARNAT (Secretaría de Medio Ambiente y Recursos Naturales). 2011. *Bi-national Recovery Plan for the Kemp's Ridley Sea Turtle (*Lepidochelys kempii*), Second Revision.* Silver Spring, MD: National Marine Fisheries Service.

NOAA (National Oceanic and Atmospheric Administration). 1994. *Tidal Characteristics and Datums of Laguna Madre, Texas.* Silver Spring, MD: Office of Ocean and Earth Sciences, Ocean and Lake Levels Division.

Noel, B. L., and C. R. Chandler. 2008. Spatial distribution and site fidelity of non-breeding Piping Plovers on the Georgia coast. *Waterbirds* 31:241–251.

Nolet, B. A., V. N. Fuld, and M. E. C. Van Rijswijk. 2006. Foraging costs and accessibility as determinants of giving-up densities in a swan-pondweed system. *Oikos* 112:353–362.

Norris, R. M. 1953. Buried oyster reefs in some Texas Bays. *Journal of Paleontology* 27:569–576.

Norwine, J., and R. Bingham. 1986. Frequency and severity of droughts in South Texas: 1900–1983. Pp. 1–17 in R. D. Brown, ed., *Livestock and Wildlife Management during Drought.* Kingsville, TX: Caesar Kleberg Wildlife Research Institute.

Norwine, J., R. Bingham, and R. V. Zepeda. 1977. Twentieth-century semi-arid and subhumid climates of Texas and northeastern Mexico. Pp. 30–41 in J. Norwine, ed., *Climate and Human Ecology.* Houston: D. Armstrong.

Norwine, J., R. Hariss, J. Yu, C. Tebaldi, and R. Binhhan. 2007. The problematic climate of South Texas 1900–2100. Pp. 15–55 in J. Norwine and K. John, eds., *The Changing Climate of South Texas 1900–2100: Problems, Impacts, Implications.* Kingsville: CREST-RESSACA [Center for Research Excellence in Science and Technology–Research on Environmental Sustainability in Semi-Arid Coastal Areas] and Texas A&M University–Kingsville.

Norwine, J., and K. John, eds. 2007. *The Changing Climate of South Texas 1900–2100: Problems and Prospects, Impacts and Implications.* Kingsville: CREST-RESSACA, Texas A&M University–Kingsville.

NPS (National Park Service). 2005. *Padre Island: An Administrative History.* http://npshistory.com/publications/pais/adhi/contents.htm.

Oberholser, H. C. 1974. *The Bird Life of Texas.* 2 vols. Austin: University of Texas Press.

O'Conner, R. J., and R. A. Brown. 1977. Prey depletion and foraging strategy in the oystercatcher *Haematopus ostralegus. Oecologia* 27:75–92.

Odum, E. D. 1971. *Fundamentals of Ecology.* Philadelphia: W. B. Saunders.

Odum, H. T., and R. F. Wilson. 1962. Further studies on reaeration and metabolism of Texas bays, 1958–1960. *Publications of the Institute of Marine Science, University of Texas* 8:23–55.

Ogren, L. H. 1989. Distribution of juvenile and sub-adult Kemp's Ridley turtles: Preliminary results from the 1984–1987 surveys. Pp. 116–123 in C. W. Caillouet Jr. and A. M. Landry, eds., *Proceedings of the First International Symposium on Kemp's Ridley Sea Turtle Biology, Conservation, and Management.* Texas A&M University Sea Grant College Program TAMU-SG-89-105, College Station.

Ohlendorf, S. M., J. M. Bigelow, and M. M. Standifer. 1980. *Journey to Mexico during the Years 1826 to 1834.* Vol. 2. Austin: Texas State Historical Association in Cooperation with the Center for Studies in Texas History, University of Texas at Austin.

Olney, J. E., and G. W. Boehlert. 1988. Nearshore ichthyoplankton associated with seagrass beds in the lower Chesapeake Bay. *Marine Ecology Progress Series* 45:33–43.

Olsen, Z. T. 2014. Potential impacts of extreme salinity and surface temperature events on population dynamics of Black Drum, *Pogonias cromis,* in the Upper Laguna Madre. *Gulf of Mexico Science* 32:60–68.

Olsen, Z. T. 2016. Emaciated Black Drum (*Pogonias cromis*) in the Upper Laguna Madre, Texas: Tracking the recovery of the population over two years. *Texas Journal of Science* 68:79–90.

Olsen, Z. T. 2017. Emaciated Black Drum (*Pogonias cromis*) in the Upper Laguna Madre, Texas: Tracking the recovery of the population over two years. *Texas Journal of Science* 68:79–90.

Olsen, Z. T., F. P. Grubbs, M. D. Morris, and J. M. Tolan. 2014. Reports of emaciated Black Drum (*Pogonias cromis*) in the Upper Laguna Madre, Texas. *Texas Journal of Science* 66:1–7.

Onuf, C. P. 1994. Seagrasses, dredging and light in Laguna Madre, Texas, USA. *Estuarine, Coastal and Shelf Science* 39:75–91.

Onuf, C. P. 1995. Seagrass meadows of the Laguna Madre of Texas. Pp. 275–277 in E. T. LaRoe, G. S. Farris, C. E. Puckett, P. D. Doran, and M. J. Mac, eds., *Our Living Resources: A Report to the Nation on the Distribution, Abundance, and Health of US Plants, Animals, and Ecosystems*. Washington, DC: US Department of the Interior National Biological Service.

Onuf, C. P. 1996a. Biomass patterns in seagrass meadows of the Laguna Madre, Texas. *Bulletin of Marine Science* 58:404–420.

Onuf, C. P. 1996b. Seagrass responses to long-term light reduction by brown tide in Upper Laguna Madre, Texas: Distribution and biomass patterns. *Marine Ecology Progress Series* 138:219–231.

Onuf, C. P. 1999. Seagrass response to and recovery (?) from seven years of brown tide. *Pacific Conservation Biology* 5:36–313.

Onuf, C. P. 2006. Aspects of the biology of *Salicornia bigelovii* Torr. in relation to a proposed restoration of a wind-tidal flat system on the South Texas, USA coast. *Wetlands* 26:649–658.

Onuf, C. P. 2007. Laguna Madre. Pp. 29–40 in L. Handley, D. Altsman, and R. DeMay, eds., *Seagrass Status and Trends in the Northern Gulf of Mexico: 1940–2002*. US Geological Survey Scientific Investigations Report 2006–5287 and US Environmental Protection Agency 855-R-04-003, Washington, DC.

Oppenheimer, C. H., and E. J. F. Wood. 1965. Quantitative aspects of the unicellular algal populations of the Texas bay systems. *Bulletin of Marine Science* 15:571–588.

Orlando, S. P., Jr., L. P. Rozas, G. H. Ward, and C. J. Klein. 1991. *Analysis of Salinity Structure and Stability for Texas Estuaries*. Rockville, MD: Strategic Assessment Branch, NOS/NOAA.

Orr, M., M. Zimmer, D. E. Jelinski, and M. Mews. 2005. Wrack deposition on different beach types: Spatial and temporal variation in the pattern of subsidy. *Ecology* 86:1496–1507.

Orth, R. J., and K. A. Moore. 1983. Chesapeake Bay: An unprecedented decline in submerged aquatic vegetation. *Science* 222:51–53.

Ortiz, A. R. 1976. The effect of human activity on the insect fauna of Padre Island. MS thesis Texas A&I University, Kingsville.

Osland, M. J., N. Enwright, R. H. Day, and T. W. Doyle. 2013. Winter climate change and coastal wetland foundation species: Salt marshes vs. mangrove forests in the southeastern United States. *Global Change Biology* 19:1482–1494.

Owen, D. E. 2008. Geology of the Chenier Plain of Cameron Parish, southwestern Louisiana. Pp. 27–38 in G. Moore, ed., *Geological Society of America Field Guide 14*, 2008 Joint Annual Meeting, Houston, TX, 5–9 October 2008.

Page, G. W., L. E. Stenzel, and C. A. Ribic. 1985. Nest site selection and clutch predation in the Snowy Plover. *Auk* 102:347–353.

Palmer, R. S. 1962. *Handbook of North American Birds*. Vol. 1. New Haven, CT: Yale University Press.

Palmer, T. A., P. A. Montagna, and R. D. Kalke. 2008. Benthic indicators of the initial effect of opening a channel. *Environmental Bioindicators* 3:205–206.

Palmer, T. A., P. A. Montagna, and R. D. Kalke. 2013. The effects of opening an artificial tidal inlet on hydrography and estuarine macrofauna in Corpus Christi, Texas. *Environmental Monitoring and Assessment* 185:5917–5935.

Parker, P. L., R. S. Scalan, B. A. Trust, and M. C. Woodin. 1992. *Stable Isotopic Composition of Redhead Ducks in South Texas and South Dakota*. Final report, US Fish Wildlife Service, Coop. Agreement No. 14-16-0009-89-917, National Wetlands Research Center Task-Order No. 2.

Parker, R. H. 1959. Macro-invertebrate assemblages of Central Texas coastal bays and Laguna Madre. *Bulletin of the American Association of Petroleum Geologists* 43:2100–2166.

Parkinson, R. W., R. D. DeLaune, and J. R. White. 1994. Holocene sea-level rise and the fate of mangrove forests within the wider Caribbean region. *Journal of Coastal Research* 10:1077–1086.

Patillo, M. E., T. E. Czapla, D. M. Nelson, and M. E. Monaco. 1997. *Distribution and Abundance of Fishes and Invertebrates in Gulf of Mexico Estuaries.* Vol. 2, *Species Life History Summaries.* ELMR Report No. 11, NOAA/NOS Strategic Environmental Assessments Division, Silver Spring, MD.

Patterson, C. S., I. A. Mendelssohn, and E. M. Swanson. 1993. Growth and survival of *Avicennia germinans* seedlings in a mangal/salt marsh community in Louisiana, U.S.A. *Journal of Coastal Research* 9:801–810.

Pearson, J. C. 1929. Natural history and conservation of the Redfish and other commercial sciaenids on the Texas coast. *Bulletin of the US Bureau of Fisheries* 54:129–214.

Pearson, T. G. 1921. Notes on the bird-life of southeastern Texas. *Auk* 38:513–523.

Pemberton, J. R. 1922a. A large tern colony in Texas. *Condor* 24:37–48.

Pemberton, J. R. 1922b. The Reddish Egrets of Cameron County, Texas. *Condor* 24:3–12.

Perales Flores, L. E., and A. J. Contreras-Balderas. 1986. Aves acuaticas y semiacuaticas de la Laguna Madre, Tamaulipas, México. *Ciencia* (Mexico) 3:39–46.

Pérez-Arteaga, A., K. J. Gaston, and M. Kershaw. 2002. Undesignated sites in Mexico qualifying as wetlands of international importance. *Biological Conservation* 107:47–57.

Pérez-Castañeda, R., Z. Blanco-Martínez, J. G. Sánchez-Martínez, J. L. Rábago-Castro, G. Aguirrre-Guzmán, and M. de la L. Vázquez-Sauceda. 2010. Distribution of *Farfantepenaeus aztecus* and *F. duorarum* on submerged aquatic vegetation habitats along a subtropical coastal lagoon (Laguna Madre, Mexico). *Journal of the Marine Biological Association of the United Kingdom* 90:445–452.

Pérez-Castañeda, R., and O. Defeo. 2005. Growth and mortality of transient shrimp populations (*Farfantepenaeus* spp.) in a coastal lagoon of Mexico: Role of the environment and density-dependence. *ICES Journal of Marine Science* 62:14–24.

Pérez-Castañeda, R., M. de la L. Vázquez-Sauceda, J. G. Sánchez-Martínez, and J. A. González-Salcaña. 2012. Estructura poblacional de camarones peneidos explotados en dos zonas pesqueras de la Laguna Madre, Tamaulipas. *Ciencia y Mar* 26:25–30.

Perry, C. L., and I. A. Mendelssohn. 2009. Ecosystem effects of expanding populations of *Avicennia germinans* in a Louisiana salt marsh. *Wetlands* 29:396–406.

Peterson, B., and B. Fry. 1987. Stable isotopes in ecosystem studies. *Annual Review of Ecology and Systematics* 18:293–320.

Peterson, J. M., and S. S. Bell. 2015. Saltmarsh boundary modulates dispersal of mangrove propagules: Implications for mangrove migration with sea-level rise. *PLoS ONE* 10 (3): e0119128.

Pfeffer, W. T., J. T. Harper, and S. O'Neel. 2008. Kinematic constraints on glacier contributions to 21st-century sea-level rise. *Science* 321:1340–1343.

Pfeiffer W. J., and R. G. Wiegert. 1981. Grazers on *Spartina* and their predators. Pp. 87–112 in L. R. Pomeroy and R. G. Wiegert, eds., *The Ecology of a Saltmarsh.* New York: Springer-Verlag.

Phillips, D. L., R. Inger, S. Bearhop, A. L. Jackson, J. W. Moore, A. C. Parnell, B. X. Semmens, and E. J. Ward. 2014. Best practices for use of stable isotope mixing models in food-web studies. *Canadian Journal of Zoology* 92:823–835.

Phillips, J. C. 1911. A year's collecting in the State of Tamaulipas, Mexico. *Auk* 28:67–89.

Phillips, R. C. 1960. Observations on the ecology and distribution of the Florida sea grasses. *Professional Papers Series, Florida Board of Conservation* 2:1–72.

Phillips, R. C. 1974. Temperate grass flats. Pp. 244–299 in H. T. Odum, B. J. Copeland, and E. A. McMahan, eds., *Coastal Ecological Systems of the United States.* Washington, DC: Conservation Foundation.

Phillips, R. C., and E. G. Meñez. 1988. *Seagrasses.* Smithsonian Contribution in Marine Science No. 34. Washington, DC: Smithsonian Institution Press.

Phlips, E. J., S. Badylak, M. A. Lasi, R. Chamberlain, W. C. Green, L. M. Hall, J. A. Hart, et al. 2015. From Red Tides to green and brown tides: Bloom dynamics in a restricted subtropical lagoon under shifting climatic conditions. *Estuaries and Coasts* 38:886–904.

Pickens, C., and M. Hester. 2011. Temperature tolerance of early life history stages of Black Mangrove, *Avicennia germinans*: Implications for range expansion. *Estuaries and Coasts* 34:824–830.

PINS (Padre Island National Seashore). 1984. *Checklist of Reptiles and Amphibians: Padre Island National Seashore*. National Park Service Information Bulletin No. 12, Corpus Christi.

PINS. 2018. *Amphibians*. Park booklet, National Park Service, Corpus Christi.

Pitelka, F. A., J. P. Myers, and P. G. Connors. 1980. Spatial and resource-use patterns in wintering shorebirds: The Sanderling in central coastal California. Pp. 1021–1044 in R. Nohring, ed., *Acta XVII Congressus Internationalis Ornithologici*, vol. 2. Berlin: Verlag der Deutschen Ornithologen-Gesellschaft.

Plotkin, P. T., and J. Bernardo. 2014. Sea turtle funding dries up. *Science* 343:484.

Plotkin, P. T., M. K. Wicksten, and A. F. Amos. 1993. Feeding ecology of the Loggerhead Sea Turtle *Caretta caretta* in the northwestern Gulf of Mexico. *Marine Biology* 115:1–15.

Poiani, K. A., and B. D. Richter. 1999. *Functional Landscapes and the Conservation of Biodiversity*. Working Papers in Conservation Science No. 1. Arlington, VA: The Nature Conservancy.

Pollack, J. B., J. W. Kinsey, and P. A. Montagna. 2009. Freshwater inflow biotic index (FIBI) for the Lavaca-Colorado estuary. *Texas Environmental Bioindicators* 4:153–169.

Ponwith, B., and Q. R. Dokken. 1996. Fisheries resources. Pp. 484–543 in J. W. Tunnell Jr., Q. R. Dokken, E. H. Smith, and K. Withers, eds., *Current Status and Historical Trends of the Estuarine Living Resources within the Corpus Christi Bay National Estuary Program Study Area*. Corpus Christi Bay National Estuary Program, CCBNEP-06A. Corpus Christi: Texas Natural Resource Conservation Commission.

Poole, J. M., W. R. Carr, D. M. Price, and J. R. Singhurst. 2007. *Rare Plants of Texas: A Field Guide*. College Station: Texas A&M University Press.

Powell, E. N., R. J. Stanton Jr., H. Cummins, and G. Staff. 1982. Temporal fluctuations in bay environments—the death assemblage as a key to the past. Pp. 203–232 in J. R. Davis, ed., *Proceedings of the Symposium on Recent Benthological Investigations in Texas and Adjacent States*. Austin: Texas Academy of Science.

Powell, E., M. C. Tyrrell, A. Milliken, J. M. Tirpak, and M. D. Staudinger. 2017. A synthesis of thresholds for focal species along the U.S. Atlantic and Gulf Coasts: A review of research and applications. *Ocean & Coastal Management* 148:75–88.

Presley, A. L. 2003. Antiquity and paleoenvironment of the Tamaulipan Biotic Province of southern Texas: The zooarchaeological perspective. MA thesis, Texas A&M University, College Station.

Price, W. A. 1933. Role of diastrophism in topography of Corpus Christi area, South Texas. *American Association of Petroleum Geologists Bulletin* 17:907–962.

Price, W. A. 1958. Sedimentology and quaternary geomorphology of South Texas. *Transactions of the Gulf Coast Association of Geological Societies* 8:41–75.

Price, W. A. 1968. *Abatement of Blowing Salt Conditions Inland from Laguna Madre, Texas*. Report to Coastal Bend Regional Planning Commission, Corpus Christi.

Price, W. A., and G. Gunter. 1942. Certain recent and geological and biological changes in South Texas, with consideration of probable causes. *Texas Academy of Science Proceedings and Transactions* 26:138–156.

Price, W. A., and L. S. Kornicker. 1961. Marine and lagoonal deposits in clay dunes, Gulf Coast, Texas. *Journal of Sedimentary Petrology* 31:245–255.

Proffitt, C. E., and D. J. Devlin. 2005a. Grazing by the intertidal gastropod *Melampus coffeus* greatly increases mangrove leaf litter degradation rates. *Marine Ecology Progress Series* 296:209–218.

Proffitt, C. E., and D. J. Devlin. 2005b. Long-term growth and succession in restored and natural mangrove forests in southwestern Florida. *Wetlands Ecology and Management* 13:531–5[illegible].

Proffitt, C. E., and S. Travis. 2014. Red Mangrove life history variables along latitudinal and anthropogenic stress gradients. *Ecology and Evolution* 4:2352–2359.

Prouty, J. S. 1996. Late Pleistocene–early Holocene karst features, Laguna Madre, South Texas: A record of climate change. *Transactions of the Gulf Coast Association of Geological Societies* 46:345–351.

Prouty, J. S., and D. W. Lovejoy. 1992. Remarkable cylindrical solution pipes in coquina south of Baffin Bay, Texas. *Transactions of the Gulf Coast Association Geological Societies* 42:599–606.

Prouty, J. S., and D. B. Prouty. 1989. Historical back barrier shoreline changes, Padre Island National Seashore, Texas. *Transactions of the Gulf Coast Association Geological Societies* 39:481–490.

Pulich W., Jr. 1980. Ecology of a hypersaline lagoon: The Laguna Madre. Pp. 103–122 in P. L. Fore and R. D. Peterson, eds., *Proceedings of the Gulf of Mexico Coastal Ecosystem Workshop*. FWS/OBS 80/30 Albuquerque, NM: US Fish and Wildlife Service.

Pulich, W., Jr. 1985. Seasonal growth dynamics of *Ruppia maritima* L. s.l. and *Halodule wrightii* Aschers. in southern Texas and evaluation of sediment fertility status. *Aquatic Botany* 23:53–66.

Pulich, W., Jr. 1991. *Effects of Freshwater Inflows on Estuarine Vegetation of San Antonio Bay System*. Special project report, Texas Parks and Wildlife Department, Resource Protection Division, Austin.

Pulich, W., Jr. 1998. *Seagrass Conservation Plan for Texas*. Austin: Texas Parks and Wildlife Department, Resource Protection Division.

Pulich, W., Jr. 2001. *Seagrass Inventory of Christmas and Drum Bays: Webster, Tex*. Project report submitted in support of US Environmental Protection Agency grant to Galveston Bay National Estuary Program.

Pulich, W., Jr., C. Blair, and W. A. White. 1997. *Current Status and Historical Trends of Seagrasses in the Corpus Christi Bay National Estuary Program Study Area*. Corpus Christi Bay National Estuary Program CCBNEP-20. Corpus Christi: Texas National Resource Conservation Commission.

Pulich, W. M., Jr., and T. Calnan, eds. 1999. *Seagrass Conservation Plan for Texas*. Austin: Texas Parks and Wildlife Department, Resource Protection Division.

Pulich, W., Jr., and J. Hinson. 1996. *Development of Geographic Information System Data Sets on Coastal Wetlands and Land Cover*. Coastal Studies Technical Report No. 1. Austin: Texas Parks and Wildlife Department, Resource Protection Division.

Pulich, W., Jr., and C. Onuf. 2007. Statewide summary for Texas. Pp. 7–17 in L. Handley, D. Altsman, and R. De May, eds., *Seagrass Status and Trends in the Northern Gulf of Mexico: 1940–2002*. US Geological Survey Scientific Investigations Report 2006-1587 and US Environmental Protection Agency 855-R-04-003, Washington, DC.

Pulich, W., Jr., and S. Rabalais. 1982. *Primary Production and Nutrient Fluxes on South Texas Tidal Flats*. Report to US Fish and Wildlife Service, Office of Environment, Region 2, Albuquerque, NM.

Pulich, W., Jr., and S. Rabalais. 1986. Primary production potential of blue-green algal mats on southern Texas tidal flats. *Southwestern Naturalist* 31:39–47.

Pulich, W., Jr., S. Rabalais, and S. Wellso. 1982. *Food Chain Components on Laguna Madre Tidal Flats*. University of Texas Marine Science Institute Contribution No. 572, Port Aransas.

Pulich, W., Jr., and R. S. Scalan. 1987. Organic carbon and nitrogen flow from marine cyanobacteria to semiaquatic insect food webs. *Contributions in Marine Science* 30:27–37.

Pulich, W., Jr., and W. White. 1991. Decline of submerged vegetation in the Galveston Bay system: Chronology and relationship to physical processes. *Journal of Coastal Research* 7:1125–1138.

Pullen, E. J. 1960. *A Study of the Marsh and Marine Plants in Upper Galveston and Trinity Bays*. Texas Game and Fish Commission, Division of Marine Fisheries, Progress Report 1960–1961, Austin.

Pye, K., and H. Tsoar. 1990. *Aeolian Sand and Sand Dunes*. Boston: Unwin Hyman.

Quammen, M. L. 1984. Predation by shorebirds, fish and crabs on invertebrates in intertidal mudflats: An experimental test. *Ecology* 65:529–537.

Quammen, M. L., and C. P. Onuf. 1993. Laguna Madre seagrass changes continue decades after salinity reduction. *Estuaries* 16:302–310.

Rabalais, N. N. 1975. *Checklist of Mammals*. Padre Island National Seashore Informational Bulletin 3. Corpus Christi: Padre Island National Seashore.

Rabalais, N. N. 1977. *Gulf Beach National Environmental Study Area*. Report on file at Padre Island National Seashore, Corpus Christi.

Rabalais, N. N. 1982. The ascidians of the Aransas Pass inlet jetties, Port Aransas, Texas. Pp. 65–74 in B. R. Chapman and J. W. Tunnell, eds., *South Texas Fauna: A Symposium Honoring Dr. Allan H. Chaney*. Kingsville, TX: Caesar Kleberg Wildlife Research Institute.

Rabalais, N. N. 1983. Adaptations of Fiddler Crabs, *Uca subcylindrica* (Stimpson, 1859), to semi-arid environments. PhD diss., University of Texas at Austin.

Rabalais, N. N., and J. N. Cameron. 1985a. The effects of factors important in semi-arid environments on the early development of *Uca subcylindrica*. *Biological Bulletin* 168:147–160.

Rabalais, N. N., and J. N. Cameron. 1985b. Physiological and morphological adaptations of adult *Uca subcylindrica* to semi-arid environments. *Biological Bulletin* 168:135–146.

Rabalais, S. C., and R. W. Flint. 1983. IXTOC-I effects on intertidal and subtidal infauna of South Texas Gulf beaches. *Contributions in Marine Science* 26:23–35.

Rabalais, S. C., and N. N. Rabalais. 1980. The occurrence of sea turtles on the South Texas coast. *Contributions in Marine Science* 23:123–129.

Rabinowitz, D. 1978. Dispersal properties of mangrove propagules. *Biotropica* 10:47–57.

Rafalski, A. 2012. Evaluation of brevetoxin accumulation and degradation in coastal mammals, birds, and fish found moribund on Texas beaches during Red Tide blooms. MS thesis, Texas A&M University–Corpus Christi.

Ramirez-Yanez, L. E., J. A. Ortega-S, L. A. Brennan, and G. A. Rasmussen. 2007. Use of prescribed fire and cattle grazing to control Guineagrass. Pp. 240–245 in R. E. Masters and K. E. M. Galley, eds., *Proceedings of the 23rd Tall Timbers Fire Ecology Conference: Fire in Grassland and Shrubland Ecosystems*. Tallahassee, FL: Tall Timbers Research Station.

RAMSAR. 2018. *The List of Wetlands of International Importance*. The Ramsar Convention on Wetlands, Gland, Switzerland. https://www.ramsar.org/sites/default/files/documents/library/sitelist.pdf.

Rappole, J. H. 1993. Personal affidavit, Exhibit 7. In *Environmental Impacts of the Plan to Dump Dredged Spoil from the Gulf Intracoastal Waterway on Lands Bordering the Laguna Madre*. Kingsville, TX: King Ranch and John G. Kenedy Charitable Trust.

Rappole, J. H. 1995. *The Ecology of Migrant Birds: A Neotropical Perspective*. Washington, DC: Smithsonian Institution Press.

Rappole, J. H., and G. W. Blacklock. 1985. *Birds of the Texas Coastal Bend: Abundance and Distribution*. W. L. Moody Natural History Series No. 7. College Station: Texas A&M University Press.

Rappole, J. H., and G. W. Blacklock. 1994. *Birds of Texas*. College Station: Texas A&M University Press.

Rappole, J. H., M. A. Ramos, R. J. Oehlenschlager, D. W. Warner, and C. H. Barkan. 1979. Timing of migration and route selection in North American songbirds. Pp. 199–214 in D. L. Drawe, ed., *Proceedings of the First Welder Wildlife Foundation Symposium*. Sinton, TX: Welder Wildlife Foundation.

Raun, G. G. 1959. An annotated checklist of the mammals of Mustang and Padre Islands. *Texas Ornithological Society Newsletter* 7.

Raymond, W. W., M. A. Albins, and T. J. Pusack. 2015. Competitive interactions for shelter between invasive Pacific Red Lionfish and native Nassau Grouper. *Environmental Biology of Fishes* 98:57–65.

Raz-Guzmán, A., and L. Huidobro. 2002. Fish communities in two environmentally different estuarine systems of Mexico. *Journal of Fish Biology* 61 (Suppl. A): 182–195.

Raz-Guzmán, A., A. J. Sánchez, P. Peralta, and R. Florido. 2004. Zoogeography of hermit crabs (Decapoda: Diogenidae, Paguridae) from four coastal lagoons in the Gulf of Mexico. *Journal of Crustacean Biology* 24:625–636.

Raz-Guzmán, A., and L. A. Soto. 2017. Updated checklist and zoogeographic remarks of benthic amphipods (Crustacea: Peracarida: Amphipoda) of two coastal lagoons in the western Gulf of Mexico. *Revista Mexicana de Biodiversidad* 88:715–734.

Raz-Guzmán, A., and A. L. Villegas. 2018. Gammarid and corophiid amphipods (Crustacea: Peracarida, Amphipoda) of Laguna de Tamiahua, Veracruz and Laguna Madre, Tamaulipas, Mexico: Spatial and temporal distribution. *Gulf of Mexico Science* 34 (1): 2–18.

Reagan, R. E. 1985. *Species Profiles: Life Histories and Environmental Requirements of Coastal Fishes and Invertebrates (Gulf of Mexico)—Red Drum*. US Fish and Wildlife Service Biological Report 82 (11.36), US Army Corps of Engineers Technical Report EL-82-4, Washington, DC.

Rebel, T. P. 1974. *Sea Turtles and the Turtle Industry of the West Indies, Florida and the Gulf of Mexico*. Coral Gables, FL: University of Miami Press.

Recher, H. F. 1966. Some aspects of the ecology of migrant shorebirds. *Ecology* 47:393–407.

Reese, M. M., G. W. Stunz, and S. M. Bushon. 2008. Recruitment of estuarine-dependent nekton through a new tidal inlet: The opening of Packery Channel in Corpus Christi, TX, USA. *Estuaries and Coasts* 31:1143–1157.

Reese, P. 1938. The history of Padre Island. MA thesis, Texas College of Arts and Industries, Kingsville.

Reguero, M., and A. Raz-Guzmán. 2018. Molluscs (Mollusca: Gastropoda, Bivalvia, Polyplacophora) of Laguna Madre, Tamaulipas, Mexico: Spatial and temporal distribution. *Gulf of Mexico Science* 34:32–55.

Reid, G. K., and R. D. Wood. 1976. *Ecology of Inland Waters and Estuaries*. 2nd ed. New York: D. Van Nostrand.

Reid, N., D. M. Stafford Smith, P. Beyer-Munzel, and J. Marroquin. 1990. Floristic and structural variation in the Tamaulipan thornscrub, northeastern Mexico. *Journal of Vegetation Science* 1:529–538.

Renaud, M. L. 1995. Movements and submergence patterns of Kemp's Ridley turtles (*Lepidochelys kempii*). *Journal of Herpetology* 29:370–374.

Renaud, M. L., J. A. Carpenter, S. A. Manzella, and J. A. Williams. 1993. *Telemetric Tracking Green Sea Turtles (*Chelonia mydas*) in Relation to Dredged Channels at South Padre Island, Texas, July through September 1992*. Report submitted to US Army Corps of Engineers (Galveston and New Orleans Districts), Galveston.

Renaud, M. L., J. A. Carpenter, and J. A. Williams. 1995. Activities of juvenile Green Turtles, *Chelonia mydas*, at a jettied pass in South Texas. *Fishery Bulletin* 93:586–593.

Renaud, M. L., J. A. Carpenter, J. A. Williams, and A. M. Landry Jr. 1996. Kemp's Ridley Sea Turtle (*Lepidochelys kempii*) tracked by satellite telemetry from Louisiana to nesting beach at Rancho Nuevo, Tamaulipas, Mexico. *Chelonian Conservation and Biology* 2:108–109.

Renaud, M. L., J. A. Carpenter, J. A. Williams, and S. A. Manzella-Tirpak. 1995. Activities of juvenile Green Turtles, *Chelonia mydas*, at a jettied pass in South Texas. *Fishery Bulletin* 93:586–593.

Renaud, M., G. Gitschlag, E. Klima, S. Manzella, and J. Williams. 1992. *Tracking of Green (*Chelonia mydas*) and Loggerhead (*Caretta caretta*) Sea Turtles Using Radio and Sonic Telemetry at South Padre Island, Texas*. Report to Southeast Fisheries Science Center, National Marine Fisheries Service, Galveston.

Renaud, M. L., J. M. Nance, E. Scott-Denton, and G. R. Gitschlag. 1997. Incidental capture of sea turtles in shrimp trawls with and without TEDs in U.S. Atlantic and Gulf waters. *Chelonian Conservation and Biology* 2:425–427.

Rhoads, D. C. 1974. Organism-sediment relations in the muddy seafloor. *Oceanography and Marine Biology Annual Review* 12:263–300.

Rhoads, D. C., and D. K. Young. 1970. The influence of deposit-feeding organisms on sediment stability and community trophic structure. *Journal of Marine Research* 28:150–178.

Rhudy, K. B., V. K. Sharma, R. L. Lehman, and D. A. McKee. 1999. Seasonal variability of the Texas "Brown Tide" (*Aureoumbra lagunensis*) in relation to environmental parameters. *Estuarine, Coastal and Shelf Science* 48:565–574.

Richard, J. 2016. Mangrove musings. *Texas Saltwater Fishing Magazine*, December. www.texassaltwaterfishingmagazine.com/fishing/features/joe-richard/mangrove-musings.

Ricklis, R. A. 1995. *Environmental and Human Adaptive Change on the Nueces Bay Shoreline: Phase I: Archaeological Data Recovery at Koch Refining Company Middle Plant, Nueces County, Texas*. Coastal Archaeological Research Inc. Report to Koch Refining Company, Corpus Christi.

Ricklis, R. A. 1996. *The Karankawa Indians of Texas: An Ecological Study of Cultural Tradition and Change*. Austin: University of Texas Press.

Ricklis, R. A., and M. D. Blum. 1997. The geoarchaelogical record of Holocene sea level change and human occupation of the Texas Gulf Coast. *Geoarchaeology* 12:287–314.

Rickner, J. A. 1979. The influence of dredged material islands in Upper Laguna Madre, Texas on selected seagrasses and macro-benthos. PhD diss., Texas A&M University, College Station.

Roberts, K., J. Collins, C. H. Paxton, R. Hardy, and J. Downs. 2014. Weather patterns associated with Green Turtle hypothermic stunning events in St. Joseph Bay and Mosquito Lagoon, Florida. *Physical Geography* 35:134–150.

Robertson, B. 1985. *Wild Horse Desert: The Heritage of South Texas*. Edinburg, TX: New Santander Press.

Robillard, M., and G. Stunz. 2012. Trout have tails . . . and know how to use them. *Texas Saltwater Fishing Magazine*, October. https://www.texassaltwaterfishingmagazine.com/fishing/by-type/inshorenearshore-fishing/trout-tails-them.

Rocha, D. D., R. L. Lehman, S. A. Alvarado, J. W. Tunnell Jr., and P. A. Montagna. 1995. *A Comparison of the Benthic Community North and South of the John F. Kennedy Causeway, Upper Laguna Madre, Texas*. Technical Report TAMU-CC-9504-CCS. Corpus Christi: Texas A&M University–Corpus Christi, Center for Coastal Studies.

Rodríguez, W., I. C. Feller, and C. Cavanaugh. 2016. Spatio-temporal changes of a mangrove–saltmarsh ecotone in the northeastern coast of Florida, USA. *Global Ecology and Conservation* 7:245–261.

Rodríguez-Almaraz, G. A., and M. del S. García-Madrigal. 2014. Crustáceos exóticos invasores. Pp. 347–371 in R. Mendoza and P. Koleff, eds., *Especies acuáticas invasoras en México*. Mexico City: Comisión Nacional para el Conocimiento y Uso de la Biodiversidad (CONABIO).

Rodríguez-Almaraz, G. A., and V. M. Ortega-Vidales. 2013. Primer registro de *Caprella scaura* y *Caprella penantis* (Crustacea: Peracarida: Amphipoda) en la Laguna Madre, Tamaulipas, México. *Revista Mexicana de Biodiversidad* 84:989–993.

Rogers, J. A. 1987. On the antipredator advantages of coloniality: A word of caution. *Wilson Bulletin* 99:269–270.

Rogers, J. P., and L. J. Korschgen. 1966. Foods of Lesser Scaups on breeding, migration, and wintering areas. *Journal of Wildlife Management* 30:258–264.

Romero-Lankao, P., J. B. Smith, D. J. Davidson, N. S. Diffenbaugh, P. L. Kinney, P. Kirshen, P. Kovacs, and L. V. Fuíz. 2014. North America. Pp. 1439–1498 in V. R. Barros, C. B. Field, D. J. Dokken, M. D. Mastrandrea, and K. J. Mach, eds., *Climate Change 2014: North America. Impacts, Adaptation, and Vulnerability. Part B: Regional Aspects*. Contribution of Working Group II to the Fifth Assessment Report of the Intergovernmental Panel on Climate Change. Cambridge: Cambridge University Press.

Rooker, J. R., G. J. Holt, and S. A. Holt. 1998. Vulnerability of newly settled Red Drum (*Sciaenops ocellatus*) to predatory fish: Is early-life survival enhanced by seagrass meadows? *Marine Biology* 131:145–151.

Ropicki, A., D. Hanselka, R. Dudensing, and T. Resinger. 2016. *The Economic Impacts of Recreational Fishing in the Lower Laguna Madre*. Texas A&M University Sea Grant Report, TAMU-SG-16-512, College Station.

Rose, F. L., and F. W. Judd. 2014. *The Texas Tortoise: A Natural History*. Norman: University of Oklahoma Press.

Rostal, D. C., J. S. Grumbles, R. A. Byles, R. Marquez-M, and D. W. Owens. 1997. Nesting physiology of Kemp's Ridley Sea Turtles, *Lepidochelys kempii*, at Rancho Nuevo, Tamaulipas, Mexico, with observations on population estimates. *Chelonian Conservation and Biology* 2:538–547.

Rowe, J. 1953. History of the King Ranch. Pp. 15–24 in *King Ranch: 100 Years of Ranching*. Corpus Christ: Corpus Christi Caller-Times.

Rubio, K. S., M. Ajemian, G. W. Stunz, T. A. Palmer, B. Lebreton, and J. B. Pollack. 2018. Dietary composition of Black Drum (*Pogonias cromis*) in a hypersaline estuary reflects water quality and prey availability. *Journal of Fish Biology* 93 (2): 250–262. DOI: 10.1111/jfb.13654.

Rupert, J. R. 1997. The brood-rearing habitat, brood home range, and fecundity of the Snowy Plover (*Charadrius alexandrinus*) in coastal southern Texas. MS thesis, University of Texas–Pan American, Edinburg.

Rusnak, G. A. 1958. Laguna Madre. Pp. 71–72 in *Sedimentology of South Texas: Field Trip Guidebook,* Gulf Coast Association of Geological Societies annual meeting, October 27–November 1, 1958, Corpus Christi.

Rusnak, G. A. 1960. Sediments of the Laguna Madre. Pp. 153–196 in F. P. Shepard, F. B. Phelger, and T. J. van Andel, eds., *Recent Sediments, Northwestern Gulf of Mexico*. Tulsa, OK: American Association of Petroleum Geologists.

Rzedowski, J. 1978. *Vegetación de México*. Mexico City: Limusa.

Rzedowski, J. 2006. *Vegetación de México*. 1ra. Edición digital. Mexico City: Comisión Nacional para el Conocimiento y Uso de la Biodiversidad (CONABIO). https://www.biodiversidad.gob.mx/publicaciones/librosDig/pdf/VegetacionMx_Cont.pdf.

Safina, C., and J. Burger. 1983. Effects of human disturbance on reproductive success in the Black Skimmer. *Condor* 85:164–171.

Saintilan N., N. C. Wilson, K. Rogers, A. Rajkaran, and K. W. Krauss. 2014. Mangrove expansion and salt marsh decline at mangrove poleward limits. *Global Change Biology* 20:147–157.

Salinas, M. 1990. *Indians of the Rio Grande Delta: Their Role in the History of Southern Texas and Northeastern Mexico*. Austin: University of Texas Press.

Salinas, S. 2013. Lift, drift, troll and pole: Seagrass protection helps sustain coastal resources. *Texas Parks and Wildlife Magazine*, November. https://tpwmagazine.com/archive/2013/nov/scout1_seagrass.

Sanchez, L. 1978. *Variables superficiales fisico-quimicia con prospección acuacultura en la Laguna Madre de Tamaulipas*. Centro de Acuacultura (in-house report). Valle Hermosa, Tamaulipas.

Sánchez-Ramírez, M., and A. Ocaña-Luna. 2015. Estructura y variación estacional de la comunidad ictioplanctónica en una laguna hipersalina del oeste del Golfo de México: Laguna Madre, Tamaulipas. *Hidrobiológica* 25:175–186.

Sanders, H. L. 1968. Marine benthic diversity: A comparative study. *American Naturalist* 102:243–282.

Sanger, P. 2002. *Mangrove Ecology, Silviculture and Conservation*. Dordrecht, Netherlands: Kluwer Academic Publishers.

Santander-Monsalvo, J., I. López-Huerta, A. Aguilar-Perera, and A. Tuz-Sulub. 2012. First record of the Red Lionfish (*Pterois volitans* [Linnaeus, 1758]) off the coast of Veracruz, Mexico. *Bio-Invasions Records* 1:121–124.

Saunders, G. B., and D. C. Saunders. 1981. *Waterfowl and Their Wintering Grounds in Mexico, 1937–64*. US Fish and Wildlife Service Resource Publication 138, Washington, DC.

Scharf, F. S. 2000. Patterns in abundance, growth, and mortality of juvenile Red Drum across estuaries on the Texas coast with implications for recruitment and stock enhancement. *Transactions of the American Fisheries Society* 296:1207–1222.

Scheffel, W. A., K. L. Heck Jr., and M. W. Johnson. 2018. Tropicalization of the northern Gulf of Mexico: Impacts of salt marsh transition to Black Mangrove dominance on faunal communities. *Estuaries and Coasts* 41:1193–1205.

Scheffel, W. A., L. H. Kenneth Jr., J. Cebrian, M. Johnson, and D. Byron. 2013. Range expansion of Black Mangroves (*Avicennia germinans*) to the Mississippi barrier islands. *Gulf of Mexico Science* 31:79–82.

Schmidly, D. J. 1981. *Marine Mammals of the Southeastern United States Coast and the Gulf of Mexico*. US Fish and Wildlife Service, Office of Biological Service FWS/OBS-80/41, Washington, DC.

Schmidly, D. J., and R. D. Bradley. 2016. *The Mammals of Texas*. Austin: University of Texas Press.

Schmidly, D. J., and S. H. Shane. 1978. *A Biological Assessment of the Cetacean Fauna of the Texas Coast*. Texas A&M University and Texas Agricultural Extension Service Report No. MMC074/05, College Station.

Schmidt-Nielsen, K., and Y. T. Kim. 1964. The effect of salt intake on the size and function of the salt gland in ducks. *Auk* 81:160–172.

Schneider, D. 1973. Equalization of prey numbers by migratory shorebirds. *Nature* 271:353–354.

Schneider, D. C., and B. A. Harrington. 1981. Timing of shorebird migration in relation to prey depletion. *Auk* 98:801–811.

Schoenbaechler, C., and C. G. Guthrie. 2011a. *Coastal Hydrology for the Laguna Madre Estuary, with Emphasis on the Lower Laguna Madre*. Austin: Texas Water Development Board. http://www.twdb.texas.gov/surfacewater/bays/major_estuaries/laguna_madre/doc/TWDB_Hydrology_LowerLagunaM_20110921.pdf?d=5874.365.

Schoenbaechler, C., and C. G. Guthrie. 2011b. *Coastal Hydrology for the Laguna Madre Estuary, with Emphasis on the Upper Laguna Madre*. Austin: Texas Water Development Board. http://www.twdb.texas.gov/surfacewater/bays/major_estuaries/laguna_madre/doc/TWDB_Hydrology_UpperLagunaM_20110921.pdf?d=3910.15.

Scott, A. J., R. A. Hoover, and J. H. McGowen. 1969. Effects of Hurricane "Beulah," 1967 on Texas coastal lagoons and barriers. Pp. 221–236 in A. A. Castañares and F. B. Phleger, eds., *Coastal Lagoons: A Symposium*. Mexico City: Universidad Nacional Autónoma de México.

Scott, F. J. 1966. *Historical Heritage of the Lower Rio Grande*. Waco: Texian Press.

Seager, R., M. Ting, I. Held, Y. Kushnir, J. Lu, G. Vecchi, H.-P. Huang, et al. 2007. Model projections of an imminent transition to a more arid climate in southwestern North America. *Science* 316:1181–1184.

SEDAR (Southeast Data, Assessment, and Review). 2009. *SEDAR 18 Stock Assessment Report, Atlantic Red Drum*. The South Atlantic Fishery Management Council, North Charleston, SC. https://sedarweb.org/docs/sar/SEDAR%2018-%20Red%20drum%20SAR%20Final.pdf

Segers, J. C., and B. R. Chapman. 1984. Ecology of the Spotted Ground Squirrel, *Spermophilus spilosoma* (Merriam), on Padre Island, Texas. *Special Publications, Museum of Texas Tech University* 22:105–112.

Selander, R. K., R. F. Johnston, B. J. Wilks, and G. G. Raun. 1962. Vertebrates from the barrier island of Tamaulipas, México. *University of Kansas Museum of Natural History Publications* 12:309–345.

SEMARNAT (Secretaría de Medio Ambiente y Recursos Naturales). 2015. *Programa de Manejo, Área de Protección de Flora y Fauna Laguna Madre y Delta del Río Bravo*. Mexico City: CONANP, SEMARNAT. https://simec.conanp.gob.mx/pdf_libro_pm/122_libro_pm.pdf.

Seminoff, J. A., C. D. Allen, G. H. Balazs, P. H. Dutton, T. Eguchi, H. Haas, S. A. Hargrove, et al. 2015. *Status Review of the Green Turtle (*Chelonia mydas*) under the Endangered Species Act*. NOAA-TM-NMFS-SWFSC-539. La Jolla, CA: National Marine Fisheries Service.

Semmens, B. X., E. R. Buhle, A. K. Salomon, and C. V. Pattengill-Semmens. 2004. A hotspot of non-native marine fishes: Evidence for the aquarium trade as an invasion pathway. *Marine Ecology Progress Series* 266:239–244.

Sengupta, R., B. Middleton, C. Yan, M. Zuro, and H. Hartman. 2005. Landscape characteristics of *Rhizophora mangle* forests and propagule deposition in coastal environments of Florida. *Landscape Ecology* 20:1–10.

Senner, S. E., and M. A. Howe. 1984. Conservation of Nearctic shorebirds. Pp. 370–415 in J. Burger and B. L. Olla, eds., *Shorebirds: Breeding Behaviour and Population*. New York: Plenum Press.

Sennett, G. B. 1878. Notes on the ornithology of the Lower Rio Grande of Texas, from observations made during the season of 1877. *Bulletin of the U.S. Geological and Geographical Survey of Territories* 4:1–66.

Sennett, G. B. 1879. Further notes on the ornithology of the Lower Rio Grande of Texas, from observations made during the season of 1878. *Bulletin of the U.S. Geological and Geographical Survey of Territories* 5:371–440.

Seward, N. W., K. C. VerCauteren, G. W. Witmer, and R. M. Engeman. 2004. Feral swine impacts on agriculture and the environment. *Sheep and Goat Research Journal* 19:34–40.

Shamblin, B. M., A. B. Bolten, F. A. Abreu-Grobois, K. A. Bjorndal, L. Cardona, C. Carreras, and M. Clusa. 2014. Geographic patterns of genetic variation in a broadly distributed marine vertebrate: New insights into Loggerhead Turtle stock structure from expanded mitochondrial DNA sequences. *PLoS ONE* 9 (1): e85956. doi:10.1371/journal.pone.0085956.

Shamblin, B. M., A. B. Bolten, K. A. Bjorndal, P. H. Dutton, J. T. Nielsen, F. A. Abreu-Grobois, K. J. Reich, et al. 2012. Expanded mitochondrial control region sequences increase resolution of stock structure among North Atlantic Loggerhead Turtle rookeries. *Marine Ecology Progress Series* 469:145–160.

Shamblin, B. M., M. G. Dodd, D. A. Bagley, L. M. Ehrhart, A. D. Tucker, C. Johnson, R. R. Carthy, et al. 2010. Genetic structure of the southeastern United States Loggerhead Turtle nesting aggregation: Evidence of additional structure within the peninsular Florida recovery unit. *Marine Biology*. DOI 10.1007/s00227-010-1582-6.

Shamblin, B. M., P. H. Dutton, D. J. Shaver, D. A. Bagley, N. F. Putman, K. L. Mansfield, L. M. Ehrhart, L. J. Peña, and C. J. Nairn. 2016. Mexican origins for the Texas Green Turtle foraging aggregation: A cautionary tale of incomplete baselines and poor marker resolution. *Journal of Experimental Marine Biology and Ecology* 488:111–120. https://doi.org/10.1016/j.jembe.2016.11.009.

Shane, S. H., and D. J. Schmidly. 1978. *The Population Biology of the Atlantic Bottlenose Dolphin,* Tursiops truncatus, *in the Aransas Pass Area of Texas*. US Marine Mammal Committee National Technical Information Service PB-283-393, Washington, DC.

Shaver, D. J. 1989. Green Sea Turtle geographic distribution. *Herpetological Review* 20:14.

Shaver, D. J. 1990a. Hypothermic stunning of sea turtles in Texas. *Marine Turtle Newsletter* 48:25–27.

Shaver D. J. 1990b. Sea turtles in South Texas inshore waters. Pp. 131–132 in T. H. Richardson, J. I. Richardson, and M. Donnelly, eds., *Proceedings of the Tenth Annual Workshop on Sea Turtle Biology and Conservation*, NOAA Technical Memorandum NMFS-SEFC-278. Miami, FL: NOAA Southeast Fisheries Center.

Shaver, D. J. 1990c. *Sea Turtles in South Texas Inshore Waters*. Padre Island National Seashore Report to US Fish and Wildlife Service, Contract No. 14-16-0002-89-919, Albuquerque, NM.

Shaver, D. J. 1991. Feeding ecology of wild and head-started Kemp's Ridley Sea Turtles in South Texas waters. *Journal of Herpetology* 25:327–334.

Shaver, D. J. 1992. Kemp's Ridley research continues at Padre Island National Seashore. *Park Science* 12:26–27.

Shaver, D. J. 1994a. Relative abundance, temporal patterns, and growth of sea turtles at the Mansfield Channel, Texas. *Journal of Herpetology* 28:491–497.

Shaver, D. J. 1994b. Sea turtle strandings along the Texas coast reach alarming levels. *Marine Turtle Newsletter* 66:8–9.

Shaver, D. J. 1995. Sea turtle strandings along the Texas coast again cause concern. *Marine Turtle Newsletter* 70:2–4.

Shaver, D. J. 1998a. *Padre Island National Seashore Kemp's Ridley Sea Turtle Project and Texas Sea Turtle Strandings 1997 Report*. Report to National Park Service, Washington, DC.

Shaver, D. J. 1998b. Sea turtle strandings along the Texas coast, 1980–94. Pp. 57–72 in R. Zimmerman, ed., *Characteristics and Causes of Texas Marine Strandings*, NOAA Technical Reports NMFS 143. Miami, FL: US Department of Commerce.

Shaver, D. J. 2000. Distribution, residency, and seasonal movements of the Green Sea Turtle, *Chelonia mydas* (Linnaeus, 1758) in Texas. PhD diss., Texas A&M University, College Station.

Shaver, D. J. 2005. Analysis of the Kemp's Ridley imprinting and headstart project at Padre Island National Seashore, Texas, 1978–88, with subsequent Kemp's Ridley nesting and stranding records on the Texas coast. *Chelonian Conservation and Biology* 4:846–859.

Shaver, D. J., and C. W. Caillouet Jr. 2015. Reintroduction of Kemp's Ridley (*Lepidochelys kempii*) Sea Turtle to Padre Island National Seashore, Texas and its connection to head-starting. *Herpetological Conservation and Biology* 10:378–435.

Shaver, D. J., K. M. Hart, I. Fujisaki, D. Bucklin, A. R. Iverson, C. Rubio, T. F. Backof, et al. 2017. Inter-nesting movements and habitat use of adult female Kemp's Ridley turtles in the Gulf of Mexico. *PLoS ONE*. https://doi.org/10.1371/journal.pone.0174248.

Shaver, D. J., K. M. Hart, I. Fujisaki, C. Rubio, and A. R. [Sartain-]Iverson. 2013. Movement mysteries unveiled: Spatial ecology of juvenile Green Sea Turtles. Pp. 463–483 in W. I. Lutterschmidt, ed., *Reptiles in Research*. Hauppauge, NY: Nova Science Publishers.

Shaver, D. J., K. M. Hart, I. Fujisaki, C. Rubio, A. R. Iverson, J. Peña, D. Gomez Gamez, et al. 2016a. Corrigendum to "Migratory corridors of adult female Kemp's Ridley turtles in the Gulf of Mexico." *Biological Conservation* 195:297.

Shaver, D. J., K. M. Hart, I. Fujisaki, C. Rubio, A. R. [Sartain-]Iverson, J. Peña, D. Gomez Gamez et al. 2016b. Migratory corridors of adult female Kemp's Ridley turtles in the Gulf of Mexico. *Biological Conservation* 194 (2016):158–167.

Shaver, D. J., K. Hart, C. Rubio, A. R. Iverson, J. Peña, P. M. Burchfield, D. Gomez Gamez, and J Ortiz. 2013. Foraging area fidelity for Kemp's Ridleys in the Gulf of Mexico. *Ecology and Evolution* 3:2002–2012. doi: 10.1002/ece3.594.

Shaver, D. J., and P. T. Plotkin. 1998. Marine debris ingestion by sea turtles in South Texas: Pre- and post-MARPOL Annex V. P. 124 in R. Byles and Y. Fernandez, comps., *Proceedings of the 16th Annual Symposium on Sea Turtle Biology and Conservation*, NOAA Technical Memorandum NMFS-SEFSC-412, Washington, DC.

Shaver, D. J., and C. Rubio. 2008. Post-nesting movement of wild and head-started Kemp's Ridley Sea Turtles (*Lepidochelys kempii*) in the Gulf of Mexico. *Endangered Species Research* 4:43–55.

Shaver, D. J., C. Rubio, J. S. Walker, J. George, A. F. Amos, K. Reich, C. Jones, and T. Shearer. 2016. Kemp's Ridley Sea Turtle (*Lepidochelys kempii*) nesting on the Texas coast: Geographic, temporal, and demographic trends through 2014. *Gulf of Mexico Science* 33:158–178.

Shaver, D. J., B. A. Schroeder, R. A. Byles, P. M. Burchfield, J. Peña, R. Márquez, and H. J. Martinez. 2005. Movements and home ranges of adult male Kemp's Ridley Sea Turtles (*Lepidochelys kempii*) in the Gulf of Mexico investigated by satellite telemetry. *Chelonian Conservation and Biology* 4:817–827.

Shaver, D. J., and W. G. Teas. 1999. Stranding and salvage networks. Pp. 152–155 in K. Eckert, K. A. Bjorndal, F. A. Abreu-Grobois, and M. Donnelly, eds., *Research and Management Techniques for the Conservation of Sea Turtles*. Washington, DC: IUCN/SSC Marine Turtle Specialist Group.

Shaver, D. J., P. E. Tissot, M. M. Streich, J. S. Walker, C. Rubio, A. F. Amos, J. A. George, and M. R. Pasawicz. 2017. Hypothermic stunning of Green Sea Turtles in a western Gulf of Mexico foraging habitat. *PLoS ONE*. https://doi.org/10.1371/journal.pone.0173920.

Shaver, D. J., and T. Wibbels. 2007. Headstarting of Kemp's Ridley turtles. Pp. 297–324 in P. T. Plotkin, ed., *Biology and Conservation of Ridley Sea Turtles*. Washington, DC: Smithsonian Institution Press.

Shaw, R. B. 2012. *Guide to Texas Grasses*. College Station: Texas A&M University Press.

Sheire, J. W. 1971. *Padre Island National Seashore, Historic Resource Study*. National Technical Information Service PB-206 132. Washington, DC: US Department of Commerce, Office of Archeology and Historic Preservation.

Shelford, V. E. 1974. *The Ecology of North America*. Urbana: University of Illinois Press.

Shepard, F. P., and G. A. Rusnak. 1957. Texas bay sediments. *Publications of the Institute of Marine Science, University of Texas* 4:5–13.

Sheridan, P. 2004. Recovery of floral and faunal communities after placement of dredged material on seagrasses in Laguna Madre, Texas. *Estuarine, Coastal and Shelf Science* 59:441–458.

Sheridan, P., and T. J. Minello. 2003. Nekton use of different habitat types in seagrass beds of Lower Laguna Madre, Texas. *Bulletin of Marine Science* 72:37–61

Sherrod, C. L., D. L. Hockaday, and C. McMillan. 1986. Survival of Red Mangrove, *Rhizophora mangle*, on the Gulf of Mexico coast of Texas. *Contributions in Marine Science* 29:27–56.

Sherrod, C. L., and C. McMillan. 1981. Black Mangrove, *Avicennia germinans*, in Texas: Past and present distribution. *Contributions in Marine Science* 24:115–131

Sherrod, C. L., and C. McMillan. 1985. The distributional history and ecology of mangrove vegetation along the northern Gulf of Mexico coastal region. *Contributions in Marine Science* 28:129–140.

Short, F. T. 1987. Effects of sediment nutrients on seagrasses: Literature review and mesocosm experiment. *Aquatic Botany* 27:47–57.

Short, F. T., D. M. Burdick, and J. E. Kaldy III. 1995. Mesocosm experiments quantify the effects of eutrophication on Eelgrass, *Zostera marina*. *Limnology and Oceanography* 40:740–749.

Short, F. T., R. G. Coles, and P. Pergent-Mantini. 2001. Global seagrass distribution. Pp. 5–30 in F. T. Short and R. G. Coles, eds., *Global Seagrass Research Methods*. Amsterdam: Elsevier Science.

Sieburth, J. McN., P. W. Johnson, and P. E. Hargraves. 1988. Ultrastructure and ecology of *Aureococcus anophageferens* Gen. et Sp. Nov. (Chrysophyceae): The dominant picoplankter during a bloom in Narragansett Bay, Rhode Island, summer 1985. *Journal of Phycology* 24:416–425.

Siegel-Causey, D., and S. P. Kharitonov. 1990. The evolution of coloniality. *Current Ornithology* 7:285–330.

Sikes, D. 2018. What are nurdles, where do they come from, and are they killing our wildlife? *Corpus Christi Caller-Times*, November 27. https://www.caller.com/story/news/local/2018/11/27/what-you-need-know-nurdles-and-threat-marine-life-texas-beaches/1807039002/.

Simersky, B. L. 1971. Competition and nesting success of four species of herons on four spoil islands in the Laguna Madre. MS thesis, Texas A&I University, Kingsville.

Simmons, E. G. 1957. An ecological survey of the Upper Laguna Madre of Texas. *Publications of the Institute of Marine Science, University of Texas* 4:156–200.

Simmons, E. G., and J. P. Breuer. 1962. A study of Redfish, *Sciaenops ocellatus* (Linnaeus) and Black Drum, *Pogonias cromis* (Linnaeus). *Publications of the Institute of Marine Science, University of Texas* 8:184–211.

Simms, A. R., N. Aryal, L. Miller, and Y. Yokoyama. 2010. The incised valley of Baffin Bay, Texas: A tale of two climates. *Sedimentology* 57:642–669. DOI: 10.1111/j.1365-3091.2009.01111.x.

Simpson, R. H. and M. B. Lawrence. 1971. *Atlantic Hurricane Frequencies along the US Coastline*. NOAA Technical Memorandum NWS SR-58, Washington, DC.

Sissom, S. L. 1990. *A Baseline Study of Three Ponds within the Padre Island National Park*. Southwest Texas State University, Department of Biology, Report to US Department of the Interior, Padre Island National Seashore, Corpus Christi.

Skoruppa, M. K., and M. C. Woodin. 2000. Impact of wintering Redhead ducks on pond water quality in southern Texas. Pp. 31–41 in F. A. Comin, J. A. Herrera, and J. Ramirez, eds., *Limnology and Aquatic Birds: Monitoring, Modelling and Management*. Mérida, Mexico: Universidad Autónoma de Yucatán.

Smee, D. L., J. A. Sanchez, M. Diskin, and C. Trettin. 2017. Mangrove expansion into salt marshes alters associated faunal communities. *Estuarine, Coastal and Shelf Science* 187:306–313.

Smeins, F. E., D. D. Diamond, and W. Hanselka. 1991. Coastal prairie. Pp. 269–290 in R. T. Coupland, ed., *Ecosystems of the World: Natural Grasslands*. Amsterdam: Elsevier Press.

Smith, D. S. 1986. *The Armstrong Chronicle: A Ranching History*. San Antonio: Corona Publishing.

Smith, E. H. 1986. Morphological and karyotypic variation of the Gulf Coast Kangaroo Rat, *Dipodomys compactus* (Rodentia: Heteromyidae). MS thesis, Corpus Christi State University, Corpus Christi.

Smith, E. H. 2002. Colonial waterbirds and rookery islands. Pp. 182–197 in J. W. Tunnell Jr. and F. W. Judd, eds., *The Laguna Madre of Texas and Tamaulipas*. College Station: Texas A&M University Press.

Smith, E. H., and S. A. Cox. 1998. *Status and Trends of Rookery Islands in the Corpus Christi Bay National Estuary Program Study Area*. Report to Corpus Christi Bay National Estuary Program, Texas Natural Resource Conservation Commission, Austin.

Smith, H. M. 1939. *The Mexican and Central American Lizards of the Genus* Sceloporus. Zoological Series, Field Museum of Natural History, vol. 26, Publication 445. Chicago: Field Museum of Natural History.

Smith, N. F., C. Wilcox, and J. Lessmann. 2009. Fiddler crab burrowing affects growth and production of the White Mangrove (*Lagunaularia racemose*) in a restored Florida coastal marsh. *Marine Biology* 156:2255–2266. DOI:10.1007/s00227-009-1253-7.

Smith, N. P. 1978. Intracoastal tides of Upper Laguna Madre, Texas. *Texas Journal of Science* 30:85–95.

Smith, P. W., and N. T. Houghton. 1984. Fidelity of Semipalmated Plovers to a migratory stopover area. *Journal of Field Ornithology* 55:247–249.

Smith, R. 1996. Temporal variation in intertidal macroinvertebrate assemblages associated with the invasive mussel, *Perna perna* (Linnaeus, 1758), on a South Texas jetty. MS thesis, Texas A&M University–Corpus Christi.

Smith, S. M., and M. C. Tyrell. 2012. Effects of Mud Fiddler Crab (*Uca pugnax*) on the recruitment of halophyte seedlings in a salt marsh dieback area of Cape Cod (Massachusetts, USA). *Ecological Research* 27:233–237.

Smith, S. V., S. Ibarra-Obando, P. R. Boudreau, and V. F. Camacho-Ibar. 1997. *Comparison of Carbon, Nitrogen, and Phosphorus Fluxes in Mexican Coastal Lagoons.* LOICZ Reports and Studies No. 10. Texel, Netherlands: Land-Ocean Interactions in the Coastal Zone (LOICZ) Core Project of the IGBP.

Smylie, V. 1964a. *Conquistadores and Cannibals.* Corpus Christi: Texas News Syndicate Press.

Smylie, V. 1964b. *The Secrets of Padre Island.* Corpus Christi: Texas News Syndicate Press.

Sobrado, M. A. 1999. Leaf photosynthesis of the mangrove *Avicennia germinans* as affected by NaCl. *Photosynthetica* 36:547–555.

Soots, R. F., and M. C. Landin. 1978. *Development and Management of Avian Habitat on Dredged Material Islands.* US Army Corps of Engineers Waterways Experiment Station Technical Report DS-78-18, Vicksburg, MS.

Soots, R. F., and J. F. Parnell. 1975. *Ecology Succession of Breeding Birds in Relation to Plant Succession on Dredge Islands in North Carolina Estuaries.* North Carolina Sea Grant Program Publication UNC-SC-75–27, Raleigh, NC.

Sorenson, L. O., and J. T. Conover. 1962. Algal mat communities of *Lyngbya confervoides* (C. Agardh.) Gioment. *Publications of the Institute of Marine Science University of Texas* 8:61–74.

Southeast Atlantic Fishery Management Council. 2009. SEDAR (*Southeast Data, Assessment, and Review) 18: Stock Assessment Report: Atlantic Red Drum.* North Charleston, SC: South Atlantic Fishery Management Council. https://sedarweb.org/docs/sar/SEDAR%2018-%20Red%20drum%20SAR%20Final.pdf.

Spotila, J. R., M. P. O'Connor, and F. V. Paladino. 1997. Thermal biology. Pp. 297–314 in P. L. Lutz and J. A. Musick, eds., *The Biology of Sea Turtles.* Boca Raton, FL: CRC Press.

Spotila, J. R., and J. R. Standora. 1985. Environmental constraints on the thermal energetics of sea turtles. *Copeia* 1985:694–702.

Sprunt, A., and C. E. Knoder. 1980. Populations of wading birds and other colonial nesting species on the Gulf and Caribbean coasts of Mexico. Pp. 3–16 in P. P. Schaeffer and S. M. Ehlers, eds., *Proceedings of the National Audubon Society's Symposium on the Birds of Mexico, Their Ecology and Conservation.* Tiburon, CA: National Audubon Society Western Education Center.

Starek, M. 2016. *High-Resolution Lidar Observations of Rookery Islands in the Upper Laguna Madre to Define a Monitoring [Program].* Corpus Christi: Conrad Blucher Institute, Texas A&M University–Corpus Christi.

Stewart, R. H. 2008. Temperature, salinity, and density. Pp. 73–102 in R. H. Stewart, *Introduction to Physical Oceanography.* Gainesville: University Press of Florida.

Stienen, H. 1990. The agroforestry potential of combined production systems in north-eastern Mexico. *Agroforestry Systems* 11:45–69.

Still, B. M., C. R. Griffin, and R. Prescott. 2005. Climatic and oceanographic factors affecting daily patterns of juvenile sea turtle cold-stunning in Cape Cod Bay, Massachusetts. *Chelonian Conservation and Biology* 4:883–890.

Stockwell, D. A., E. J. Buskey, and T. E. Whitledge. 1993. Studies on conditions conducive to the development and maintenance of a persistent "brown tide" in Laguna Madre, Texas. Pp. 693–698 in T. J. Smayda and Y. Shimizu, eds., *Toxic Phytoplankton Blooms in the Sea: Proceedings of the 5th International Conference on Toxic Marine Phytoplankton.* Amsterdam: Elsevier.

Stokes, G. M. 1974. *The Distribution and Abundance of Penaeid Shrimp in the Lower Laguna Madre of Texas, with a Description of the Live Bait Shrimp Fishery.* Texas Parks and Wildlife Department Technical Series No. 15, Austin.

Street, G. T., P. A. Montagna, and P. L. Parker. 1997. Incorporation of brown tide into an estuarine food web. *Marine Ecology Progress Series* 152:67–78.

Stunz, G. W., and T. J. Minello. 2001. Habitat-related predation on juvenile wild-caught and hatchery-reared Red Drum *Sciaenops ocellatus* (Linnaeus). *Journal of Experimental Marine Biology and Ecology* 260:13–25.

Suárez, N., and E. Medina. 2005. Salinity effect on plant growth and leaf demography of the mangrove, *Avicennia germinans* L. *Trees* 19:721–727.

Suárez, N., M. A. Sobrado, and E. Medina. 1998. Salinity effects on the leaf water relations components and ion accumulation patterns in *Avicennia germinans* (L.) L. seedlings. *Oecologia* 114:299–304.

Sullivan, M. J., and C. A. Moncreiff. 1990. Edaphic algae are an important component of salt marsh food webs: Evidence from multiple stable isotope analyses. *Marine Ecology Progress Series* 62:149–159.

Sun, M. M., J. Sun, J. W. Qiu, H. Jing, and H. Liu. 2012. Characterization of the proteomic profiles of the brown tide alga *Aureoumbra lagunensis* under phosphate- and nitrogen-limiting conditions and of its phosphate limitation-specific protein with alkaline phosphatase activity. *Applied and Environmental Microbiology* 78:2025–2033.

Tamayo, J. 1949. *Geografia general de México.* Vol. 2. Mexico City: Talleres Gráficos de la Nacion.

Tamez, C. A. 2014. Gauging wetland restoration status using benthic-based structural and functional metrics. MS thesis, University of Texas at Brownsville.

Tanaka, K., H. Takada, R. Yamashita, K. Mizukawa, M. Fukuwaka, and Y. Watanuki. 2013. Accumulation of plastic-derived chemicals in tissues of seabirds ingesting marine plastics. *Marine Pollution Bulletin* 69:219–222.

TCEQ (Texas Commission on Environmental Quality). 2003. *Twelve Total Maximum Daily Loads for Legacy Pollutants in the Arroyo Colorado above Tidal and the Donna Reservoir and Canal System.* Austin: TCEQ Strategic Assessment Division.

TCWC (Texas Colonial Waterbird Census). 1998. Texas colonial waterbird census: 1972–1996, database summary, March 1998. Unpublished manuscript on file at Texas A&M University–Corpus Christi, Center for Coastal Studies Library.

TCWS (Texas Colonial Waterbird Society). 1982. *An Atlas and Census of Texas Waterbird Colonies 1972–1980.* Kingsville, TX: Caesar Kleberg Wildlife Research Institute.

TCWS. 2017. Texas colonial waterbird census: 1972–2017, database summary. Unpublished manuscript on file at Texas A&M University–Corpus Christi, Center for Coastal Studies Library.

TDA (Texas Department of Agriculture). 1983. *Texas Family Land Heritage: 1975 Registry.* Vol. 2. Austin: TDA.

TDA. 1985. *Texas Family Land Heritage: 1983 Registry.* Vol. 9. Austin: TDA.

TDWR (Texas Department of Water Resources). 1983. *Laguna Madre Estuary: A Study of the Influence of Freshwater Inflows.* Texas Department of Water Resources Report LP-182, Austin.

Teas, W. G. 1993. *Species Composition and Size Class Distribution of Marine Turtles: Strandings on the Gulf of Mexico and Southeast United States Coast, 1985–1991.* NOAA Tech. Memo. NMFS-SEFSC-315. Miami, FL: US Department of Commerce.

Telfair, R. C., II, D. A. McCrimmon Jr., and S. T. Fryska. 2000. Population dynamics of the Cattle Egret in Texas, 1954–1999. *Waterbirds* 23:187–195.

Tella, J. L. 2002. The evolutionary transition to coloniality promotes higher blood parasitism in birds. *Journal of Evolutionary Biology* 15:32–41.

Téllez, S. J., M. Oliva, J. A. Ramírez de León, and M. Vázquez. 1999. Evaluación de la calidad microbiológica del ostión de "la Laguna Madre" de Tamaulipas (México). *Ciencia y Tecnología Alimentaria* 2:152–157.

Terán-Juárez, S. A., E. García-Padilla, V. Mata-Silva, J. D. Johnson, and L. D. Wilson. 2016. The herpetofauna of Tamaulipas, Mexico: Composition, distribution, and conservation status. *Mesoamerican Herpetology* 3:43–111.

Tewes, M. E. 1993. Personal affidavit, Exhibit 9. In *Environment Impacts of the Plan to Dump Dredged Spoil from the Gulf Intracoastal Waterway on Lands Bordering the Laguna Madre.* Kingsville, TX: King Ranch and John G. Kenedy Charitable Trust.

TGLO (Texas General Land Office). 1988. *Guide to Spanish and Mexican Land Grants in South Texas.* Austin: TGLO.

TGLO. 1997. *Oil Spill Planning and Response Atlas: Lower Coast of Texas.* Austin: TGLO.

Thayer, G. W., S. M. Adams, and M. W. LaCroix. 1975. Structural and functional aspects of a recently established *Zostera marina* community. Pp. 517–540 in L. E. Cronin, ed., *Estuarine Research*, vol. 1. New York: Academic Press.

Thomas, C., and L. Blum. 2010. Importance of the fiddler crab *Uca pugnax* to salt marsh soil organic matter accumulation. *Marine Ecology Progress Series* 414:167–177.

Thomas, G. W. 1975. Texas plants and ecological summary. Pp. 7–14 in F. W. Gould, ed., *Texas Plants: A Checklist and Ecological Summary*. College Station: Texas Agricultural Experiment Station MP-585, Texas A&M University.

Thomas, W. D. 1972. The mammals of Padre Island. Pp. 59–61 in Corpus Christi Geological Society, ed., *Padre Island National Seashore Field Guide*. Corpus Christi: Golden Banner Press.

Thompson, J. 1997. *A Wild and Vivid Land: An Illustrated History of the South Texas Border*. Austin: Texas State Historical Association.

Thornthwaite, C. W. 1948. An approach toward a rational classification of climate. *Geographic Review* 38:55–94.

Thronson, A., and A. Quigg. 2008. Fifty-five years of fish kills in coastal Texas. *Estuaries and Coasts* 31:802–813.

Thrush, S. F., J. E. Hewitt, G. A. Funnell, V. J. Cummings, J. Ellis, D. Schultz, D. Talley, and A. Norkko. 2001. Fishing disturbance and marine biodiversity: The role of habitat structure in simple soft-sediment systems. *Marine Ecology Progress Series* 223:277–286.

Thurman, C. L., III. 1984. Ecological notes on fiddler crabs of South Texas, with special reference to *Uca subcylindrica*. *Journal of Crustacean Biology* 4:665–681.

Tinnin, R. K. 1974. A trammel net survey of a disturbed hypersaline environment. MS thesis, Texas A&I University, Kingsville.

TNC (The Nature Conservancy). 1996. *Conservation by Design*. Arlington, VA: TNC.

TNC. 2000. *The Five-S Framework for Site Conservation: A Practitioner's Handbook for Site Conservation Planning and Measuring Conservation Success*. Vol. 1. Arlington, VA: TNC.

TNC. 2001. *Conservation Action Plan for the Texas Portion of the Laguna Madre*. Arlington, VA: TNC.

TNC. 2002. *The Gulf Coast Prairies and Marshes Ecoregional Conservation Plan*. San Antonio: Gulf Coast Prairies and Marshes Ecoregional Planning Team, TNC.

TNC. 2016. *Conservation by Design 2.0*. Guidance Document, Version 1.0. Arlington, VA: TNC.

Tolan, J. M. 1994. Habitat selection by larval fishes immigrating into estuarine nursery grounds. MS thesis, Texas A&M University–Corpus Christi.

Tolan, J. M. 2007. El Niño–Southern Oscillation impact translated to the watershed scale: Estuarine salinity patterns along the Texas Gulf Coast, 1982–2004. *Estuarine, Coastal and Shelf Science* 72:247–260.

Tolan, J. M. 2013. Estuaries fisheries community-level response to freshwater inflows. Pp. 73–94 in R. Wurbs, ed., *Water Resources Planning, Development and Management*. IntechOpen. http://dx.doi.org/10.5772/52313.

Tolan, J. M., S. A. Holt, and C. P. Onuf. 1997. Distribution and community structure of ichthyoplankton in Laguna Madre seagrass meadows: Potential impact of seagrass species change. *Estuaries* 20:450–464.

Toland, J. 1996. Assault, a look back at Texas' greatest racehorse. *Corpus Christi Caller-Times* 114 (122): 18–19.

Tomasko, D. A., C. J. Dawes, and M. O Hall. 1996. The effects of anthropogenic nutrient enrichment on Turtle Grass (*Thalassia testudinum*) in Sarasota Bay, Florida. *Estuaries* 19 (2), Part B: 448–456.

Tomasko, D. A., and K. H. Dunton. 1995. Primary productivity in *Halodule wrightii*: A comparison of techniques based on daily carbon budgets. *Estuaries* 18:271–278.

Tomlinson, P. B. 2016. *The Botany of Mangroves*. Cambridge Tropical Biology Series. New York: Cambridge University Press.

Tompkins, I. R. 1933. Ways of the Black Skimmer. *Wilson Bulletin* 45:147–151.

TPWD (Texas Parks and Wildlife Department). 1998a. *Special Species List, Kenedy County; Revised 3/27/98*. Austin: TPWD, Endangered Resources Branch.

TPWD. 1998b. *Special Species List, Willacy County; Revised 3/31/98*. Austin: TPWD, Endangered Resources Branch.

TPWD. 1999a. *Annotated County Lists of Rare Species, Cameron County; Revised 8/26/99*. Austin: TPWD, Endangered Resources Branch.

TPWD. 1999b. *Annotated County Lists of Rare Species, Kleberg County; Revised 8/26/99*. Austin: TPWD, Endangered Resources Branch.

TPWD. 1999c. *Annotated County Lists of Rare Species, Nueces County; Revised 8/26/99*. Austin: TPWD, Endangered Resources Branch.

TPWD. 2003. Texas threatened and endangered species, January 2003. https://tpwd.texas.gov/publications/pwdpubs/media/pwd_lf_w7000_0017.pdf.

TPWD. 2019. *Outdoor Annual*. https://tpwd.texas.gov/regulations/outdoor-annual/fishing/shellfish-regulations/shrimp-regulations.

Trewartha, G. T. 1961. *The Earth's Problem Climates*. Oshkosh: University of Wisconsin Press.

Trewartha, G. T. 1968. *An Introduction to Climate*. New York: McGraw-Hill.

Tristan, T., D. J. Shaver, J. Kimbro, T. deMaar, T. Metz, J. George, and A. Amos. 2010. Identification of fibropapillomatosis in Green Sea Turtles (*Chelonia mydas*) on the Texas coast. *Journal of Herpetological Medicine and Surgery* 20:109–112.

True, F. W. 1889. Description of *Geomys personatus* and *Dipodomys compactus*, two new species of rodents from Padre Island, Texas. *Proceedings of the United States National Museum* 11:159–160.

Tunnell, J. W., Jr. 2002. The environment. Pp. 73–84 in J. W. Tunnell and F. W. Judd, eds., *The Laguna Madre of Texas and Tamaulipas*. College Station: Texas A&M University Press.

Tunnell, J. W., Jr. 2017. Shellfish of the Gulf of Mexico. Pp. 769–839 in C. H. Ward, ed., *Habitats and Biota of the Gulf of Mexico: Before the Deepwater Horizon Oil Spill*, vol. 1. New York: Springer.

Tunnell, J. W., Jr., and S. A. Alvarado, eds. 1996. *Checklist of Species within Corpus Christi Bay National Estuary Program Study Area: References, Habitats, Distribution, and Abundance*. Corpus Christi National Estuary Program CCBNEP-06D. Corpus Christi: Texas Natural Resources Conservation Commission.

Tunnell, J. W., Jr., J. Andrews, N. Barrera, and F. Moretzsohn. 2010. *Encyclopedia of Texas Seashells*. College Station: Texas A&M University Press.

Tunnell, J. W., Jr., E. A. Chávez, and Kim Withers, eds. 2007. *Coral Reefs of the Southern Gulf of Mexico*. College Station: Texas A&M University Press.

Tunnell, J. W., Jr., Q. R. Dokken, M. E. Kindinger, and L. C. Thebeau. 1981. Effects of the IXTOC I oil spill on the intertidal and subtidal infaunal populations along lower Texas coast barrier island beaches. Pp. 467–475 in *Proceedings of the 1981 Oil Spill Conference*. Washington, DC: American Petroleum Institution.

Tunnell, J. W., Jr., and F. W. Judd, eds. 2002. *The Laguna Madre of Texas and Tamaulipas*. College Station: Texas A&M University Press.

Tunnell, J. W., Jr., and J. W. Tunnell. 2015. *Pioneering Archeology in the Texas Coastal Bend: The Pape-Tunnell Collection*. College Station: Texas A&M University Press.

Turner, J. W., F. Hernandez, C. W. Boal, B. M. Ballard, F. C. Bryant, and D. B. Wester. 2014. Raptor abundance and Northern Bobwhite survival and habitat use. *Wildlife Society Bulletin* 38:689–696.

Turner, R. E. 1977. Intertidal vegetation and commercial yields of penaeid shrimp. *Transactions of the American Fisheries Society* 106:411–416.

Twilley, R. R., E. J. Barron, H. L. Gholz, M. A. Harwell, R. L. Miller, D. J. Reed, J. B. Rose, E. H. Siemann, R. G. Wetzel, and R. J. Zimmerman. 2001. *Confronting Climate Change in the Gulf Coast Region*. Cambridge, MA: Union of Concerned Scientists; Washington, DC: Ecological Society of America.

UCS (Union of Concerned Scientists). 2013. *Causes of Sea Level Rise: What the Science Tells Us*. Cambridge, MA: UCS.

UCS. 2018. *Underwater: Rising Seas, Chronic Floods, and the Implications for US Coastal Real Estate*. Cambridge, MA: UCS.

UNESCO (United Nations Educational, Scientific and Cultural Organization). 2006. Laguna Madre y Delta Río Bravo Biosphere Reserve, Mexico. Updated September 2018. https://en.unesco.org/biosphere/lac/laguna-madre_rio-bravo.

US Census Bureau. 2020a. Cameron County, Texas. https:/data.census.gov/cedsci/profile?g=0500000US48061.

US Census Bureau. 2020b. Kenedy County, Texas. https://data.census.gov/cedsci/profile?g=0500000US48261.

US Census Bureau. 2020c. Texas. https://data.census.gov/cedsci/profile?g=0400000US48.

US Climate Data. 2017. Average weather for Sarita, TX, 78385: 1981–2010 normals. https://www.usclimatedata.com/climate/sarita/texas/united-states/ustx2663.

USACE (US Army Corps of Engineers) and ICT (Interagency Coordination Team). 2002. *Laguna Madre GIWW Dredged Material Management Plan.* https://www.swg.usace.army.mil/Portals/26/docs/Navigation/GIWW-N-ICT/LagunaMadreDMMP.pdf.

USDA (US Department of Agriculture). 2012. *USDA Census of Agriculture.* https://www.agcensus.usda.gov/Publications/2012/Full_Report/Census_by_State/Texas/index.asp.

USFWS (US Fish and Wildlife Service). 1983. *Land Protection Plan: Lower Rio Grande Valley National Wildlife Refuge.* Albuquerque, NM: US Fish and Wildlife Service.

USFWS. 2016. *Recovery Plan for the Ocelot (*Leopardus pardalis*): First Revision.* Albuquerque, NM: US Fish and Wildlife Service, Region 2.

USFWS and NMFS. 1992. *Recovery Plan for the Kemp's Ridley Sea Turtle (*Lepidochelys kempii*).* St. Petersburg, FL: USFWS and NMFS.

Valderrama-Landeros, L. H., M. T. Rodríguez-Zúñiga, C. Troche, S. Velázquez-Salazar, E. Villeda, A. Alcántara, B. Vazquez, M. I. Cruz López, and R. A. Ressl. 2017. *Manglares de México: Actualización y exploración de los datos del sistema de monitoreo 1970/1980–2015.* Mexico City: CONABIO.

Valiela, I., and C. S. Rietsma. 1995. Disturbance of salt marsh vegetation by wrack mats in Great Sippewissett Marsh. *Oecologia* 102:106–112.

Van Diggelen, A. D., and P. A. Montagna. 2016. Is salinity variability a benthic disturbance in estuaries? *Estuaries and Coasts* 39:967–980.

van Montfrans, J., R. L. Wetzel, and R. J. Orth. 1984. Epiphyte-grazer relationships in seagrass meadows: Consequences for seagrass growth and production. *Estuaries* 7:289–309.

Varney, R. L., C. E. Galindo-Sanchez, P. Cruz, and P. M. Gaffney. 2009. Population genetics of the Eastern Oyster *Crassostrea virginica* (Gmelin, 1791) in the Gulf of Mexico. *Journal of Shellfish Research* 28:855–864.

Velázquez de la Cruz, G., J. A. Ramírez de León, R. Pérez-Castañeda, M. A. Reyes-López, and A. V. Martínez Vázquez. 2012. *Aprovechamiento de la jaiba azul (*Callinectes sapidus*) en la Laguna Madre de Tamaulipas.* Mexico City: Plaza y Valdés, S. A. de C. V.

Vermillion, W. G., B. Eley, B. C. Wilson, S. A. Heath, and M. W. Parr. 2012. *Gulf Coast Joint Venture Landbird Conservation Plan, Bird Conservation Regions 25, 26, & 27, Gulf Coast Joint Venture Region.* http://www.gcjv.org/docs/Conservation_Plans/GCJV_BCR25_26_27_landbird_plan_Vers_1.pdf.

Villarreal, J. M. 2000. *Proyecto de ordenamiento ecológico de la Laguna Madre—Tamaulipas: Ordenamiento del clima regional.* Technical Report, Centro de Calidad Ambiental, ITESM Campus Monterrey—Pronatura Noreste, Monterrey, Nuevo León, Mexico.

Villareal, T. A., A. Mansfield, and E. J. Buskey. 1998. Growth and chemical composition of the Texas Brown Tide–forming pelagophyte *Aureoumbra lagunensis*. Pp. 359–362 in B. Reguera, J. Blanco, M. L. Fernandez, and T. Wyatt, eds., *Harmful Algae*. Vigo, Spain: Xunta de Galicia and Intergovernmental Oceanographic Commission of UNESCO.

Vora, R. S. 1989. Seed germination characteristics of selected native plants of the Lower Rio Grande Valley, Texas. *Journal of Range Management* 42:36–40.

Vora, R. S. 1990. Plant communities of the Santa Ana National Wildlife Refuge, Texas. *Texas Journal of Science* 42:115–128.

Wakida-Kusunoki, A. T. 2009. The Bay Scallop, *Argopecten irradians amplicostatus*, in northeastern Mexico. *Marine Fisheries Review* 71:17–19.

Wakida-Kusunoki, A. T., and L. E. Amador del Ángel. 2015. First record of the lionfish, *Pterois volitans*, on the coast of Tabasco, Mexico. *Hidrobiológica* 25:307–309.

Wakida-Kusunoki, A. T., L. García-Solorio, and N. G. Vázquez-Benavides. 2008. Abundancia de juveniles de camarones peneidos comerciales en la zona norte de Laguna Madre, México. *Hidrobiológica* 18:85–88.

Wakida-Kusunoki, A. T., R. I. Rojas-González, A. González-Cruz, L. E. Amador-del Ángel, J. L. Cruz-Sanchez, and N. A. Lopez-Tellez. 2013. Presence of Giant Tiger Shrimp *Penaeus monodon* Fabricius, 1798 on the Mexican coast of the Gulf of Mexico. *Bioinvasions Records* 2 (4): 325–328.

Wallace, B. P., A. D. Di Matteo, B. J. Hurley, E. M. Finkbeiner, A. B. Bolten, M. Y. Chaloupka, B. J. Hutchinson, et al. 2010. Regional management units for marine turtles: A novel framework for prioritizing conservation and research across multiple scales. *PLoS ONE* 5:e15465.

Walsh, G. E. 1974. Mangroves: A review. Pp. 51–174 in R. J. Reimhold and W. H. Queens, eds., *Ecology of Halophytes*. New York: Academic Press.

Walsh, S. C., B. M. Ballard, D. B. Wester, W. P. Kuvlesky Jr., L. A. Brennan, M. L. Morrison, and K. Boydston. 2017. High passage rates and different seasonal migration strategies of birds along the lower Texas coast. *International Journal of Biodiversity and Conservation* 9:183–199.

Ward, G. H., and N. E. Armstrong. 1997. *Current Status and Historical Trends of Ambient Water, Sediment, Fish and Shellfish Tissue Quality in the Corpus Christi Bay National Estuary Program Study Area*. Corpus Christi Bay National Estuary Program, CCBNEP 13. Corpus Christi: Texas Natural Resources Conservation Commission.

Ward, L. A., P. A. Montagna, R. D Kalke, and E. J. Buskey. 2000. Sublethal effects of Texas Brown Tide on *Streblospio benedicti* (Polychaeta) larvae. *Journal of Experimental Marine Biology and Ecology* 248:121–129.

Ward, P., and A. Zahavi. 1973. The importance of certain assemblages of birds as "information centres" for food-finding. *Ibis* 115:517–534.

Ward, R., E. G. Zimmerman, and T. L. King. 1990. Multivariate analyses of terrestrial reptilian distribution in Texas: An alternative view. *Southwestern Naturalist* 35:441–445.

Warshaw, S. 1975. *Water Quality Segment Report for Segment No. 2491 Laguna Madre*. Texas Water Quality Board Report WQS-14, Austin.

Warwick, R. M., and R. J. Uncles. 1980. Distribution of benthic macrofauna associations in the Bristol Channel in relation to tidal stress. *Marine Ecology Progress Series* 3:97–103.

Watson, R. L. 1971. Origin of shell beaches, Padre Island, Texas. *Journal of Sedimentary Petrology* 41:1105–1111.

Watson, R. L. 1989. Sedimentation rates, Kenedy Ranch supratidal flats. Unpublished manuscript on file at Texas A&M University–Corpus Christi, Center for Coastal Studies Library.

Weaver, C. A., and A. R. Armitage. 2018. Nutrient enrichment shifts mangrove height distribution: Implications for coastal woody encroachment. *PLoS ONE* 13 (3): e0193617. https://doi.org/10.1371/journal.pone.0193617.

Weddle, R. S. 1968. *San Juan Bautista Mission: Gateway to Spanish Texas*. Austin: University of Texas Press.

Weddle, R. S. 1985. *Spanish Sea: The Gulf of Mexico in North American Discovery*. College Station: Texas A&M University Press.

Weise, B. R., and W. A. White. 1980. *Padre Island National Seashore: A Guide to the Geology, Natural Environments, and History of a Texas Barrier Island*. Guidebook 17. Austin: Bureau of Economic Geology, University of Texas at Austin.

Weller, M. W. 1964. Distribution and migration of the Redhead. *Journal of Wildlife Management* 28:64–103.

Wells, F. C., G. A. Jackson, and W. J. Rogers. 1988. *Reconnaissance Investigation of Water Quality, Bottom Sediment, and Biota Associated with Irrigation Drainage in the Lower Rio Grande Valley and Laguna Atascosa National Wildlife Refuge, Texas 1986–87*. US Geological Service Water-Resources Investigation No. 87-4277, Austin.

Weniger, D. 1997. *The Explorers' Texas*. Vol. 2, *The Animals They Found*. Austin: Eakin Press.

Werler, J. E., and J. R. Dixon. 2000. *Texas Snakes: Identification, Distribution, and Natural History*. Austin: University of Texas Press.

Wern, J. O. 1995. Nemerteans of the northwestern Gulf of Mexico. PhD diss., Texas A&M University, College Station.

Wester, D. B., F. C. Bryant, A. D. Tjelmeland, J. L. Grace, S. L. Mitchell, J. T. Edwards, F. Hernández et al. 2018. Tanglehead in southern Texas: A native grass with invasive behavior. *Rangelands* 40:37–44.

Wetz, M. S. 2014. *Baffin Bay Volunteer Water Quality Monitoring Study: Synthesis of May 2013–July 2014 Data.* Final report, Publication CBBEP-95, Project No. 1413, Coastal Bend Bays and Estuaries Program, Corpus Christi.

Wetz, M. S., E. K Cira, B. Sterba-Boatwright, P. A. Montagna, T. A. Palmer, and K. C. Hayes. 2017. Exceptionally high organic nitrogen concentrations in a semi-arid South Texas estuary susceptible to brown tide blooms. *Estuarine, Coastal and Shelf Science* 188:27–37.

Whelan, T. 1975. Low-altitude aerosol distribution along a barrier island coast. *Contribution of Marine Science* 19:3–12.

White, D. H., and D. James. 1978. Differential use of freshwater environments by wintering waterfowl of coastal Texas. *Wilson Bulletin* 90:99–111.

White, D. H., C. A. Mitchell, and T. E. Kaiser. 1983. Temporal accumulations of organochlorine pesticides in shorebirds wintering on the South Texas coast. *Texas Archives of Environmental Contamination and Toxicology* 12:241–250.

White, D. H., C. A. Mitchell, H. D. Kennedy, A. J. Krynitsky, and M. A. Ribick. 1983. Elevated DDE and toxaphene residues in fishes and birds reflect local contamination in the Lower Rio Grande Valley, Texas. *Southwestern Naturalist* 28:325–333.

White, D. H., C. A. Mitchell, and D. M. Swineford. 1984. Reproductive success of Black Skimmers in Texas relative to environmental pollutants. *Journal of Field Ornithology* 55:18–30.

White, W. A., and W. B. Brogden. 1977. Descriptions of land and water resources. Appendix A in R. S. Kier and E. G. Fruh, eds., *Environmental and Economic Impacts of Recreational Community Development, Mustang Island and North Padre Island*, vol. 1. Austin: University of Texas.

White, W. A., T. R. Calnan, R. A. Morton, R. S. Kimble, T. G. Littleton, J. H. McGowen, H. S. Nance, et al. 1983. *Submerged Lands of Texas, Corpus Christi Area: Sediments, Geochemistry, Benthic Macroinvertebrates, and Associated Wetlands.* Austin: Bureau of Economic Geology, University of Texas at Austin.

White, W. A., T. R. Calnan, R. A. Morton, R. S. Kimble, T. G. Littleton, J. H. McGowen, H. S. Nance, et al. 1986. *Submerged Lands of Texas, Brownsville-Harlingen Area: Sediments, Geochemistry, Benthic Macroinvertebrates, and Associated Wetlands.* Austin: Bureau of Economic Geology, University of Texas at Austin at Austin.

White, W. A., T. R. Calnan, R. A. Morton, R. S. Kimble, T. G. Littleton, J. H. McGowen, H. S. Nance, et al. 1989. *Submerged Lands of Texas, Kingsville Area: Sediments, Geochemistry, Benthic Macroinvertebrates, and Associated Wetlands.* Austin: Bureau of Economic Geology, University of Texas at Austin.

White, W. A., and W. E. Galloway. 1977. *Guide to Modern Barrier Environments of Mustang and North Padre Islands and Jackson (Eocene) Barrier/Lagoon Facies of South Texas Uranium District.* Research Note 7. Austin: Bureau of Economic Geology, University of Texas at Austin.

White, W. A., R. A. Morton, and C. W. Holmes. 2002. A comparison of factors controlling sedimentation rates and wetland loss in fluvial-deltaic systems. *Texas Gulf Coast Geomorphology* 44:47–66.

White, W. A., T. A. Tremblay, R. L. Waldinger, and T. R. Calnan. 2006. *Status and Trends of Wetland and Aquatic Habitats on Texas Barrier Islands, Coastal Bend.* Final report, No. NA04NOS4190058, Coastal Coordination Division, Texas General Land Office, Austin.

White, W. A., T. A. Tremblay, R. L. Waldinger, and T. R. Calnan. 2007. *Status and Trends of Wetland and Aquatic Habitats on Barriers: Upper Coast Strandplain-Chenier System and Southern Coast Padre Island National Seashore.* Final report for the Texas General Land Office and National Oceanic and Atmospheric Administration (GLO Contract 06-44). Austin: Bureau of Economic Geology, University of Texas at Austin.

White, W. A., T. A. Tremblay, R. L. Waldinger, T. L. Hepner, and T. R. Calnan. 2005. *Status and Trends of Wetland and Aquatic Habitats on Barrier Islands, Freeport to East Matagorda Bay, and South Padre Island.* Final report for the Texas General Land Office and National Oceanic

and Atmospheric Administration (GLO Contracts: 04-044 and 04-045). Austin: Bureau of Economic Geology, University of Texas at Austin.
Whitledge, T. E. 1993. The nutrient and hydrographic conditions prevailing in the Laguna Madre, Texas before and during a brown tide bloom. Pp. 711–716 in T. J. Smayda and Y. Shimizu, eds., *Toxic Phytoplankton Blooms in the Sea: Proceedings of the Fifth International Conference on Toxic Marine Phytoplankton*. Amsterdam: Elsevier.
Whitledge, T. E., D. A. Stockwell, E. J. Buskey, K. C. Dunton, G. J. Holt, S. A. Holt, and P. A. Montagna. 1999. Persistent brown tide bloom in Laguna Madre, Texas. Pp. 338–359 in H. Kumpf, K. Steidinger, and K. Sherman, eds., *The Gulf of Mexico Large Marine Ecosystem: Assessment, Sustainability, and Management*. London: Blackwell.
Whitten, H. L., H. F. Rosene, and J. W. Hedgpeth. 1950. The invertebrate fauna of Texas coast jetties: A preliminary survey. *Publications of the Institute of Marine Science, University of Texas* 1:53–87.
Whorff, J. S. 1992. Physical and biological interactions in the midlittoral zone along a Central Texas inlet. PhD diss., Texas A&M University, College Station.
Whyte, R. J. 1978. The effects of cattle on shoreline vegetation of ponds and tanks in South Texas. MS thesis, Texas A&M University, College Station.
Wilhite, H. S., T. C. Allison, and J. A. Rickner. 1982. The diversity and distribution of living mollusks in the Lower Laguna Madre of Texas. Pp. 233–247 in J. R. Davis, ed., *Proceedings of the Symposium on Recent Benthological Investigations in Texas and Adjacent States*. Austin: Texas Academy of Science.
Wilkinson, B. H., J. H. McGowen, and C. R. Lewis. 1975. Ingleside strandplain sand of central Texas coast. *American Association of Petroleum Geologists Bulletin* 59:347–352.
Williams, D. D., and N. C. Kraus. 2011. Seasonal change in nearshore and channel morphology at Packery Channel, a new inlet serving Corpus Christi, Texas. *Journal of Coastal Research* SI59:86–97.
Williams, D. D., N. C. Kraus, and C. M. Anderson. 2007. Morphologic response to a new inlet, Packery Channel, Corpus Christi, Texas. Pp. 11529–11542 in N. C. Kraus and J. D. Rosati, eds., *Coastal Sediments '07: Proceedings of the ASCE Coastal Sediments 2007 Conference*. Reston, VA: American Society of Civil Engineers.
Williams, N. C., K. A. Bjorndal, M. M. Lamont, and R. R. Carthy. 2013. Winter diets of immature Green Turtles (*Chelonia mydas*) on a northern feeding ground: Integrating stomach contents and stable isotope analyses. *Estuaries and Coasts* 37:986–994.
Williams, T. C., J. M. Williams, L. C. Ireland, and J. M. Teal. 1977. Autumnal bird migration over the western North Atlantic Ocean. *American Birds* 31:251–267.
Williamson, C. J. 1980. Population dynamics of molluscs in a seagrass bed surrounding a dredged material island, Upper Laguna Madre, Texas. MS thesis, Corpus Christi State University, Corpus Christi.
Williamson, E. A., K. B. Strychar, and K. Withers. 2011. Populations of the gorgonian genus *Leptogorgia* at two jetties in the northwestern Gulf of Mexico. *Gulf of Mexico Science* 29:126–130.
Williges, K. 1989. The avian fauna of two oak mottes on the sand sheet of South Texas. MS thesis, Texas A&I University, Kingsville.
Wilson, S. S., and K. H. Dunton. 2017. Hypersalinity during regional drought drives mass mortality of the seagrass *Syringodium filiforme* in a subtropical lagoon. *Estuaries and Coasts* 41:855–865. doi: 10.1007/s12237-017-0319-x.
Wilson, W. H., Jr. 1990. Relationship between prey abundance and foraging site selection by Semipalmated Sandpipers on a Bay of Fundy mudflat. *Journal of Field Ornithology* 61:9–19.
Wiser, R., and M. Bolinger. 2017. *2016 Wind Technologies Market Report*. https://www.energy.gov/sites/default/files/2017/10/f37/2016_Wind_Technologies_Market_Report_101317.pdf.
Witherington, B. E., and L. E. Ehrhart. 1989. Hypothermic stunning and mortality of marine turtles in the Indian River Lagoon system, Florida. *Copeia* 1989:696–703.
Witherington, B. E., S. Hirama, and R. Hardy. 2012. Young sea turtles of the pelagic *Sargassum*-dominated drift community: Habitat use, population density, and threats. *Marine Ecology Progress Series* 463:1–22.

Withers, K. 1994. The relationship of macrobenthic prey availability to shorebird use of blue-green algal flats in the Upper Laguna Madre. PhD diss., Texas A&M University, College Station.

Withers, K. 1996a. *An Evaluation of Recovery of Benthic Invertebrate Communities in Vehicle Tracks and Restored Oil and Gas Impacted Areas on Wind-Tidal Flats in the Upper Laguna Madre, Padre Island National Seashore, Texas*. Report to Padre Island National Seashore, Contract No. 1443PX749050162, Corpus Christi.

Withers, K. 1996b. Seagrass meadows. Pp. 175–249 in J. W. Tunnell Jr., Q. R. Dokken, E. H. Smith, and K. Withers, eds., *Status and Trends of the Estuarine Living Resources within the Corpus Christi Bay National Estuary Program Study Area*. Corpus Christi Bay National Estuary Program CCBNEP-06A. Corpus Christi: Texas Natural Resource Conservation Commission.

Withers, K. 1996c. Tidal flats. Pp. 294–356 in J. W. Tunnell Jr., Q. R. Dokken, E. H. Smith, and K. Withers, eds., *Status and Trends of the Estuarine Living Resources within the Corpus Christi Bay National Estuary Program Study Area*. Corpus Christi Bay National Estuary Program CCBNEP-06A. Corpus Christi: Texas Natural Resource Conservation Commission.

Withers, K. 1998. *Biological Productivity of Southerly Wind-Tidal Flats within Padre Island National Seashore*. Technical Report TAMU-CC-9810-CCS. Corpus Christi: Texas A&M University–Corpus Christi, Center for Coastal Studies.

Withers, K. 2002a. Seagrass meadows. Pp. 85–101 in J. W. Tunnell and F. W. Judd, eds., *The Laguna Madre of Texas and Tamaulipas*. College Station: Texas A&M University Press.

Withers, K. 2002b. Shorebird use of coastal wetland and barrier island habitat in the Gulf of Mexico. *Scientific World Journal* 2:514–536.

Withers, K. 2002c. Wind-tidal flats. Pp. 114–126 in J. W. Tunnell Jr. and F. W. Judd, eds., *The Laguna Madre of Texas and Tamaulipas*. College Station: Texas A&M University Press.

Withers, K., and T. S. Brooks. 2004. Diet of Double-crested Cormorants (*Phalacrocorax auritus*) wintering on the central Texas coast. *Southwestern Naturalist* 49:48–53.

Withers, K., and B. R. Chapman. 1993. Seasonal abundance and habitat use of shorebirds on an Oso Bay mudflat, Corpus Christi, Texas. *Journal of Field Ornithology* 64:382–392.

Withers, K., and J. W. Tunnell Jr. 1998. *Identification of Tidal Flat Alterations and Determination of Effects on Biological Productivity of These Habitats within the Coastal Bend*. Corpus Christi Bay National Estuary Program, CCBNEP-26. Corpus Christi: Texas Natural Resource Conservation Commission.

Witzell, W. N. 1994a. The origin, evolution, and demise of the U.S. sea turtle fisheries. *Marine Fisheries Review* 56:8–23.

Witzell, W. N. 1994b. *The U.S. Commercial Sea Turtle Landings*. NOAA Technical Memorandum NMFS-SEFSC-350. Miami, FL: US Department of Commerce.

Wolff, W. J. 1969. Distribution of non-breeding waders in an estuarine area in relation to the distribution of their food organisms. *Ardea* 57:1–28.

Woodin, M. C. 1994. Use of saltwater and freshwater habitats by wintering Redheads in southern Texas. *Hydrobiologia* 279/280:279–287.

Woodin, M. C. 1996. Wintering ecology of Redheads (*Aythya americana*) in the western Gulf of Mexico region. *Gibier Faune Sauvage* 13:653–665.

Woodin, M. C., M. K. Skoruppa, and G. C. Hickman. 2000. *Surveys of Night Birds along the Rio Grande in Webb County, Texas*. Final report, US Fish and Wildlife Service, Corpus Christi.

Wright, S. L., R. C. Thompson, and T. S. Galloway. 2013. The physical impacts of microplastics on marine organisms: A review. *Environmental Pollution* 178:483–492. http://dx.doi.org/10.1016/j.envpol.2013.02.031.

WRT (Writer's Round Table). 1950. *Padre Island*. San Antonio: Naylor.

Wunderle, J. M., Jr., R. B. Waide, and J. Fernandez. 1989. Seasonal abundance of shorebirds in the Hobos Bay estuary in southern Puerto Rico. *Journal of Field Ornithology* 60:329–339.

Wynne, M. 2009. A checklist of benthic marine algae of the coast of Texas. *Gulf of Mexico Science* 1:64–87.

Xavier, R., A. Barata, L. P. Cortez, N. E. Queiroz, and E. Cuevas. 2006. Hawksbill Turtle (*Eretmochelys imbricata* Linnaeus 1766) and Green Turtle (*Chelonia mydas* Linnaeus 1754) nesting activity (2002–2004) at El Cuyo beach, Mexico beach, Mexico. *Amphibia-Reptilia* 27:539–547.

Xiong, Y., B. Liao, E. Proffitt, W. Guan, Y. Sun, F. Wang, and X. Liul. 2018. Patterns of soil carbon in mangroves are primarily controlled by soil properties: A landscape-scale study. *Science of the Total Environment* 619–620:1226–1235.

Yañez, A., and C. J. Schlaepfer. 1968. Composicion y distribucion de los sedimentos recientes de la Laguna Madre, Tamaulipas. *Universidad Nacional Autónoma de México, Instituto de Geologia Boletín* 84 (Parte 1): 5–44.

Yáñez-Arancibia, A., and J. W. Day. 2004. Environmental sub-regions of the Gulf of Mexico coastal zone: The ecosystem approach as an integrated management tool. *Ocean & Coastal Management* 47:727–757.

Yáñez-Arancibia, A., J. W. Day, and R. H. Day. 2014. Manglares: Ecosistema centinela frente al cambio climático, Golfo de México. *Madera y Bosques* 20 (número especial): 39–75.

Yates, T. L., and D. J. Schmidly. 1977. Systematics of *Scalopus aquaticus* (Linnaeus) in Texas and adjacent states. *Occasional Papers, Museum of Texas Tech University* 45:1–36.

Yeager, V., K. R. Summy, F. W. Judd, R. I. Lonard, and R. A. Mazariegos. 2010. Factors affecting the spatial distribution of Black Mangrove on the dredged-material or "spoil" islands in the Lower Laguna Madre of Texas. *Subtropical Plant Science* 62:44–55.

Yzaguirre, G. A. 1974. Distribution and variation of rodents in different vegetation-terrain types on Padre Island. MS thesis, Texas A&I University, Kingsville.

Zamora-Tovar, C. 2002. Algas de importancia económica en Tamaulipas, México: Una síntesis de su conocimiento. *BIOTAM*, n.s., 13 (2): 51–70.

Zamora-Tovar, C. 2004. Laguna Madre: Where the sea and land blend together. Pp. 83–103 in P. Robles Gil, E. Ezcurra, E. Peters, E. Pallares, and A. Ezcurra, eds., *The Great Tamaulipan Natural Province*. Tamaulipas: Tamaulipas State Government, Agrupación Sierra Madre, Ciudad Victoria.

Zamora-Tovar, C. 2007. *Restauración de la cuenca hidrográfica de la Laguna Madre*. Universidad Autónoma de Tamaulipas, Instituto de Ecología Aplicada. Informe final SNIB-CONABIO proyecto No. CJ069, Mexico City.

Zamora-Tovar, C., P. J. L. Jiménez, E. A. Cardona, C. E. González-Romo, H. A. Garza-Torres, G. Herrera-Patiño, and G. Sánchez-Ramos. 2011. Participación comunitaria en la restauración ecológica de la Laguna Madre. *CienciaUAT* 21 (3): 38–47.

Zamora-Tovar, C., and A. Mora-Olivo. 2014. Vegetación acuática sumergida de la Laguna Madre, Tamaulipas, México. In *1er Simposio de Investigación Científico de Laguna Madre y Delta del Río Bravo, "Biodiversidad y Cambio Climático."* Ciudad Victoria, Tamaulipas: CONANP.

Zavaleta-Lizárraga, L., and J. E. Morales-Mávil. 2013. Nest site selection by the Green Turtle (*Chelonia mydas*) in a beach of the north of Veracruz, Mexico. *Revista Mexicana de Biodiversidad* 84:927–937.

Zehner, W. 1985. First record of *Pipistrellus subflavus* (Chiroptera: Vespertilionidae) on Padre Island, Texas. *Southwestern Naturalist* 30:306–307.

Zeitzschel, G. 1980. Sediment-water interactions in nutrient dynamics. Pp. 195–218 in K. R. Tenore and B. C. Coull, eds., *Marine Benthic Dynamics*. Columbia: University of South Carolina Press.

Zhang, Q.-C., L.-M. Qiu, R.-C. Yu, F.-Z. Kong, Y.-F. Wang, T. Yan, C. J. Gobler, and M.-J. Zhou. 2012. Emergence of brown tides caused by *Aureococcus anophageferens* Hargraves et Sieburth in China. *Harmful Algae* 19:117–124.

Zhou, L., Y. Tian, S. Baidya Roy, C. Thorncroft, L. F. Bosart, and Y. Hu. 2012. Impacts of wind farms on land surface temperature. *Nature Climate Change* 2:539–543.

Ziegler, M. 2018. Facilitation plays a key role in the survival and growth of Black Mangrove, *Avicennia germinans*, seedlings in a marsh-mangrove ecotone under moderate and hypersaline conditions. MS thesis, Texas A&M University–Corpus Christi.

Zieman, J. C. 1975. Seasonal variation of Turtle-grass, *Thalassia testudinum* (König), with reference to temperature and salinity effects. *Aquatic Botany* 1:107–123.

Zieman, J. C. 1976. The ecological effects of physical damage from motor boats on Turtle-grass beds in southern Florida. *Aquatic Botany* 2:127–139.

Zieman, J. C. 1982. *The Ecology of the Seagrasses of South Florida: A Community Profile*. US Fish and Wildlife Service, Office of Biological Services FWS/OBS-82/25, Washington, DC.

Zieman, J. C., and R. T. Zieman. 1989. *The Ecology of the Seagrass Meadows of the West Coast of Florida: A Community Profile*. US Fish and Wildlife Service Biological Report 85 (7.25), Washington, DC.

Zonick, C. 1997a. *Snowy Plover Breeding Ecology along the Texas Gulf Coast*. Report to Texas Parks and Wildlife Department, Resource Protection and US Fish and Wildlife Service, Ecological Services, Corpus Christi.

Zonick, C. 1997b. *The Use of Texas Barrier Island Washover Pass Habitat by Piping Plovers and Other Coastal Waterbirds*. Report to Texas Parks and Wildlife Department, Resource Protection and US Fish and Wildlife Service, Ecological Services, Corpus Christi.

Zonick, C., K. R. Drake, K. Drake, J. Thompson, and L. Elliott. 1998. *The Effects of Dredged Material on Piping Plovers (*Charadrius melodus*) and Snowy Plovers (*C. alexandrinus*) in the Lower Laguna Madre of Texas*. Final report for the 1997/1998 Season to US Army Corps of Engineers, Galveston District, Laguna Madre Interagency Coordination Team, Galveston.

Zonick, C., and M. Ryan. 1994. *Ecology and Conservation of Piping and Snowy Plovers Wintering along the Texas Gulf Coast*. Report to Texas Parks and Wildlife Department, Resource Protection and US Fish and Wildlife Service, Ecological Services, Corpus Christi.

Zupan, A. W. 1971. *Surficial Sediments and Sedimentary Structures: Middleground, Padre Island, Texas*. Texas A&M University Department of Oceanography Report 71-12-T, College Station.

Index

Note: page numbers in *italics* indicate figures and page numbers in **bold** indicate maps.